COMPUTING EQUILIBRIA AND FIXED POINTS

THEORY AND DECISION LIBRARY

*General Editors:* W. Leinfellner (*Vienna*) and G. Eberlein (*Munich*)

Series A: Philosophy and Methodology of the Social Sciences

Series B: Mathematical and Statistical Methods

Series C: Game Theory, Mathematical Programming and Operations Research

# SERIES C: GAME THEORY, MATHEMATICAL PROGRAMMING AND OPERATIONS RESEARCH

VOLUME 21



*Scope:* Particular attention is paid in this series to game theory and operations research, their formal aspects and their applications to economic, political and social sciences as well as to socio-biology. It will encourage high standards in the application of game-theoretical methods to individual and social decision making.

*The titles published in this series are listed at the end of this volume.*

# COMPUTING EQUILIBRIA AND FIXED POINTS

## The Solution of Nonlinear Inequalities

*by*

ZAIFU YANG

*Yokohama National University, Japan*

KLUWER ACADEMIC PUBLISHERS

BOSTON / DORDRECHT / LONDON

A C.I.P. Catalogue record for this book is available from the Library of Congress.

ISBN 0-7923-8395-8

---

Published by Kluwer Academic Publishers,
P.O. Box 17, 3300 AA Dordrecht, The Netherlands.

Sold and distributed in North, Central and South America,
by Kluwer Academic Publishers,
101 Philip Drive, Norwell, MA 02061, U.S.A.

In all other countries, sold and distributed
by Kluwer Academic Publishers,
P.O. Box 322, 3300 AH Dordrecht, The Netherlands.

*Printed on acid-free paper*



Printed in the Netherlands.

*To the memory of my grandparents*

## Table of Contents

# Preface

This book is an introduction to the computation of equilibria, fixed points and stationary points in finite dimension spaces. We wrote this book with three goals in mind:

(i) To give a comprehensive introduction to fixed point methods and to the definition and construction of Gröbner bases;

(ii) To discuss several interesting applications of these methods in the fields of general equilibrium theory, game theory, mathematical programming, algebra and symbolic computation;

(iii) To introduce several advanced fixed point and stationary point theorems.

These methods and topics are of interest not only to economists and game theorists concerned with the computation and existence of equilibrium outcomes in economic models and cooperative and non-cooperative games, but also to applied mathematicians, computer scientists and engineers dealing with models of highly nonlinear systems of equations (or polynomial equations). Many of the results in this book are quite new and only available in research papers. We have also made the mathematical prerequisites to the book very modest so that it should be suitable for an advanced undergraduate or a first year graduate student.

For a long time the arduous problem of solving a system of nonlinear equations or inequalities has been a challenge. In fact, this problem dates back to the beginning of human civilization when the ancient Babylonians, Chinese, Egyptians and Greeks were studying specific examples of linear and quadratic equations with very few variables (see Midonick [1968] and van der Waerden [1983]). However, it was not until the appearance of Hilbert's Nullstellensatz (Hilbert's zero point theorem) in the late nineteenth century, and Brouwer's fixed point theorem in the early twentieth century, that a breakthrough was made in quest of the existence of a solution to a general system of nonlinear (polynomial) equations. Generally speaking, Hilbert's zero point theorem gives a sufficient and necessary condition for the existence of a solution to a system of nonlinear polynomial equations. Furthermore Brouwer's fixed point theorem states that every continuous function from a nonempty, convex and compact subset of the Euclidean space into itself has at least one fixed point, i.e., a point in the set being mapped into itself. For many years these theorems and related results appeared to be remote from any practical use because of their mathematical abstractness and the non-constructive nature of their proofs. However, in the 1960s, Scarf developed a finite algorithm for finding an approximate fixed point of a continuous function from the unit simplex into itself. This algorithm provided the first constructive proof to Brouwer's fixed point theorem. At the same time, Buchberger discovered a finite algorithm for manipulating systems of nonlinear polynomial equations, which can actually determine whether a system of polynomial equations has a solution or not, (though it does not show how to find a solution); this gave the first constructive proof to Hilbert's zero point theorem. The former algorithm laid the foundation for fixed point methods; the latter one, the foundation for Gröbner basis methods. These two pioneering studies have stimulated many researchers to extend the applicability and improve the efficiency of these methods. More specifically, efficient fixed point algorithms have been designed for applications in general equilibrium theory, game theory, mathematical programming and engineering problems, whereas Gröbner basis algorithms have found applications in computational algebra, symbolic computation and computer science. This book is intended to introduce the recent fixed point algorithms, the basic Gröbner basis algorithm, and their applications, to a wider audience.

Note that the aim of this book is the solution of highly nonlinear systems of equations which are required only to be continuous, or upper semicontinuous. Therefore, the book does not deal with conceptually simpler methods to solve smooth systems that require first or even higher order derivatives, like Newton methods and continuation methods. The interested reader is referred to Ortega and Rheinbolt [1970], Kellogg, Li and Yorke [1976], Smale [1976], Allgower and Georg [1990], Harker and Pang [1990], Yu and Lin [1996].

The book is organized as follows. Chapter 1 describes the mathematical tools needed to understand the rest of the book. It includes basic notation and concepts which will be used throughout the book, as well as several classic theorems from polyhedral theory, definitions of graphs, simplices, and triangulations. We also present Brouwer's fixed point theorem, Kakutani's fixed point theorem, Browder's fixed point theorem, Knaster-Kuratowski-Mazurkiewicz's (KKM) lemma and Sperner's lemma which will be demonstrated in constructive manners in the following chapters. Furthermore, some other types of fixed point theorems such as Tarski's theorem, Caristi's theorem, and Ekeland's theorem are introduced and proved.

It is a rather common belief that the ultimate goal of even the most abstract and theoretical research work should be its practical applicability. Therefore, Chapter 2 discusses several applications of fixed point theorems. In fact, the topics under consideration have been the major source of motivation and inspiration for the development of fixed point algorithms. These topics include the core of cooperative games, the Nash, perfect Nash and proper Nash equilibria of non-cooperative finite games, the Walrasian equilibrium of a pure exchange economy, the Drèze equilibrium and supply-constrained (or unemployment) equilibrium of a general exchange economy under price rigidities, the equilibrium of an exchange economy with increasing returns to scale, the competitive equilibrium of an exchange economy with indivisibilities and money, and the Radner equilibrium of an exchange economy under uncertainty.

In Chapters 3, 4 and 5 we systematically introduce the classic fixed point algorithms as developed over the last thirty years and familiarize the reader with the basic theory and techniques of these algorithms. These algorithms will be presented approximately in the order of the events. In doing so, we hope that the reader will easily see how these methods were developed and improved, and where there is room for further improvement. We discuss these methods in an accessible way but with no loss of mathematical rigor. More precisely, these algorithms include the first algorithm of Scarf based on primitive sets, the algorithms of Kuhn based on simplices, Merrill's (and Kuhn-MacKinnon's) restart algorithm, the homotopy algorithms of Eaves, and Eaves-Saigal, and the variable dimension restart algorithms of van der Laan-Talman.

Chapter 6 presents an algorithm for solving a class of integer programming problems which are known to be intractable, as they cannot be shown to be solvable in polynomial time. The algorithm is based on both a specific integer labeling rule and a triangulation of the Euclidean space. Starting with an arbitrarily chosen integer vector, the algorithm terminates within a finite number of steps either yielding an optimal solution or proving that the problem has no feasible integer solution. Prominent features of the algorithm are its computational simplicity, flexibility and stability. Numerical results are also reported.

In Chapter 7 the stability and refinement of stationary points is studied. There is growing evidence that a continuous function may have many stationary points and some of them are undesirable with regard to stability. Hence, it is very important to eliminate these stationary points. To achieve this goal, we introduce a new solution to the stationary point problem. This solution has certain stable properties and is called a robust stationary point. The concept of robust stationary point is a refinement of that of stationary point. Although a stationary point need not be robust, it has been shown that every continuous function on polytopes has at least one robust stationary point. A simplicial algorithm is developed to compute such a robust stationary point. An important application of the algorithm is the computation of proper Nash equilibria and robust Walrasian equilibria. This theory provides a unifying framework for studying the stability and refinement of Walrasian equilibria and Nash equilibria.

Chapter 8 introduces a new fixed point theorem which states that under some mild conditions there exists a continuum of fixed points for a point-to-set mapping. Although many fixed point theorems have been established following the appearance of Brouwer's fixed point theorem, to date the existence of results for multiple fixed points is very rare. This theorem is proved in a constructive manner by means of a simplicial algorithm. Applying the algorithm to the model of an exchange economy under price rigidities, it is proved that there exists a connected set of constrained equilibria linking together two trivial no-trade equilibria. The algorithm is also applied so as to provide a first constructive proof for Browder's fixed point theorem. This chapter is based on the work of Herings, Talman and Yang [1996].

In Chapter 9, two algorithms are discussed for computing zero points, respectively, stationary points of a continuous function on a general polytope. These methods were first developed by Talman and Yamamoto [1989] and later modified by Fujishige and Yang [1998]. Talman and Yamamoto's method is the first to compute a stationary point of any continuous function on a general polytope. In simplicial algorithms we often have to resolve degeneracy problems. Here, a general method is presented for dealing with these problems. Moreover, a new existence theorem of zero points on general polytopes is introduced which generalizes the well-known multivariate mean value theorem.

In Chapter 10 an algorithm is presented to compute an antipodal fixed point of a continuous function on a symmetric cubical set. As a result, constructive proofs of the well-known Borsuk-Ulam's fixed point theorem and Tucker's combinatorial theorem are obtained. This algorithm is designed to find a simplex in a given triangulation, which contains an approximate antipodal fixed point. Furthermore, we present a generalization of Borsuk-Ulam's theorem to a point-to-set mapping. This chapter largely reflects the work of Todd and Wright [1980], Freund and Todd [1981], and van der Laan [1984].

In Chapter 11, we discuss Kuhn's algorithm for computing all roots of a univariate polynomial with complex coefficients. Consequently, we also obtain a constructive proof of the Fundamental Theorem of Algebra. Kuhn's algorithm is the first to compute all roots of a univariate polynomial. The algorithm is built upon both a particular integer labeling rule and a triangulation of the complex plane. The complexity of the algorithm is analyzed and numerical results are reported.

In Chapter 12, we provide the definition of a Gröbner basis and then show how to compute it by using Buchberger's algorithm. Then, we proceed to discuss the famous Hilbert's basis theorem and Hilbert's zero point theorem and solve the ideal membership problem in a constructive manner. Furthermore, we give a method to solve systems of multivariate polynomial equations with complex coefficients. To do so, we first determine whether such a system even has solutions. If a solution exists, we determine whether it has a finite number of solutions. If the system has a finite number of solutions, then all solutions in the system can be found by using Kuhn's algorithm successively. If the system has an infinite number of solutions, we can apply a homotopy algorithm to find as many solutions as we wish to have. Thus, this procedure for solving a system of polynomial equations is a remarkable extension of Gauss-Jordan elimination method for solving a system of linear equations. Meanwhile, we also discuss how to apply the above theories to solve a large class of linear and nonlinear complementarity problems.

Finally, in Chapters 13 and 14 we discuss both the theory on intersection points and Sperner's theory. These theories are closely related to fixed point theory and provide geometric intuitions to fixed point algorithms. In particular, three unifying combinatorial theorems on general polytopes are presented. The first is a general intersection theorem, while the second and third are two general integer labeling theorems. Many well-known results, such as the lemmas of Sperner, KKM, Scarf, Shapley, among others, will be derived from these three general theorems. We also discuss intersection theorems of Freidenfelds and Herings-Talman which state conditions for the existence of a continuum of intersection points. Furthermore, we present several permutation-based generalizations of Brouwer's fixed point theorem, KKM lemma and Sperner's lemma. By applying such generalization we prove the existence of an equilibrium in an economy with multiple indivisibilities and money.

## Acknowledgments

I would like to thank the many people who have helped me write this book. To the aforementioned individuals who created this exciting research field, my gratitude is obvious. To my teachers, friends and co-researchers, Dolf Talman, Gerard van der Laan, Yoshi Yamamoto, Kai-zhou Chen, I have come to incur an incalculable intellectual debt. I cannot find the words to pay adequate tribute to them. Over the years, I have had many interesting and stimulating discussions with my friends Antoon van den Elzen and Jean-Jacques Herings to whom I owe a great deal. I am also indebted to Walter Forster, Satoru Fujishige, Pieter Ruys, Stef Tijs, Ze-ke Wang, and Xinhui Xing for their encouragement and help. I would also like to express my gratitude to my colleagues at Tilburg, Tsukuba and Yokohama for creating stimulating research environments and offering me the opportunity to complete this work. I am also grateful to Allard Winterink and Siep Kroonenberg of Kluwer Academic Publishers for Allard's advice and for Siep's technical assistance. Finally, I send my very special thanks to my wife and my parents for their understanding and support.

CHAPTER 1

# Mathematical Preliminaries

## 1.1. Introduction

This chapter presents mathematical preliminaries which will be used throughout the book. The reader is presumed to be familiar with elementary linear algebra, linear programming and real analysis.

This chapter is organized as follows. Section 2 introduces some basic notation and terminology. Section 3 provides several basic results from polyhedral theory. In Section 4 we review the basic concept of a simplicial subdivision and acquaint the readers with some well-known triangulations. We also recall some elementary concepts from graph theory. In Section 5 we introduce the concepts of fixed points and stationary points, and discuss some existence results, such as Brouwer's fixed point theorems, and the lemmas of Sperner, and Knaster, Kuratowski and Mazurkiewicz. In Section 6 Kakutani's fixed point theorem and its extension are discussed. Finally in Section 7 we introduce three different types of fixed point theorems due to Tarski, Caristi, and Ekeland.

## 1.2. Basic notation and terminology

The symbols $\mathbb{Z}$, $\mathbb{R}$, $\mathbb{Q}$ and $\mathbb{C}$ denote the sets of integer, real, rational and complex numbers, respectively. $\mathbb{Z}_+$ ($\mathbb{Z}_{++}$) and $\mathbb{R}_+$ ($\mathbb{R}_{++}$) are the restrictions of the sets $\mathbb{Z}$ and $\mathbb{R}$ to the non-negative values (positive values), respectively. The set of positive integers is also denoted by $\mathbb{N}$. For $n \in \mathbb{N}$, $I_n$ denotes the index set $\{1, \cdots, n\}$. Given a real number $\beta$, then

$$\lfloor \beta \rfloor \text{ and } \lceil \beta \rceil$$

denote the lower integer part and the upper integer part, respectively, of $\beta$.

For a vector $x$ (matrix $A$), $x^\top$ ($A^\top$) denotes its *transpose*. Let $n \in \mathbb{N}$. The $n$-dimensional Euclidean space is specified as $\mathbb{R}^n = \{ x = (x_1, \cdots, x_n)^\top \mid x_i \in \mathbb{R}, i = 1, \cdots, n\}$. An element or point $x = (x_1, \cdots, x_n)$ in $\mathbb{R}^n$ is called a vector or n-vector with components $x_i$, $i = 1, \cdots, n$. It is taken for granted that all elements in $\mathbb{R}^n$ are column vectors, unless stated otherwise. Accordingly, $\mathbb{R}^n_+$ ($\mathbb{R}^n_{++}$) and $\mathbb{Z}^n_+$ ($\mathbb{Z}^n_{++}$) denote the sets of the non-negative (positive) elements in $\mathbb{R}^n$ and $\mathbb{Z}^n$, respectively. Given a number $n \in \mathbb{N}$, we denote the $i$-th unit vector in the Euclidean space by $e(i)$ for each $i \in I_n$, where the dimension is usually clear from the context. The vector $0^n$ ($e^n$)

denotes the all-zero (all-one) vector in $\mathbb{R}^n$. In addition, the symbol $\mathbf{0}$ stands for an all-zero vector (matrix), $e$ for an all-one vector, and $\mathbf{I}$ for an identity matrix, all of appropriate dimension. The *inner product* $\sum_{i=1}^{n} x_i y_i$ of two vectors $x$ and $y$ in $\mathbb{R}^n$ is denoted by $x^\top y$. Given two elements $x$ and $y$ of $\mathbb{R}^n$, $x \leq y$, $x < y$, and $x \ll y$ are defined respectively if $x_i \leq y_i$ for all $i \in I_n$, if $x \leq y$ and $x \neq y$, and if $x_i < y_i$ for all $i \in I_n$. The relations $x \geq y$, $x > y$, and $x \gg y$ are defined similarly.

Let two sets $A$ and $B$ in $\mathbb{R}^n$ be given. Then $A \subset B$ means that $A$ is a *proper subset* of $B$, whereas $A \subseteq B$ means that $A$ is a *proper subset* of $B$ or *equal* to $B$. By $A \bigcap B$, $A \bigcup B$ and $A \backslash B$ we denote the *intersection*, *union* and *set difference* of the sets $A$ and $B$, respectively. Furthermore, $A + B$ and $A - B$ represent the *algebraic sum* and *difference* of the sets $A$ and $B$, i.e., $A + B = \{ x + y \,|\, x \in A, y \in B \}$ and $A - B = \{ x - y \,|\, x \in A,\ y \in B \}$.

We use $||\cdot||$ or $||\cdot||_2$ for the *Euclidean norm*, i.e.,

$$||x|| = ||x||_2 = \sqrt{x^\top x}$$

for any vector $x$ in $\mathbb{R}^n$. The open and closed balls with center $x \in \mathbb{R}^n$ and radius $r \in \mathbb{R}_{++}$ are the sets

$$B(x, r) = \{ y \in \mathbb{R}^n \,|\, ||x - y|| < r \},$$

and

$$\bar{B}(x, r) = \{ y \in \mathbb{R}^n \,|\, ||x - y|| \leq r \},$$

respectively. Let $C$ be a subset of $\mathbb{R}^n$ and $r \geq 0$. Then we define

$$B(C, r) = C + B(\mathbf{0}, r) \quad \text{and} \quad \bar{B}(C, r) = C + \bar{B}(\mathbf{0}, r).$$

Another norm occurring in the book is the maximum norm defined by

$$||x||_\infty = \max_{1 \leq i \leq n} |x_i|,$$

for any vector $x \in \mathbb{R}^n$. For two vectors $x$ and $y$ in $\mathbb{R}^n$, $d(x, y)$ and $d_\infty(x, y)$ denote the Euclidean and $||\cdot||_\infty$ norm distances between $x$ and $y$, respectively.

Given a finite set $I$, we denote the number of elements of $I$ by $|I|$. A vector $s$ in $\mathbb{R}^n$ is called a *sign vector* if each of its components is either $-1$, $0$ or $+1$. For a vector $x \in \mathbb{R}^n$, the following notation is used throughout the book,

$$I^-(x) = \{ h \in I_n \,|\, x_h < 0 \},$$

$$I^0(x) = \{ h \in I_n \,|\, x_h = 0 \},$$

$$I^+(x) = \{ h \in I_n \,|\, x_h > 0 \}.$$

For $x \in \mathbb{R}$, the *sign* of $x$ is given by

$$sign(x) = \begin{cases} +1 & \text{if } x > 0, \\ 0 & \text{if } x = 0, \\ -1 & \text{if } x < 0. \end{cases}$$

The sign vector of $x$ in $\mathbb{R}^n$ is given by $sign(x) = (sign(x_1), \cdots, sign(x_n))^\top$.

Let $x^1, \cdots, x^m$ be vectors in $\mathbb{R}^n$. The point $x = \sum_{i=1}^m \lambda_i x^i$ is a *linear combination* of $x^1, \cdots, x^m$ if $\lambda_i \in \mathbb{R}$, $i \in I_m$. If the sum of all $\lambda_i$ is equal to one, the point $x$ is an *affine combination* of $x^1, \cdots, x^m$. If all $\lambda_i$ are restricted to be non-negative with sum equal to 1, then the point $x$ is a *convex combination* of $x^1, \cdots, x^m$. The *affine hull* of the vectors $x^1, \cdots, x^m$, denoted by $aff(\{x^1, \cdots, x^m\})$, is given by

$$aff(\{x^1, \cdots, x^m\}) = \{x \in \mathbb{R}^n \mid x = \sum_{i=1}^m \lambda_i x^i, \sum_{i=1}^m \lambda_i = 1, \lambda_i \in \mathbb{R}, i \in I_m\}.$$

Thus, the affine hull of $x^1, \cdots, x^m$ consists of all affine combinations of these vectors. In general, for any subset $C$ of $\mathbb{R}^n$, the affine hull of $C$, denoted by $aff(C)$, is defined as the set of all affine combinations of points in $C$. The *convex hull* of the vectors $x^1, \cdots, x^m$, denoted by $co(\{x^1, \cdots, x^m\})$, is given by

$$co(\{x^1, \cdots, x^m\}) = \{x \in \mathbb{R}^n \mid x = \sum_{i=1}^m \lambda_i x^i, \sum_{i=1}^m \lambda_i = 1, \lambda_i \geq 0, i \in I_m\}.$$

Thus, the convex hull of $x^1, \cdots, x^m$ consists of all convex combinations of these vectors. Moreover, for any subset $C$ of $\mathbb{R}^n$, the convex hull of $C$, denoted by $co(C)$, is defined as the set of all convex combinations of points in $C$. A set $C$ of $\mathbb{R}^n$ is *convex* if it contains all the convex combinations of its elements. The *dimension* of a convex set $C$ is measured by the dimension of the linear subspace parallel to $aff(C)$ and denoted by $dim(C)$, i.e., $dim(C) = dim(aff(C) - \{x\})$ for any $x \in C$. By convention, $dim(C) = -1$ if $C$ is the empty set. For a set $C$ in $\mathbb{R}^n$, by $bd(C)$, $cl(C)$, and $int(C)$ we mean the *boundary*, *closure*, and *interior* of $C$ with respect to $aff(C)$, respectively. Boundedness, closedness and openness of a set $C$ in $\mathbb{R}^n$ are defined in the usual sense. A set $C$ in $\mathbb{R}^n$ is *compact* if it is both bounded and closed. In particular, a nonempty compact set in $\mathbb{R}^n$ is said to be *connected* if it can not be expressed as the union of two or more disjoint nonempty closed subsets. Given a subset $B$ of $\mathbb{R}^n$, $B$ is *comprehensive* if for any $y \in B$, $x \leq y$ implies $x \in B$.

**Example 1.2.1** *Let $x^1$ and $x^2$ be two elements in $\mathbb{R}^2$ not lying on the same line through the origin. Then the set of all linear combinations of $x^1$*

*and $x^2$ is the whole plane $\mathbb{R}^2$. The affine hull of $x^1$ and $x^2$ is the straight line through $x^1$ and $x^2$, and the convex hull of $x^1$ and $x^2$ is the closed line segment between $x^1$ and $x^2$.*

A function $f : C \mapsto \mathbb{R}$, defined on the convex set $C \subseteq \mathbb{R}^n$, is *convex* if

$$f(\beta x + (1-\beta)y) \leq \beta f(x) + (1-\beta) f(y)$$

for all $x$, $y \in C$ and all $\beta \in [0,1]$. If the above inequality is strict for all $x \neq y$ and all $\beta \in (0,1)$, then we say that $f$ is *strictly convex*. Analogously, the function $f$ is said to be *(strictly) concave* if $-f$ is (strictly) convex.

A function $f : C \mapsto \mathbb{R}$, defined on the convex set $C \subseteq \mathbb{R}^n$, is *quasi-convex* if the set $\{x \in C \mid f(x) \leq \beta\}$ is convex for any $\beta \in \mathbb{R}$. Analogously, we say that the function is *quasi-concave* if the set $\{x \in C \mid f(x) \geq \beta\}$ is convex for any $\beta \in \mathbb{R}$.

A binary relation $\preceq$ on a set $X$ is called to be a *partial order* if it is
(a) reflexive, i.e., $x \preceq x$ for all $x \in X$;
(b) anti-symmetric, i.e., $x \preceq y$ and $y \preceq x$ imply $x = y$; and
(c) transitive, i.e., $x \preceq y$ and $y \preceq z$ imply $x \preceq z$.

A set $X$ equipped with a partial order is a *partially order set*, sometimes called a *poset*. A *total order* or *linear order* is a partial order with the property that if $x \neq y$, then either $x \preceq y$ or $y \preceq x$. A *chain* in a poset is a subset on which the order is total. In other words, any two distinct elements of a chain are ranked by the partial order.

Let $X$ be a partially order set. An *upper bound* for a set $Y \subseteq X$ is an element $x \in X$ satisfying $y \preceq x$ for all $y \in Y$. An element $x \in X$ is a *maximal element* of $X$ if there is no $y \neq x$ in $X$ such that $x \preceq y$. An element $x \in X$ is a *greatest (biggest, best) element* of $X$ if $y \preceq x$ for all $y \in X$. Similarly, we can define *lower bound*, *minimal element* and *least (smallest, worst) element.* A total order on $X$ is a *well-ordering* if each nonempty subset of $X$ contains a smallest element. The *supremum* $z$ of a subset $Y$ of $X$ is its least upper bound and we write $z = \sup Y$. The *infimum* $z$ of a set $Y$ of $X$ is its greatest lower bound and we write $z = \inf Y$. A *lattice* is a partially ordered set in which every pair of elements has an infimum and a supremum. Note that the infimum and supremum of a set need not exist.

A sequence $\{a_m \mid m \in \mathbb{N}\}$ in $\mathbb{R}^n$ is called a *Cauchy sequence* if for any given $\epsilon > 0$ there is a positive integer $M$ such that $||x_k - x_l|| \leq \epsilon$ for all $k, l \geq M$. Note that a sequence in $\mathbb{R}^n$ has a limit if and only if it is Cauchy. Let $\{a_n \mid n \in \mathbb{N}\}$ be a sequence in $\mathbb{R}$. We define

$$\limsup_{n\to\infty} a_n = \lim_{k\to\infty} \sup \{a_n \mid n \geq k\},$$

and

$$\liminf_{n\to\infty} a_n = \lim_{k\to\infty} \inf \{a_n \mid n \geq k\}.$$

Here we adopt the following conventions. If $\{a_n \mid n \in \mathbb{N}\}$ is not bounded above, we decree $\limsup_{n\to\infty} a_n = +\infty$. Likewise, if $\{a_n \mid n \in \mathbb{N}\}$ is not bounded below, we decree $\liminf_{n\to\infty} a_n = -\infty$.

Given a subset $C$ of $\mathbb{R}^n$, the *diameter* of $C$ is defined to be

$$diam(C) = \sup\{||x - y|| \mid x, y \in C\}.$$

Given a subset $C$ of $\mathbb{R}$, a function $f : C \mapsto \mathbb{R} \cup \{-\infty, +\infty\}$ is said to be *lower semi-continuous* at $x^*$ if $f(x^*) \leq \liminf_{x\to x^*} f(x)$; and *upper semi-continuous* at $x^*$ when $f(x^*) \geq \limsup_{x\to x^*} f(x)$. The function is said to be lower (upper) semi-continuous if it is lower (upper) semi-continuous at every point in $C$.

Given a subset $C$ of $\mathbb{R}^n$ and a subset $X$ of $[-\infty, +\infty]$, a function $f : C \mapsto X$ is said to be a *proper function* if there exists at least one point $x \in C$ such that $f(x)$ is finite.

Finally, we will give an intuitive definition of an algorithm. An *algorithm* or a *method* is a list of instructions to solve a problem that can be programmed and implemented on a computer. For precise definitions we refer to Knuth [1968]. Algorithms or methods are important building blocks of this book.

## 1.3. Basic polyhedral theory

We recall some well-known concepts and theorems from polyhedral theory. For more details, we refer to Dantzig [1963], Stoer and Witzgall [1970], Schrijver [1986], Nemhauser and Wolsey [1988]. Let us first give some standard results from linear algebra. A set of points $x^1, \cdots, x^m$ in $\mathbb{R}^n$ is *linearly independent* if the unique solution of $\sum_{i=1} \beta_i x^i = \mathbf{0}$ is $\beta_i = 0$, for all $i \in I_m$.

**Definition 1.3.1** *A set of points $x^1, \cdots, x^m$ in $\mathbb{R}^n$ is affinely independent if $\sum_{i=1}^m \beta_i x^i = \mathbf{0}$ and $\sum_{i=1}^m \beta_i = 0$ imply $\beta_i = 0$ for all $i \in I_m$.*

It is easy to verify the following result.

**Lemma 1.3.2** *The following statements are equivalent:*
*(i) $x^1, \cdots, x^m$ are affinely independent;*
*(ii) $x^2 - x^1, \cdots, x^m - x^{m-1}$ are linearly independent;*
*(iii) $(1, x^{1\top})^\top, \cdots, (1, x^{m\top})^\top$ are linearly independent;*
*(iv) $dim(co(\{x^1, \cdots, x^m\})) = m - 1$.*

Observe that the maximum number of affinely independent vectors in $\mathbb{R}^n$ equals $n + 1$.

Given a nonzero vector $a \in \mathbb{R}^n$ and a number $d \in \mathbb{R}$, the sets $\{x \in \mathbb{R}^n \mid a^\top x \leq d\}$, $\{x \in \mathbb{R}^n \mid a^\top x \leq 0\}$, and $\{x \in \mathbb{R}^n \mid a^\top x = d\}$ are called an *affine half-space*, a *linear half-space*, and a *hyperplane* in $\mathbb{R}^n$, respectively.

A *polyhedron* $P$ in $\mathbb{R}^n$ is the intersection of finitely many affine half-spaces. In other words, $P = \{ x \in \mathbb{R}^n \mid Ax \leq b \}$, where $A$ is an $m \times n$ real matrix and $b$ is a vector in $\mathbb{R}^m$. A bounded polyhedron is called a *polytope*. It is clear that a polyhedron is a convex set. Given a polyhedron $P = \{ x \in \mathbb{R}^n \mid Ax \leq b \}$, we say that a constraint $a^{i\top} x \leq b_i$ from the system $Ax \leq b$ is *redundant* if it can be removed from the system without affecting the set $P$. A system $Ax \leq b$ is *irredundant* if it has no redundant constraints.

A set $C$ in $\mathbb{R}^n$ is called a *cone* if $x \in C$ implies $\beta x \in C$ for all $\beta \in \mathbb{R}_+$. A cone is polyhedral if it can be expressed as $P = \{ x \in \mathbb{R}^n \mid Ax \leq \mathbf{0} \}$ for some matrix $A$. A cone generated by a finite number of vectors $x^1, \cdots, x^m$ in $\mathbb{R}^n$ is defined as the set

$$Cone(\{ x^1, \cdots, x^m \}) = \{ x \in \mathbb{R}^n \mid x = \sum_{i=1}^{m} \beta_i x^i,\ \beta_i \geq 0,\ i = 1, \cdots, m \}.$$

It is called a *finitely generated cone.*

An $n \times n$ matrix $A$ with entries $a_{i,j}$ is called *doubly stochastic* if

$$\begin{array}{ll} \sum_{i=1}^n a_{i,j} = 1, & j = 1, \cdots, n, \\ \sum_{j=1}^n a_{i,j} = 1, & i = 1, \cdots, n, \\ a_{i,j} \geq 0, & i,j = 1, \cdots, n. \end{array}$$

An $n \times n$ matrix is a *permutation matrix* if each entry is 0 or 1 and there is exactly one 1 in each row and in each column. An $n \times n$ matrix $A$ is *regular* or *non-singular* if its inverse $A^{-1}$ exists.

We can now introduce several classical results related to the concepts above.

**Lemma 1.3.3 (Farkas-Minkowski-Weyl Theorem)** *A cone is polyhedral if and only if it is finitely generated.*

**Lemma 1.3.4 (Minkowski Theorem)** *A set $P$ in $\mathbb{R}^n$ is a polyhedron if and only if $P = Q + C$ for some polytope $Q$ and some polyhedral cone $C$ in $\mathbb{R}^n$. In other words, a polyhedron $P = \{x \in \mathbb{R}^n \mid Ax \leq b\}$ can be expressed as*

$$\begin{array}{lll} P = \{x \in \mathbb{R}^n & \mid & x = \sum_{i=1}^m \lambda_i x^i + \sum_{j=1}^k \mu_j y^j,\ \sum_{i=1}^m \lambda_i = 1, \\ & & \lambda_i \geq 0 \text{ for all } i \in I_m;\ \mu_j \geq 0 \text{ for all } j \in I_k \} \end{array}$$

*where $x^1, \cdots, x^m, y^1, \cdots, y^k$ are elements of $\mathbb{R}^n$.*

**Lemma 1.3.5 (Farkas Lemma)** *Let $A$ be an $(m \times n)$-matrix and let $b$ be an $m$-vector. Then there exists a vector $x \in \mathbb{R}^n_+$ with $Ax = b$ if and only if $y^\top b \leq 0$ for each $m$-vector $y$ with $y^\top A \leq \mathbf{0}$.*

The first two theorems are quite obvious to geometric intuition. The content of Farkas lemma geometrically says that if a vector $b$ is not an element of the cone generated by the column vectors $a^1, \cdots, a^n$ of $A$, then there exists a hyperplane $\{ x \mid y^\top x = \delta \}$ for some nonzero $m$-vector $y$ and some number $\delta$ separating $b$ strictly from $a^1, \cdots, a^n$.

**Theorem 1.3.6 (Birkhoff-von Neumann Theorem)** *For $n \in \mathbb{N}$, an $n \times n$ matrix $A$ is doubly stochastic if and only if $A$ is a convex combination of $n \times n$ permutation matrices.*

Let $c$ be a nonzero element of $\mathbb{R}^n$, $P$ a polyhedron $\{ x \in \mathbb{R}^n \mid Ax \leq b \}$ and $\delta = \max\{ c^\top x \mid x \in P \}$. The hyperplane $\{ x \in \mathbb{R}^n \mid c^\top x = \delta \}$ is called a *supporting hyperplane* of $P$ if $\delta$ is finite. We say that a subset $F$ of $P$ is a *face* of $P$ if $F$ is either $P$ itself or the intersection of $P$ with a supporting hyperplane of $P$. We will specially reserve the geometric terminology for the following cases. If a face $F$ of $P$ consists of a single point then it is called a *vertex* of $P$. If $P$ has a vertex, $P$ is called *pointed*. When $P$ is pointed, a face $F$ of $P$ is called an *edge* if $F$ consists of a half line or a line segment, while a face $F$ of dimension one lower than the dimension of $P$ is called a *facet* of $P$.

**Example 1.3.7** *Let a polyhedron $P = \{ x \in \mathbb{R}^2 \mid a^{i\top} x \leq b_i,\ i \in I_3 \}$ be given, where $a^1 = (1, -2)^\top$, $a^2 = (-1, -1)^\top$, and $a^3 = (-1, 1)^\top$, $b_1 = 2$, $b_2 = 1$, and $b_3 = 1$. The points $x^1 = (-1, 0)^\top$ and $x^2 = (0, -1)^\top$ are the vertices of $P$, and the line segment connecting $x^1$ and $x^2$ and the two half-lines initiating at $x^1$ and $x^2$, are the edges of $P$. Moreover, the system $Ax \leq b$ is irredundant.*

Let $P = \{ x \in \mathbb{R}^n \mid Ax \leq b \}$ be a nonempty polyhedron where $A$ is an $(m \times n)$-matrix and $b$ is an $m$-vector. An inequality $a^i x \leq b_i$ from $Ax \leq b$ is called an *implicit equality* if $a^i x = b_i$ for all $x$ satisfying $Ax \leq b$. Throughout the book, unless stated otherwise, we shall assume that a nonempty polyhedron $P$ is given as

$$P = \{ x \in \mathbb{R}^n \mid Ax \leq b,\, Cx = d \},$$

where $A$ is an $m \times n$ matrix and $C$ a $k \times n$ matrix, $b$ an $m$-vector and $d$ a $k$-vector, under the condition that no constraint of the system $Ax \leq b$, $Cx = d$ is redundant, and the system $Ax \leq b$ has no implicit equalities. For $I \subseteq I_m$, we define a subset $F(I)$ of $P$ as

$$F(I) = \{ x \in P \mid a^{i\top} x = b_i \quad \text{for all} \quad i \in I \}.$$

Define $F(\emptyset) = P$. In case $F(I)$ is nonempty, it is obviously a face of $P$. A polyhedron $P$ is *simple* if the dimension of any face $F(I)$ of $P$ is equal to

$n-k-|I|$. Notice that the polytope $P$ given by $P = \{x \in \mathbb{R}^3 \mid -2x_1+x_3 \leq 0,\ -2x_2+x_3 \leq 0,\ 2x_2+x_3 \leq 4,\ 2x_1+x_3 \leq 4,\ -x_3 \leq 0\}$ is not simple since the vertex $v = (1,1,2)^\top$ has dimension zero but $n-|I| = -1$, where $I$ is the set of the indices of the first four constraints. Note that the system $Ax \leq b$ is irredundant.

Consider a linear programming problem $\max\{c^\top x \mid Ax \leq b\}$. Now we introduce the following fundamental theorem.

**Lemma 1.3.8 (Duality Theorem of Linear Programming)** *Let $A$ be an $m \times n$ matrix, let $b$ be an $m$-vector and let $c$ be an $n$-vector. Then*

$$\max\{c^\top x \mid Ax \leq b\} = \min\{b^\top y \mid y \geq 0,\ y^\top A = c^\top\}$$

*provided that both sets above are nonempty.*

Another equivalent form of the duality theorem can be stated as follows:

$$\max\{c^\top x \mid x \geq 0,\ Ax = b\} = \min\{b^\top y \mid y^\top A \geq c^\top\}$$

provided that both sets are nonempty. If the above maximization problem is called the *primal* problem, then the corresponding minimization problem is referred to as the *dual* problem, or vice versa.

## 1.4. Simplices, triangulations and graphs

In this section we start with the notion of triangulations and introduce some particular triangulations to be used for the development of simplicial algorithms. Triangulations or simplicial subdivisions originally appeared in the field of topology. They are now one of the indispensable ingredients of simplicial algorithms.

To describe a triangulation of a convex set in $\mathbb{R}^n$, we first give the definition of a simplex.

**Definition 1.4.1** *If $x^1, \cdots, x^{t+1}$ in $\mathbb{R}^n$ are affinely independent, the convex hull of these points is a $t$-dimensional simplex or $t$-simplex.*

A $t$-simplex $\sigma = co(\{x^1, \cdots, x^{t+1}\})$ is often denoted by $\sigma(x^1, \cdots, x^{t+1})$, or simply by $< x^1, \cdots, x^{t+1} >$. It easily follows from Lemma 1.3.2 that no simplex in $\mathbb{R}^n$ has dimension more than $n$.

**Definition 1.4.2** *A $k$-simplex $\tau$ is a $k$-dimensional face or $k$-face of a $t$-simplex $\sigma(x^1, \cdots, x^{t+1})$ if all vertices of $\tau$ are vertices of $\sigma$.*

It is clear from the previous section that in particular, $x^1, \cdots, x^{t+1}$ are the vertices of the $t$-simplex $\sigma(x^1, \cdots, x^{t+1})$, and that a $(t-1)$-face of $\sigma$ is a facet of $\sigma$. Naturally, $\sigma$ is a face of itself. A facet $\tau$ of the $t$-simplex $\sigma$ is called the facet of $\sigma$ *opposite* the vertex $y$ if $y$ is a vertex of $\sigma$ not being a vertex of $\tau$.

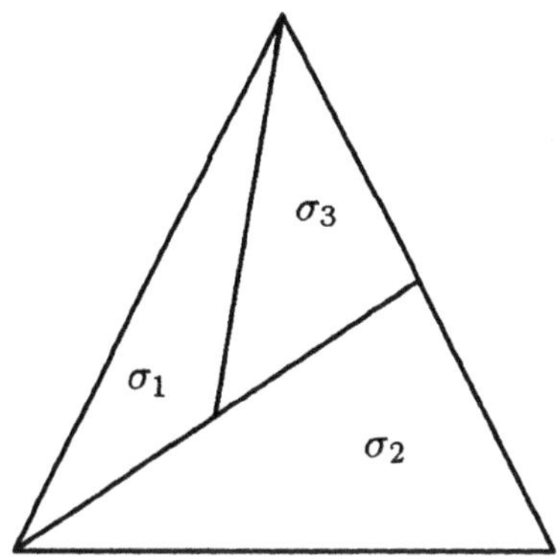


*Figure 1.1.* Illustration of a non-triangulation.

**Definition 1.4.3** *Let $C$ be a $t$-dimensional convex set in $\mathbb{R}^n$. A collection $\mathcal{T}$ of $t$-simplices is a triangulation or simplicial subdivision of $C$ if*
*(a) $C$ is the union of all simplices in $\mathcal{T}$;*
*(b) for any $\sigma_1$, $\sigma_2$ in $\mathcal{T}$, $\sigma_1 \bigcap \sigma_2$ is either the empty set or a common face of both;*
*(c) each element $x$ in $C$ has a neighborhood intersecting only a finite number of simplices in $\mathcal{T}$.*

Denote by $\mathcal{T}^k$ the collection of all $k$-faces of simplices of $\mathcal{T}$. It is trivial to see that $\mathcal{T}^t = \mathcal{T}$, and that the members of $\mathcal{T}^0$ are the vertices of simplices of $\mathcal{T}$. Finally, we denote by $\mathcal{T}^+$ the collection of all faces of simplices of $\mathcal{T}$.

Let us give some examples to illustrate the definition of a triangulation. In Figure 1.1, $C$ is the union of the 2-simplices $\sigma_1$, $\sigma_2$ and $\sigma_3$. However, the collection of these simplices is not a triangulation since $\sigma_1 \bigcap \sigma_2$ is not a common face of both simplices. In this case Condition (b) is violated. In Figure 1.2, $C$ is the union of the 2-simplices $\sigma_1$, $\sigma_2$, $\sigma_3$ and $\sigma_4$. The collection of these simplices is indeed a triangulation of $C$ as one can easily check.

**Example 1.4.4** *Let $C = [0, 2]$, and let $\mathcal{T}$ be the collection of 1-simplices given by $\mathcal{T} = \{ [\frac{i-1}{i+1}, \frac{i}{i+1}] \mid i \in \mathbb{N} \} \bigcup \{ [1, 2] \}$. It is easy to check that the collection $\mathcal{T}$ is not a triangulation of $C$ since any neighborhood of the point 1 meets an infinite number of simplices of $\mathcal{T}$. However, $\mathcal{T}' = \{ [\frac{i-1}{i+1}, \frac{i}{i+1}] \mid i \in \mathbb{N}\}$ is a triangulation of the interval $[0, 1)$.*

To compare simplices and triangulations, we introduce the diameter of a simplex and the mesh or grid size of a triangulation. For other measures of triangulation, the reader is referred to Todd [1976a], Talman [1980], Eaves and Yorke [1984].

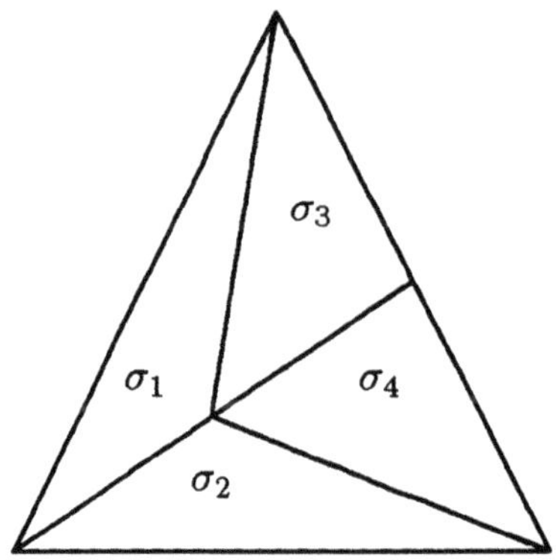


*Figure 1.2.* Illustration of a triangulation.

**Definition 1.4.5** *Let $\mathcal{T}$ be a triangulation of a convex set in $\mathbb{R}^n$. The diameter of a simplex $\sigma \in \mathcal{T}$ is given by*

$$diam(\sigma) = \max\{\, ||x - y|| \,|\, x,\, y \in \sigma \,\}.$$

*The mesh size of the triangulation $\mathcal{T}$ is given by*

$$mesh(\mathcal{T}) = \sup\{\, diam(\sigma) \,|\, \sigma \in \mathcal{T} \,\}.$$

Similarly, we can define the diameter and the mesh size for the maximum norm and denote these by $diam_\infty(\sigma)$ and $mesh_\infty(\mathcal{T})$, respectively.

The most important properties of triangulations are given below.

**Theorem 1.4.6** *Let $\mathcal{T}$ be a triangulation of a $t$-dimensional convex set $C$ in $\mathbb{R}^n$.*
*(i) If $D \subseteq bd(C)$ and $D = aff(D) \bigcap C$ with $dim(D) = t - 1$, then $\mathcal{T}' = \{\, \tau \,|\, \tau \subseteq D,\, \tau \in \mathcal{T}^{t-1} \,\}$ is a triangulation of $D$;*
*(ii) If $\tau \in \mathcal{T}^{t-1}$, then either $\tau \subseteq bd(C)$ and $\tau$ is a facet of exactly one simplex of $\mathcal{T}$, or $\tau \not\subseteq bd(C)$ and $\tau$ is a common facet of just two simplices of $\mathcal{T}$;*
*(iii) If $D$ is a compact subset of $C$, then $D$ meets only a finite number of simplices of $\mathcal{T}$.*

We are now ready to present some special triangulations. The first one was proposed by Freudenthal [1942] and rediscovered by Tucker in 1945 (see Lefschetz [1949, p. 140]). It is called $K_1$-triangulation and described below.

**Definition 1.4.7** *The $K_1$-triangulation of $\mathbb{R}^n$ is the collection of all $n$-simplices $\sigma(x^1, \pi)$ with vertices $x^1, \cdots, x^{n+1}$ such that*

(*a*) $\pi = (\pi(1), \cdots, \pi(n))$ *is a permutation of the elements of the set* $I_n$;
(*b*) $x^1 \in \mathbf{Z}^n$ *and* $x^{i+1} = x^i + e(\pi(i))$ *for each* $i \in I_n$.

The following theorem shows that $K_1$ is indeed a triangulation of $\mathbb{R}^n$. The proof is standard and can also be found in Todd [1976a]. The idea here can be adapted when considering other triangulations.

**Theorem 1.4.8** $K_1$ *is a triangulation of* $\mathbb{R}^n$.

Proof: It is clear that $K_1$ satisfies Condition (c) of Definition 1.4.3. It remains to show that each point $x$ in $\mathbb{R}^n$ lies in the interior of just one face of a simplex of $K_1$. Take an arbitrary point $x$ from $\mathbb{R}^n$. Let $y_i^1 = \lfloor x_i \rfloor$ for each $i \in I_n$. Thus $y^1 \in K_1^0$. Let $z = x - y^1$. Clearly, $0 \le z_i < 1$ for all $i \in I_n$. Let $\pi = (\pi(1), \cdots, \pi(n))$ be a permutation of $I_n$ such that $1 > z_{\pi(1)} \ge \cdots \ge z_{\pi(n)} \ge 0$. For convenience, denote these terms correspondingly as

$$\beta_1 = 1 > \beta_2 \ge \cdots \ge \beta_{n+1} \ge \beta_{n+2} = 0.$$

Let $\gamma_i = \beta_i - \beta_{i+1}$ for each $i \in I_{n+1}$. Obviously, $\sum_{i \in I_{n+1}} \gamma_i = \beta_1 - \beta_{n+2} = 1$, and $\gamma_i \ge 0$ for all $i \in I_{n+1}$. Take the $n$-simplex $\sigma = \sigma(y^1, \pi) =< y^1, \cdots, y^{n+1} >$ of $K_1$ and consider $\sum_{i \in I_{n+1}} \gamma_i y^i$. It follows that

$$\begin{aligned}
\sum_{i \in I_{n+1}} \gamma_i y^i &= y^1 + \sum_{i=2}^{n+1} \gamma_i (y^i - y^1) \\
&= y^1 + \sum_{i=2}^{n+1} \gamma_i \left(\sum_{j=2}^{i} (y^j - y^{j-1})\right) \\
&= y^1 + \sum_{i=2}^{n+1} (y^i - y^{i-1}) \left(\sum_{j=i}^{n+1} \gamma_j\right) \\
&= y^1 + \sum_{i=2}^{n+1} e(\pi(i-1)) \beta_i \\
&= y^1 + \sum_{i \in I_n} z_{\pi(i)} e(\pi(i)) \\
&= y^1 + z \\
&= x.
\end{aligned}$$

Therefore $x$ belongs to $\sigma$. Let $J = \{ j \in I_{n+1} \mid \gamma_j > 0 \}$. Let $\tau$ be a face of $\sigma$ whose vertices are $y^j$, $j \in J$. It is readily seen that $x$ lies in the interior of $\tau$. Now we need to show that all $y^j$, $j \in J$, are uniquely determined by $x$ independently of $y^1$ and $\pi$. This can be done as follows. For each $t \in (0, 1]$, define $y(t) \in \mathbf{Z}^n$ by

$$y_i(t) = \begin{cases} \lfloor x_i \rfloor + 1, & \text{if } x_i - \lfloor x_i \rfloor \ge t, \\ \lfloor x_i \rfloor, & \text{if } x_i - \lfloor x_i \rfloor < t. \end{cases}$$

Then $(0, 1]$ is divided into $|J|$ disjoint intervals, each of which corresponds to exactly one vertex $y^j$, $j \in J$. Clearly, this construction is proceeded irrespective to $y^1$ and $\pi$. □

The $K_1$-triangulation is illustrated in Figure 1.3 for $n = 2$. To comprehend the above proof, let us furnish an example.

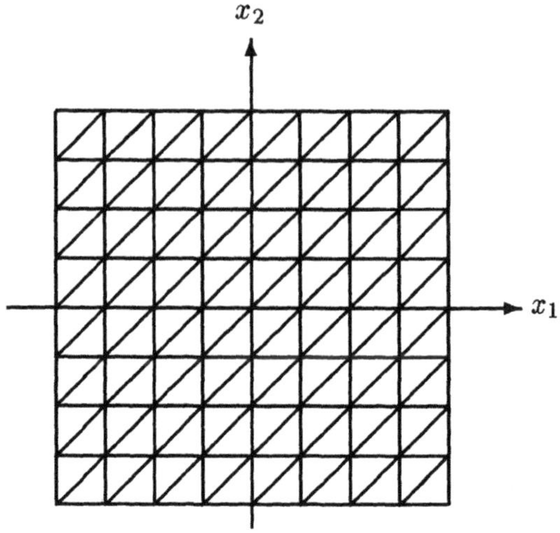


*Figure 1.3.* Illustration of $K_1$-triangulation with $n = 2$.

**Example 1.4.9** *Consider the $K_1$-triangulation of $\mathbb{R}^n$ for $n = 3$. Let $x = (2.5, -1.2, 0.5)^\top$. We have $y^1 = (2, -2, 0)^\top$ and $z = (0.5, 0.8, 0.5)^\top$. Since $1 > z_2 > z_1 = z_3 > 0$, we can choose $\pi^1 = (2, 3, 1)$ and $\pi^2 = (2, 1, 3)$. In either case, $\beta = (1, 0.8, 0.5, 0.5)$ and $\gamma = (0.2, 0.3, 0, 0.5)$. Then the two possible simplices containing $x$ are $\sigma^1(x^1, \pi^1) =< x^1, \cdots, x^4 >$ and $\sigma^2(z^1, \pi^2) =< z^1, \cdots, z^4 >$, where $x^1 = z^1 = y^1 = (2, -2, 0)^\top$, $x^2 = z^2 = (2, -1, 0)^\top$, $x^3 = (2, -1, 1)^\top$, $z^3 = (3, -1, 0)^\top$, and $x^4 = z^4 = (3, -1, 1)^\top$. It is easy to check that $x = \sum_{i=1}^4 \gamma_i x^i = \sum_{i=1}^4 \gamma_i z^i$. The point $x$ lies in the interior of the common face $< x^1, x^2, x^4 >$. Next, we put aside $y^1$ and $\pi$ to determine the vertices $x^1$, $x^2$ and $x^4$ directly from $x$. Recall that for each $t \in (0, 1]$, $y(t) \in \mathbb{Z}^n$ is defined by*

$$y_i(t) = \begin{cases} \lfloor x_i \rfloor + 1 & \textit{if } \ x_i - \lfloor x_i \rfloor \geq t, \\ \lfloor x_i \rfloor, & \textit{if } \ x_i - \lfloor x_i \rfloor < t. \end{cases}$$

*In this case, it follows that*

$$y_1(t) = \begin{cases} 3, & \textit{if } \ t \leq 0.5, \\ 2, & \textit{if } \ t > 0.5, \end{cases}$$

$$y_2(t) = \begin{cases} -1, & \textit{if } \ t \leq 0.8, \\ -2, & \textit{if } \ t > 0.8, \end{cases}$$

$$y_3(t) = \begin{cases} 1, & \textit{if } \ t \leq 0.5, \\ 0, & \textit{if } \ t > 0.5. \end{cases}$$

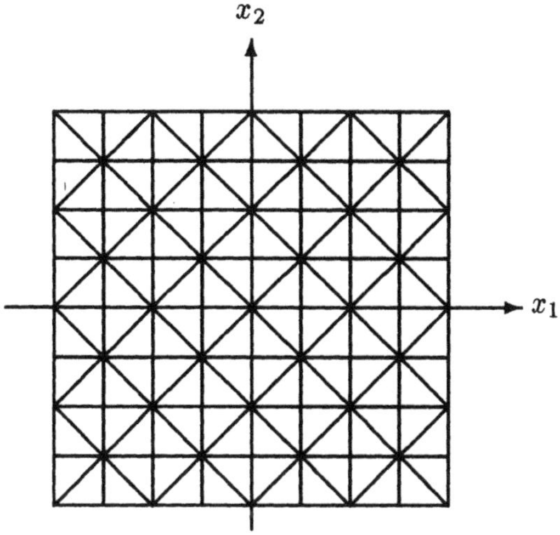


*Figure 1.4.* Illustration of $J_1$-triangulation for $n = 2$.

*In summary, we have*

$$y(t) = \begin{cases} (3,-1,1)^\top = x^4, & \textit{for} \quad 0 < t \le 0.5, \\ (2,-1,0)^\top = x^2, & \textit{for} \quad 0.5 < t \le 0.8, \\ (2,-2,0)^\top = x^1, & \textit{for} \quad 0.8 < t \le 1. \end{cases}$$

Now we introduce the $J_1$-triangulation of $\mathbb{R}^n$ originally due to Tucker (see also Todd [1977]).

**Definition 1.4.10** *Let $O^n = \{x \in \mathbb{Z}^n \mid x_i$ is odd for each $i \in I_n\}$. The $J_1$-triangulation of $\mathbb{R}^n$ is the collection of all simplices $\sigma(y^1, s, \pi)$ with vertices $y^1, \cdots, y^{n+1}$ such that*
*(a) $\pi = (\pi(1), \cdots, \pi(n))$ is a permutation of the elements of the set $I_n$;*
*(b) $s$ is a sign vector in $\{-1, +1\}^n$;*
*(c) $y^1 \in O^n$, and $y^{i+1} = y^i + s_{\pi(i)} e(\pi(i))$ for each $i \in I_n$.*

The $J_1$-triangulation of $\mathbb{R}^n$ is illustrated in Figure 1.4 for $n = 2$.

The next simplicial subdivision of $\mathbb{R}^n$ due to Todd [1978a] is called $K'$-triangulation. It is illustrated in Figure 1.5 for $n = 2$.

**Definition 1.4.11** *The $K'$-triangulation of $\mathbb{R}^n$ is the collection of all simplices $\sigma(y^1, s, \pi)$ with vertices $y^1, \cdots, y^{n+1}$ such that*
*(a) $\pi = (\pi(1), \cdots, \pi(n))$ is a permutation of the elements of the set $I_n$;*
*(b) $s$ is a sign vector in $\mathbb{R}^n$ such that $s_i = -1$ for $i \in I^-(y^1)$, $s_i = +1$ for $i \in I^+(y^1)$ and $s_i \in \{-1, +1\}$ for $i \in I^0(y^1)$;*
*(c) $y^1 \in \mathbb{Z}^n$, and $y^{i+1} = y^i + s_{\pi(i)} e(\pi(i))$ for each $i \in I_n$.*

Now we describe a method of obtaining a triangulation with any desired mesh size. For any subset $C$ of $\mathbb{R}^n$ and $\delta$, define $\delta C = \{\delta x \mid x \in C\}$. For any family $F$ of subsets of $\mathbb{R}^n$ and $\delta \neq 0$, define $\delta F = \{\delta C \mid C \in F\}$.

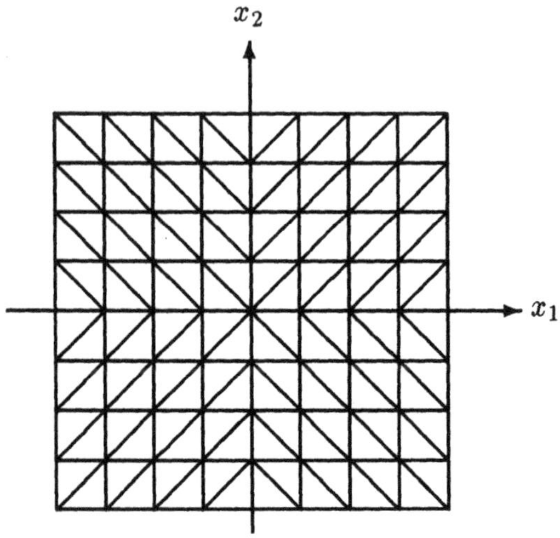


*Figure 1.5.* Illustration of $K'$-triangulation for $n = 2$.

So if $\mathcal{T}$ is a triangulation of $C$ and $\delta > 0$, then $\delta\mathcal{T}$ is a triangulation of $\delta C$. In this way we can get $\delta K_1$ and $\delta J_1$ triangulations for $\mathbb{R}^n$ with $mesh(\delta K_1) = mesh(\delta J_1) = \delta\sqrt{n}$ provided that $\delta > 0$.

Let us now introduce the following four basic convex sets which underlie many problems. These sets are the unit cube, the symmetric cube, the unit simplex and the simplotope. The $n$-dimensional unit cube is given by

$$U^n = \{x \in \mathbb{R}^n \mid 0 \leq x_i \leq 1,\ i \in I_n\}.$$

The $n$-dimensional symmetric cube is given by

$$C^n = \{x \in \mathbb{R}^n \mid -1 \leq x_i \leq 1,\ i \in I_n\}.$$

The $(n-1)$-dimensional unit simplex is given by

$$S^n = \{\, x \in \mathbb{R}^n_+ \mid \sum_{i=1}^{n} x_i = 1 \,\}.$$

It is easy to see that $S^n$ is the convex hull of all unit vectors $e(i)$, $i \in I_n$, in $\mathbb{R}^n$. For $h \in I_n$, $S^n_h$ denotes the facet $S^n_h = \{\, x \in S^n \mid x_h = 0 \,\}$, and for $T \subset I_n$, $S^n(T) = \bigcap_{h \in T} S^n_h$.

The *simplotope* $\mathcal{S}$ is defined to be the Cartesian product of, say $N$, $N \in \mathbb{N}$, unit simplices $S^{n_j}$, $n_j \in \mathbb{N}$, $j \in I_N$, i.e.,

$$\mathcal{S} = \prod_{j=1}^{N} S^{n_j} = \{\, x = (x_1, \cdots, x_N) \mid x_j = (x_{j,1}, \cdots, x_{j,n_j})^\top \in S^{n_j}, j \in I_N \,\}.$$

Now we present the $V$-triangulation of the unit simplex $S^n$ which is introduced by Doup and Talman [1987b] (see also Doup [1988]). For $\emptyset \subset K \subset I_n$, we define the subset $AS^n(K)$ of $S^n$ by $AS^n(K) = \{ x \in S^n \mid x_h = 0, h \notin K \}$, i.e., $AS^n(K) = S^n(I_n \backslash K)$.

**Definition 1.4.12** *Let $v$ be an arbitrary point in $S^n$ and let $K$ be a nonempty, proper subset of $I_n$. Then the projection $p(K)$ of $v$ on $AS^n(K)$ is given by*

$$p_i(K) = \begin{cases} 0, & \text{if } i \notin K, \\ \frac{(1-\sum_{h\in K} v_h)}{(|K_0|+\sum_{h\in K} v_h)}, & \text{if } i \in K_0, \\ \frac{v_i(|K_0|+1)}{(|K_0|+\sum_{h\in K} v_h)}, & \text{if } i \in K\backslash K_0, \end{cases}$$

*where $K_0$ is the set $\{ j \in K \mid v_j = 0 \}$. For $K = \emptyset$, we define $p(\emptyset) = v$.*

Let $T(n-1)$ be a subset of $I_n$ containing $n-1$ elements and let $\gamma = (\gamma_1, \cdots, \gamma_{n-1})$ be a permutation of the elements of $T(n-1)$. A subset $A(T(n-1); \gamma)$ of $S^n$ is defined by

$$A(T(n-1); \gamma) = \{ x \in S^n \mid x = v + \sum_{i=1}^{n-1} \beta_i q(\gamma_i), \ 0 \le \beta_{n-1} \le \cdots \le \beta_1 \le 1 \}$$

where $q(\gamma_i)$, $i = 1, \cdots, n-1$, is defined by

$$q(\gamma_i) = p(\{ \gamma_1, \cdots, \gamma_i \}) - p(\{ \gamma_1, \cdots, \gamma_{i-1} \}).$$

The sets $A(T(n-1); \gamma)$ are illustrated in Figure 1.6 for $n = 3$ where $B_1 = A(\{2,3\}; (3,2))$, $B_2 = A(\{2,3\}; (2,3))$, $B_3 = A(\{1,2\}; (2,1))$, $B_4 = A(\{1,2\}; (1,2))$, $B_5 = A(\{1,3\}; (1,3))$, and $B_6 = A(\{1,3\}; (3,1))$.

Let $m$ be a positive integer. The $V$-triangulation with grid size $m^{-1}$ subdivides $A(T(n-1); \gamma)$ into $(n-1)$-simplices as follows:

**Definition 1.4.13** *Let $m$ be a positive integer. Let $T(n-1)$ be a subset of $I_n$ and let $\gamma = (\gamma_1, \cdots, \gamma_{n-1})$ be a permutation of the elements in $T(n-1)$. Then the $V(T(n-1); \gamma)$-triangulation with grid size $m^{-1}$ of $A(T(n-1); \gamma)$ is the collection of all simplices $\sigma(y^1, \pi)$ with vertices $y^1, \cdots, y^n$ such that*
*(a) $y^1 = v + \sum_{h=1}^{n-1} \beta(\gamma_h) q(\gamma_h)/m$ for integers $\beta(\gamma_h)$, $h = 1, \cdots, n-1$, such that $0 \le \beta(\gamma_1) \le \cdots \le \beta(\gamma_{n-1}) \le m-1$;*
*(b) $\pi = (\pi(1), \cdots, \pi(n-1))$ is a permutation of the elements in $T(n-1)$ satisfying that for some $h$, $1 < h \le n-1$, it holds that $t > s$ if $\beta(\pi(t)) = \beta(\pi(s))$, where $\pi(t) = \gamma_h$ and $\pi(s) = \gamma_{h-1}$;*
*(c) $y^{i+1} = y^i + q(\pi(i))/m$ for each $i \in I_{n-1}$.*

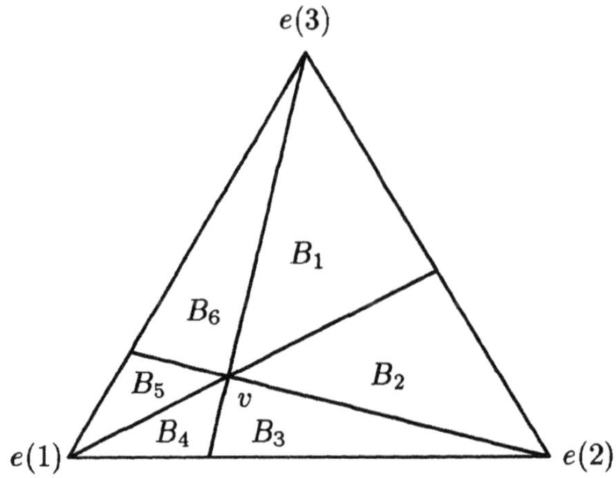


*Figure 1.6.* The sets $A(T(n-1);\gamma)$ of $S^n$ for $n = 3$.

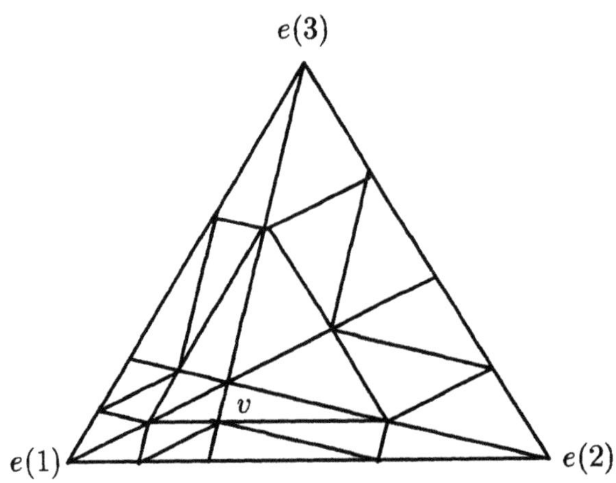


*Figure 1.7.* Illustration of $V$-triangulation of $S^n$ for $n = 3$.

It is easy to show that the unit simplex $S^n$ is the union of $A(T(n-1);\gamma)$ over all permutations $\gamma$ and over all sets $T(n-1) \subset I_n$. The $V$-triangulation with grid size $m^{-1}$ of $S^n$ is obtained by taking the union of the collections $V(T(n-1);\gamma)$ over all permutations $\gamma$ of $T(n-1)$ and over all subsets $T(n-1)$ of $I_n$. It is illustrated in Figure 1.7 for $n = 3$ and $m = 2$. In a similar way, a simplotope can be triangulated; see Doup [1988].

At the end of this section we give a brief review of some elementary concepts and results in graph theory.

A *graph* is a pair $G = (V, A)$, where $V$ is a finite or countably infinite set, and $A$ is a family of unordered pairs of elements of $V$. The elements of $V$ are called the *nodes or vertices* of $G$, and the elements of $A$ the *arcs or edges* of $G$. An arc is denoted by a pair $\{u, v\}$, $u \neq v$, and is said to connect the nodes $u$ and $v$. The nodes $u$ and $v$ are adjacent if there is an arc connecting them. An arc $\{u, v\}$ is said to be incident to the node $u$ and to the node $v$. The nodes $u$ and $v$ are called the ends or end points of the arc $\{u, v\}$. The number of arcs incident to a node $v$ is called the *degree* of $v$, denoted by $deg(v)$. If a node has degree zero, we say it is *isolated.*

A *path* in a graph $G = (V, A)$ from node $v_0$ to node $v_t$ is defined as a sequence of the form

$$(v_0, a_1, v_1, a_2, \cdots, v_{t-1}, a_t, v_t) \tag{1.4.1}$$

where $v_0, \cdots, v_t$ are nodes and $a_1, \cdots, a_t$ are arcs, such that $a_h = \{v_{h-1}, v_h\}$ for $h = 1, \cdots, t$. A path of form (1.4.1) is called a $(v_0, v_t)$-path. A $(v_0, v_t)$-path connects $v_0$ and $v_t$. The nodes $v_0$ and $v_t$ are called the starting point and the end point of a $(v_0, v_t)$-path, respectively, or simply the end points of the path. A path is *simple* if all its nodes and arcs are different. The *length* of a simple $(v_0, v_t)$-path is $t$. If $v_0 = v_t$, the $(v_0, v_t)$-path is *closed.* A closed path of length at least three and without repeated arcs or nodes (except for the end points) is called a *circuit.* A path goes to infinity from $v_0$ if it is an infinite sequence of the form

$$(v_0, a_1, v_1, a_2, \cdots, v_{t-1}, a_t, v_t, \cdots).$$

Similarly, we can define a circuit with length of infinity. A path or circuit of a graph is called a *connected component* of the graph.

The following theorem is very useful for proving the convergence of simplicial algorithms.

**Theorem 1.4.14** *Let $G = (V, A)$ be a graph of which each node has at most degree two. Then every path in the graph is simple. There may exist some circuits and isolated nodes. For any node $v$ with degree one, there exists a unique path from $v$ which either goes to infinity or leads to another node with degree one.*

The theorem is illustrated in Figure 1.8. This graph is finite. There are one path, one circuit, and three isolated nodes.

## 1.5. Brouwer's fixed point theorem, Sperner's lemma and KKM lemma

In this section we are concerned with fixed point problems, other equivalent problems and some existence theorems. The fixed point problem states that

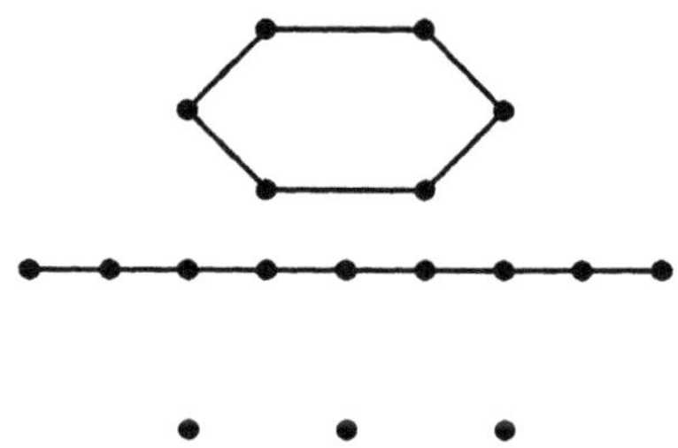

*Figure 1.8.* Illustration of a finite graph with degree at most two.

if a function $f$ maps from a nonempty subset $C$ of $\mathbb{R}^n$ into a subset $D$ of $\mathbb{R}^n$, does there exist any point in $C$ which is mapped by $f$ into itself? If so, such a point, say, $x^*$ is called a fixed point, i.e., $f(x^*) = x^*$. In general, it is a question which can not be answered affirmatively. However, in 1912 the Dutch mathematician L.E.J. Brouwer demonstrated that any continuous function from a nonempty, convex and compact subset of $\mathbb{R}^n$ into itself has at least one fixed point. Brouwer [1912] proved his theorem using degree theory. It is a folklore that Brouwer himself eventually became to dislike the theorem because the proof was too abstract and was not constructive. However, Brouwer's fixed point theorem, with no doubt, is one of the most important and elegant results in modern mathematics. The interested reader is recommended to, for example, Smart [1974] and Istratescu [1981]. Following Todd [1976a], we will provide a proof of Brouwer theorem via the famous lemmas of Sperner [1928], and Knaster, Kuratowski and Mazurkiewicz [1929] (KKM lemma).

**Theorem 1.5.1 (Brouwer Theorem)** *Let $C$ be a nonempty, compact and convex subset of $\mathbb{R}^n$, and let $f : C \mapsto C$ be a continuous function. Then there exists at least one point $x^*$ in $C$ such that $f(x^*) = x^*$.*

A more general statement of Brouwer theorem would be: if a compact subset $C$ of $\mathbb{R}^n$ is homeomorphic to a convex and compact nonempty subset of $\mathbb{R}^n$, then any continuous function $f : C \mapsto C$ has a fixed point. Brouwer theorem can be vividly displayed in the following way. As Franklin [1980, p. 232] explains: "Sit down with a cup of coffee. Gently and continuously swirl the coffee about in the cup. Put the coffee down, and let the motion subside. When the coffee is still, Brouwer says there is at least one point in the coffee that has returned to the exact spot in the cup where it was when you first sat down." It is easy to see that the theorem is true for $n = 1$. But for $n > 1$ the proof of the theorem is much more complicated. Without loss of generality, we can confine ourselves to continuous functions on the unit simplex $S^n$.

**Theorem 1.5.2 (KKM Lemma)** *Let $\{C^1, \cdots, C^n\}$ be a collection of closed subsets of $S^n$ such that*
*(a) $S^n = \bigcup_{i=1}^n C^i$;*
*(b) $S^n(T) \subseteq \bigcup_{j \in I_n \setminus T} C^j$ for any $T$, $\emptyset \subset T \subset I_n$.*
*Then $\bigcap_{i=1}^n C^i \neq \emptyset$.*

We will prove the lemma using Sperner lemma later. First, we show that the KKM lemma implies Brouwer theorem.

**Theorem 1.5.3** *Brouwer theorem is implied by the KKM lemma.*

Proof: Define $C^i = \{x \in S^n \mid f_i(x) \leq x_i\}$ for all $i \in I_n$. Then it can easily be verified that the sets $C^1$, $\cdots$, $C^n$ satisfy the two hypotheses of KKM lemma. Thus $\bigcap_{i=1}^n C^i \neq \emptyset$. Choose $x^* \in \bigcap_{i=1}^n C^i$. It follows from the definition of $C^i$ that

$$f_i(x) \leq x_i, \; i = 1, \cdots, n.$$

Since $\sum_{i=1}^n f_i(x) = \sum_{i=1}^n x_i = 1$ we must have $f_i(x^*) = x_i^*$, $i = 1$, $\cdots$, $n$. Hence $x^*$ is a fixed point of $f$. □

Before proceeding to Sperner lemma we first introduce the following definition.

**Definition 1.5.4** *Let $\mathcal{T}$ be a triangulation of a convex set $C$ in $\mathbb{R}^n$ with $dim(C) = m$, and let $l : \mathcal{T}^0 \mapsto I_{m+1}$ be a labeling function. An $m$-simplex of $\mathcal{T}$ with vertices $y^1$, $\cdots$, $y^{m+1}$, is completely labelled if all its vertices are differently labelled, i.e., $\{l(y^i) \mid i \in I_{m+1}\} = I_{m+1}$.*

The following lemma due to Sperner [1928] provides a condition for the existence of a completely labelled simplex in a triangulation of the unit simplex.

**Theorem 1.5.5 (Sperner Lemma)** *Let $\mathcal{T}$ be a triangulation of $S^n$ and let $l : \mathcal{T}^0 \mapsto I_n$ be a labeling function such that $x_i = 0$ implies $l(x) \neq i$ for any vertex $x$ of $\mathcal{T}^0$. Then there exists at least one completely labelled simplex in $\mathcal{T}$.*

The influence of Sperner lemma is far beyond its beauty and simplicity in its form[1] .

[1]For example, Herbert Scarf [1991, p. 114] says: "I finally met Professor Sperner at a conference on fixed point methods at the University of Southampton in July of 1979. The two of us alighted from the same train, and after a few moments of hesitation, we introduced ourselves. We talked about the consequences of the theorem he had demonstrated some fifty years earlier, and he seemed very pleased. Afterwards, reflecting on

In Cohen [1967] a stronger form of Sperner lemma is proved by induction, which claims the existence of an odd number of completely labelled simplices. Here we do not want to give a proof for Sperner lemma because later we will introduce several algorithms to actually find such a completely labelled simplex, which lead to constructive proofs of Sperner lemma.

The following theorem then proves the KKM lemma.

**Theorem 1.5.6** *Sperner lemma implies the KKM lemma.*

Proof: Let the collection $\{C^1, \cdots, C^n\}$ of closed subsets of $S^n$ satisfy the conditions of the KKM lemma. Furthermore, let $\{\mathcal{T}_k \mid k \in \mathbb{N}\}$ be a sequence of triangulations of $S^n$ with $mesh(\mathcal{T}_k)$ converging to zero as $k$ goes to infinity. For each $k$, assign a label to each vertex $y$ of $\mathcal{T}_k$ by

$$l(y) = \min\{j \in I_n \mid y \in C^j \text{ and } y_j > 0\}.$$

It is easy to check that there always exists such an index. Moreover, since the labeling function satisfies the condition of Sperner lemma, there exists a completely labelled simplex $\sigma_k(y^1(k), \cdots, y^n(k))$ in $\mathcal{T}_k$ such that $y^j(k) \in C^j$ for each $k \in \mathbb{N}$. Since, say, $y^1(k)$, $k \in \mathbb{N}$, lies in the compact set $S^n$, we have that for some subsequence $\{k_j \mid j \in \mathbb{N}\}$ with $k_j \to \infty$ if $j \to \infty$, the sequence $\{y^1(k_j) \mid j \in \mathbb{N}\}$ converges to a point $x^*$ in $S^n$. Since $\lim_{k\to\infty} mesh(\mathcal{T}_k) = 0$, it follows that $\lim_{j\to\infty} y^i(k_j) = x^*$ for all $i \in I_n$. Since every $C^i$ is closed, we deduce that $x^* \in C^i$ for all $i \in I_n$. □

It is clear from the proof that if the sets $C^j$, $j \in I_n$, are taken as in the proof of Theorem 1.5.3, Brouwer theorem is also proved. Therefore, in order to find a fixed point as in Brouwer theorem, it suffices to have a method which can find a completely labelled simplex in Sperner lemma.

Now we show that Sperner lemma is implied by Brouwer theorem.

**Theorem 1.5.7** *Brouwer theorem implies Sperner lemma.*

Proof: For any $x \in S^n$, there exists an $(n-1)$-simplex $\sigma(x^1, \cdots, x^n)$ containing $x$. So, there exist non-negative numbers $\beta_i$ with sum equal to 1 such that $x = \sum_{i=1}^n \beta_i x^i$. Define a function $f : S^n \mapsto S^n$ by

$$f(x) = \sum_{i=1}^{n} \beta_i e(l(x^i) + 1),$$

where $l(x^i) + 1 = 1$ if $l(x^i) = n$. It is easy to see that the function $f$ is continuous. By Brouwer theorem, there exists a fixed point $x^* \in S^n$, i.e.,

our talk, I thought about that mysterious feature of mathematics, always perplexing and enchanting to me: the way in which an act of pure imagination can lead to unexpected practical applications in areas that are totally removed from the originator's wholly abstract concerns."

$f(x^*) = x^*$. Since for any vertex $x$ in $S^n$, $x_i = 0$ implies $l(x) \neq i$, this means that $x^*$ does not lie on the boundary of $S^n$. Thus, $x^*$ lies in the interior of $S^n$. From the definition of $f$, there must exist a completed labelled simplex containing $x^*$. □

Most problems disguise in one fashion or another. So does the fixed point problem. In this book we often meet the following equivalent forms of the fixed point problem. The first problem is called the stationary point problem. This problem is also often called the variational inequality problem. See Hartman and Stampacchia [1966] and Eaves [1971a].

**Definition 1.5.8** *Let $C$ be a subset of $\mathbb{R}^n$, and let $f : C \mapsto \mathbb{R}^n$ be a function. Then a point $x^*$ in $C$ is a stationary point of $f$ on $C$ if*

$$(x^* - x)^\top f(x^*) \geq 0$$

*for all $x \in C$.*

The problem of finding a stationary point is called *stationary point problem.* Before giving an existence theorem of stationary points, we introduce the following result.

**Lemma 1.5.9** *For any nonempty, convex and compact subset $C$ of $\mathbb{R}^n$, define a function $r : \mathbb{R}^n \mapsto C$ by*

$$r(x) = \arg\min\{||y - x||^2 \mid y \in C\}.$$

*Then for any $x, y \in \mathbb{R}^n$, it holds that*

$$||r(x) - r(y)|| \leq ||x - y||.$$

Proof: Consider the problem

$$\begin{array}{c} \min \ ||y - x||^2 \\ s.t. \ y \in C. \end{array}$$

Since $C$ is a nonempty compact set, the problem has an optimal solution, say, $y^1$. Suppose it is not unique. Then there exists $y^2 \in C$ such that $||y^1 - x||^2 = ||y^2 - x|| = \alpha$ and $y^1 \neq y^2$. Take $y^0 = (y^1 + y^2)/2$. Since $C$ is convex, $y^0 \in C$. By using the relation

$$||x^1 - x^2||^2 + ||x^1 + x^2||^2 = 2||x^1||^2 + 2||x^2||^2,$$

we have

$$\begin{array}{rcl} ||y^0 - x||^2 & = & ||(y^1 - x)/2 + (y^2 - x)/2||^2 \\ & = & ||y^1 - x||^2/2 + ||y^2 - x||^2/2 - ||y^1 - y^2||^2/4 \\ & = & \alpha - ||y^1 - y^2||^2/4 \\ & < & \alpha. \end{array}$$

So, $y^1$ and $y^2$ are not optimal solutions, yielding a contradiction.

Now consider two points $x^1, x^2 \in \mathbf{R}^n$, and let $r(x^1)$, $r(x^2)$ be the respective optimal solutions in $C$. For $0 \le t \le 1$, the point $r(x^1)+t(r(x^2)-r(x^1))$ is also in $C$. Further, we have

$$\begin{aligned}||x^1 - (r(x^1) + t(r(x^2) - r(x^1))||^2 &= ||x^1 - r(x^1)||^2 \\ &\quad -2t(x^1 - r(x^1))^\top (r(x^2) - r(x^1)) \\ &\quad +t^2||r(x^2) - r(x^1)||^2.\end{aligned}$$

This implies $(x^1 - r(x^1))^\top (r(x^2) - r(x^1)) \le 0$. Suppose to the contrary that $(x^1 - r(x^1))^\top (r(x^2) - r(x^1)) > 0$. Then for sufficiently small $t > 0$, $r(x^1) + t(r(x^2) - r(x^1))$ has a shorter distance from $x^1$ than $r(x^1)$, contradicting the definition of $r(x^1)$. By a similar argument, we have $(x^2 - r(x^2))^\top (r(x^1) - r(x^2)) \le 0$. Now it follows that

$$\begin{aligned}||x^1 - x^2||^2 - ||r(x^1) - r(x^2)||^2 &= ||(x^1 - r(x^1) - (x^2 - r(x^2))||^2 \\ &\quad -2(x^1 - r(x^1))^\top (r(x^2) - r(x^1)) \\ &\quad -2(x^2 - r(x^2))^\top (r(x^1) - r(x^2)) \\ &\ge 0.\end{aligned}$$

That is, $||r(x^1) - r(x^2)|| \le ||x^1 - x^2||$. □

From Lemma 1.5.9 it follows that the function $r$ is a Lipschitz continuous function. Moreover, the above lemma is also true if the function $r : \mathbf{R}^n \mapsto C$ is given by

$$r(x) = \arg\min\{||y - x|| \mid y \in C\}.$$

The following theorem asserts that any continuous function from a nonempty, convex and compact set of $\mathbf{R}^n$ to $\mathbf{R}^n$ has a stationary point.

**Theorem 1.5.10 (Stationary Point Theorem I)** *Any continuous function $f$ from a nonempty, convex and compact subset $C$ of $\mathbf{R}^n$ to $\mathbf{R}^n$ has at least one stationary point.*

Proof: The proof goes via defining a function $g : C \mapsto C$ by

$$g(x) = \arg\min\{\, ||x + f(x) - y||^2 \mid y \in C\}$$

for $x \in C$. It follows from Lemma 1.5.9 that $g$ is a continuous function since $x + f(x)$ is a continuous function with respect to $x$. By Brouwer theorem there exists a point $x^* \in C$ such that $x^* = g(x^*)$. We shall prove that $x^*$ must be a stationary point of $f$. Suppose to the contrary that it is not a stationary point. Then there exists a point $\bar{x}$ such that $\bar{x}^\top f(x^*) > x^{*\top} f(x^*)$. Let $z$ be a convex combination of $x^*$ and $\bar{x}$, i.e., $z = \beta\bar{x} + (1 - \beta)x^*$ for some $\beta \in [0, 1]$. From the definition of $g$ we have

$$||x^* + f(x^*) - z||^2 \ge ||f(x^*)||^2.$$

It follows that for any $\beta \in [0,1]$,

$$\beta(\beta(x^* - \bar{x})^\top (x^* - \bar{x}) + 2(x^* - \bar{x})^\top f(x^*)) \geq 0. \tag{1.5.2}$$

Notice that $2(x^* - \bar{x})^\top f(x^*)$ is a constant negative real number. Let

$$\beta_0 = \frac{-2(x^* - \bar{x})^\top f(x^*)}{||x^* - \bar{x}||^2}.$$

Obviously, $\beta_0 > 0$. Let $\beta_1 = \min\{1, \beta_0\}$. It is clear that for $\beta \in (0, \beta_1)$,

$$\beta(\beta(x^* - \bar{x})^\top (x^* - \bar{x}) + 2(x^* - \bar{x})^\top f(x^*)) < 0$$

which contradicts equation ( 1.5.2). So $x^*$ must be a stationary point of $f$ on $C$. □

So Brouwer theorem implies Theorem 1.5.10. For the reverse, let $g$ be a continuous mapping from a nonempty, convex and compact subset $C$ of $\mathbb{R}^n$ into itself. Define a function $f : C \mapsto \mathbb{R}^n$ by $f(x) = g(x) - x$ for each $x \in C$. By the above theorem there exists $x^* \in C$ such that $(g(x^*) - x^*)^\top (x^* - x) \geq 0$ for all $x \in C$. Since $g(x^*)$ is in $C$, we must have $x^* = g(x^*)$.

Another equivalent problem concerns the finding of equilibrium points.

**Definition 1.5.11** *Let $h : S^n \mapsto \mathbb{R}^n$ be a function satisfying $p^\top h(p) = 0$ for any $p \in S^n$. Then a point $p^*$ in $S^n$ is an equilibrium point of the function $h$ if $h(p^*) \leq \mathbf{0}$.*

The problem of finding an equilibrium point is called *equilibrium problem.* We can also establish an existence theorem for the equilibrium problem. See Arrow and Hahn [1971].

**Theorem 1.5.12** *Let $h : S^n \mapsto \mathbb{R}^n$ be a continuous function satisfying $p^\top h(p) = 0$ for any $p \in S^n$. Then $h$ has an equilibrium point $p^*$ in $S^n$.*

Proof: Let $\beta$ be a positive real number. Define a function $f : S^n \mapsto S^n$ by

$$f_i(p) = \max\{0, p_i + \beta h_i(p)\}/(\sum_{j \in I_n} \max\{0, p_j + \beta h_j(p)\})$$

for all $i \in I_n$. We will show that $f$ is well-defined and continuous. Note that

$$\sum_{j \in I_n} \max\{0, p_j + \beta h_j(p)\} \geq 0$$

for all $p \in S^n$. Suppose that there exists a point $y \in S^n$ such that

$$\sum_{j \in I_n} \max\{0, y_j + \beta h_j(y)\} = 0.$$

Then we have $y + \beta h(y) \leq \mathbf{0}$. Furthermore, $0 \leq (y + \beta h(y))^\top y = y^\top y > 0$, yielding a contradiction. Hence $f$ is well-defined. It is easy to see that $f$ is continuous. It follows from Brouwer theorem that there is $p^* \in S^n$ such that $p^* = f(p^*)$. Thus for all $i \in I_n$ we have $\max\{0, p_i^* + \beta h_i(p^*)\} = \gamma p_i^*$ for some $\gamma > 0$. Let $K = \{i \in I_n \mid p_i^* > 0\}$. We have the following two cases.

(i) For all $i \in K$, $\gamma p_i^* = \max\{0, p_i^* + \beta h_i(p^*)\} = p_i^* + \beta h_i(p^*)$.

(ii) For all $i \in I_n \setminus K$, $0 = \gamma p_i^* = \max\{0, p_i^* + \beta h_i(p^*)\} \geq \beta h_i(p^*)$.

If $i \in K$, $h_i(p^*) = \alpha p_i^*$ with $\alpha = (\gamma - 1)/\beta$. Since $(p^*)^\top h(p^*) = 0$, we have $0 = \sum_{i \in I_n} p_i^* h_i(p^*) = \sum_{i \in K} p_i^* \alpha p_i^* = \alpha (p^*)^\top p^*$. Since $(p^*)^\top p^* > 0$, this means that $\alpha = 0$ and hence $h_i(p^*) = 0$ for all $i \in K$. This together with $(ii)$ implies that $h(p^*) \leq \mathbf{0}$. □

Another way of proving the above result is given in Scarf [1967b, 1973]. Scarf defined a function $g : S^n \mapsto S^n$ by

$$g_i(p) = (p_i + \max\{0, h_i(p)\})/(1 + \sum_{i=1}^{n} \max\{0, h_i(p)\}), i = 1, \cdots, n$$

for $p \in S^n$. It is again easy to show that $p^*$ is an equilibrium point if $p^*$ is a fixed point of $g$. However, Todd [1976a] pointed out that for computational purpose, the function defined by Arrow and Hahn is better than that defined by Scarf. Uzawa [1962] showed that the implication of the above theorem can be reversed. This can be done as follows. Given a continuous function $g : S^n \mapsto S^n$, define a function $f : S^n \mapsto \mathbb{R}^n$ by $f(x) = g(x) - \frac{x^\top g(x)}{x^\top x} x$ for each $x \in S^n$. Then $f$ satisfies the hypothesis of the above theorem. It is easy to see that if $f(x^*) \leq \mathbf{0}$ then $g(x^*) = x^*$.

Given a function $f : C \mapsto D$, a point $x^*$ in $C$ is called *a zero point* if $f(x^*) = \mathbf{0}$. Finding a zero point is referred to as the *zero point problem.*

Another important problem related to fixed point problems is the so-called *complementarity problem.* It is a problem of finding a point $x^* \in \mathbb{R}^n_+$ such that $f(x^*) \in \mathbb{R}^n_+$ and

$$(x^*)^\top f(x^*) = 0,$$

where $f : \mathbb{R}^n \mapsto \mathbb{R}^n$ is a function.

The problems above can easily be extended to the simplotope and the existence theorems can also be easily obtained.

**Theorem 1.5.13** *Any continuous function $f$ from the simplotope $\mathcal{S}$ to $\prod_{j \in I_N} \mathbb{R}^{n_j}$ has at least one stationary point $p^*$ in $\mathcal{S}$, i.e., $(p_j^*)^\top f_j(p^*) \geq (p_j)^\top f_j(p^*)$, $j = 1, \cdots, N$, for all $p \in \mathcal{S}$.*

Define a function $g_j : \mathcal{S} \mapsto S^{n_j}$ for $j = 1, \cdots, N$ by

$$g_j(x) = \arg\min\{ \|y_j - (x_j + f_j(x))\| \mid y_j \in S^{n_j} \}.$$

Again, $x^*$ is a stationary point of $f$ on $\mathcal{S}$ if $x^*$ is a fixed point of $(g_1, \cdots, g_N)$ in $\mathcal{S}$.

## 1.6. Fixed point theorems of Kakutani and Browder

Many practical problems motivate the necessity of more general fixed point theorems. This leads us to weaken the continuity of the function to be addressed, and to relax the requirements on the domain and range of the function. A natural extension of the continuity of a function is the important concept of upper semi-continuity of a point-to-set mapping, and a natural generalization of Brouwer theorem is a fixed point theorem due to Kakutani [1941].

**Definition 1.6.1** *Let $\phi$ be a point-to-set mapping from a set $C$ in $\mathbb{R}^n$ to a subset $D$ of $\mathbb{R}^m$, i.e., for each $x \in C$, $\phi(x) \subseteq D$. Then $\phi$ is upper semi-continuous (u.s.c.) at the point $x^*$ in $C$ if for any convergent sequence $\{x^k \mid k \in \mathbb{N}\}$ of points in $C$ with limit $x^*$ and any convergent sequence $\{y^k \mid k \in \mathbb{N}\}$ with $y^k \in \phi(x^k)$ and limit $y^*$, it holds that $y^* \in \phi(x^*)$. The point-to-set mapping $\phi$ is lower semi-continuous (l.s.c.) at a point $x^*$ if for any convergent sequence $\{x^k \mid k \in \mathbb{N}\}$ of points in $C$ with limit $x^*$ and for any $y^* \in \phi(x^*)$ there exists a sequence $\{y^k \mid k \in \mathbb{N}\}$ of points in $D$ such that $y^k \in \phi(x^k)$ for all $k$ and* $\lim_{k\to\infty} y^k = y^*$.

A point-to-set mapping $\phi : C \mapsto D$ is called *upper semi-continuous (lower semi-continuous)* if $\phi$ is u.s.c. (l.s.c.) at every point of $C$. A point-to-set mapping is called to be continuous if this mapping is both u.s.c. and l.s.c.(see Berge [1963] and Debreu [1959] for the above definition). Note that a function $f : C \mapsto D$ is continuous if and only if the point-to-set mapping $\phi : C \mapsto D$ defined by $\phi(x) = \{f(x)\}$ for each $x \in C$ is u.s.c. Often we call a point-to-set mapping $\phi$ a *correspondence.* Here we give a warning that *the reader should distinguish the upper (lower) semi-continuity of a point-to-set mapping with the upper (lower) semi-continuity of a function.* The reason is that when a function $f$ is considered, the upper (lower) semi-continuity of $f$ is weaker than the upper (lower) semi-continuity of the point-to-set mapping $\{f\}$.

Let us consider an example. Let $C = [0, 1]$ and $D = [0, 1]$. We define $\phi : C \mapsto D$ by

$$\phi(x) = \begin{cases} [0.5, 1], & \text{if} \quad x \in [0, 0.5], \\ [x - 0.5, 1], & \text{if} \quad x \in (0.5, 1]. \end{cases}$$

This correspondence is u.s.c. except for the point $x^* = 0.5$.

The following theorems characterize some of the most important properties of u.s.c. mappings. Their proofs can be found in Debreu [1959], Berge [1963], or Todd [1976a].

**Theorem 1.6.2** *Let $C$ and $D$ be two subsets in $\mathbb{R}^n$ and $\mathbb{R}^m$, respectively, and let $D$ be a compact set. Then a correspondence $\phi : C \mapsto D$ is u.s.c. if and only if the graph of $\phi$*

$$\{(x, y) \in C \times D \mid y \in \phi(x)\}$$

*is closed in $C \times D$.*

**Theorem 1.6.3** *Let $C$ and $D$ be two subsets in $\mathbb{R}^n$ and $\mathbb{R}^m$, respectively, and let $D$ be a compact set. Let $\phi : C \mapsto D$ be a point-to-set mapping such that $\phi(x)$ is compact for every $x \in C$. Then $\phi$ is u.s.c. if and only if for every $x \in C$, and every $\epsilon > 0$, there is $\delta > 0$ such that if $y \in B(x, \delta) \cap C$, $\phi(y) \subseteq B(\phi(x), \epsilon)$.*

**Theorem 1.6.4** *All correspondences below are from a set $C$ in $\mathbb{R}^n$ to a subset $Y$ of $\mathbb{R}^m$, unless otherwise specified.*
*(i) If $\phi$ is u.s.c. and $\varphi$ is defined by $\varphi(x) = co(\phi(x))$ for each $x \in C$, then $\varphi$ is u.s.c.*
*(ii) If $\phi_1$ and $\phi_2$ are u.s.c. and $\phi$ is defined by $\phi(x) = \phi_1(x) + \phi_2(x)$ for each $x$ in $C$, then $\phi$ is u.s.c.*
*(iii) If $\phi_1$ and $\phi_2$ are u.s.c. and $\phi$ is defined by $\phi(x) = \phi_1(x) \bigcup \phi_2(x)$ for each $x$ in $C$, then $\phi$ is u.s.c.*
*(iv) If $\phi_1$ and $\phi_2$ are u.s.c. and $\phi$ is defined by $\phi(x, y) = \phi_1(x) \times \phi_2(y)$ for each $(x, y) \in C \times C$, then $\phi$ is u.s.c. from $C \times C$ to $Y \times Y$.*
*(v) If $\phi$ is u.s.c. and $C$ is closed with $C \subseteq D$, then $\varphi : D \mapsto Y$ is also u.s.c., where $\varphi(x) = \phi(x)$ for $x \in C$, and $\varphi(x) = \emptyset$ for $x \in D \backslash C$.*

**Theorem 1.6.5** *Let $X$, $Y$ and $Z$ be three subsets of $\mathbb{R}^l$, $\mathbb{R}^m$ and $\mathbb{R}^n$, respectively. Let $\phi_1 : X \mapsto Y$ and $\phi_2 : Y \mapsto Z$ be u.s.c. point-to-set mappings. Then the point-to-set mapping $\phi_2 \circ \phi_1 : X \mapsto Z$ is u.s.c.*

The following concept is often used in simplicial algorithms.

**Definition 1.6.6** *Let $\phi$ be a point-to-set mapping from a t-dimensional convex set $C$ in $\mathbb{R}^n$ to a set $D$ in $\mathbb{R}^m$. Let $\mathcal{T}$ be a triangulation of $C$. A function $g : C \mapsto \mathbb{R}^m$ is a piecewise linear (P.L.) approximation to the mapping $\phi$ with respect to the triangulation $\mathcal{T}$ if for each vertex $x$ of any $\sigma \in \mathcal{T}$; $g(x) \in \phi(x)$ and $g$ is affine on each simplex of $\mathcal{T}$, i.e., when $x = \sum_{i=1}^{t+1} \lambda_i x^i \in \sigma(x^1, \cdots, x^{t+1})$ with non-negative numbers $\lambda_i$ then $g(x) = \sum_{i=1}^{t+1} \lambda_i g(x^i)$.*

It is trivial to see that $g$ is a continuous function. Similarly, we define a piecewise linear approximation $g$ to a function $f$ with respect to a triangulation by taking $\phi(\cdot) = \{f(\cdot)\}$.

In 1941 Kakutani gave a sufficient condition which guarantees that a u.s.c. mapping has a fixed point, i.e., a point $x^*$ with $x^* \in \phi(x^*)$.

**Theorem 1.6.7 (Kakutani Theorem)** *Let $C$ be a nonempty, convex, and compact subset of $\mathbb{R}^n$, and let $\phi$ be a u.s.c. mapping from $C$ to the collection of nonempty and convex subsets of $C$. Then $\phi$ has a fixed point in $C$.*

Proof: We follow here a proof of Eaves [1971b]. Let $t = dim(C)$ and let $D$ be a $t$-simplex containing $C$. Take an interior point $v$ of $C$. Define a correspondence $\psi : D \mapsto D$ by

$$\psi(x) = \begin{cases} \phi(x) & if \;\; x \in int(C), \\ co(\phi(x) \cup \{v\}) & if \;\; x \in bd(C), \\ \{v\} & if \;\; x \in D \setminus C. \end{cases}$$

It is easy to see that $\psi$ is a u.s.c. mapping with nonempty, convex and compact values. Suppose that $x^*$ is a fixed point of $\psi$. Clearly, $x^* \in C$. If $x^* \in int(C)$, then $x^* \in \phi(x^*)$ and $x^*$ is a fixed point of $\phi$. If $x^* \in bd(C)$, then for some $y \in \phi(x^*)$, $x^* = \beta v + (1 - \beta)y$ for some $\beta$, $0 \leq \beta \leq 1$. Since $v \in int(C)$, we have $\beta > 0$ which implies $x^* \in int(C)$. This is a contradiction. Hence $\beta = 0$. Again we have $x^* \in \phi(x^*)$. Therefore, $x^*$ is a fixed point of $\phi$ if $x^*$ is a fixed point of $\psi$.

We now prove that $\psi$ has a fixed point. Let $G_k$, $k = 1, 2, \cdots$, be a sequence of triangulations of $D$ with mesh$(G_k)$ tending to zero. For each $k$, let $f^k$ be the P.L. approximation of $\psi$ with respect to $G_k$. By Brouwer theorem, $f^k$ has a fixed point, say $x^k$. From Definition 1.6.6 there exists a simplex $\sigma_k$ of $G_k$ with vertices $w^{1,k}, \cdots, w^{t+1,k}$ such that

$$\begin{array}{l} x^k = \sum_{i=1}^{t+1} \beta_{i,k} w^{i,k} = \sum_{i=1}^{t+1} \beta_{i,k} f^{i,k}, \\ \sum_{i=1}^{t+1} \beta_{i,k} = 1, \;\; \beta_{i,k} \geq 0, \;\; i \in I_{t+1}, \end{array} \tag{1.6.3}$$

where $f^{i,k} \in \psi(w^{i,k})$ for all $i \in I_{t+1}$ and all $k \in \mathbb{N}$. Since $D$ is compact, there is a sub-sequence (without loss of generality, we may assume the whole sequence) such that $x^k \to x^*$, $\beta_{i,k} \to \beta_i$ and $f^{i,k} \to f^i$ as $k \to \infty$. Since mesh$(G_k) \to 0$, we have $w^{i,k} \to x^*$ for all $i \in I_{t+1}$. Because $\psi$ is u.s.c., $f^i \in \psi(x^*)$ for all $i \in I_{t+1}$. Taking limits in ( 1.6.3), we have $x^* = \sum_{i \in I_{t+1}} \beta_i f^i$, $\sum_{i \in I_{t+1}} \beta_i = 1$ and $\beta_i \geq 0$ for all $i \in I_{t+1}$. Since $\psi(x^*)$ is convex, it follows that $x^* \in \psi(x^*)$. □

We illustrate the theorem with an example. Let $C = [0, 1]$. Define $\phi : C \mapsto C$ by

$$\phi(x) = \begin{cases} \{x + 0.5\}, & \text{if} \;\; x \in [0, 0.5), \\ [0, 1], & \text{if} \;\; x = 0.5, \\ \{x - 0.5\}, & \text{if} \;\; x \in (0.5, 1]. \end{cases}$$

The correspondence satisfies the conditions of the theorem. It is easy to see that the point $x^* = 0.5$ is a fixed point.

The following theorem is known as *the maximum theorem*. See, for example, Hildenbrand [1974, p.30] for a proof.

**Theorem 1.6.8** *Let $C$ be a subset of $\mathbb{R}^n$, let $D$ be a subset of $\mathbb{R}^m$, and let $\psi : C \mapsto D$ be a continuous point-to-set mapping with compact values. Let $f : C \times D \mapsto \mathbb{R}$ be a continuous function. Define the mapping $\phi : C \mapsto D$ by*

$$\phi(x) = \{z \in \psi(x) \mid f(x,z) \geq f(x,y) \text{ for all } y \in \psi(x)\}$$

*for each $x \in C$ and define $g : C \mapsto \mathbb{R}$ by*

$$g(x) = \{f(x,y) \mid y \in \phi(x)\}$$

*for each $x \in C$. Then $\phi$ is a u.s.c. point-to-set mapping with compact values and $g$ is a continuous function.*

**Definition 1.6.9** *Let $C$ be a subset of $\mathbb{R}^n$, and let $\phi : C \mapsto \mathbb{R}^n$ be a point-to-set mapping. Then a point $x^*$ in $C$ is a stationary point of $\phi$ on $C$ if there exists a point $y^* \in \phi(x^*)$ such that*

$$(x^* - x)^\top y^* \geq 0$$

*for all $x \in C$.*

The following existence theorem of stationary points can be found in Mas-Colell [1974]. Here we give a proof due to van der Laan and Talman [1993] using Kakutani theorem.

**Theorem 1.6.10 (Stationary Point Theorem II)** *Let $C$ be a nonempty, convex and compact subset of $\mathbb{R}^n$, let $D$ be a compact subset of $\mathbb{R}^n$, and let $\psi : C \mapsto D$ be a u.s.c. point-to-set mapping with nonempty, compact and convex values. Then $\psi$ has a stationary point $x^* \in C$.*

Proof: Let $Y$ be a convex and compact subset of $\mathbb{R}^n$ containing the set $\cup_{x \in C} \psi(x)$. Define the point-to-set mapping $\phi : Y \mapsto C$ by

$$\phi(y) = \{z \in C \mid y^\top z \geq y^\top x \text{ for all } x \in C\}$$

for each $y \in Y$. By applying Theorem 1.6.8, it is easy to see that $\phi$ is u.s.c. Moreover, since $C$ is convex, $\phi(x)$ is nonempty, convex and compact for every $x \in Y$. For each $(x,y) \in C \times Y$, let $\chi(x,y)$ be defined by

$$\chi(x,y) = \phi(y) \times \psi(x).$$

Clearly, $\chi : C \times Y \mapsto C \times Y$ is a u.s.c. point-to-set mapping with nonempty, convex and compact values. By Kakutani theorem there exists a pair $(x^*, y^*)$

$\in C \times Y$ such that $(x^*, y^*) \in \chi(x^*, y^*) = \phi(y^*) \times \psi(x^*)$. That is, $x^* \in \phi(y^*)$ and $y^* \in \psi(x^*)$. So we have that $y^* \in \psi(x^*)$ and

$$(x^* - x)^\top y^* \geq 0$$

for all $x \in C$. $x^*$ is a stationary point of $\psi$. □

It is easy to show that the above theorem implies Kakutani theorem. Now we introduce one more fundamental fixed point theorem due to Browder [1960]. Browder proved this result using a rather sophisticated machinery. In Chapter 8 we will give a constructive proof for Browder theorem.

**Theorem 1.6.11 (Browder Theorem)** *Let $X$ be a nonempty, convex and compact subset of $\mathbb{R}^n$. If $\phi : X \times [0,1] \mapsto X$ is a u.s.c. point-to-set mapping with nonempty, convex and compact values, then the set $C_\phi = \{(x,t) \in X \times [0,1] \mid x \in \phi(x,t)\}$ contains a connected set $D$ such that*

$$D \cap (X \times \{0\}) \neq \emptyset \text{ and } D \cap (X \times \{1\}) \neq \emptyset.$$

Now we have the following theorem which states the existence of a connected set of stationary points for a continuous function.

**Theorem 1.6.12** *Let $f : C \mapsto \mathbb{R}^n$ be a continuous function where $C$ is a nonempty, convex and compact subset of $\mathbb{R}^n$, and let $\psi : [0,1] \mapsto C$ be a continuous point-to-set mapping with nonempty, convex and compact values. Then the set*

$$C_\psi = \{(x,t) \in C \times [0,1] \mid (x - z)^\top f(x) \geq 0 \text{ for all } z \in \psi(t)\}$$

*contains a connected set $D$ such that*

$$D \cap (C \times \{0\}) \neq \emptyset \quad \text{and} \quad D \cap (C \times \{1\}) \neq \emptyset.$$

Proof: Define a function $h : C \times [0,1] \mapsto C$ by

$$h(x,t) = \arg\min\{||x + f(x) - y||^2 \mid y \in \psi(t)\}.$$

We prove that $h(x,t)$ is a continuous function with respect to $x$ and $t$. It follows from Lemma 1.5.9 that for any given $t \in [0,1]$, $h(\cdot, t)$ is a continuous function. We now prove that for any given $x \in C$, $h(x, \cdot)$ is a continuous function. Define a function $q : [0,1] \mapsto [0, \infty)$ by

$$q(t) = \min_{y \in \psi(t)} ||x + f(x) - y||^2.$$

Let $Q(t) = \{y \in \psi(t) \mid ||x + f(x) - y||^2 \leq q(t)\}$. We first show that $q$ is a continuous function. Take any $t \in [0,1]$ and any $\epsilon > 0$. Let $\{t^k \mid k \in \mathbb{N}\}$

with $t^k \in [0,1]$ be any sequence converging to $t^*$. Choose any $y^* \in \psi(t^*)$ such that $||x + f(x) - y^*||^2 \leq q(t^*) + \epsilon$. Since $\psi$ is continuous, there exists a sequence $\{y^k \mid k \in \mathbb{N}\}$ with $y^k \in \psi(t^k)$ converging to $y^*$. Then we have

$$\begin{aligned} \limsup_{k\to\infty} q(t^k) &\leq \limsup_{k\to\infty} ||x + f(x) - y^k||^2 \\ &\leq ||x + f(x) - y^*||^2 \\ &\leq q(t^*) + \epsilon. \end{aligned}$$

The second inequality comes from the u.s.c. property of the function $||x + f(x) - y||^2$ with respect to $y$. Hence $q$ is u.s.c. Now we are going to demonstrate that $q$ is also l.s.c. Take any $\epsilon > 0$ and $t^* \in [0,1]$. Let $\{t^k \mid k \in \mathbb{N}\}$ with $t^k \in [0,1]$ be any sequence converging to $t^*$. Choose $y^k \in \psi(t^k)$ such that $||x+f(x)-y^k||^2 \leq q(t^k)+\epsilon$. Then the sequence $\{y^k \mid k \in \mathbb{N}\}$ must have a convergent subsequence because $C$ is compact. For simplicity, we may assume that the subsequence is the same as the original sequence. Because $\psi$ is u.s.c., we have that $\lim_{k\to\infty} y^k = y^*$ and $y^* \in \psi(t^*)$. Furthermore, it follows that

$$\begin{aligned} q(t^*) &\leq ||x + f(x) - y^*||^2 \\ &\leq \liminf_{k\to\infty} ||x + f(x) - y^k||^2 \\ &\leq \liminf_{k\to\infty} q(t^k) + \epsilon. \end{aligned}$$

The second inequality comes from the l.s.c. property of the function $||x + f(x) - y||^2$ with respect to $y$. This implies that $q$ is l.s.c. Now we can conclude from the above discussions that $q$ is a continuous function.

Notice that for each $t \in [0,1]$, $Q(t)$ only contains a single element because $\psi(t)$ is convex. Now we prove that $Q$ is u.s.c. Choose any $t^* \in [0,1]$. Assume that there are two sequences $\{y^k \mid k \in \mathbb{N}\}$ and $\{t^k \mid k \in \mathbb{N}\}$ with $y^k \in Q(t^k)$, converging to $y^*$ and $t^*$, respectively. Since $\psi$ is u.s.c., $y^* \in \psi(t^*)$. On the other hand because it holds

$$||x + f(x) - y^k||^2 \leq q(t^k)$$

for all $k \in \mathbb{N}$, and $q(t)$ and $||x + f(x) - y||^2$ are continuous functions with respect to $t$ and $y$, respectively, then

$$||x + f(x) - y^*||^2 \leq q(t^*).$$

It is clear that $y^* \in Q(t^*)$. Hence $Q$ is u.s.c. Since $Q$ is u.s.c. and single-valued, $Q$ is a continuous function.

For any $(x^1, t^1) \in C \times [0,1]$ and $\epsilon > 0$, there exists a $\delta_1 > 0$ such that for all $t \in [0,1]$ with $|t - t^1| < \delta_1$ and for all $y, z \in C$ with $||y - z|| < \delta_1$, it holds

$$||h(x^1, t) - h(x^1, t^1)|| \leq \epsilon/3 \text{ and } ||f(y) - f(z)|| \leq \epsilon/3.$$

Let $\delta = min\{\delta_1, \epsilon/3\}$. Then for all $t \in [0,1]$ with $|t - t^1| < \delta$ and for all $x \in C$ with $||x - x^1|| < \delta$, we have

$$\begin{aligned} ||h(x,t) - h(x^1,t^1)|| &= ||h(x,t) - h(x^1,t) + h(x^1,t) - h(x^1,t^1)|| \\ &\leq ||h(x,t) - h(x^1,t)|| + ||h(x^1,t) - h(x^1,t^1)|| \\ &\leq ||x + f(x) - x^1 - f(x^1)|| + ||h(x^1,t) - h(x^1,t^1)|| \\ &\leq ||x - x^1|| + ||f(x) - f(x^1)|| + \epsilon/3 \\ &\leq \epsilon. \end{aligned}$$

The second inequality follows from Lemma 1.5.9. Since $(x^1, t^1)$ is chosen arbitrarily, $h$ is continuous.

Since $h : C \times [0,1] \mapsto C$ is continuous, all the conditions of Theorem 1.6.11 are satisfied. So the set $K_h = \{(x,t) \in X \times [0,1] \mid x = h(x,t)\}$ contains a connected set $L$ such that

$$L \cap (X \times \{0\}) \neq \emptyset \text{ and } L \cap (X \times \{1\}) \neq \emptyset.$$

It follows from Theorem 1.6.11 that for each $t \in [0,1]$, $x = h(x,t)$ if and only if $x$ is a stationary point of $f$ on $\psi(t)$. Therefore, $(x,t) \in K_h$ if and only if $(x,t) \in C_\psi$. Hence the set $C_\psi$ contains a connected set $D$ such that

$$D \cap (C \times \{0\}) \neq \emptyset \quad \text{and} \quad D \cap (C \times \{1\}) \neq \emptyset.$$

□

This theorem implies the following corollary.

**Corollary 1.6.13** *Let $C$ be a nonempty, convex and compact subset of $\mathbb{R}^n$, and let $\psi : [0,1] \mapsto C$ be a continuous point-to-set mapping with nonempty, convex and compact values. Then the function $h : C \times [0,1] \mapsto C$ defined by*

$$h(x,t) = \arg\min\{||x - y||^2 \mid y \in \psi(t)\},$$

*is a continuous function with respect to $x$ and $t$.*

## 1.7. Fixed point theorems of Tarski, Caristi and Ekeland

In this section we introduce three different types of fixed point theorems. We remark that neither these theorems are implied by the fixed point theorems in the previous two sections nor the reverse implication holds true. We first quote a famous result due to Zorn in 1923.

**Lemma 1.7.1 (Zorn Lemma)** *If every chain in a partially ordered set $X$ has an upper bound, then $X$ has a maximal element in $X$.*

The next fixed point theorem was due to Tarski [1955]. Let $(X, \preceq)$ be a partially ordered set. A function $f : X \mapsto X$ is called to be *monotone* if $x \preceq y$ implies $f(x) \preceq f(y)$.

**Theorem 1.7.2 (Tarski Theorem)** *Let $(X, \preceq)$ be a partially ordered set with the property that every chain in $X$ has a supremum. Let $f : X \mapsto X$ be monotone such that there is a point $b \in X$ with $b \preceq f(b)$. Then the set of fixed points of $f$ is nonempty and has a maximal fixed point.*

Proof: Consider the following partially ordered subset

$$Y = \{x \in X \mid x \preceq f(x)\}.$$

The set $Y$ is nonempty since $Y$ contains $b$. Suppose $Z$ is a chain in $Y$, and $q$ is its supremum in $X$. Since $p \preceq q$ for all $p \in Z$, we see that $f(p) \preceq f(q)$. Because $p \preceq f(p)$ for all $p \in Z$, it is easy to see that $f(q)$ is an upper bound for $Z$. Since $q$ is the least such upper bound, we have $q \preceq f(q)$. This implies $q \in Y$. Hence the supremum of any chain in $Y$ belongs to $Y$. Then by Zorn lemma, $Y$ has a maximal element, say, $x^*$. Now we have $x^* \preceq f(x^*)$, since $x^*$ is in $Y$. Since $f$ is monotone, $f(x^*) \preceq f(f(x^*))$. But this means that $f(x^*)$ belongs to $Y$ by the definition of $Y$. Since $x^*$ is a maximal element of $Y$, we see that $x^* = f(x^*)$.

Furthermore, if $x$ is a fixed point of $f$, then $x$ is in $Y$. This shows that $x^*$ is a maximal fixed point of $f$. □

Let $T$ be a set and let $d$ be a function defined for all pairs $(x, y)$ of elements from $T$. We say that $d$ is a *distance function* or *metric* on $T$ if the following conditions are satisfied

(a) for any $x, y \in T$, $d(x, y) = 0$ if and only if $x = y$;

(b) $d(x, y) = d(y, x)$ for all $x, y \in T$;

(c) $d(x, z) \leq d(x, y) + d(y, z)$ for all $x, y, z \in T$ (*triangle inequality*).

Now we are going to introduce Caristi theorem and Ekeland theorem. There are several ways of proving both results. Here we adopt a proof due to Takahashi [1991]. We first introduce a result of Takahashi.

**Lemma 1.7.3** *Let $d$ be a distance function defined on $\mathbf{R}^n$. Let $g : \mathbf{R}^n \mapsto (-\infty, +\infty]$ be a proper lower semi-continuous function, bounded from below. Suppose that for each $a \in \mathbf{R}^n$ with $\inf_{x \in \mathbf{R}^n} g(x) < g(a)$, there exists $b \in \mathbf{R}^n$ such that $a \neq b$ and $g(b) + d(a, b) \leq g(a)$. Then there exists $x^* \in \mathbf{R}^n$ such that $g(x^*) = \inf_{x \in \mathbf{R}^n} g(x)$.*

Proof: Suppose to the contrary that $\inf_{x \in \mathbf{R}^n} g(x) < g(y)$ for all $y \in \mathbf{R}^n$. Since $g$ is proper, there exists a point $a \in \mathbf{R}^n$ such that $g(a)$ is finite. Define a sequence $\{ b_n \mid n \in \mathbf{Z}_+ \}$ with $b_0 = a$ as follows. Let

$$S_n = \{x \in \mathbf{R}^n \mid g(x) + d(b_{n-1}, x) \leq g(b_{n-1})\}.$$

Note that $S_n$ is nonempty since $b_{n-1} \in S_n$. Moreover, since $\inf_{x \in \mathbf{R}^n} g(x) < g(b_{n-1})$, there exists a point $y \neq b_{n-1}$ by assumption such that $g(y) +$

$d(b_{n-1}, y) \leq g(b_{n-1})$. Hence $g(y) < g(b_{n-1})$ since $d(b_{n-1}, y) > 0$. Note that $y \in S_n$. This means that

$$g(b_{n-1}) - \inf_{x \in S_n} g(x) > 0.$$

Take $\epsilon = \frac{1}{2}(g(b_{n-1}) - \inf_{x \in S_n} g(x))$. By definition of infimum there exists a point $b_n \in S_n$ such that

$$\inf_{x \in S_n} g(x) \leq g(b_n) \leq \inf_{x \in S_n} g(x) + \epsilon. \tag{1.7.4}$$

Since $\{g(b_n) \mid n \in \mathbb{Z}_+\}$ is a decreasing sequence by definition and has a lower bound by assumption, it has a limit. We now show that $\{b_n \mid n \in \mathbb{Z}_+\}$ is a Cauchy sequence. Indeed, when $m < n$, we have

$$\begin{aligned} 0 \leq d(b_m, b_n) &\leq \sum_{k=m}^{n-1} d(b_k, b_{k+1}) \\ &\leq \sum_{k=m}^{n-1} (g(b_k) - g(b_{k+1})) \\ &= g(b_m) - g(b_n). \end{aligned} \tag{1.7.5}$$

Let $\lim_{n\to\infty} b_n = q$. By the definition of lower semi-continuity we have

$$\lim_{n\to\infty} g(b_n) \geq g(\lim_{n\to\infty} b_n) = g(q).$$

Then it follows from (1.7.5) that

$$d(b_m, q) \leq g(b_m) - \lim_{n\to\infty} g(b_n) \leq g(b_m) - g(q).$$

On the other hand, by assumption, there exists $p \in \mathbb{R}^n$ such that $p \neq q$ and $g(p) + d(q, p) \leq g(q)$. It follows that $g(p) < g(q)$ and

$$\begin{aligned} g(p) &\leq g(q) - d(p, q) \\ &\leq g(q) - d(p, q) + g(b_n) - g(q) - d(b_n, q) \\ &= g(b_n) - (d(p, q) + d(b_n, q)) \\ &\leq g(b_n) - d(p, b_n). \end{aligned} \tag{1.7.6}$$

This means that $p$ is in $S_{n+1}$. Since $n$ is arbitrary, we have $p \in S_n$. Using (1.7.4), we have

$$2g(b_n) - g(b_{n-1}) \leq \inf_{x \in S_n} g(x) \leq g(p).$$

By taking the limit, it follows that

$$\lim_{n\to\infty} g(b_n) = \lim_{n\to\infty} (2g(b_n) - g(b_{n-1})) \leq \lim_{n\to\infty} \inf_{x \in S_n} g(x) \leq g(p).$$

Hence we obtain that

$$g(p) < g(q) \leq \lim_{n\to\infty} g(b_n) \leq g(p).$$

This is a contradiction. Therefore there exists $x^* \in \mathbb{R}^n$ such that $g(x^*) = \inf_{x\in\mathbb{R}^n} g(x)$. □

By using the above lemma, we prove the theorems of Caristi [1976] and Ekeland [1979].

**Theorem 1.7.4 (Caristi Theorem)** *Let $d$ be a distance function defined on $\mathbb{R}^n$ and let $g : \mathbb{R}^n \mapsto \mathbb{R}$ be a proper lower semi-continuous function, bounded from below. Suppose that a function $f : \mathbb{R}^n \mapsto \mathbb{R}^n$ satisfies*

$$d(x, f(x)) \le g(x) - g(f(x))$$

*for every $x \in \mathbb{R}^n$. Then there exists a point $x^* \in \mathbb{R}^n$ such that $x^* = f(x^*)$.*

Proof: Suppose that $f(x) \neq x$ for all $x \in \mathbb{R}^n$. So for each $x \in \mathbb{R}^n$ (with $\inf_{z\in\mathbb{R}^n} g(z) < g(x)$ since $g(x) \ge d(x, f(x)) + g(f(x)) > g(f(x))$) there is a point $y \in \mathbb{R}^n$ (for example take $y = f(x)$) such that $y \neq x$ and $g(y) + d(x, y) \le g(x)$. Applying Lemma 1.7.3, we have a point $x^* \in \mathbb{R}^n$ with $g(x^*) = \inf_{x\in\mathbb{R}^n} g(x)$. For such point $x^*$, it follows that

$$0 < d(x^*, f(x^*)) \le g(x^*) - g(f(x^*)) \le g(f(x^*)) - g(f(x^*)) = 0.$$

This leads to a contradiction. □

We remark that Caristi theorem is an important generalization of the fixed point theorem for contractive mappings.

**Theorem 1.7.5 (Ekeland $\epsilon$ Principle)** *Let $d$ be a distance function defined on $\mathbb{R}^n$ and let $g : \mathbb{R}^n \mapsto (-\infty, +\infty]$ be a proper lower semi-continuous function, bounded from below. Let $\epsilon > 0$ be given. Suppose that $u$ is a point in $\mathbb{R}^n$ such that*

$$g(u) \le \inf_{x\in\mathbb{R}^n} g(x) + \epsilon.$$

*Then there exists a point $v \in \mathbb{R}^n$ such that $g(v) \le g(u)$, $d(u, v) \le 1$ and $g(w) > g(v) - \epsilon d(v, w)$ for any $w \neq v$.*

Proof: Define a set

$$X = \{x \in \mathbb{R}^n \mid g(x) \le g(u) - \epsilon d(x, u)\}.$$

It is easy to verify that $X$ is nonempty and closed. Moreover, for each $x \in X$ we have

$$\epsilon d(u, x) \le g(u) - g(x) \le g(u) - \inf_{w\in\mathbb{R}^n} g(w) \le \epsilon.$$

Hence $d(u, x) \le 1$. It also holds that $g(x) \le g(u)$.

Suppose to the contrary that for any $x \in X$, there exists $w \in \mathbb{R}^n$ such that $w \neq x$ and $g(w) \leq g(x) - \epsilon d(x, w)$. Then we have

$$\begin{aligned} \epsilon d(w,u) &\leq \epsilon d(w,x) + \epsilon d(x,u) \\ &\leq g(x) - g(w) + g(u) - g(x) \\ &= g(u) - g(w). \end{aligned}$$

This implies that $w \in X$. By applying Lemma 1.7.3, there exists $x^0 \in X$ such that $g(x^0) = \inf_{x \in X} g(x)$. Hence there is a point $y^0 \in X$ with $g(y^0) < g(x^0)$. This is a contradiction. □

Note that although the above two theorems look so different, they are in fact equivalent to each other. See Ekeland [1979] and its relevant references.

# CHAPTER 2

# Applications in Game Theory and Economics

## 2.1. Introduction

This chapter selects several important topics from game theory and economic theory to which the methodology introduced in this book will be applied. To study and analyze the outcomes of an economic system or a society in a broader sense, we may put the underlying problems into the framework of game theory or deal with them by means of general equilibrium theory. Both theories can be regarded as complementary methodologies rather than substitutes. As we see in the literature, there is a close relationship between both methods of analysis.

The organization of this chapter is summarized next. Section 2 introduces non-transferable utility cooperative games and discusses some central concepts, such as the core of a game and the concept of balancedness. Section 3 reviews the non-cooperative game in normal form and three fundamental equilibrium concepts: Nash equilibrium, perfect Nash equilibrium, and proper Nash equilibrium. Section 4 captures the well-known model of a pure exchange economy and the concept of a Walrasian equilibrium. Section 5 presents a typical model of an exchange economy with price rigidities and a rationing system and discusses the concept of a constrained equilibrium. Section 6 deals with an exchange economy with linear production technologies. Section 7 presents an exchange economy in which production technologies exhibit increasing returns to scale. Section 8 introduces an exchange economy in which commodities are indivisible. Finally Section 9 sets up an exchange economy under uncertainty.

## 2.2. The core of a game

Since the appearance of the fundamental book by von Neumann and Morgenstern [1944], game theory has been spread into two streams: cooperative game theory and non-cooperative game theory. First, let us look at the cooperative game theory. Loosely speaking, the theme of cooperative game theory is concerned with what each group of players or coalition can achieve by the cooperation of its members and how coalitions should be formed. For $N \in \mathbb{N}$, let $I_N$ be the set of players. A nonempty subset $S$ of $I_N$ is called a *coalition*. Let $\mathcal{N}$ be the collection of all coalitions of $I_N$. The simplest game of cooperative games is a so-called *transferable utility (TU) game in characteristic function form* and defined below.

**Definition 2.2.1** *A transferable utility game in characteristic function form is an ordered pair* $< I_N, v >$, *where* $v : \mathcal{N} \mapsto \mathbb{R}$ *is a function which assigns to each coalition* $S \in \mathcal{N}$ *a real number* $v(S)$. *The number* $v(S)$ *is the worth of coalition* $S$.

The interpretation of the game is as follows. There is usually assumed to be a commodity, called money, which players can transfer without costs among themselves and the payoff of any player is increased by one unit when he gets one unit of money. The worth $v(S)$ is the maximal total payoff that the coalition $S$ can earn if the members of $S$ cooperate regardless of the actions from the players outside of $S$. In the sequel, we will simply call the above game a transferable utility game. One of the central problems is how to divide the payoffs among the players. An important solution concept for that would be the concept of the core which is a generalization of the contract curve first introduced by Edgeworth [1881] in the context of a pure exchange economy. The core of a transferable utility game is introduced below.

**Definition 2.2.2** *The core of a transferable utility game* $< I_N, v >$, *denoted by* $C(I_N, v)$, *is given by the set of vectors* $x \in \mathbb{R}^N$ *such that*
*(a)* $x$ *is a feasible allocation for the grand coalition* $I_N$, *i.e.,* $\sum_{i=1}^N x_i = v(I_N)$;
*(b) no coalition can do strictly better by acting alone, i.e.,* $\sum_{i \in S} x_i \geq v(S)$ *for any* $S \in \mathcal{N}$.

Thus, $x \in C(I_N, v)$ can be regarded as a stable outcome of the game, where each component of $x$ is interpreted as the utility payoff to some player. Although the above concept is very appealing, unfortunately the core may be empty. But the following condition gives us an affirmative answer for the existence of the core of a game. This condition is known as balancedness, introduced by Bondareva [1962]. For $S \subseteq I_N$, we define an $N$-vector $m^S$ by $\sum_{i \in S} \frac{1}{|S|} e(i)$, where $e(i)$ is the $i$-th unit vector in $\mathbb{R}^N$ for $i \in N$. Note that $m^S = e(i)$ if $S = \{i\}$.

**Definition 2.2.3** *A family* $\mathcal{B} = \{B_1, \cdots, B_k\}$ *of* $k$ *elements of* $\mathcal{N}$ *is coalitionally balanced if there exist positive numbers* $\lambda_j$, $j = 1, \cdots, k$ *such that*

$$\sum_{j=1}^{k} \lambda_j m^{B_j} = m^{I_N},$$

*where* $\lambda_1, \cdots, \lambda_k$ *are called balancing coefficients. A transferable utility game* $< I_N, v >$ *is coalitionally balanced if for every coalitionally balanced family* $\mathcal{B} = \{B_1, \cdots, B_k\}$ *of* $\mathcal{N}$ *with balancing coefficients* $\lambda_1, \cdots, \lambda_k$, *it holds that* $\sum_{j=1}^{k} \lambda_j v(B_j) \leq v(I_N)$.

The following theorem reveals the significance of the concept of balancedness.

**Theorem 2.2.4** *Let* $< I_N, v >$ *be a transferable utility game. The core of the game is nonempty if and only if the game is coalitionally balanced.*

The proof follows immediately from Lemma 1.3.8.

Sometimes it may be very difficult to give a plausible interpretation of the worth $v(S)$ of a coalition $S$ for a transferable utility game, since the ultimate goal of each player is to pursue his own maximal satisfaction. In order to get rid of this flaw, a more general cooperative game has been developed by Aumann and Peleg [1960]. For each coalition $S$, define

$$\mathbb{R}^S = \{ x \in \mathbb{R}^N \mid x_j = 0, \ j \notin S \}.$$

**Definition 2.2.5** *A non-transferable utility (NTU) game in characteristic form is an ordered pair* $< I_N, V >$, *where* $V : \mathcal{N} \mapsto \mathbb{R}^N$ *is a correspondence satisfying*
(*a*) *for each coalition* $S$, $V(S)$ *is nonempty, comprehensive and closed;*
(*b*) *for each coalition* $S$, $V(S) \bigcap \mathbb{R}^S_+$ *is bounded;*
(*c*) *for each coalition* $S$, $V(S)$ *is a cylinder in the sense that for two vectors* $x$, $y$ *in* $\mathbb{R}^N$ *with* $x_i = y_i$ *for all* $i \in S$, $x \in V(S)$ *implies* $y \in V(S)$.

We will simply call the above game a non-transferable utility game. If $u \in V(S)$ for some coalition $S$, then each member $i$ of $S$ can guarantee for himself a payoff $u_i$ if they act cooperatively, independently of other players outside $S$. The core of a non-transferable utility game is defined as follows.

**Definition 2.2.6** *The core of a non-transferable utility game* $< I_N, V >$, *denoted by* $C(I_N, V)$, *is given by the set of vectors* $x \in \mathbb{R}^N$ *such that*
(*a*) $x$ *is an attainable payoff vector for the grand coalition* $I_N$, *i.e.,* $x \in V(I_N)$;
(*b*) *no coalition* $S$ *can do strictly better by acting alone, i.e., there is no feasible allocation* $y \in V(S)$ *such that* $y_i > x_i$ *for all members* $i \in S$.

From the definition a payoff vector lies in the core of a game if it can be attained by the grand coalition and no coalition can improve upon. The following important concept is introduced in Scarf [1967a].

**Definition 2.2.7** *A non-transferable utility game* $< I_N, V >$ *is coalitionally balanced if for every coalitionally balanced family* $\mathcal{B} = \{B_1, \cdots, B_k\}$ *of* $\mathcal{N}$, *it holds that* $\bigcap_{S \in \mathcal{B}} V(S) \subseteq V(I_N)$.

Moreover, Scarf [1967a] established the first and probably most important theorem for the non-emptiness of the core of a non-transferable utility game.

**Theorem 2.2.8 (Scarf Theorem)** *Let* $< I_N, V >$ *be a non-transferable utility game. The core of the game is nonempty if the game is coalitionally balanced.*

To illustrate coalitional balancedness of a game, let us consider the three-person game given by

$$\begin{aligned} V(\{i\}) &= \{x \in \mathbf{R}^3 \,|\, x_i \leq 0\},\ i = 1,\ 2,\ 3, \\ V(\{1,2\}) &= \{x \in \mathbf{R}^3 \,|\, x_1 \leq 1,\ x_2 \leq 1\}, \\ V(\{2,3\}) &= \{x \in \mathbf{R}^3 \,|\, x_2 \leq 1,\ x_3 \leq 1\}, \\ V(\{3,1\}) &= \{x \in \mathbf{R}^3 \,|\, x_3 \leq 1,\ x_1 \leq 1\}, \\ V(\{1,2,3\}) &= \{x \in \mathbf{R}^3 \,|\, x_1 + x_2 + x_3 \leq 3\}. \end{aligned}$$

Note that the balanced collections are $\{\{1\},\{2\},\{3\}\}$, $\{\{1,2\},\{2,3\},\{3,1\}\}$, $\{\{1,2\},\{3\}\}$, $\{\{2,3\},\{1\}\}$, $\{\{3,1\},\{2\}\}$, and $\{\{1,2,3\}\}$. It is easy to check that the game is coalitionally balanced. The core of the game is given by

$$\begin{aligned} C(I_3, V) &= \{x \in \mathbf{R}^3_+ \,|\, x_1 \geq 1,\ x_2 \geq 1,\ x_1 + x_2 + x_3 = 3\} \\ &\cup \{x \in \mathbf{R}^3_+ \,|\, x_2 \geq 1,\ x_3 \geq 1,\ x_1 + x_2 + x_3 = 3\} \\ &\cup \{x \in \mathbf{R}^3_+ \,|\, x_3 \geq 1,\ x_1 \geq 1,\ x_1 + x_2 + x_3 = 3\}. \end{aligned}$$

## 2.3. Nash, perfect Nash, and proper Nash equilibria

Now we turn to the non-cooperative game theory. Non-cooperative game theory is mainly concerned with multi-person decision problems in which each player acts individually to maximize his interests according to the rules of the game. Here we will restrict ourselves to non-cooperative games in normal (or strategic) form. We are concerned with the concept of Nash equilibrium and its refinements. The refinement and stability of Nash equilibria has been one of the major preoccupations of finite non-cooperative game theory during the last three decades.

For $N \in \mathbf{N}$, let $I_N$ be the set of players. A *non-cooperative $N$-person game in normal form* is characterized by a $2N$-tuple $\Gamma = (\Phi_1, \cdots, \Phi_N; U_1, \cdots, U_N)$, where $\Phi_j$ denotes a nonempty finite set of player $j$'s pure strategies, indexed by $(j, 1)$, $\cdots$, $(j, n_j)$, and $U_j : \Phi \mapsto \mathbf{R}$ is the payoff function of player $j$ with $\Phi = \prod_{j=1}^N \Phi_j$. The set of all mixed strategies of player $j$ is the $(n_j - 1)$-dimensional unit simplex $S^{n_j}$ and the mixed strategy space of the game is the Cartesian product $\mathcal{S} = \prod_{j=1}^N S^{n_j}$. A point $x \in \mathcal{S}$ is called a *completely mixed strategy* if $x$ lies in the interior of $\mathcal{S}$, i.e., $x \in int(\mathcal{S})$. Given a mixed strategy $x = (x_1, \cdots, x_N) \in \mathcal{S}$ the probability that a pure strategy combination

$$\phi = ((1, l_1), \cdots, (N, l_N)) \in \Phi$$

occurs is given by $x(\phi) = \prod_{h=1}^{N} x_{h,l_h}$, so $x_{j,k}$ denotes the probability with which player $j$ chooses his $k$-th pure strategy $(j,k)$. Then the expected payoff for player $j$, $j \in I_N$, at strategy $x \in \mathcal{S}$ is equal to $U_j(x) = \sum_{\phi \in \Phi} x(\phi) U_j(\phi)$. The expected marginal payoff for player $j$ at $x \in \mathcal{S}$ when he plays his pure strategy $(j,k) \in \Phi_j$ is given by

$$U_j^k(x) = \sum_{\{\phi \in \Phi | (j,l_j)=(j,k)\}} U_j(\phi) \prod_{h \neq j} x_{h,l_h}.$$

It is readily seen that $U_j(x) = \sum_{h=1}^{n_j} x_{j,h} U_j^h(x)$ for every $j \in I_N$ and $x \in \mathcal{S}$.

The following equilibrium concept is proposed in Nash [1950], which is the most fundamental idea in non-cooperative game theory.

**Definition 2.3.1** *A mixed strategy $x \in \mathcal{S}$ is a Nash equilibrium if for all $j \in I_N$ it holds that*

$$U_j(x) \geq U_j^k(x), \text{ for all } k \in I_{n_j}.$$

The definition says that a strategy $x \in \mathcal{S}$ is a Nash equilibrium if no player has an incentive to deviate from the prediction of the strategy. Nash [1950] established the following milestone theorem.

**Theorem 2.3.2 (Nash Theorem)** *Every non-cooperative $N$-person game in normal form has at least one Nash equilibrium in the mixed strategy space $\mathcal{S}$.*

As argued by Selten [1975], it may happen that some Nash equilibria are not consistent with the intuitive notions about what should be the outcome of the game. This is illustrated in the next example.

**Example 2.3.3** *Consider the non-cooperative 2-person game of Myerson [1978]. Each player has three pure strategies and the payoffs are given in Table 2.1.*

*Player 2*

| | | $\phi_1$ | $\phi_2$ | $\phi_3$ |
|---|---|---|---|---|
| Player 1 | $\psi_1$ | $(1,1)$ | $(0,0)$ | $(-9,-9)$ |
| | $\psi_2$ | $(0,0)$ | $(0,0)$ | $(-7,-7)$ |
| | $\psi_3$ | $(-9,-9)$ | $(-7,-7)$ | $(-7,-7)$ |

TABLE 2.1. Payoffs of the game.

*This game has three Nash equilibria in pure strategies: $(\phi_1, \psi_1)$, $(\phi_2, \psi_2)$, and $(\phi_2, \psi_3)$.*

In this example the equilibrium $(\phi_3, \psi_3)$ leads to the outcome $(-7, -7)$ which is not very reasonable. To eliminate undesirable Nash equilibria, Selten introduced the concept of a perfect equilibrium. For $\epsilon > 0$, a completely mixed strategy $x \in \mathcal{S}$ is called an *$\epsilon$-perfect equilibrium* if $x_{j,k} \leq \epsilon$ in case $U_j^k(x) < U_j^l(x)$ for all $j \in I_N$ and $k, l \in I_{n_j}$. It means that every pure strategy gets a positive probability, but only optimal strategies can get more than $\epsilon$ probability. A mixed strategy $x \in \mathcal{S}$ is a perfect equilibrium if it is the limit point of a sequence of $\epsilon$-perfect equilibria when $\epsilon$ goes to zero.

**Definition 2.3.4** *A mixed strategy $x \in \mathcal{S}$ is a perfect (Nash) equilibrium if there exist some sequences $\{ \epsilon_h \mid h \in \mathbb{N} \}$ of positive numbers and $\{ x^h \mid h \in \mathbb{N} \}$ of $\epsilon_h$-perfect equilibria such that*

$$\lim_{h \to \infty} \epsilon_h = 0 \quad \textit{and} \quad \lim_{h \to \infty} x^h = x.$$

The following result was due to Selten [1975].

**Theorem 2.3.5 (Selten Theorem)** *Every non-cooperative $N$-person game in normal form has at least one perfect equilibrium in the mixed strategy space $\mathcal{S}$. The set of perfect equilibria is a subset of the set of Nash equilibria of the game.*

This theorem implies that each perfect equilibrium is a Nash equilibrium. The game in Example 2.3.3 has two perfect equilibria: $(\phi_1, \psi_1)$ and $(\phi_2, \psi_2)$.

In Myerson [1978] it is demonstrated that even perfect equilibria are not always reasonable. In Example 2.3.3, the perfect equilibrium $(\phi_2, \psi_2)$ is clearly not as desirable as the perfect equilibrium $(\phi_1, \psi_1)$. So Myerson proposed the notion of a proper equilibrium to further refine the perfect equilibrium concept. For $\epsilon > 0$, a completely mixed strategy $x \in \mathcal{S}$ is called an *$\epsilon$-proper equilibrium* if $x_{j,k} \leq \epsilon x_{j,l}$ in case $U_j^k(x) < U_j^l(x)$, for all $j \in I_N$ and $k, l \in I_{n_j}$. It implies that a player gives to any strategy at least $1/\epsilon$ times more probability than all other worse strategies. A mixed strategy $x \in \mathcal{S}$ is a proper equilibrium if it is the limit point of a sequence of $\epsilon$-proper equilibria when $\epsilon$ goes to zero.

**Definition 2.3.6** *A mixed strategy $x \in \mathcal{S}$ is a proper (Nash) equilibrium if there exist some sequences $\{ \epsilon_h \mid h \in \mathbb{N} \}$ of positive numbers and $\{ x^h \mid h \in \mathbb{N} \}$ of $\epsilon_h$-proper equilibria such that*

$$\lim_{h \to \infty} \epsilon_h = 0 \quad \textit{and} \quad \lim_{h \to \infty} x^h = x.$$

The following theorem assures that every non-cooperative $N$-person game has a proper equilibrium.

**Theorem 2.3.7 (Myerson Theorem)** *Every non-cooperative $N$-person game in normal form has at least one proper equilibrium in the mixed strategy space $\mathcal{S}$. The set of proper equilibria is a subset of the set of perfect equilibria of the game.*

This result implies that each proper equilibrium is a perfect equilibrium and hence is also a Nash equilibrium. The game in Example 2.3.3 has exactly one proper equilibrium: $(\phi_1, \psi_1)$.

Now we define the expected marginal payoff function $f : \mathcal{S} \mapsto \prod_{h=1}^{N} \mathbb{R}^{n_h}$ by

$$\begin{aligned} f(x) &= (f_1(x), \cdots, f_N(x)), \\ f_j(x) &= (f_{j,1}(x), \cdots, f_{j,n_j}(x))^\top \text{ for } j \in I_N, \\ f_{j,l}(x) &= U_j^l(x) \text{ for } j \in I_N \text{ and } l \in I_{n_j}. \end{aligned} \tag{2.3.1}$$

Clearly, $f$ is a continuous function from the strategy space $\mathcal{S}$ to $\prod_{h=1}^{N} \mathbb{R}^{n_h}$. It follows from Theorem 1.5.13 that $f$ has a stationary point $x^* \in \mathcal{S}$, i.e., $(x_j^*)^\top f_j(x^*) \geq (x_j)^\top f_j(x^*)$ for any $j \in I_N$ and for any $x \in \mathcal{S}$. Recalling that $U_j(x) = \sum_h x_{j,h} U_j^h(x)$ for every $j \in I_N$ and $x \in \mathcal{S}$, it is easy to derive the following result.

**Theorem 2.3.8** *Let a non-cooperative N-person game in normal form be given. Then $x^* \in \mathcal{S}$ is a Nash equilibrium if and only if $x^*$ is a stationary point of the expected marginal payoff function defined by* (2.3.1) *on $\mathcal{S}$.*

For Example 2.3.3, let $P$ be given by

$$P = \begin{bmatrix} 1 & 0 & -9 \\ 0 & 0 & -7 \\ -9 & -7 & -7 \end{bmatrix}.$$

Then $f_1$ and $f_2$ of (2.3.1) are given by

$$f_1(x) = Px_2 \quad \text{and} \quad f_2(x) = Px_1.$$

It is easy to check that indeed only $(1,0,0,1,0,0)^\top$, $(0,1,0,0,1,0)^\top$ and $(0,0,1,0,0,1)^\top$ are stationary points of $f$ on $\mathcal{S}$.

## 2.4. A pure exchange economy

In this and the next four sections we introduce economic equilibrium theory. The theme of this theory is to investigate the complex economic system as an equilibrium resulting from the interaction of many economic agents with different interests.

In the current section we focus on the typical model of a pure exchange economy. This model was first elaborated by Walras [1874] but became studied rigorously with the publication of the classic book by Debreu in 1959. It should be mentioned that the model addressed here excludes production, but incorporating production with non-increasing returns to scale to the model does not cause additional difficulties.

The model of a pure exchange economy is given by $\mathcal{E} = (\{ (X^i, u^i, \omega^i) \mid i \in I_m \})$. There are $n$ types of commodities, indexed by $j = 1, \cdots, n$, and $m$

consumers, indexed by $i = 1, \cdots, m$. Each consumer $i \in I_m$ is characterized by his consumption set $X^i$, his utility function $u^i$ and his initial endowment vector $\omega^i$. The *consumption set* $X^i$ is a subset of the commodity space $\mathbb{R}^n$ and represents the set of possible commodity bundles. The *utility function* $u^i : X^i \mapsto \mathbb{R}$ represents the *preference (or taste)* of consumer $i$ for all commodity bundles in his consumption set. Consumer $i$ is endowed with $\omega^i_j$ of commodity $j \in I_n$ before trade takes place.

Concerning a consumer's consumption set, his initial endowment and his utility function, the following assumptions are often made.

**Assumption 2.4.1** *For each consumer $i$, the consumption $X^i$ is a non-empty, closed, and convex subset of $\mathbb{R}^n_+$ such that $X^i + \mathbb{R}^n_+ \subseteq X^i$ and $\omega^i \in int(X^i)$.*

**Assumption 2.4.2** *For each consumer $i$, the utility function $u^i : X^i \mapsto \mathbb{R}$ satisfies*
*(a) continuity;*
*(b) monotonicity, i.e., $x^i$, $y^i \in X^i$ and $x^i > y^i$ imply $u^i(x^i) > u^i(y^i)$;*
*(c) quasi-concavity.*

Given a price vector $p \in \mathbb{R}^n_+ \setminus \{\mathbf{0}\}$, the budget set of consumer $i \in I_m$ is defined by

$$B^i(p) = \{\, x \in X^i \mid p^\top x \leq p^\top \omega^i \,\}.$$

Then the *demand mapping* $D^i : \mathbb{R}^n_+ \setminus \{\mathbf{0}\} \mapsto \mathbb{R}^{\mathbf{n}}$ of consumer $i$ is given by

$$D^i(p) = \{\, x \in B^i(p) \mid u^i(x) = \max_{y \in B^i(p)} u^i(y) \,\}.$$

It has been shown (see Debreu [1959]) that for each $i \in I_m$, $D^i$ is a u.s.c. mapping on $\mathbb{R}^n_{++}$ and $D^i(p)$ is convex and compact. Moreover, $D^i$ is homogeneous of degree zero in $p$. The *aggregated demand mapping* $D : \mathbb{R}^n_+ \setminus \{\mathbf{0}\} \mapsto \mathbb{R}^{\mathbf{n}}$ is given by $D(p) = \sum_{i=1}^m D^i(p)$. Let $\omega$ be the total initial endowments, i.e., $\omega = \sum_{i=1}^m \omega^i$. The *excess demand mapping* $Z : \mathbb{R}^n_+ \setminus \{\mathbf{0}\} \mapsto \mathbb{R}^n$ is given by

$$Z(p) = D(p) - \{\omega\}.$$

Now we obtain the following result.

**Lemma 2.4.3** *Let a pure exchange economy $\mathcal{E} = (\{X^i, u^i, \omega^i\}_{i=1}^m)$ be given, which satisfies Assumptions 2.4.1 and 2.4.2. Then for the excess demand mapping $Z : \mathbb{R}^n_+ \setminus \{\mathbf{0}\} \mapsto \mathbb{R}^n$, it holds that*
*(i) $Z$ is u.s.c. on $\mathbb{R}^n_{++}$;*
*(ii) $Z(p)$ is convex and compact, for any $p \in \mathbb{R}^n_{++}$;*
*(iii) $Z$ is homogeneous of degree zero;*
*(iv) $p^\top z = 0$ for all $z \in Z(p)$ and all $p \in \mathbb{R}^n_+ \setminus \{\mathbf{0}\}$.*

The property (iv) is known as *Walras' law*. An element $(x^{1*}, \cdots, x^{m*}, p^*) \in \prod_{i=1}^m X^i \times (\mathbb{R}^n_+ \setminus \{\mathbf{0}\})$ is called a *Walrasian equilibrium* if $x^{i*} \in D^i(p^*)$ for all $i \in I_m$ and $\sum_{i=1}^m x^{i*} = \omega$. This implies that $\mathbf{0} \in Z(p^*)$. The vector $p^*$ is called a *Walrasian equilibrium price* and the vector $(x^{1*}, \cdots, x^{m*})$ a *Walrasian equilibrium allocation*. The fundamental question arises here: Given an economy $\mathcal{E}$, does it have a Walrasian equilibrium price? The next theorem answers the question in an affirmative manner.

**Theorem 2.4.4** *Let a pure exchange economy $\mathcal{E} = (\{(X^i, u^i, \omega^i) \mid i \in I_m\})$ be given. Then the economy has a Walrasian equilibrium if it satisfies Assumptions 2.4.1 and 2.4.2.*

Note that we can normalize the price vectors $p$ to lie in the unit simplex $S^n$ due to the homogeneity of $Z$. Moreover, if the condition (c) of Assumption 2.4.2 is strengthened to *strict quasi-concavity* (i.e., $u^i(tx + (1-t)y) > \min\{u^i(x), u^i(y)\}$ for all $x, y \in X^i$ with $x \neq y$ and all $t$ such that $0 < t < 1$), then the u.s.c. demand mapping $Z$ is reduced to a function $z$ and the Walrasian equilibrium price is a zero point of $z$.

We conclude with an example of a pure exchange economy.

**Example 2.4.5** *The economy consists of two agents and two commodities. The initial endowments of agent 1 and agent 2 are $(3,2)^\top$ and $(1,3)^\top$, respectively. Both agents have $\mathbb{R}^2_+$ as consumption sets and have Cobb-Douglas utility functions given by*

$$\begin{aligned} u^1(x^1) &= (x_1^1)^{\frac{1}{3}}(x_2^1)^{\frac{2}{3}}, \; x^1 \in \mathbb{R}^2_+, \\ u^2(x^2) &= (x_1^2)^{\frac{2}{3}}(x_2^2)^{\frac{1}{3}}, \; x^2 \in \mathbb{R}^2_+, \end{aligned}$$

*respectively.*

It is easy to calculate the demand functions of both consumers,

$$\begin{aligned} d^1(p) &= \left(\tfrac{3p_1+2p_2}{3p_1}, \tfrac{6p_1+4p_2}{3p_2}\right)^\top, \; p \in \mathbb{R}^2_{++}, \\ d^2(p) &= \left(\tfrac{2p_1+6p_2}{3p_1}, \tfrac{p_1+3p_2}{3p_2}\right)^\top, \; p \in \mathbb{R}^2_{++}. \end{aligned}$$

It should be observed that for any price vector $p \in \mathbb{R}^2_+ \setminus \{\mathbf{0}\}$, the demand $d^i_j(p)$ tends to infinity if $p_j$ converges to zero. The excess demand at $p \in \mathbb{R}^2_{++}$ is given by

$$z(p) = d^1(p) + d^2(p) - \omega = d^1(p) + d^2(p) - (4,5)^\top.$$

The unique Walrasian equilibrium price of the economy on $S^2$ is equal to

$$p^* = (8/15, 7/15)^\top.$$

The Walrasian equilibrium allocations for the two agents are given by

$$x^{1*} = (19/12, 76/21)^\top \quad \text{and} \quad x^{2*} = (29/12, 29/21)^\top.$$

## 2.5. An exchange economy with price rigidities

As we have seen, the assumption of perfect competition is essential in the neoclassical economic theory (see e.g., Debreu [1959], Arrow and Hahn [1971]). However, the assumption is invalid in many economic situations, such as, where the minimum wage law is implemented in the presence of underemployment, or the grain price is subject to a maximum level in case of a bad harvest. In such cases, a Walrasian equilibrium may not be achieved. To cope with these more realistic and sophisticated economic problems, Drèze [1975] and Bénassy [1975] independently proposed to clear the markets by imposing quantity constraints on net trades. In their theories both authors developed the notion of quantity rationing in general equilibrium models with prices subject to inequality constraints. In this framework each agent chooses his most preferred commodity bundle subject not only to his budget constraint as usual, but also to quantity constraints on net trades. An important feature of a rationing scheme is that rationing may affect either supply or demand but it never affects both simultaneously. Moreover, quantity constraints come into effect only if price rigidities are binding. The work of Drèze [1975] and Bénassy [1975] marked one of the major achievements of mathematical economics during the last two decades and has spawned a lot of literature. Here we will discuss one of the widely-used models in this field as introduced by Drèze [1975].

A general exchange economic equilibrium model with price rigidities is defined by $\mathcal{E} = (\{ (X^i, u^i, \omega^i) \mid i \in I_m \}, P)$. In the economy there are $n$ types of commodities indexed by $j = 1, \cdots, n$, and $m$ consumers indexed by $i = 1, \cdots, m$. Each consumer $i \in I_m$ is characterized by his consumption set $X^i$, his utility function $u^i$ and his initial endowment $\omega^i$. $P$ is the set of admissible price vectors. With respect to $X^i$, $u^i$ and $\omega^i$, we assume that Assumptions 2.4.1 and 2.4.2 are satisfied here.

Due to price rigidities the price vector is usually assumed to be in a proper subset of $\mathbb{R}^n_+$. We restrict our attention to the case that the set of admissible prices is given by

$$P = \{ p \in \mathbb{R}^n_+ \mid \underline{p}_j \leq p_j \leq \bar{p}_j,\ j \in I_n \},$$

for certain $0 < \underline{p}_j \leq \bar{p}_j < \infty$, $j \in I_n$. Because of the lower and upper bounds on the prices, a Walrasian equilibrium price $p^*$ does not necessarily exist. Thus, in case of excess demands or excess supplies, rationing may be needed to clear the markets. In particular, $l^i \in -\mathbb{R}^n_+$ stands for the constraints on the net supplies of consumer $i$, and $L^i \in \mathbb{R}^n_+$ for the constraints on the net demands of consumer $i$. The constrained budget set of consumer $i$ at price $p \in P$ and constraint $(l^i, L^i) \in -\mathbb{R}^n_+ \times \mathbb{R}^n_+$ is given by

$$B^i(p, l^i, L^i) = \{ x \in X^i \mid p^\top x \leq p^\top \omega^i,\ l^i \leq x^i - \omega^i \leq L^i \}.$$

The demand of consumer $i$ at price $p \in P$ and constraints $(l^i, L^i)$ is then defined by

$$D^i(p, l^i, L^i) = \{ x \in B^i(p, l^i, L^i) \mid u^i(x) = \max_{y \in B^i(p, l^i, L^i)} u^i(y) \}.$$

Now we are able to introduce the concept of a constrained equilibrium.

**Definition 2.5.1** *A constrained equilibrium is an element* $(x^{1*}, \cdots, x^{m*}, l^{1*}, \cdots, l^{m*}, L^{1*}, \cdots, L^{m*}, p^*)$ *of the set* $\prod_{i=1}^m X^i \times \prod_{j=1}^m -\mathbb{R}_+^n \times \prod_{k=1}^m \mathbb{R}_+^n \times P$ *such that*

*(a) for all* $i \in I_m$, $x^{i*}$ *is an element of* $D^i(p^*, l^{i*}, L^{i*})$;

*(b)* $\sum_{i=1}^m x^{i*} = \sum_{i=1}^m \omega^i$;

*(c) for all* $j \in I_n$, $x_j^{i*} - \omega_j^i = l_j^{i*}$ *for some* $i \in I_m$ *implies* $x_j^{h*} - \omega_j^h < L_j^{h*}$ *for all* $h \in I_m$, *and* $x_j^{i*} - \omega_j^i = L_j^{i*}$ *for some* $i \in I_m$ *implies* $x_j^{h*} - \omega_j^h > l_j^{h*}$ *for all* $h \in I_m$;

*(d) for all* $j \in I_n$, $p_j^* < \bar{p}_j$ *implies* $L_j^{i*} > x_j^{i*} - \omega_j^i$ *for all* $i \in I_m$, *and* $p_j^* > \underline{p}_j$ *implies* $l_j^{i*} < x_j^{i*} - \omega_j^i$ *for all* $i \in I_m$.

The first two conditions need no explanation. Condition (c) says that rationing will never simultaneously affect both sides of any market. Condition (d) states that no quantity rationing is allowed unless price rigidities are binding. Observe that there are two *trivial constrained equilibria or no-trade equilibria*: $x^{i*} = \omega^i$ for all $i \in I_m$, $l^{i*} = \mathbf{0}$ for all $i \in I_m$, $L_j^{i*} > 0$ for all $i \in I_m$ and all $j \in I_n$, and $p^* = \underline{p}$; $x^{i*} = \omega^i$ for all $i \in I_m$, $L^{i*} = \mathbf{0}$ for all $i \in I_m$, $l_j^{i*} < 0$ for all $i \in I_m$ and all $j \in I_n$, and $p^* = \bar{p}$. In the first case there is full rationing on all excess supplies. In the second case there is full rationing on all excess demands.

Two special equilibrium concepts need to be distinguished. The first one is usually called a Drèze equilibrium.

**Definition 2.5.2** *A constrained equilibrium is a Drèze equilibrium with respect to commodity* $j \in I_n$, *if* $l_j^{i*} = -\infty$ *and* $L_j^{i*} = \infty$ *for all* $i \in I_m$.

Notice that in a Drèze equilibrium there is no rationing on the a priori chosen market of commodity $j$.

The second special equilibrium concept is often called a supply-constrained (or unemployment) equilibrium and is introduced in van der Laan [1980a] and Kurz [1982].

**Definition 2.5.3** *A constrained equilibrium is a supply-constrained (or unemployment) equilibrium if for all* $i \in I_m$, $L_j^{i*} = \infty$ *for all* $j \in I_n$ *and* $l_k^{i*} = -\infty$ *for at least one* $k \in I_n$.

In an unemployment equilibrium, there is no rationing on the demand sides of all markets and at least one commodity is not constrained on the supply side.

The existence of a Drèze equilibrium was shown by Drèze [1975].

**Theorem 2.5.4** *Let $\mathcal{E} = (\{X^i, u^i, \omega^i\}_{i=1}^m, P)$ be an exchange economy with price rigidities. Then the economy has a Drèze equilibrium if it satisfies Assumptions 2.4.1 and 2.4.2.*

The existence theorem of a supply-constrained equilibrium was independently established by van der Laan [1980a] and Kurz [1982].

**Theorem 2.5.5** *Let $\mathcal{E} = (\{X^i, u^i, \omega^i\}_{i=1}^m, P)$ be an exchange economy with price rigidities. Then the economy has a supply-constrained (or unemployment) equilibrium if it satisfies Assumptions 2.4.1 and 2.4.2.*

For computational purposes, it is convenient to transform the above problems on the $n$-dimensional unit cube $U^n$. This approach was first employed by Drèze [1975]. Here we follow Herings [1995]. The *rationing scheme* $(l, L) = (l^1, \cdots, l^m, L^1, \cdots, L^m)$ is specified as follows. For all $i \in I_m$, $l^i$ and $L^i$ are continuous functions and satisfy for all $q \in U^n$ and all $j \in I_n$

$$\begin{array}{rcll} l^i_j(q) & = & l^i_j(r), & \text{if } r \in U^n \text{ and } q_j = r_j, \\ l^i_j(q) & = & 0, & \text{if } q_j = 0, \\ l^i_j(q) & < & -\omega^i_j, & \text{if } q_j \geq 1/3, \end{array}$$

and

$$\begin{array}{rcll} L^i_j(q) & = & L^i_j(r), & \text{if } r \in U^n \text{ and } q_j = r_j, \\ L^i_j(q) & = & 0, & \text{if } q_j = 1, \\ L^i_j(q) & > & \sum_{h \neq i} \omega^h_j, & \text{if } q_j \leq 2/3. \end{array}$$

It should be emphasized that the above rationing functions include most of the well-known rationing schemes in the literature. The price function $p : U^n \mapsto P$ is defined by

$$p_j(q) = \max\{\underline{p}_j, \min\{(2 - 3q_j)\underline{p}_j + (3q_j - 1)\bar{p}_j, \bar{p}_j\}\}$$

for each $j \in I_n$. Let us give some explanation on the rationing scheme. For every $j \in I_n$ the variable $q_j \in [0, 1]$ determines the state of the market of commodity $j$. In case $0 \leq q_j \leq 1/3$ there is no rationing on the demand side of the market of commodity $j$ and $p_j(q) = \underline{p}_j$. When $2/3 \leq q_j \leq 1$ there is no rationing on the supply side of the market of commodity $j$ and $p_j(q) = \bar{p}_j$. If $1/3 \leq q_j \leq 2/3$ there is no rationing on the market of commodity $j$ at all and $\underline{p}_j \leq p_j(q) \leq \bar{p}_j$.

Now we can define the excess demand correspondence $Z : U^n \mapsto \mathbf{R}^n$ by

$$Z(q) = \sum_{i=1}^m D^i(p(q), l^i(q), L^i(q)) - \sum_{i=1}^m \{\omega^i\}.$$

The properties of $Z$ are summarized in the following theorem; see Herings [1995] for its proof.

**Theorem 2.5.6** *Let $\mathcal{E} = (\{ (X^i, u^i, \omega^i) \mid i \in I_m \}, P)$ be an exchange economy with price rigidities and rationing scheme $(l, L)$. If the economy satisfies Assumptions 2.4.1 and 2.4.2, then the excess demand correspondence $Z : U^n \mapsto \mathbb{R}^n$ has the following properties:*
*(i) $Z$ is a u.s.c. correspondence with nonempty, convex and compact values;*
*(ii) for any $q \in U^n$, any $j \in I_n$ and any $z \in Z(q)$, $q_j = 0$ implies $z_j \geq 0$;*
*(iii) for any $q \in U^n$, any $j \in I_n$ and any $z \in Z(q)$, $q_j = 1$ implies $z_j \leq 0$;*
*(iv) for any $q \in U^n$ and any $z \in Z(q)$, $z^\top p(q) = 0$.*

Note that all zero points of $Z$ correspond to constrained equilibria and vice versa. More precisely, when $\mathbf{0} \in Z(q^*)$ for some $q^* \in U^n$ then there exists an $x^{i*} \in D^i(p(q^*), l^i(q^*), L^i(q^*))$ for all $i \in I_m$ such that $(x^{1*}, \cdots, x^{m*}, l^1(q^*), \cdots, l^m(q^*), L^1(q^*), \cdots, L^m(q^*), p(q^*))$ is a constrained equilibrium. Two constrained equilibria are said to be *equivalent* if they have the same equilibrium price, the same equilibrium allocations, and only non-binding rationing schemes are allowed to be different. Notice that $q = \mathbf{0}$ and $q = e$ correspond to the two trivial constrained equilibria, respectively, where $p = \underline{p}$ and $p = \bar{p}$. Furthermore, a Drèze equilibrium with respect to commodity $j \in I_n$ corresponds to an element $q^* \in U^n$ such that $\mathbf{0} \in Z(q^*)$ and $1/3 \leq q_j^* \leq 2/3$. A supply-constrained equilibrium corresponds to an element $q^* \in U^n$ such that $\mathbf{0} \in Z(q^*)$ and $1/3 \leq \max_{j \in I_n} q_j^* \leq 2/3$.

The following theorem follows immediately from Theorem 4.3 of Herings, Talman and Yang [1996]. It says that there exists a *continuum of constrained equilibria* in the above economy.

**Theorem 2.5.7** *Let $\mathcal{E} = (\{ (X^i, u^i, \omega^i) \mid i \in I_m \}, P)$ be an exchange economy with price rigidities and rationing scheme $(l, L)$. If the economy satisfies Assumptions 2.4.1 and 2.4.2, then there exists a connected subset $C$ in $U^n$ such that $0^n, 1^n \in C$ and $0^n \in Z(q)$ for every $q \in C$.*

In Chapter 8 an algorithm will be developed to compute a continuum of constrained equilibria linking the two trivial constrained equilibria. Now we exhibit the above results with an example.

**Example 2.5.8** *The economy consists of two agents and two commodities. The set of admissible prices is given by $P = \{ p \in \mathbb{R}^2_+ \mid \frac{1}{6} \leq p_1 \leq 2,\ p_2 = 1 \}$. The initial endowments of agent 1 and agent 2 are $(1, 4)^\top$ and $(2, 1)^\top$, respectively. Both agents have $\mathbb{R}^2_+$ as consumption sets and have Cobb-Douglas utility functions given by*

$$\begin{aligned} u^1(x^1) &= (x_1^1)^{\frac{3}{4}} (x_2^1)^{\frac{1}{4}}, \ x^1 \in \mathbb{R}^2_+, \\ u^2(x^2) &= (x_1^2)^{\frac{1}{4}} (x_2^2)^{\frac{3}{4}}, \ x^2 \in \mathbb{R}^2_+, \end{aligned}$$

*respectively.*

We first consider the case without price rigidities. It is easy to calculate the demand functions of both consumers

$$\begin{aligned} d^1(p) &= (\tfrac{3p_1+12p_2}{4p_1}, \tfrac{p_1+4p_2}{4p_2})^\top, \ p \in \mathbb{R}^2_{++}, \\ d^2(p) &= (\tfrac{2p_1+p_2}{4p_1}, \tfrac{6p_1+3p_2}{4p_2})^\top, \ p \in \mathbb{R}^2_{++}. \end{aligned}$$

The excess demand function is given by

$$z(p) = d^1(p) + d^2(p) - \omega = d^1(p) + d^2(p) - (3,5)^\top$$

for all $p \in \mathbb{R}^2_{++}$. The unique Walrasian equilibrium price of the economy on $S^2$ is

$$p^* = (13/7, 1)^\top.$$

The Walrasian equilibrium allocations for the two agents are given by

$$x^{1*} = (123/52, 41/28)^\top \ \text{ and } \ x^{2*} = (33/52, 99/28)^\top.$$

Now we consider the case with price rigidities. Clearly, $\underline{p}_1 = 1/6$, $\bar{p}_1 = 2$, and $\underline{p}_2 = \bar{p}_2 = 1$. The rationing scheme $(l, L)$ is defined by

$$\begin{aligned} l^1_1(q) = l^2_1(q) &= -3\min\{1, 3q_1\}, \ q \in U^2, \\ l^1_2(q) = l^2_2(q) &= -5\min\{1, 3q_2\}, \ q \in U^2, \\ L^1_1(q) = L^2_1(q) &= 18\min\{1, 3-3q_1\}, \ q \in U^2, \\ L^1_2(q) = L^2_2(q) &= 5\min\{1, 3-3q_2\}, \ q \in U^2. \end{aligned}$$

The price function $p : U^2 \mapsto \mathbb{R}^2_+$ is defined by

$$\begin{aligned} p_1(q) &= \max\{1/6, \min\{11q_1/2 - 5/3, 2\}\}, \ q \in U^2, \\ p_2(q) &= 1, \ q \in U^2. \end{aligned}$$

The excess demand function of agent 1 is given by

$$z^1(q) = \begin{cases} (90q_2, -15q_1)^\top, & 0 \le q_1 \le 1/3, \ 0 \le q_2 \le 71/360, \\ (\frac{74}{4}, -\frac{71}{24})^\top, & 0 \le q_1 \le 1/3, \ 71/360 \le q_2 \le 1, \\ (\frac{90q_2}{33q_1-10}, -15q_2)^\top, & 1/3 \le q_1 \le 2/3, \ 0 \le q_2, \ 33q_1 + 360q_2 \le 82, \\ (\frac{82-33q_1}{132q_1-40}, \frac{33q_1-82}{24})^\top, & 1/3 \le q_1 \le 2/3, \ q_2 \le 1, \ 33q_1 + 360q_2 \ge 82, \\ (\frac{15q_2}{2}, -15q_2)^\top, & 2/3 \le q_1, \ 0 \le q_2 \le 1/6, \ 36q_1 + 5q_2 \le 36, \\ (\frac{5}{4}, -\frac{5}{2})^\top, & 2/3 \le q_1 \le 211/216, \ 1/6 \le q_2 \le 1, \\ (54 - 54q_1, 108q_1 - 108)^\top, & 211/216 \le q_1 \le 1, \ q_2 \le 1, \ 36q_1 + 5q_2 \ge 36. \end{cases}$$

The excess demand function of agent 2 is given by

$$z^2(q) = \begin{cases} (0,0)^\top, & 0 \le q_1 \le 1/3, \ 0 \le q_2 \le 1, \\ (\frac{33-99q_1}{66q_1-20}, \frac{33q_1-11}{4})^\top, & 1/3 \le q_1 \le 2/3, \ 0 \le q_2, \ 33q_1 + 60q_2 \le 71, \\ (\frac{90q_2-90}{33q_1-10}, 15 - 15q_2)^\top, & q_1 \le 2/3, \ q_2 \le 1, \ 33q_1 + 60q_2 \ge 71, \\ (-\frac{11}{8}, \frac{11}{4})^\top, & 2/3 \le q_1 \le 1, \ 0 \le q_2 \le 49/60, \\ (\frac{15q_2-15}{2}, 15 - 15q_2)^\top, & 2/3 \le q_1 \le 1, \ 49/60 \le q_2 \le 1. \end{cases}$$

The total excess demand function $z : U^2 \mapsto \mathbb{R}^2$ is given by $z = z^1 + z^2$. The zero points of $z$ can be analytically determined. It can be verified that the points

$$\begin{pmatrix} 0 \\ 0 \end{pmatrix}, \begin{pmatrix} \frac{1}{3} \\ 0 \end{pmatrix}, \begin{pmatrix} \frac{148}{231} \\ \frac{71}{420} \end{pmatrix}, \begin{pmatrix} \frac{148}{231} \\ \frac{349}{420} \end{pmatrix}, \begin{pmatrix} \frac{2}{3} \\ \frac{5}{6} \end{pmatrix}, \begin{pmatrix} \frac{211}{216} \\ \frac{5}{6} \end{pmatrix}, \begin{pmatrix} 1 \\ 1 \end{pmatrix}$$

and the convex combinations of any two subsequent points are all zero points of $z$. It is clear that in this example the set of constrained equilibria is a connected path whose end points are the two trivial equilibria. Notice that $q^* = (\frac{1}{2}, \frac{11}{120})^\top$ induces a Drèze equilibrium with respect to commodity 1, where $p(q^*) = (\frac{13}{12}, 1)^\top$, $x^{1*}(q^*) = (\frac{59}{26}, \frac{21}{8})^\top$, and $x^{2*}(q^*) = (\frac{19}{26}, \frac{19}{8})^\top$. In this case commodity 1 is not rationed. This equilibrium is also a supply-constrained equilibrium. A Drèze equilibrium with respect to commodity 2 corresponds to $\bar{q} = (\frac{148}{231}, \frac{1}{2})^\top$, where $p(\bar{q}) = (\frac{13}{7}, 1)^\top$, $x^{1*}(\bar{q}) = (\frac{123}{52}, \frac{41}{28})^\top$, and $x^{2*}(\bar{q}) = (\frac{33}{52}, \frac{99}{28})^\top$. In this case commodity 2 is not rationed.

## 2.6. An exchange economy with linear technologies

In this section we discuss another important economic equilibrium model in which an explicit production sector is embodied in the model of a pure exchange economy that we have studied in Section 4.

Let $\mathcal{E} = (\{ (X^i, u^i, \omega^i) \mid i \in I_m\}, A)$ be an exchange economy with linear technologies. In the economy there are $n$ types of commodities, indexed by $j = 1, \cdots, n$, and $m$ consumers, indexed by $i = 1, \cdots, m$, and $l$ firms, indexed by $k = 1, \cdots, l$. Firms are owned by consumers. Each consumer $i \in I_m$ is characterized by his consumption set $X^i$, his utility function $u^i$, and his initial endowment vector $\omega^i$. Firm $k \in I_l$ is represented by the $k$-th column $a^k$ of the $n \times l$ real matrix $A$. The vector $a^k$ is called firm $k$'s production activity vector at a unit level. More precisely, the interpretation of $a^k$ is as follows. A positive component $a^k_j$ of $a^k$ implies that firm $k$ produces $a^k_j$ units of commodity $j$, while a negative component $a^k_j$ means that firm $k$ needs to use $-a^k_j$ units of commodity $j$ as an input, when the activity level of firm $k$ is fixed at a unit level. A vector $y \in \mathbb{R}^l_+$ denotes an activity level of all firms with component $y_k$ the activity level of firm $k$. Thus, given $y \in \mathbb{R}^l_+$, $Ay$ is the *aggregate net input-output vector* for activity level $y$, i.e., $(Ay)_j > 0$ means commodity $j$ serving as a net output to the economy as a whole and $(Ay)_j < 0$ implies commodity $j$ serving as a net input to the economy as a whole. Given a nonzero price vector $p \in \mathbb{R}^n_+$, the $l$-vector $A^\top p$ is the unit level profit vector of all firms. In other words, component $(A^\top p)_k$ is the profit of firm $k$ when it operates at its unit level. With reference to $X^i$, $u^i$ and $\omega^i$, for all $i \in I_m$, we assume that Assumptions 2.4.1 and 2.4.2 are also satisfied here.

As usual, each consumer maximizes his utility subject to the budget constraint and each firm maximizes its profit by choosing an activity level. As discussed in Section 4, $D^i : \mathbb{R}^n_+\backslash\{\mathbf{0}\} \mapsto \mathbb{R}^n$ denotes the demand correspondence of consumer $i \in I_m$ and $Z : \mathbb{R}^n_+\backslash\{\mathbf{0}\} \mapsto \mathbb{R}^n$ denotes the excess demand correspondence. Now we can introduce the concept of a Walrasian equilibrium.

**Definition 2.6.1** *A vector* $(x^{1*}, \cdots, x^{m*}, y^*, p^*) \in \prod_{i=1}^m X^i \times \mathbb{R}^l_+ \times \mathbb{R}^n_+\backslash\{\mathbf{0}\}$ *is a Walrasian equilibrium if*

(a) *for each* $i \in I_m$, $x^{i*}$ *is an element of* $D^i(p^*)$;

(b) $\sum_{i=1}^m x^{i*} \leq Ay^* + \sum_{i=1}^m \omega^i$;

(c) $A^\top p^* \leq \mathbf{0}$.

Condition (b) states that the total supply meets the total demand. Condition (c) says that at an equilibrium state no firm makes a positive profit. As a matter of fact, all firms make zero profit in equilibrium. To guarantee the existence of a Walrasian equilibrium, we make a very natural and economically appealing assumption on the production sector.

**Assumption 2.6.2 (No Production without Input Assumption)**
*If* $Ay \geq \mathbf{0}$ *and* $y \in \mathbb{R}^l_+$, *then* $y = \mathbf{0}$.

The following theorem assures us that the economy has a Walrasian equilibrium.

**Theorem 2.6.3** *Let an exchange economy with linear technologies* $\mathcal{E} = (\{(X^i, u^i, \omega^i) \mid i \in I_m\}, A)$ *be given. Then the economy has a Walrasian equilibrium, if it satisfies Assumptions 2.4.1, 2.4.2, and 2.6.2.*

We remark that due to the homogeneity of $D^i(\cdot)$, $i \in I_m$, the price vector $p \in \mathbb{R}^n_+\backslash\{\mathbf{0}\}$ is permitted to lie in the unit simplex $S^n$. We give an example to demonstrate the theory above.

**Example 2.6.4** *The economy consists of three commodities, two consumers and two firms. The initial endowments of consumer 1 and consumer 2 are* $(1,2,3)^\top$ *and* $(3,2,1)^\top$, *respectively. Both consumers have* $\mathbb{R}^3_+$ *as consumption sets and have Cobb-Douglas utility functions*

$$\begin{aligned} u^1(x^1) &= (x^1_1)^{\frac{1}{3}}(x^1_2)^{\frac{1}{3}}(x^1_3)^{\frac{1}{3}}, \ x^1 \in \mathbb{R}^3_+, \\ u^2(x^2) &= (x^2_1)^{\frac{1}{3}}(x^2_2)^{\frac{1}{3}}(x^2_3)^{\frac{1}{3}}, \ x^2 \in \mathbb{R}^3_+, \end{aligned}$$

*respectively. The production activity vectors are given by* $a^1 = (-3/2, 1, 1)^\top$ *and* $a^2 = (-1, -1, 3/2)^\top$.

The demand functions are given by

$$\begin{aligned} d^1(p) &= \left(\tfrac{p_1+2p_2+3p_3}{3p_1}, \tfrac{p_1+2p_2+3p_3}{3p_2}, \tfrac{p_1+2p_2+3p_3}{3p_3}\right)^\top, \ p \in \mathbb{R}^3_{++}, \\ d^2(p) &= \left(\tfrac{3p_1+2p_2+p_3}{3p_1}, \tfrac{3p_1+2p_2+p_3}{3p_2}, \tfrac{3p_1+2p_2+p_3}{3p_3}\right)^\top, \ p \in \mathbb{R}^3_{++}. \end{aligned}$$

The total excess demand function is

$$z(p) = (\frac{4}{3p_1} - 4, \frac{4}{3p_2} - 4, \frac{4}{3p_3} - 4)^\top, \quad p \in \mathbf{R}^3_{++}.$$

The Walrasian equilibrium for the economy without production is given by

$$p^* = (1/3, 1/3, 1/3)^\top,$$

$$x^{1*} = (2,2,2)^\top \text{ and } x^{2*} = (2,2,2)^\top.$$

In the economy with production the Walrasian equilibrium is $(x^{1*}, x^{2*}, y^*, p^*)$ with

$$x^{1*} = (7/4, 7/3, 7/3)^\top \text{ and } x^{2*} = (19/12, 19/9, 19/9)^\top,$$

$$p^* = (4/10, 3/10, 3/10)^\top,$$

$$y^* = (4/9, 0)^\top.$$

We close this section with one more result, where $S^n_A$ is defined to be the set $\{ p \in S^n \mid A^\top p \leq \mathbf{0} \}$.

**Theorem 2.6.5** *Let an exchange economy with linear technologies* $\mathcal{E} = (\{ X^i, u^i, \omega^i \}_{i=1}^m, A)$ *be given, which satisfies Assumptions 2.4.1, 2.4.2, and 2.6.2. Then* $p^* \in S^n_A$ *is a Walrasian equilibrium price if and only if there exists some* $z \in Z(p^*)$ *such that* $z^\top p \leq z^\top p^*$ *for all* $p \in S^n_A$.

## 2.7. An exchange economy with non-convex technologies

In this section we present an exchange economy model in which the production sets may exhibit increasing returns to scale or more generally, non-convexities. It has been widely recognized since Marshall [1920] that because of the presence of increasing returns to scale the competitive mechanism may fail. In term of equilibrium analysis, due to non-convexities in the production sets, the supply correspondence might not be upper semi-continuous, convex-valued or even well defined. Starting with the seminal work of Guesnerie [1975], there is now a substantial literature on the existence of equilibria in economies with increasing returns to scale. See, for example, Beato [1976], Bonnisseau and Cornet [1988], Kamiya [1988], Vohra [1988], and many others. Here we introduce the model of Bonnisseau and Cornet [1988] which is suitable for computational purpose.

Let $\mathcal{E} = (\{ (X^i, u^i, r_i, \omega^i) \mid i \in I_m\}, \{(Y^k, \phi_k) \mid k \in I_l\})$ be an exchange economy with non-convex technologies. In the economy there are $n$ types of commodities, indexed by $j = 1, \cdots, n$, and $m$ consumers, indexed by $i = 1, \cdots, m$, and $l$ firms, indexed by $k = 1, \cdots, l$. Firms are owned by consumers. Firm $k \in I_l$ is represented by the production possibilities set

$Y^k$. Let $y$ be an element in $Y^k$. Then a positive component $y_j$ of $y$ implies that firm $k$ produces $y_j$ units of commodity $j$, while a negative component $y_j$ means that firm $k$ needs to use $-y_j$ units of commodity $j$ as an input. The behavior of each firm is characterized by a *pricing rule* . A pricing rule for firm $k$ is a point-to-set mapping $\phi_k : bd(Y^k) \mapsto \mathbf{R}^n_+$ such that for each $y^k \in bd(Y^k)$, $\phi_k(y^k)$ is a closed cone. Each consumer $i \in I_m$ is characterized by his consumption set $X^i$, his utility function $u^i$, his wealth distribution function $r_i : \mathbf{R}^n_+ \times \prod_{k=1}^l bd(Y^k) \mapsto \mathbf{R}$, and his initial endowment vector $\omega^i$.

Let $\omega = \sum_{i=1}^m \omega^i$. Now we postulate the following assumptions.

**Assumption 2.7.1** *For each $k \in I_l$, $Y^k$ satisfies*

*(a) $Y^k$ is nonempty and closed;*

*(b) $Y^k - \mathbf{R}^n_+ \subseteq Y^k$. (free disposal condition)*

**Assumption 2.7.2**

*(a) For each $i \in I_m$, $X^i$ is a nonempty, closed and convex subset of $\mathbf{R}^n$, bounded from below;*

*(b) For each $i \in I_m$, $u^i : X^i \mapsto \mathbf{R}$ is continuous, quasi-concave and local non-satiated; i.e., local non-satiation means that given any $x \in X^i$ and any $\epsilon > 0$, then there is some vector $z \in X^i$ with $||x - z|| < \epsilon$ such that $u^i(z) > u^i(x)$;*

*(c) For each $i \in I_m$, $r_i : \mathbf{R}^n_+ \times \prod_{k=1}^l bd(Y^k) \mapsto \mathbf{R}$ is continuous, $\sum_{i=1}^m r_i(p, y^1, \cdots, y^l) = p^\top(\sum_{k=1}^l y^k + \omega)$, and $r_i(\gamma p, y^1, \cdots, y^l) = \gamma r_i(p, y^1, \cdots, y^l)$ for all $(p, y^1, \cdots, y^l, \gamma) \in \mathbf{R}^n_+ \times \prod_{k=1}^l bd(Y^k) \times \mathbf{R}_+$.*

**Assumption 2.7.3** *For every $\omega' \geq \omega$, the set*

$$A(\omega') = \{(x^1, \cdots, x^m, y^1, \cdots, y^l) \in \prod_{i=1}^m X^i \times \prod_{k=1}^l Y^k \mid \sum_{i=1}^m x^i \leq \sum_{k=1}^n y^k + \omega'\}$$

*is bounded.*

The above assumption says that it is impossible for any firm to produce an unlimited amount of goods out of a finite amount of inputs.

Given a price $p \in \mathbf{R}^n$ and a production plan $y = (y^1, \cdots, y^l) \in \prod_{k=1}^l Y^k$, the demand set of consumer $i \in I_m$ is defined by

$$D_i(p, y) = \{x^i \mid \begin{array}{ll} & u^i(x^i) = \max u^i(x), \\ s.t. & x \in X^i,\ p^\top x \leq r_i(p, y^1, \cdots, y^l)\}. \end{array}$$

Now we can introduce the concept of an equilibrium.

**Definition 2.7.4** *A vector $(x^{1*}, \cdots, x^{m*}, y^{1*}, \cdots, y^{l*}, p^*)$ in $\prod_{i=1}^m X^i \times \prod_{k=1}^l Y^k \times \mathbf{R}^n$ is an equilibrium if* (*a*) *for each* $i \in I_m$, $x^{i*}$ *is an element of* $D^i(p^*, y^{1*}, \cdots, y^{l*})$;

(*b*) *for each* $k \in I_l$, $y^{k*} \in bd(Y^k)$ *and* $p^* \in \phi_k(y^{k*})$;

(*c*) $\sum_{i=1}^m x^{i*} = \sum_{i=1}^m \omega^i + \sum_{k=1}^l y^{k*}$.

The above homogeneity assumptions of $r_i$ and $\phi_k$ and the local non-satiation allow us to normalize the price vector in the unit simplex $S^n$. Let us define $\psi_k : bd(Y^k) \mapsto S^n$ by

$$\psi_k(y^k) = \phi_k(y^k) \cap S^n.$$

Define

$$PE = \{(p, y^1, \cdots, y^l) \in S^n \times \prod_{k=1}^l bd(Y^k) \mid p \in \prod_{k=1}^l \phi_k(y^k)\}.$$

To guarantee the existence of an equilibrium, Bonnisseau and Cornet introduced the following assumptions.

**Assumption 2.7.5** *For each* $k \in I_l$, *the mapping* $\psi_k$ *is an upper semi-continuous point-to-set mapping with nonempty, convex and compact values.*

**Assumption 2.7.6**

(*a*) *For each* $k \in I_l$, *there exists a real number* $\alpha_k$ *such that for all* $(p, y^k) \in S^n \times bd(Y^k)$, $p \in \psi_k(y^k)$ *implies* $p^\top y^k \geq \alpha_k$; (*bounded losses assumption*)

(*b*) $(p, y^1, \cdots, y^l) \in PE$ *implies* $p^\top(\sum_{k=1}^l y^k + \omega) > \inf_{x \in \sum_{i=1}^m X^i} p^\top x$; (*survival assumption*)

(*c*) $(p, y^1, \cdots, y^l) \in PE$ *and* $p^\top(\sum_{k=1}^l y^k + \omega) > \inf_{x \in \sum_{i=1}^m X^i} p^\top x$ *imply* $r_i(p, y^1, \cdots, y^l) > \inf_{x \in X^i} p^\top x$ *for all* $i \in I_m$.

The following theorem assures us that the economy has an equilibrium.

**Theorem 2.7.7** *Let an exchange economy with non-convex technologies*

$$\mathcal{E} = (\{(X^i, u^i, r_i, \omega^i) \mid i \in I_m\}, \{(Y^k, \phi_k) \mid k \in I_l\})$$

*be given. Then the economy has an equilibrium, if it satisfies Assumptions 2.7.1, 2.7.2, 2.7.3, 2.7.5, and 2.7.6.*

## 2.8. An exchange economy with indivisibilities and money

The importance and difficulties of the equilibrium problems on indivisibilities have been realized since 1950s by Koopmans and Beckman [1957] and Debreu [1959, p.36]. Starting with the seminal article of Gale and Shapley [1962], over the last twenty years or so there has been a good deal of progress made in understanding certain classes of these economic problems such as two-sided matchings. See, for example, Shapley and Shubik [1972], Shapley and Scarf [1974], Kelso and Crawford [1982], Quinzii [1984], Gale [1984], Curiel and Tijs [1985], Kaneko and Yamamoto [1986], Yamamoto [1987a], Roth and Sotomayor [1990], Scarf [1994], and van der Laan, Talman and Yang [1997], Yang [1997], among others. Here we present a general model due to van der Laan et al [1997] and Gale [1984].

Consider an economy in which there are finitely many agents and finitely many different types of indivisible goods. We may think of these different types of indivisible goods as houses, cars, trucks, horses, and so on. Goods of the same type have all the same function for agents but may be different in quality. So for example houses will be classified into several different types of indivisible goods when they vary in function. Each agent initially owns at most one item of each type of indivisible goods and some amount of money. Agents have preferences over goods and money with the constraint that no agent desires more than one item of each type of indivisible goods. More precisely, the model is as follows. We denote the set of agents and the set of different types of indivisible goods by $I_n$ and $I_m$, respectively. It is assumed that each agent initially owns one item of each type of indivisible goods and some amount of money. Notice that this involves no loss of generality since if some agent does not own one item of some type of goods we may assume that he has a dummy good of that type which is of no value to any of the agents. We denote that agent $i$ owns one item of the indivisible goods of type $j$ by the ordered pair $(j, i)$. Then $I_m \times I_n$ represents the set of all indivisible goods in the economy, i.e., $I_m \times I_n = \{ (j,i) \mid j \in I_m, i \in I_n \}$. A price vector is a vector in $\prod_{j=1}^m \mathbb{R}_+^n$. A vector $p \in \prod_{h=1}^m \mathbb{R}_+^n$ is represented by $p = (p_1, \cdots, p_m)$ where $p_j = (p_{j,1}, \cdots, p_{j,n})^\top$ for each $j \in I_m$. Let $\Phi = \{ \rho \mid \rho = (\rho(1), \cdots, \rho(n))$ is a permutation of $(1, \cdots, n)\}$. An element $\pi \in \prod_{j=1}^m \Phi$ is written as $\pi = (\pi_1, \cdots, \pi_m)$ where $\pi_j = (\pi_j(1), \cdots, \pi_j(n))$ for each $j \in I_m$. Furthermore, for a positive integer $k$, let $I_n^k$ denote $I_n \times \cdots \times I_n$ where $I_n$ is repeated $k$ times.

For each $i \in I_n$ and each $j \in I_m$, the demand of agent $i$ for the indivisible goods of type $j$ is specified by a covering $C_i^j = \{ C_i^{(j,0)}, C_i^{(j,1)}, \cdots, C_i^{(j,n)} \}$ of $\prod_{h=1}^m \mathbb{R}_+^n$. The interpretation is that if $p \in C_i^{(j,k)}$ for some $k \in I_n$, then agent $i$ will demand the item of type $j$ owned by agent $k$ at price vector $p$, while if $p \in C_i^{(j,0)}$, then agent $i$ will demand no item of type $j$. We remark that each demand set $C_i^{(j,l)}$ is not necessarily a connected set and could be the union of several disjoint subsets.

**Definition 2.8.1** *A pair of vectors $(p, \pi) \in \prod_{h=1}^m \mathbb{R}_+^n \times \prod_{h=1}^m \Phi$ is a competitive equilibrium if it holds that $p \in C_i^{(j,\pi_j(i))}$ for all $j \in I_m$ and all $i \in I_n$.*

Let $B^n$ denote the boundary of $\mathbb{R}^n_+$. Now we make the following conditions on demand:

**Assumption 2.8.2**

(*a*) *The set* $C_i^{(j,l)}$ *is closed for any* $j \in I_m$, $i \in I_n$, *and* $l \in I_n \cup \{0\}$;

(*b*) *For each* $i \in I_n$ *and each* $j \in I_m$, $C_i^{(j,1)}, \ldots, C_i^{(j,n)}$ *cover* $\prod_{h=1}^m B^n$;

(*c*) *There exists* $M > 0$ *such that if* $p_{j,l} \geq M$, *then* $p \notin C_k^{(j,l)}$ *for all* $k \in I_n$.

Assumption (*a*) says that the demand of every agent exhibits certain continuous behavior and that each agent has a positive amount of money. More precisely, it says that if $(z^k)$ is a sequence of price vectors in some demand set $C_i^{(j,l)}$, and if this sequence converges to some price vector $z^*$, then $z^*$ also lies in the demand set $C_i^{(j,l)}$. Assumption (*b*) is a free disposal condition saying that since $p \in \prod_{h=1}^m B^n$ implies that at least one item of type $j$ is free, so agent $i$ would demand that item without hurting himself. Finally, Assumption (*c*) implies that no agent is willing to spend a huge amount of money on any good. The following equilibrium theorem is established in van der Laan, Talman and Yang [1997]. A proof will be given in Chapter 13.

**Theorem 2.8.3** *Under Assumption 2.8.2 the economy has at least one competitive equilibrium* $(p, \pi) \in \prod_{h=1}^m B^n \times \prod_{h=1}^m \Phi$.

Note that when there is only one type of indivisible goods present in the economy, i.e., $m = 1$, the model above is reduced to the model of Gale [1984].

## 2.9. An exchange economy under uncertainty

In the previous economic equilibrium models it is assumed that everything is deterministic and that all trade takes place just one time. However, in reality, economic agents face a lot of uncertainty and trade takes place to a large extent sequentially over time. There is now a very rich literature dealing with this economic problem; see Merton [1990] and Duffie [1992]. In this section we are concerned with the existence of equilibrium and will consider the equilibrium model under uncertainty due to Werner [1985]. The interested reader is further referred to Radner [1972], Hart [1974], Duffie [1987], Werner [1987], Nielsen [1989,1990], Kim [1998], and to Duffie and Shafer [1985, 1986] and the special issue of Journal of Mathematical Economics [1990, Vol.19] for generic existence results of equilibrium in economic models under uncertainty.

Let $\mathcal{E} = (\{ (X^i, u^i) \mid i \in I_m\}, \{\omega^i(1), \omega^i(2, s) \mid s \in \Omega, i \in I_m \})$ be an exchange economy under uncertainty. In the economy there are $m$ consumers (or investors), indexed by $i = 1, \cdots, m$. The economy lasts for two dates, i.e., date 1 and date 2. Date 1 will be understood as today (or the present) and date 2 as tomorrow (or the future). At date 2 there are $T$ possible states of nature $\{s_1, \cdots, s_T\} = \Omega$. A state of the nature gives a complete description of a possible outcome of uncertainty. All futures markets are treated as financial phenomena and described by means of securities in term of monetary returns. The return of a security depends on which state of nature occurs. There are $n$ consumption commodities, indexed by $j = 1, \cdots, n$, available at each date in each state of nature. At date 1 there are spot markets for $n$ current commodities and futures markets for $L$ securities, indexed by $k = 1, \cdots, L$. Each security is characterized by its state-dependent expected return at date 2. More precisely, an investor holding one share of security $k$ receives $r_k(s)$ units of account (e.g., dollar) if state $s \in \Omega$ prevails. Let

$$r(s) = (r_1(s), \cdots, r_L(s))^{\top}$$

for every $s \in \Omega$. This means that the expected returns of securities are completely determined by nature. Hence all consumers (or investors) will agree upon them. Or one may say that all investors' probability beliefs about securities returns are identical. It is assumed that $r(s) \neq \mathbf{0}$ for all $s \in \Omega$. At date 2, in each state of the nature all $n$ commodities will be traded on spot markets. Each consumer $i \in I_m$ is characterized by his consumption set $X^i$, his utility function $u^i : X^i \mapsto \mathbb{R}$, and his initial endowment of date-1 commodities $\omega^i(1) \in \mathbb{R}^n_+$ and state-dependent endowment of date-2 commodities $\omega^i(2, s)$, where $\omega^i_j(2, s)$ denotes the endowment of commodity $j$ at date 2 in state $s$, for $j \in I_n$. There are no initial endowments of securities. It is assumed that for all $i \in I_m$, $X^i = \mathbb{R}^{n(T+1)}_+$.

In the following we denote a price vector of commodities by

$$p = (p(1), p(2, s_1), \cdots, p(2, s_T)) \in \mathbb{R}^{n(T+1)}_+,$$

a price vector of securities by $\pi \in \mathbb{R}^L_+$, a consumption plan vector by

$$x = (x(1), x(2, s_1), \cdots, x(2, s_T)) \in \mathbb{R}^{n(T+1)}_+,$$

and a portfolio of securities by $\theta \in \mathbb{R}^L$, respectively.

Now we can introduce the concept of an equilibrium, which was originally introduced by Radner [1972].

**Definition 2.9.1** *A vector*

$$(x^{1*}, \cdots, x^{m*}, \theta^{1*}, \cdots, \theta^{m*}, p^*, \pi^*) \in \mathbb{R}^{mn(T+1)}_+ \times \mathbb{R}^{mL} \times \mathbb{R}^{n(T+1)}_+ \times \mathbb{R}^L_+$$

*is a Radner equilibrium if*

*(a) for each* $i \in I_m$, $(x^{i*}, \theta^{i*})$ *is an element of the demand set*

$$\{(y,\theta) \mid \quad u^i(y) = \max_x u^i(x), \\ s.t. \quad x \in X,\ \theta \in \mathbb{R}^L,\ p^{*\top}(1)x(1) + \pi^{*\top}\theta \le p^{*\top}(1)\omega^i(1), \\ p^{*\top}(2,s)x^i(2,s) \le p^{*\top}(2,s)\omega^i(2,s) + \theta^\top r(s),\ \textit{for all } s \in \Omega\};$$

*(b)* $\sum_{i=1}^m x^{i*}(1) = \sum_{i=1}^m \omega^i(1)$, $\sum_{i=1}^m x^{i*}(2,s) = \sum_{i=1}^m \omega^i(2,s)$, *for all* $s \in \Omega$, *and* $\sum_{i=1}^m \theta^{i*} = \mathbf{0}$.

In the definition (a) says that every investor chooses his best consumption plan and his best portfolio under his sequence of budget constraints, one for each date-event; (b) states that all markets, both today's and tomorrow's, are cleared. In detail the last equation of (b) says that for each investor $i \in I_m$ who demands some security $k$, $\theta_k^{i*} > 0$, there will be some (or more than one) investor $j \in I_m$ who creates security $k$, $\theta_k^{j*} < 0$, and sells it to investor $i$.

**Assumption 2.9.2**

*(a) For each* $i \in I_m$, $u^i$ *is continuous, quasi-concave and monotonic;*

*(b) For each* $i \in I_m$, $\omega^i(1) > \mathbf{0}$ *or* $\omega^i(2,s) > \mathbf{0}$ *for all* $s \in \Omega$; *furthermore* $\sum_{i=1}^m \omega^i(1) >> \mathbf{0}$ *and for every* $j = 1, \cdots, n$, *there exists an index* $i \in I_m$ *such that* $\omega_j^i(2,s) > 0$ *for all* $s \in \Omega$.

Using Kakutani theorem Werner [1985] proved the following theorem in an elegant manner.

**Theorem 2.9.3** *Let an exchange economy under uncertainty*

$$\mathcal{E} = (\{ (X^i, u^i) \mid i \in I_m\}, \{\omega^i(1),\ \omega^i(2,s) \mid s \in \Omega,\ i \in I_m \})$$

*be given. Then the economy has a Radner equilibrium, if it satisfies Assumptions 2.9.2.*

CHAPTER 3

# First Algorithms for Approximating Fixed Points on the Unit Simplex

## 3.1. Introduction

In this chapter we consider the problem of computing fixed points of a continuous function $f$ from the unit simplex $S^n$ into itself. As we see in Chapter 1, Brouwer theorem guarantees the existence of a fixed point on $S^n$ but does not offer any insight about the location of such fixed points. Nevertheless it is clear to us that in order to approximate a fixed point, one possible way is to design a systematic procedure for finding completely labelled simplices. Such procedure did not exist until 1967. In his pioneering work Scarf [1967b] developed the first such procedure. Starting with a specific primitive set, this procedure generates a path of adjacent primitive sets and terminates within a finite number of steps with a completely labelled primitive set which yields a good approximation of a fixed point. (Completely labelled primitive sets are analogue of completely labelled simplices.) To prove the convergence of the procedure, Scarf utilized an argument of Lemke and Howson [1964] and Lemke [1965] which is a purely combinatorial argument and does not rely on any monotonicity property. Later Kuhn [1968,1969] proposed two alternative procedures by using simplices and triangulations instead of primitive sets. These very first fixed point methods are the central topic of this chapter.

This chapter is organized as follows. In Section 2 we introduce integer labeling rules and derive the accuracy of approximate fixed points induced by them for any continuous function on the unit simplex. In Section 3 we present two versions of Scarf's algorithm, one based on simplices, another based on primitive sets. In Sections 4 and 5 we introduce Kuhn's artificial start algorithm and Kuhn's variable dimension algorithm, respectively. Finally in Section 6 we conclude with some remarks.

## 3.2. Integer labeling, approximation and triangulation

In this section we introduce several integer labeling rules and analyze their accuracy. Consider the problem of finding fixed points of a continuous function $f$ from $S^n$ into itself.

**Definition 3.2.1** *Given a function $f : S^n \mapsto S^n$ and $\mathcal{T}$ a triangulation of $S^n$, a labeling rule $l^1 : \mathcal{T}^0 \mapsto I_n$ is given by*

$$l^1(y) = \min\{j \in I_n \mid f_j(y) \leq y_j > 0,\ f_{j+1}(y) \geq y_{j+1}\}$$

*where $j+1 = 1$ for $j = n$.*

**Lemma 3.2.2** *Given a function $f : S^n \mapsto S^n$ and the labeling rule $l^1$, then the set*

$$\{j \in I_n \mid f_j(y) \leq y_j > 0,\ f_{j+1}(y) \geq y_{j+1}\}$$

*is nonempty for every $y \in \mathcal{T}^0$.*

Proof: For each $y \in S^n$, define $J = \{i \in I_n \mid y_i > 0\}$. Since

$$\sum_{i \in J} y_i = \sum_{i \in I_n} y_i = 1 = \sum_{i \in I_n} f_i(y) \geq \sum_{i \in J} f_i(y),$$

there exists $k \in J$ such that $y_k \geq f_k(y)$, i.e., $y_k - f_k(y) \geq 0$. Consider the following sequence $y_{k+1} - f_{k+1}(y)$, $y_{k+2} - f_{k+2}(y)$, $\cdots$ by defining $n+1 = 1$, $n+2 = 2$, and so on. Let $l$ be the smallest non-negative integer such that $y_{k+l} - f_{k+l}(y) \geq 0$ and $y_{k+l+1} - f_{k+l+1}(y) \leq 0$. Since $\sum_{i \in I_n} y_i = 1$ and $\sum_{i \in I_n} f_i(y) = 1$, such an integer $l$ always exists. If $l = 0$, we have $f_k(y) \leq y_k > 0$, and $y_{k+1} \leq f_{k+1}(y)$. If $l > 0$, then we have $y_{k+1} - f_{k+1}(y) > 0, \cdots, y_{k+l} - f_{k+l}(y) > 0$ and $y_{k+l+1} - f_{k+l+1}(y) \leq 0$. Here we have $y_{k+1} > 0, \cdots, y_{k+l} > 0$. Therefore, we have $f_{k+l}(y) \leq y_{k+l} > 0$ and $f_{k+l+1}(y) \geq y_{k+l+1}$. The proof is completed. □

**Definition 3.2.3** *Given a triangulation $\mathcal{T}$ of $S^n$, a labeling rule $l : \mathcal{T}^0 \mapsto I_n$ is a proper labeling rule if $y_i = 0$ implies $l(y) \neq i$ for any $y \in \mathcal{T}^0$.*

**Definition 3.2.4** *Given a triangulation $\mathcal{T}$ of $S^n$ and a labeling rule $l : \mathcal{T}^0 \mapsto I_n$, an $(n-1)$-simplex $\sigma$ with vertices $y^1, \cdots, y^n$ is a completely labelled simplex if*

$$\{l(y^i) \mid i \in I_n\} = I_n.$$

The following theorem says that a completely labelled simplex yields a good approximation of a fixed point of a continuous function.

**Theorem 3.2.5** *Let $f : S^n \mapsto S^n$ be a continuous function, let $\mathcal{T}$ be a triangulation of $S^n$ and let $\epsilon > 0$, $0 < h < \frac{\epsilon}{2\sqrt{n}}$ be such that $mesh(\mathcal{T}) \leq h$, and for any $x, y \in S^n$, $\max_i |x_i - y_i| \leq h$ implies*

$$\max_i |f_i(x) - f_i(y)| \leq \frac{\epsilon}{2\sqrt{n}}.$$

*Given the labeling rule $l^1$ and a completely labelled simplex $\sigma$, then we have*

$$||x - f(x)|| \leq \epsilon$$

*for all $x \in \sigma$.*

Proof: We may assume that $\sigma =< y^1, \cdots, y^n >$ and $l^1(y^i) = i$ for all $i \in I_n$. Now it follows from the labeling rule $l^1$ that for any $i \in I_n$,

$$f_i(y^i) \leq y^i_i.$$

This leads to

$$\begin{array}{rcl} f_i(x) - x_i & = & (f_i(x) - f_i(y^i)) + (f_i(y^i) - y^i_i) + (y^i_i - x^i_i) \\ & \leq & \frac{\epsilon}{2\sqrt{n}} + 0 + h \\ & \leq & \frac{\epsilon}{\sqrt{n}} \end{array}$$

for all $i \in I_n$. On the other hand for any $i \in I_n$ it holds that

$$f_{i+1}(y^i) \geq y^i_{i+1}$$

by defining $i + 1 = 1$ if $i = n$. This leads to

$$\begin{array}{rcl} f_{i+1}(x) - x_{i+1} & = & (f_{i+1}(x) - f_{i+1}(y^i)) + (f_{i+1}(y^i) - y^i_{i+1}) + (y^i_{i+1} - x_{i+1}) \\ & \geq & -\frac{\epsilon}{2\sqrt{n}} + 0 - h \\ & \geq & -\frac{\epsilon}{\sqrt{n}} \end{array}$$

for all $i \in I_n$. Consequently, we have

$$\max_i |x_i - f_i(x)| \leq \frac{\epsilon}{\sqrt{n}}.$$

Now it is easy to see that

$$||x - f(x)|| \leq \epsilon$$

for all $x \in \sigma$. □

The following labeling rule is also often used and is called the *standard integer labeling rule on* $S^n$.

**Definition 3.2.6** *Given a function* $f : S^n \mapsto S^n$ *and* $\mathcal{T}$ *a triangulation of* $S^n$, *a labeling rule* $l^2 : \mathcal{T}^0 \mapsto I_n$ *is given by*

$$l^2(y) = \min\{j \in I_n \mid f_j(y) - y_j = \min_h (f_h(y) - y_h) \text{ and } y_j > 0\}.$$

Obviously, there always exists such an index. Again, the following theorem says that a completely labelled simplex induced by $l^2$ also yields a good approximation of a fixed point of a continuous function.

**Theorem 3.2.7** *Let $f : S^n \mapsto S^n$ be a continuous function, let $\mathcal{T}$ be a triangulation of $S^n$ and let $\epsilon > 0$, $\delta > 0$ be such that $mesh(\mathcal{T}) \leq \delta$, and for any $x, y \in S^n$, $\max_i |x_i - y_i| < \delta$ implies*

$$\max_i |f_i(x) - f_i(y)| < \epsilon.$$

*Given the labeling rule $l^2$ and a completely labelled simplex $\sigma$, then we have*

$$||x - f(x)|| < n(\epsilon + \delta)$$

*for all $x \in \sigma$.*

Proof: We may assume that $\sigma = < y^1, \cdots, y^n >$ and $l^2(y^i) = i$ for all $i \in I_n$. Clearly, for all $i \in I_n$ it holds

$$f_i(y^i) - y_i^i \leq 0.$$

Hence, for all $i$,

$$f_i(x) - x_i = (f_i(x) - f_i(y^i)) + (f_i(y^i) - y_i^i) + (y_i^i - x_i) < \epsilon + \delta.$$

On the other hand because $\sum_{i=1}^n f_i(x) - x_i = 0$, we have for all $i$,

$$f_i(x) - x_i = -\sum_{j \neq i}(f_j(x) - x_j) > -n(\epsilon + \delta).$$

By combining the above two inequalities we obtain the desired result. □

It is clear that both labeling rules $l^1$ and $l^2$ are proper. By Sperner lemma, there exists a completely labelled simplex for any triangulation of $S^n$ and for any function $f : S^n \mapsto S^n$. Then by the above two approximation estimate theorems a completely labelled simplex yields a good approximation of a fixed point for any continuos function $f : S^n \mapsto S^n$. In the coming sections and chapters we will introduce various algorithms for finding completely labelled simplices.

Now we introduce a triangulation of $S^n$ which is easily implementable for the algorithms to be introduced later. In principle, triangulations of $S^n$ can be obtained from the triangulations of $\mathbf{R}^n$ introduced in Chapter 1 by affine homeomorphism. See Todd [1976a].

**Definition 3.2.8** *Given a positive integer $m$, the $K_2(m)$-triangulation of $S^n$ with grid size $m^{-1}$ is the collection of all $(n-1)$-simplices $\sigma(y^1, \pi)$ with vertices $y^1$, $\cdots$, $y^n$ in $S^n$ such that*
*(a) each component of $y^1$ is a non-negative multiple of $m^{-1}$;*
*(b) $\pi = (\pi(1), \cdots, \pi(n-1))$ is a permutation of the elements in $I_{n-1}$;*
*(c) $y^{i+1} = y^i + m^{-1}q(\pi(i))$, $i \in I_{n-1}$, where $q(j) = e(j+1) - e(j)$, $j \in I_{n-1}$.*

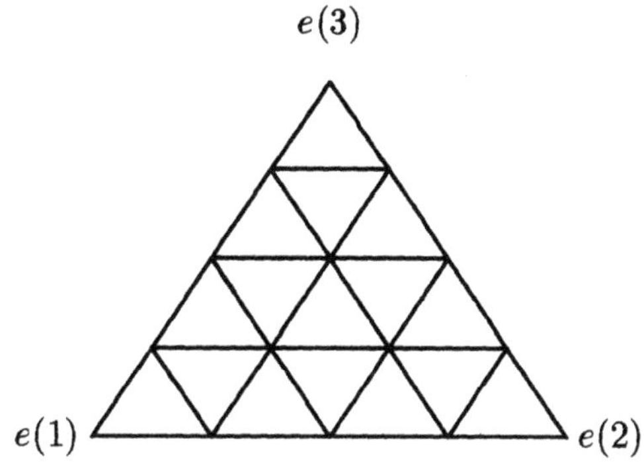


*Figure 3.1.* Illustration of $K_2(m)$-triangulation.

Let $Q$ be the $n \times (n-1)$ matrix with $j$-th column $q(j)$, $j \in I_{n-1}$. Then it is clear that the rank of this matrix is equal to $n-1$. The $K_2(m)$-triangulation of $S^n$ is illustrated in Figure 3.1 for $n = 3$ and $m = 4$. It is easy to prove that the mesh size of the $K_2(m)$-triangulation is equal to $\sqrt{n-1}/m$ if $n$ is even and $\sqrt{n}/m$ if $n$ is odd. There also exist other triangulations of $S^n$. See Todd [1976a] for detail. Those triangulations are less suitable in term of computational efficiency.

## 3.3. Scarf's algorithm

In this section we introduce the first algorithm (see Scarf [1967b, 1973]) for computing a fixed point of a continuous function $f : S^n \mapsto S^n$. We first introduce Scarf's algorithm in the framework of simplices (from Scarf [1973]) and then briefly describe this algorithm in term of primitive sets (from Scarf [1967b, 1973]).

In order to get an explicit computational procedure for finding a completely labelled simplex in $S^n$. We first imbed the $(n-1)$-dimensional unit simplex $S^n$ properly in a larger simplex $S$

$$S = \{x \in \mathbb{R}^n \mid \sum_{i=1}^{n} x_i = 1 \text{ and } \quad x_i \leq 2 \text{ for all } i\}.$$

So $S^n$ is completely contained in the interior of $S$. Let $s^1, \cdots, s^n$ be the vertices of $S$. Let $\mathcal{T}$ be a triangulation of $S^n$ and let $l : \mathcal{T}^0 \mapsto I_n$ be a proper labeling rule. We then extend the triangulation of $S^n$ to the simplex $S$ as follows. For each nonempty proper set $T$ of $I_n$, new simplices are constructed by taking a simplex with $(n - |T|)$ vertices on the face of $S^n$ defined by $x_i = 0$ for all $i \in T$ and adjoining to them the vertices $s^i$ for

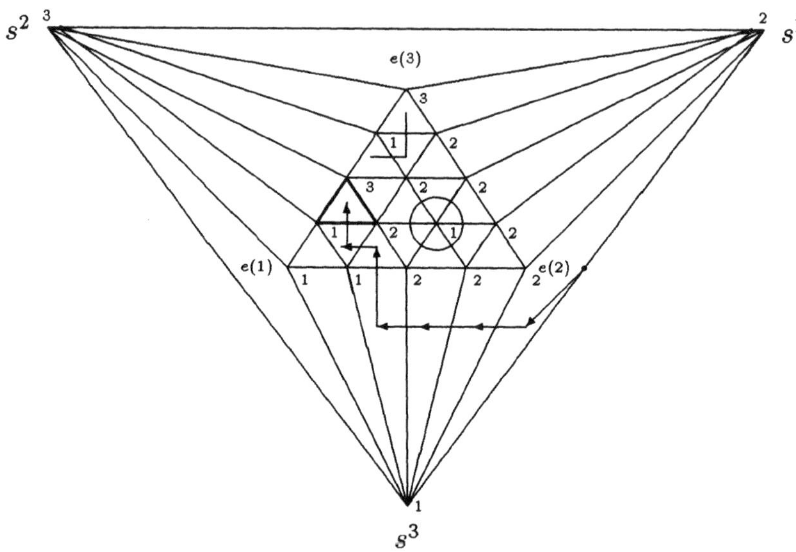


*Figure 3.2.* Illustration of Scarf's algorithm.

$i \in T$. For each $i \in I_n$, the vertex $s^i$ will receive label $i+1$ modulo $n$. In other words, $l(s^1) = 2$, $l(s^2) = 3$, $\cdots$, $l(s^n) = 1$. We demonstrate this augmentation in Figure 3.2 for $n = 3$ and $m = 4$. For a simplex $\sigma$, we denote its label set by $L(\sigma)$, i.e.,

$$L(\sigma) = \{l(y^i) \mid y^i \text{ is a vertex of } \sigma\}.$$

From the above construction, we have the following easy but important observation.

**Lemma 3.3.1** *There are no completely labelled simplices lying on $S \setminus int(S^n)$.*

Proof: Note that an $(n-1)$-simplex $\sigma$ lying on $S \setminus int(S^n)$ must be formed by $(n - |T|)$ vertices $y^j$ on the face of $S^n$ defined by $x_i = 0$ for all $i \in T$ and the vertices $s^i$ for $i \in T$, where $T$ is some nonempty proper set $T$ of $I_n$. Since $l$ is a proper labeling rule, this implies that all $(n - (T|)$ vertices $y^j$ have labels different from the members of $T$. In order for $\sigma$ to be completely labelled, the collection of vertices $s^i$, $i \in T$, must bear all of the labels in $T$. However, it is impossible, since $l(s^i) = i + 1$ modulo $n$ for $i \in T$, i.e., $\{l(s^i) \mid i \in T\} \neq T$. □

**Definition 3.3.2** *Let the simplex $S$ be triangulated and let the vertices of the triangulation be labelled as above. An $(n-1)$-simplex $\sigma$ is completely labelled (c.l.) if its vertices carry all labels in $I_n$, i.e., $L(\sigma) = I_n$, while an $(n-1)$ or $(n-2)$-simplex $\sigma$ is almost completely labelled (a.c.l.) if its vertices carry at least all labels in $I_{n-1}$, i.e., $I_{n-1} \subseteq L(\sigma)$.*

It is clear that a c.l. simplex is an a.c.l. simplex. We are now able to construct the following graph $G = (V, A)$ where $V$ is the set of nodes and $A$ is the set of arcs.

**Definition 3.3.3** *Let a.c.l. $(n-1)$-simplices in $S$ and a.c.l. $(n-2)$ simplices lying on the boundary of $S$ be the nodes of $G$. Two nodes are adjacent and connected by an arc or edge if one is a facet of the other or they share a common a.c.l. facet. The degree of a node $\sigma$ is the number of nodes adjacent to $\sigma$, denoted by $deg(\sigma)$.*

This definition is illustrated in Figure 3.2. From the definition it is also clear that the $(n-2)$-simplex $\tau^0 =< s^i \mid i \in I_n \setminus \{n-1\} >$ is the only node lying on $bd(S)$. Now we have the following lemma.

**Lemma 3.3.4** *Given the graph $G = (V, A)$ defined as above, for each node $\sigma$ of $G$, we have:*
*(i) if $\sigma = \tau^0$, then $deg(\sigma) = 1$;*
*(ii) if $\sigma$ is a c.l. simplex, then $deg(\sigma) = 1$;*
*(iii) in all other cases, $deg(\sigma) = 2$.*

Proof: (i) Note that $\tau^0$ lies on the boundary of $S$. By Theorem 1.4.6, $\tau^0$ is a facet of just one $(n-1)$-simplex $\sigma$ and clearly $\sigma$ is a.c.l. Hence $deg(\tau^0) = 1$.

(ii) $\sigma$ is a c.l. $(n-1)$-simplex. Then $\sigma$ has exactly one a.c.l. facet $\tau$. By Lemma 3.3.1 we know that $\tau$ must be lying on $S^n$. By Theorem 1.4.6 $\tau$ is a facet of exactly one other $(n-1)$-simplex $\sigma'$. Obviously, $\sigma'$ is a.c.l. Thus, $deg(\sigma) = 1$.

(iii) $\sigma$ is an a.c.l. but not c.l. $(n-1)$-simplex. Since the $n$ vertices of $\sigma$ have all labels in $I_{n-1}$, i.e., $L(\sigma) = I_{n-1}$, there are exactly two vertices, $y^1$ and $y^2$, with the same label. Therefore $\sigma$ has precisely two a.c.l. facets $\tau_1$ and $\tau_2$ opposite $y^1$ and $y^2$, respectively. Each $\tau_i$ either is $\tau^0$ (i.e., $\tau_i$ lies on $bd(S)$) and is a node of $G$ or leads to another a.c.l. $(n-1)$-simplex $\sigma^i$. In each case $deg(\sigma) = 2$. □

By using Theorem 1.4.14, Lemmas 3.3.1, 3.3.4 and the finite number of simplices in $S$, we have

**Theorem 3.3.5** *For the graph $G = (V, A)$ defined as above, each connected component of $G$ has one of the following three forms:*
*(i) a finite simple circuit whose nodes are all a.c.l. but not c.l. $(n-1)$-simplices;*
*(ii) a finite simple path whose end nodes are a.c. $(n-1)$-simplices;*

*(iii) a finite simple path starts with the a.c.l. $(n-2)$-simplex $\tau^0$ and ends with a c.l. $(n-1)$-simplex in $S^n$. Its intermediate nodes are all a.c.l. but not c.l. $(n-1)$-simplices.*

Scarf's algorithm starts with $\tau^0$ and then generates a finite sequence of adjacent a.c.l. simplices and finally leads to a c.l. simplex. As a matter of fact, we can explain the convergence of this algorithm in somewhat leisure way. Imagine a house consisting of a finite number of rooms. Each room has either one door or two doors. A room with only one door is called a *dead room.* Suppose that there is only one door leading to the outside! We label this door by *ENTER*. Imagine a situation that a bear has entered the house through ENTER. Now we claim that a hunter can always chase this bear into a dead room by a simple rule: He enters the house through ENTER and proceeds from room to room. As soon as he goes through a door, he immediately closes that door and never opens that door again. Doing so, he can chase the bear into a dead room. That bear cannot escape from the hunter!

Now we give a detailed description of Scarf's algorithm as follows. Let $f : S^n \mapsto S^n$ be a continuous function. Given a positive integer $m$, we triangulate $S^n$ by $K_2(m)$ and then extend the triangulation as above. Label the vertices of $S$ as above. Let $\tau^0$ be the unique a.c.l. $(n-2)$-simplex on $bd(S)$, i.e., $\tau^0 =< s^i \mid i \in I_n \setminus \{n-1\} >$.

Scarf's Algorithm on $S^n$

**Step (0)** Let $\sigma^0$ be the unique $(n-1)$-simplex in $S$ which has $\tau^0$ as its facet. Let $y^+$ be the vertex of $\sigma^0$ that is not a vertex of $\tau^0$. Set $k = 0$.

**Step (1)** Compute $l(y^+)$. If $l(y^+) = n$, stop and a c.l. simplex $\sigma^k$ is found. Otherwise, $l(y^+)$ is the label of exactly one other vertex of $\sigma^k$, say $y^-$. Let $\tau^{k+1}$ be the facet of $\sigma^k$ opposite $y^-$.

**Step (2)** Find a simplex $\sigma^{k+1}$ sharing the facet $\tau^{k+1}$ with $\sigma^k$, and let $y^+$ be the vertex of $\sigma^{k+1}$ not being a vertex of $\tau^{k+1}$. Set $k = k+1$ and return to Step (1).

In the algorithm, the replacement of the vertex $y^-$ by $y^+$ is called the *complementary pivoting step.* It is very important to observe from (ii) in Theorem 1.4.6 that at each step of the algorithm the vertices $y^-$ and $y^+$ are uniquely determined. The algorithm is illustrated in Figure 3.2.

**Corollary 3.3.6** *There is a finite simple path which starts with $\tau^0$ and ends with a completely labelled simplex in $S^n$.*

The reader may have realized that the proof of the convergence of Scarf's algorithm above is significantly distinct from any other types of numerical methods (such as Newton methods). That is, the above complementary pivoting argument depends purely on combinatorial reasoning to ensure finite convergence, whereas other types of numerical methods typically require the monotonic property. This novel argument was first used by Lemke and Howson [1964] in their algorithm which calculates a Nash equilibrium in a two-person finite game. Such arguments are now standard in fixed point algorithms, or more generally, in complementary pivot theory. In other algorithms to be discussed later the arguments to prove the finite convergence are quite similar to the above arguments. So, they will not be repeated each time.

Now we briefly discuss Scarf's original algorithm on the unit simplex $S^n$ which is based on the concept of primitive sets; see Scarf [1967b,1973]. Hence we will not give proofs here. The interested reader should be able to work out proofs for the following three theorems by himself or herself. To describe the algorithm we consider a finite set $P_k$ of vectors $y^1, \cdots, y^n, \cdots, y^k$ in $\mathbb{R}^n$ for some fixed integer $k$ with $k > n$. The first $n$ vectors $y^1, \cdots, y^n$, are defined by

$$\begin{aligned} y^1 &= (0, M_1, \cdots, M_1)^\top, \\ y^2 &= (M_2, 0, \cdots, M_2)^\top, \\ &\vdots \\ y^n &= (M_n, \cdots, M_n, 0)^\top, \end{aligned}$$

with the $M_i$ satisfying $M_1 > M_2 > \cdots > M_n > 1$. Clearly, $y^1, \cdots, y^n$ are not in $S^n$. The vectors $y^{n+1}, \cdots, y^k$ are selected arbitrarily from the interior of $S^n$ such that no two vectors of $y^{n+1}, \cdots, y^k$ have the same $i$-th coordinate for any $i \in I_n$.

**Definition 3.3.7** *A set of $n$ points $\{y^i \mid i \in J\}$ in $P_k$ is a primitive set if there are no vectors $y^m$ in $P_k$ such that $y_i^m > \min_{j\in J} y_i^j$ for all $i \in I_n$.*

Clearly, the set $\{y^i \mid i \in I_n\}$ does not form a primitive set. Let $y^{j_1}, \cdots, y^{j_n}$ be a set of $n$ vectors in $P_k$. Consider the subsimplex $Q$ of $S^n$ defined by

$$x_i \geq \min_{h\in I_n} y_i^{j_h}, \ i = 1, \cdots, n,$$

and $\sum_{i=1}^n x_i = 1$. If no vectors of $P_k$ are contained in the interior of this subsimplex, then the vectors $y^{j_1}, \cdots, y^{j_n}$ forms a primitive set. Such a subsimplex $Q$ is called a *primitive subsimplex*. We illustrate the above concepts by an example. Let $y^1, \cdots, y^8$ be given as

$$\begin{array}{lll} y^1 = (0,4,4)^\top, & y^2 = (3,0,3)^\top, & y^3 = (2,2,0)^\top, \\ y^4 = (7/9,1/9,1/9)^\top, & y^5 = (3/4,1/8,1/8)^\top, & y^6 = (1/3,1/3,1/3)^\top, \\ y^7 = (1/10,8/10,1/10)^\top, & y^8 = (1/11,1/11,9/11)^\top. & \end{array}$$

The sets

$$\begin{array}{lll} \{y^1,y^2,y^8\}, & \{y^1,y^3,y^7\}, & \{y^2,y^3,y^4\}, \\ \{y^6,y^7,y^8\}, & \{y^1,y^6,y^7\}, & \{y^2,y^6,y^8\}, \end{array}$$

are primitive sets. The reader may find more primitive sets.

Let $l : P_k \mapsto I_n$ be a labeling function such that $l(y^1) = 1, \cdots, l(y^n) = n$. A primitive set $\{y^h \mid h \in J\}$ is called a *completely labelled primitive set* if all of its vectors are labelled differently, i.e., $\{l(y^h) \mid h \in J\} = I_n$. Scarf proved the following result in a constructive way.

**Theorem 3.3.8 (Scarf's Combinatorial Theorem)** *Let the points $y^1, y^2, \cdots, y^k$ be defined and labelled as above. Then there exists a completely labelled primitive set.*

This theorem is very similar to Sperner lemma but not the same. Scarf developed an algorithm which will find a completely labelled primitive set within a finite number of steps. This algorithm is based on the next result.

**Theorem 3.3.9 (Scarf's Replacement Theorem)**
*(i) For each $m^- \in I_n$, there exists a unique index $m^+ \in I_k \setminus I_n$, such that the set of points $\{y^h \mid h \in (I_n \cup \{m^+\}) \setminus \{m^-\}\}$ forms a primitive set.*
*(ii) Let $\{y^j \mid j \in J\}$ be a primitive set and $m^- \in J$. If $\{y^j \mid j \in J \setminus \{m^-\}\}$ is not a subset of $\{y^i \mid i \in I_n\}$, then there exists a unique index $m^+ \in I_k \setminus J$ such that $\{y^j \mid j \in (J \cup \{m^+\}) \setminus \{m^-\}\}$ forms a primitive set.*

Now we can describe Scarf's algorithm. By Theorem 3.3.9 (i) there is a unique primitive set $\sigma^I = \{y^2, y^3, \cdots, y^n, y^+\}$ where $y^+$ is an element from $y^{n+1}, \cdots, y^k$. Starting with this unique primitive set, Scarf's algorithm searches for label 1. Thus, $l(y^+)$ is computed. Proceed steps similar to Steps (1) and (2) in Scarf's algorithm based on simplices. Because of Theorem 3.3.9 (ii) each replacement is unique. Since the number of primitive sets is finite and each primitive set is visited at most once, the algorithm must terminate with a completely labelled primitive set within a finite number of steps. This leads to the following result.

**Theorem 3.3.10 (Scarf's Convergence Theorem)** *Let the points $y^1, y^2, \cdots, y^k$ be given and labelled as above. Then starting with the primitive set $\sigma^I$, Scarf's algorithm will find a completely labelled primitive set with a finite number of steps.*

Let us illustrate the algorithm with the previous example with $l(y^1) = 1$, $l(y^2) = 2$, $l(y^3) = 3$, $l(y^4) = 2$, $l(y^5) = 3$, $l(y^6) = 2$, $l(y^7) = 3$ and $l(y^8) = 1$. The process of the algorithm goes as follows: A sequence of primitive sets is generated as $\sigma^I = \{y^2, y^3, y^4\}$, $\sigma^1 = \{y^3, y^4, y^5\}$, $\sigma^2 = \{y^2, y^4, y^5\}$, $\sigma^3 = \{y^2, y^5, y^6\}$, $\sigma^4 = \{y^3, y^5, y^6\}$, $\sigma^5 = \{y^3, y^6, y^7\}$, and $\sigma^6 = \{y^6, y^7, y^8\}$. The set $\sigma^6$ is a completely labelled primitive set.

Now we show how Brouwer theorem is proved. Consider a continuous function $f : S^n \mapsto S^n$. Let a labeling function $l : S^n \cup \{y^1, \cdots, y^n\} \mapsto I_n$ be defined as

$$l(y) = \max\{j \in I_n \mid f_j(y) - y_j = \max_h(f_h(y) - y_h)\},$$

and $l(y^i) = i$ for all $i \in I_n$. Clearly, if $y \in S^n$ and $l(y) = i$, then $f_i(y) \geq y_i$. Choose $y^{n+1}, \cdots, y^k$ from $int(S^n)$ as required, and label these vectors as above. In the sequel we show that a completely labelled primitive set $W = \{y^h \mid h \in J\}$ indeed yields an approximate fixed point of $f$. This primitive set $W$ contains some vectors from the first $n$ vectors, say $y^h$ with $h \in I$, and some vectors from the remaining vectors in $P_k$. Note that $I$ is a proper subset of $I_n$. The primitive subsimplex associated with this primitive set is bounded by an edge $x_i = 0$ for each $i \in I$, and by an edge passing through each vector $y^h$, $h \in J \setminus I$. Note that for each $h \in I$, $h \neq l(y^i)$ for all $i \in J \setminus I$. So for every $i \in I_n$ there is some vector $y$ in this primitive subsimplex for which $f_i(y) \geq y_i$. Clearly, any point in this subsimplex is an approximate fixed point of $f$. As $k$ goes to infinity, the diameter of each primitive subsimplex tends to zero. As a result, a fixed point of $f$ will be obtained by taking the limit of a sequence of the primitive subsimplices just described.

## 3.4. Kuhn's artificial start algorithm

After the work of Scarf, Kuhn [1968] introduced the first fixed point algorithm which is based on the notions of simplices and simplicial subdivisons instead of primitive sets. Since then most fixed point algorithms are built upon simplices and triangulations. It is why fixed point algorithms are often called simplicial fixed point algorithms, or simply, simplicial algorithms. Before describing Kuhn's algorithm, we recall that $S_i^n$ is the set $\{x \in S^n \mid x_i = 0\}$, a subset of the $(n-1)$-dimensional unit simplex $S^n$, for $i \in I_n$. In order to find a completely labelled simplex in $S^n$, we first extend the triangulation $K_2(m)$ of $S^n$ to a larger set $\bar{S}^n = \{x \in \mathbb{R}^n \mid x_i \geq 0, i \in I_{n-1}$ and $x_n \geq -m^{-1}\}$. That is, an extra layer is added to $S^n$ below $S_n^n$. Define

$$\bar{S}_i^n = \{x \in \bar{S}^n \mid x_i = 0\},\ i \in I_{n-1}$$

and

$$\bar{S}_n^n = \{x \in \bar{S}^n \mid x_n = -m^{-1}\}.$$

Clearly, $\bar{S}_i^n = S_i^n$ for each $i \in I_{n-1}$. $K_2(m)$ can be easily extended to $\bar{S}^n$ as follows.

**Definition 3.4.1** *Given a positive integer $m$, the $\bar{K}_2(m)$-triangulation of $\bar{S}^n$ with grid size $m^{-1}$ is the collection of all $(n-1)$-simplices $\sigma(y^1, \pi)$ with vertices $y^1, \cdots, y^n$ in $\bar{S}^n$ such that*

(a) *each component of* $y^1$ *is an integer multiple of* $m^{-1}$;
(b) $\pi = (\pi(1), \cdots, \pi(n-1))$ *is a permutation of the elements in* $I_{n-1}$;
(c) $y^{i+1} = y^i + m^{-1}q(\pi(i))$, $i \in I_{n-1}$, *where* $q(j) = e(j+1) - e(j)$, $j \in I_{n-1}$.

From the definition, we can see that each vertex of $\bar{K}_2(m)$ lies in $S^n$ or lies on $\bar{S}^n_n$. For a vertex $y \in K^0_2(m)$ we label it by $l^2$, whereas if a vertex $y \in \bar{K}^0_2(m)$ lying on $\bar{S}^n_n$, it will be artificially labelled as

$$l(y) = \min\{j \in I_n \mid y_j = \max_h y_h\}.$$

Now one can easily prove the following lemma by noticing the property of the above labeling rule.

**Lemma 3.4.2** *Given the above labeling rule, no vertex in* $\bar{S}^n_i$ *has the label* $i$ *for any* $i \in I_n$.

Since $l^2$ is a proper labeling rule, no vertex in $S^n_n$ has the label $n$, this together with Lemma 3.4.2 implies that all completely labelled simplices in effect lie in $S^n$. Now we can construct the following graph $G = (V, A)$ where $V$ is the set of nodes and $A$ is the set of arcs.

**Definition 3.4.3** *Let a.c.l.* $(n-1)$*-simplices in* $\bar{S}^n$ *and a.c.l.* $(n-2)$ *simplices lying on the boundary of* $\bar{S}^n$ *be the nodes of* $G$. *Two nodes are adjacent and connected by an arc or edge if one is a facet of the other or they share a common a.c.l. facet. The degree of a node* $\sigma$ *is the number of nodes adjacent to* $\sigma$, *denoted by* $deg(\sigma)$.

It follows from Lemma 3.4.2 that there exists no node lying on $\bar{S}^n_i$ for any $i \in I_{n-1}$. This means that a node (if any) lying on $bd(\bar{S}^n)$ must lie on $\bar{S}^n_n$. The following lemma says that there is a unique a.c.l. $(n-2)$-simplex $\tau^I$ (i.e., a unique node) of $\bar{K}_2(m)$ lying on $\bar{S}^n_n$. When we assume that $m$ is a multiple of $n-1$, say $m = k(n-1)$, then $\tau^I$ can be given explicitly. That is, $\tau^I = \tau^I(y^*, \pi^*)$ is defined in the natural way, where $y^* = (\frac{k+1}{m}, \frac{k}{m}, \cdots, \frac{k}{m}, -\frac{1}{m})^\top$ and $\pi^* = (1, 2, \cdots, n-2)$.

**Lemma 3.4.4** *Given the above labeling rule and* $m = k(n-1)$, *then* $\tau^I$ *is the only a.c.l.* $(n-2)$*-simplex of* $\bar{K}_2(m)$ *in* $\bar{S}^n_n$.

Proof: We first prove that $\tau^I = \tau^I(y^*, \pi^*)$ is a.c.l. Note that the vertices of $\tau^I$ are $y^1 = y^*$, and $y^{i+1} = y^i + m^{-1}q(\pi^*(i))$ for $i = 1, \cdots, n-2$. It follows that $y^i = (\frac{k}{m}, \cdots, \frac{k}{m}, \frac{k+1}{m}, \frac{k}{m}, \cdots, \frac{k}{m}, -\frac{1}{m})^\top$, with $\frac{k+1}{m}$ in the $i$-th coordinate. By the above labeling rule it is readily seen that $y^i$ has label $i$ for each $i \in I_{n-1}$. Thus, $\tau^I$ is a.c.l. Furthermore, $\tau^I$ is a facet of the simplex $\sigma(y^1, (\pi^*, n-1))$.

Now we need to prove the uniqueness. Let $\tau = \tau(z^1, \rho) = < z^1, \cdots, z^{n-1} >$ be an arbitrary a.c.l. simplex in $\bar{S}^n_n$. We have to show $\tau = \tau^I$. For each $i$, let $z^i = (\frac{a_{i,1}}{m}, \cdots, \frac{a_{i,n-1}}{m}, -\frac{1}{m})^\top$. We must have

$$\sum_{h=1}^{n-1} a_{i,h} - 1 = m = k(n-1), \tag{3.4.1}$$

where all $a_{i,h}$ are non-negative integers. Suppose that $z^i$ receives label $j$. Then we must have

$$\frac{\sum_{h=1}^{n-1} a_{i,h}}{n-1} = k + \frac{1}{n-1} \leq \max_h a_{i,h} = a_{i,j}.$$

Because $a_{i,j}$ is non-negative integer, we have

$$a_{i,j} \geq k+1. \tag{3.4.2}$$

So ( 3.4.2) is a necessary condition for $z^i$ to receive label $j$.

We proceed in several steps. (1) If $a_{1,1} > k+1$, then ( 3.4.1) implies that there is some $h \in I_{n-1}$ such that $a_{1,h} < k$. Thus, $a_{i,h} \leq k$ for all $i$. Consequently, no vertex $z^i$ receives label $h$ by ( 3.4.2). On the other hand, if $a_{1,1} < k+1$, then $a_{h,1} < k+1$ for every $h$ and no vertex receives label 1. Thus $a_{1,1} = k+1$. Similarly one can show $a_{1,h} = k$ for $h = 2, \cdots, n-1$. Hence $z^1 = y^1 = y^*$. (2) If $\rho(1) \neq 1$, then we have $z^2 = z^1 + m^{-1}q(\rho(1)) = (\frac{k+1}{m}, \frac{k}{m}, \cdots, \frac{k-1}{m}, \frac{k+1}{m}, \frac{k}{m}, \cdots, \frac{k}{m}, -\frac{1}{m})^\top$, where the 1-th and $(\rho(1)+1)$-th components are $(k+1)/m$. This implies that $l(z^2) = 1$. It is a contradiction. Thus, $\rho(1) = 1$. Similarly one can show $\rho(h) = h$ for $h = 2, \cdots, n-2$. So $\tau = \tau^I$. We are done. □

The proof of the following results goes similarly to that of Lemma 3.3.4 and Theorem 3.3.5.

**Lemma 3.4.5** *Given the graph $G = (V, A)$ defined as above, for each node $\sigma$ of $G$, we have:*
*(i) if $\sigma = \tau^I$, then $deg(\sigma) = 1$;*
*(ii) if $\sigma$ is a c.l. simplex, then $deg(\sigma) = 1$;*
*(iii) in all other cases, $deg(\sigma) = 2$.*

**Theorem 3.4.6** *For the graph $G = (V, A)$ defined as above, each connected component of $G$ has one of the following three forms:*
*(i) a finite simple circuit whose nodes are all a.c.l. but not c.l. $(n-1)$-simplices;*
*(ii) a finite simple path whose end nodes are a.c. $(n-1)$-simplices;*
*(iii) a finite simple path starts with the a.c.l. $(n-2)$-simplex $\tau^I$ and ends with a c.l. $(n-1)$-simplex in $S^n$. Its intermediate nodes are all a.c.l. but not c.l. $(n-1)$-simplices.*

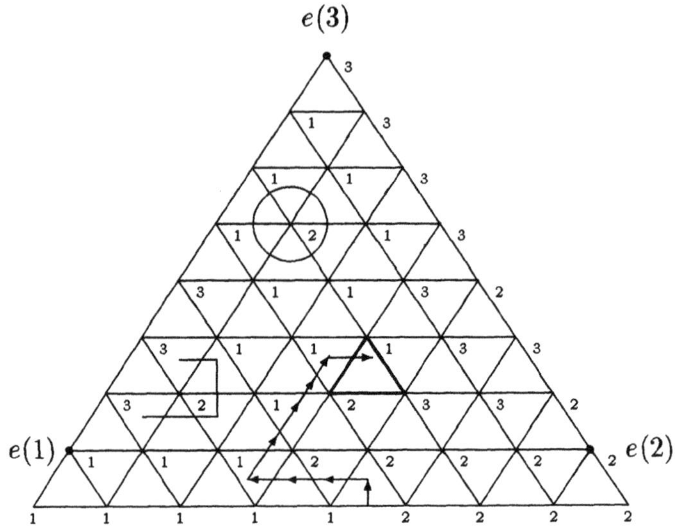


*Figure 3.3.* Illustration of Kuhn's (AS) algorithm.

Kuhn's algorithm starts with $\tau^I$ and then generates a finite sequence of adjacent a.c.l. simplices and finally leads to a c.l. simplex. We can formally describe this algorithm as follows.

Kuhn's Artificial Start (AS) Algorithm on $S^n$

**Step (0)** Set $\tau^0 = \tau^I$. Let $\sigma^0$ be the unique $(n-1)$-simplex in $\bar{S}^n$ which has $\tau^0$ as its facet. Let $y^+$ be the vertex of $\sigma^0$ that is not a vertex of $\tau^0$. Set $k = 0$.

**Step (1)** Compute $l(y^+)$. If $l(y^+) = n$, stop and a c.l. simplex $\sigma^k$ is found. Otherwise, $l(y^+)$ is the label of exactly one other vertex of $\sigma^k$, say $y^-$. Let $\tau^{k+1}$ be the facet of $\sigma^k$ opposite $y^-$.

**Step (2)** Find a simplex $\sigma^{k+1}$ sharing the facet $\tau^{k+1}$ with $\sigma^k$, and let $y^+$ be the vertex of $\sigma^{k+1}$ not being a vertex of $\tau^{k+1}$. Set $k = k+1$ and return to Step (1).

This algorithm is illustrated in Figure 3.3 for $n = 3$ and $m = 7$.

**Corollary 3.4.7** *There is a finite simple path which starts with $\tau^I$ and ends with a completely labelled simplex in $S^n$.*

## 3.5. Kuhn's variable dimension algorithm

In this section we introduce Kuhn's second fixed point algorithm (see Kuhn [1969]). This method differs from the first one in the following respects. It uses no extra layer below the unit simplex $S^n$, but starts with one of the vertices of the unit simplex. Moreover, the dimension of adjacent simplices traced by the method varies from time to time. Due to that this method is therefore called the variable dimension algorithm.

**Definition 3.5.1** *For each $k \in I_n$, let*

$$S^n(k) = \{x \in S^n \mid x_i = 0,\ i = k+1, \cdots, n\}.$$

Note that $S^n(1) = \{e(1)\}$, $S^n(n-1) = S^n_n$ and $S^n(n) = S^n$. Now we can define a graph as follows.

**Definition 3.5.2** *For $k \in I_n$, a $(k-1)$-simplex $\sigma$ in $S^n(k)$ is an $I_{k-1}$-complete if one facet $\tau$ of $\sigma$ carries all the labels in $I_{k-1}$, i.e., $L(\tau) = I_{k-1}$.*

Clearly, a c.l. $(n-1)$-simplex is $I_{n-1}$-complete.

**Definition 3.5.3** *For each $k \in I_n$, a $(k-1)$-simplex in $S^n(k)$ is a node if it is $I_{k-1}$-complete. For $k \in I_{n-1}$, two nodes $\sigma^1$ and $\sigma^2$ in $S^n(k)$ or $S^n(k+1)$ are adjacent and connected by an edge or arc if one is a facet of the other or if they have a common facet whose vertices carry all the labels in $I_{k-1}$ ($I_k$) when $\sigma^1$ and $\sigma^2$ are in $S^n(k)$ ($S^n(k+1)$). The degree of a node $\sigma$ is the number of nodes adjacent to $\sigma$, denoted by $deg(\sigma)$. Let $V$ be the set of nodes and let $A$ be the set of arcs. In this way a graph $G = (V, A)$ is obtained.*

Given a triangulation of $S^n$ and a proper labeling rule $l$ on $S^n$, then we have the following:

**Lemma 3.5.4** *Given the graph $G = (V, A)$ defined as above, for each node $\sigma$ of $G$, we have:*
*(i) if $\sigma = \{e(1)\}$, then $deg(\sigma) = 1$;*
*(ii) if $\sigma$ is a c.l. simplex, then $deg(\sigma) = 1$;*
*(iii) in all other cases, $deg(\sigma) = 2$.*

A more general result than the above one will be given in Chapter 5 and a detailed proof will also be given there. At this moment, the reader should note that due to the proper labeling rule the possible labels in $S^n(k)$ include only the elements in $I_k$.

**Theorem 3.5.5** *For the graph $G = (V, A)$ defined as above, each connected component of $G$ has one of the following three forms:*
*(i) a finite simple circuit;*
*(ii) a finite simple path whose end nodes are c.l. $(n-1)$-simplices;*
*(iii) a finite simple path starts with the node $\{e(1)\}$ and ends with a c.l. $(n-1)$-simplex in $S^n$.*

Kuhn's algorithm starts with $\{e(1)\}$ and generates a finite sequence of adjacent $I_k$-complete simplices of varying $I_k$ and finally leads to a c.l. simplex.

A natural triangulation to use is $K_2(m)$. In the following a permutation of the elements of $I_k$ is denoted by $\pi(I_k) = (\pi(1), \cdots, \pi(k))$. A $k$-simplex $\sigma(y^1, \pi(I_k))$ is defined by $K_2(m)$ in a natural way, i.e., $y^{i+1} = y^i + m^{-1}q(\pi(i))$, for $i = 1, \cdots, k$. Let $f : S^n \mapsto S^n$ be given and let $l : K_2^0(m) \mapsto I_n$ be a proper labeling rule. Then, the steps of this algorithm are as follows.

Kuhn's Variable Dimension (VD) Algorithm on $S^n$

**Step (0)** Set $y^1 = e(1)$, $k = 2$, $\sigma^1 = \sigma(y^1, \pi(I_1))$ with $\pi(I_1) = (1)$. Let $y^+ = y^2 = e(1) + m^{-1}q(1)$. Set $j = 1$.

**Step (1)** Compute $l(y^+)$. If $l(y^+) = k$, go to Step (3). Otherwise, $l(y^+)$ is the label of exactly one other vertex of $\sigma^j$, say $y^s$. Let $\tau^j$ be the facet of $\sigma^j$ opposite $y^s$.

**Step (2)** If $s = k$ and $y_k^{k-1} = 0$ (i.e., $\tau^j \subseteq S^n(k-1)$), go to Step (4). Otherwise, $\sigma(y^1, \pi(I_{k-1}))$ is adapted according to Table 3.1 by replacing $y^s$. Let $\sigma^{j+1} = \sigma(y^1, \pi(I_{k-1}))$ and let $y^+$ be the new vertex of $\sigma^{j+1}$. Set $j = j + 1$ and return to Step (1).

**Step (3)** (Increasing the dimension) If $k = n$, a c.l. $(n-1)$-simplex is found and stop. Otherwise, let $\sigma^{j+1} = \sigma(y^1, \pi(I_k))$ where $\pi(I_k) = (\pi(I_{k-1}), k)$. Let $y^+$ be the new vertex of $\sigma^{j+1}$. Set $j = j+1$, $k = k+1$ and return to Step (1).

**Step (4)** (Decreasing the dimension) Let $\sigma^{j+1} = \sigma(y^1, \pi(I_{k-2}))$ with $\pi(I_{k-2}) = (\pi(1), \cdots, \pi(k-2))$. Let $y^s$ be the vertex of $\sigma^{j+1}$ with label $k-1$. Set $k = k-1$ and $j = j+1$. Return to Step (2).

| | $y^1$ becomes | $\pi(I_{k-1})$ becomes |
|---|---|---|
| $s = 1$ | $y^1 + q(\pi(1))/m$ | $(\pi(2), \cdots, \pi(k-1), \pi(1))$ |
| $1 < s < k$ | $y^1$ | $(\pi(1), \cdots, \pi(s-2), \pi(s), \pi(s-1), \pi(s+1), \cdots, \pi(k-1))$ |
| $s = k$ | $y^1 - q(\pi(k-1))/m$ | $(\pi(k-1), \pi(1), \cdots, \pi(k-2))$ |

TABLE 3.1. Pivot rules if $y^s$ is to be replaced.

This algorithm is illustrated in Figure 3.4 for $n = 3$ and $m = 8$.

**Corollary 3.5.6** *There is a finite simple path which starts with* $\{e(1)\}$ *and ends with a completely labelled simplex in* $S^n$.

Finally we point out that Kuhn's two methods discussed above can be easily adapted to compute fixed points on $\mathbb{R}^n_+$. The interested reader should try to figure them out or refer to Section 4.5 of van der Laan [1980b].

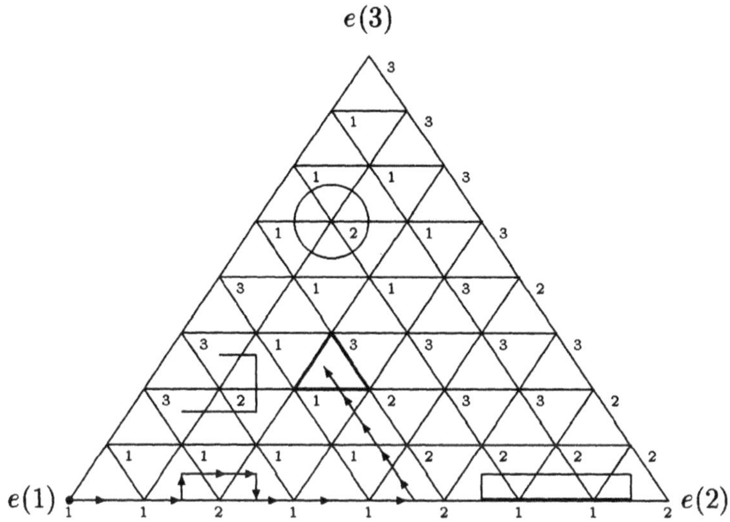


*Figure 3.4.* Illustration of Kuhn's (VD) algorithm.

## 3.6. Concluding remarks

So far we have discussed two versions of Scarf's algorithm and two algorithms of Kuhn on the unit simplex $S^n$. As the reader may observe, all these methods suffer from computational inefficiency, because they have to be started outside the region of interest while the grid size of simplicial subdivision is kept fixed during the operation of the algorithms. This implies that if an approximate solution is found and one wishes to search for a better approximate solution by refining the triangulation, one has to start again the methods outside the region of interest. So all obtained useful information is lost. In other words, if the mesh size is large, we cannot guarantee to obtain a good approximation. So if one wishes to find a satisfactory approximation, one has to use a triangulation with very small mesh size. Then the number of simplices can become very large. The cost of searching for a c.l. simplex by these methods could be very high since they have to start outside the region of interest. In the coming chapters we will introduce more efficient methods to alleviate these difficulties.

CHAPTER 4

# Simplicial Homotopy Algorithms

## 4.1. Introduction

In the previous chapter we introduced the first generation fixed point algorithms. As we see, those algorithms all suffer from computational inefficiency, caused by the following two difficulties: (1) they have to start outside the region of interest; (2) the mesh size of the underlying triangulation is kept constant throughout the whole computing process. To circumvent these impasses, several methods have been developed over the last thirty years. In this chapter we discuss two of them. The first method is that of Merrill [1972] and Kuhn and MacKinnon [1975], called the restart algorithm. This algorithm can start anywhere. When an approximate solution is found, the algorithm can restart from this approximation with a finer triangulation. The second method is that of Eaves [1972] and Eaves and Saigal [1972], called the simplicial homotopy algorithm. This algorithm can start anywhere and the grid size of simplices is automatically refined during the course of the algorithm. To grasp the idea of these methods, let us consider the fixed point problem of a continuous function $f : C \mapsto C$. Recall that the classical homotopy idea in topology goes as follows: Let $f_t(x) = (1-t)g(x) + tf(x)$ with $t \in [0,1]$ and $x \in C$, where $g : C \mapsto C$ is a constant function with a known fixed point, say, $x_0$, (i.e., $g(x) = x_0$ for all $x \in C$.) In order to find a fixed point of $f$, we begin with the constant function $f_0(x) = g(x)$ and its fixed point $x_0$, deform $f_t$ as $t$ converges to 1 and $f_t$ converges to the original function $f$, and follow the path of fixed points $x_t$ of $f_t$. Cluster points of the sequence $\{x_t \mid t \in \mathbf{Z}_+\}$ are fixed points of $f$. The reader may have realized that in this framework we need to work in a one dimension higher space, i.e., $C \times [0,1]$. It will become clear that both Merrill's algorithm and the algorithms of Eaves and (Saigal) are indeed homotopy algorithms but the latter ones are much more natural than the former one. In this chapter we present Merrill's algorithm first and then the algorithms of Eaves and (Saigal), because we believe the latter algorithms are more elegant and sophisticated than the former one, although Eaves and Merrill discovered their respective algorithms independently and simultaneously.

This chapter is organized as follows. In Section 2 we introduce the vector labeling rule and derive the approximation accuracy. Vector labeling rules are suitable for the computation of fixed points of point-to-set mappings.

In Section 3 we introduce the method of Merrill on $\mathbb{R}^n$. In Sections 4 and 5 we discuss the method of Eaves on the unit simplex $S^n$ and the method of Eaves and Saigal on $\mathbb{R}^n$, respectively.

## 4.2. Vector labeling, approximation and lexicographic system

Vector labeling rules were first introduced by Hansen and Scarf [1969] and Eaves [1971b] for u.s.c. point-to-set mappings. Compared with integer labeling rules these rules will provide a better approximation and be more suitable for designing various new algorithms.

Suppose now that we want to compute a fixed point of a u.s.c. point-to-set mapping $\psi : \mathbb{R}^n \mapsto \mathbb{R}^n$. Let $f$ be the piecewise linear approximation of $\psi$ with respect to a triangulation $\mathcal{T}$ of $\mathbb{R}^n$. We define a vector labeling rule in the following.

**Definition 4.2.1** *Let $l : \mathbb{R}^n \mapsto \mathbb{R}^n$ be given by*

$$l(x) = f(x) - x.$$

*For each vertex $x$, $l(x)$ is called the vector label of $x$. $l$ is said to be the vector labeling rule of $\psi$.*

**Definition 4.2.2** *Given an $n$-simplex $\sigma$ with vertices $y^1, \cdots, y^{n+1}$ and the vector labeling rule $l$, the following matrix*

$$L(\sigma) = \begin{bmatrix} 1 & 1 & \cdots & 1 \\ l(y^1) & l(y^2) & \cdots & l(y^{n+1}) \end{bmatrix}$$

*is said to be the label matrix of $\sigma$.*

**Definition 4.2.3** *An $n$-simplex $\sigma$ with vertices $y^1, \cdots, y^{n+1}$ is called a completely labelled simplex if the following system of linear equations has a solution*

$$L(\sigma)w = E(1), \ w \in \mathbb{R}^{n+1}_+$$

*where $E(1) = (1, 0, \cdots, 0)^\top$ is an $(n+1)$-vector.*

Then the following lemma states that a completely labelled simplex yields a fixed point of $f$.

**Lemma 4.2.4** *The piecewise linear approximation $f$ of $\psi$ with respect to $\mathcal{T}$ has a fixed point in the simplex $\sigma(y^1, \cdots, y^{n+1})$ if and only if $\sigma$ is completely labelled. If $w^*$ is the solution of the system of equations, then*

$$y^* = \sum_{i=1}^{n+1} w_i^* y^i$$

*is the fixed point of $f$ in $\sigma$.*

Proof: If $\sigma$ is completely labelled, then we have $y^* = \sum_{i=1}^{n+1} w_i^* y^i$ and $\sum_{i=1}^{n+1} w_i^* = 1$ with $w_i^* \geq 0$ for all $i \in I_{n+1}$. Thus, $y^* \in \sigma$. Moreover, since

$$\sum_{i=1}^{n+1} w_i^* l(y^i) = 0^n$$

we have $y^* = \sum_{i=1}^{n+1} w_i^* y^i = \sum_{i=1}^{n+1} w_i^* f(y^i) = f(y^*)$.

Conversely, if $y$ is a point of $\sigma(y^1, \cdots, y^{n+1})$, then there is a $w \in S^{n+1}$ such that $y = \sum_{i=1}^{n+1} w_i y^i$. Because $f$ is affine on $\sigma$, we have

$$\sum_{i=1}^{n+1} w_i l(y^i) = l(y) = f(y) - y. \tag{4.2.1}$$

If $y^*$ is a fixed point of $f$ in $\sigma$, then we can find $w^* \in S^{n+1}$ with $y^* = \sum_{i=1}^{n+1} w_i^* y^i$. It follows from equation (4.2.1) that $w^*$ is a solution of

$$L(\sigma)w = E(1), \ \ w \in \mathbb{R}_+^{n+1}.$$

Thus, $\sigma$ is completely labelled. □

The next theorem says that a completely labelled simplex yields an approximate fixed point of $\psi$.

**Theorem 4.2.5** *Let $\psi : \mathbb{R}^n \mapsto \mathbb{R}^n$ be a nonempty-valued point-to-set mapping. Let $\epsilon > 0$ and $\delta > 0$ be such that mesh$(\mathcal{T}) < \delta$ and for all $x \in \mathbb{R}^n$, $y \in B(x, \delta)$, we have $\psi(y) \subseteq B(\psi(x), \epsilon)$. Let $\sigma^*(y^1, \cdots, y^{n+1})$ be a completely labelled simplex with solution $w^*$ and $y^* = \sum_{i=1}^{n+1} w_i^* y^i$. Then*

$$y^* \in B(\psi(y^*), \epsilon).$$

Proof: Since $y^* \in \sigma^*$ and $y^i \in \sigma^*$ for all $i \in I_{n+1}$, we have

$$||y^i - y^*|| < \delta$$

for all $i \in I_{n+1}$. This means that $f(y^i) \in \psi(y^i) \subseteq B(\psi(y^*), \epsilon)$. Note that $B(\psi(y^*), \epsilon)$ is convex. Thus it contains all convex combinations of the $f(y^i)'s$. Since $y^* = \sum_{i=1}^{n+1} w_i^* y^i = \sum_{i=1}^{n+1} w_i^* f(y^i) = f(y^*)$, this means that $y^* \in B(\psi(y^*), \epsilon)$. □

Notice that when $\psi$ is reduced to a continuous function, the accuracy given above is better than the one given in Theorem 3.2.7 for an integer labeling rule with the same mesh size.

In the integer labeling case the operation of fixed point algorithms is quite easy, as we have seen in the last chapter. In the vector labeling case we have to solve a sequence of systems of linear equations in order to make the corresponding complementary pivot steps. This can be done by using the simplex method of linear programming and will somehow become more complex. Now we discuss the basic framework for using vector labeling rules.

In order to avoid the degeneracy problem, we will introduce lexicographic systems. A row vector $x \in R^n$ is *lexicographically positive*, denoted by $x \succ \mathbf{0}$, if $x$ is not a zero vector and its first nonzero entry is positive, and *lexicographically non-negative,* denoted by $x \succeq \mathbf{0}$, if $x$ is either zero or lexicographically positive. A matrix $A$ is called *lexico positive*, denoted by $A \succ \mathbf{0}$, if each row vector is lexicographically positive, and *lexico non-negative*, denoted by $A \succeq \mathbf{0}$, if each row vector is lexicographically non-negative. Given two vectors (matrices) $a$ and $b$, we write $a \succ b$ ($a \succeq b$) if $a - b \succ \mathbf{0}$ ($a - b \succeq \mathbf{0}$).

A matrix $A$ is *semilexicopositive*, denoted by $A \succ_s \mathbf{0}$, if each row vector is lexicographically positive regardless of the last row, and *semilexico non-negative*, denoted by $A \succeq_s \mathbf{0}$, if each row vector is lexicographically non-negative regardless of the last row. In the sequel, we mean "lexicographically" simply by "lexico".

Consider the system of linear equations

$$AX = \mathbf{B}, \quad X \succeq \mathbf{0} \tag{4.2.2}$$

where $A$ is an $n \times (n+1)$ matrix with $rank(A) = n$ and $B$ an $n \times k$ matrix with $rank(B) = n$. It is clear that a solution of system (4.2.2) must be an $(n+1) \times k$ matrix with rank $n$. If $X$ is a solution, then it has at most one zero row. We say that a solution $X$ with one zero row, say row $j$, is a *basic feasible solution.* Then the corresponding column $j$ of $A$ is a *non-basis vector.* The remaining columns of $A$ are called *basis vectors.*

From linear programming theory, one can easily prove

**Lemma 4.2.6** *If system* (4.2.2) *has a solution, it must have at least one basic feasible solution. Moreover, system* (4.2.2) *has at most two basic feasible solutions.*

The following basic theorem will be often used later.

**Theorem 4.2.7** *Suppose that system* (4.2.2) *has a solution. Then it has two basic feasible solutions if and only if for any non-basis column, say $A_i$, of $A$, expressed as a linear combination of the basis columns, say $A_j$ for $j \in J$, of $A$, (i.e.,*

$$A_i = \sum_{j \in J} \mu_j A_j$$

*where $\mu_j \in \mathbb{R}$ for all $j \in J$), there exists at least one positive coefficient $\mu_j$ for some $j \in J$.*

Proof: Suppose that $W$ and $X$ are the two basic feasible solutions of system (4.2.2). Without loss of generality, assume that the unique zero rows of $W$ and $X$ are the first and the second rows, respectively. Let $W_i$ and $X_i$ denote the $i$-th rows of $W$ and $X$ respectively, and $A_i$ the $i$-th column of $A$. Then we have

$$AW = \sum_{i\neq 1} A_i W_i = B = AX = \sum_{i\neq 2} A_i X_i. \tag{4.2.3}$$

Now we use basis vectors of $A$ to represent non-basis vectors of $A$. Then there is a non-zero vector $\mu$ such that

$$A_1 = \sum_{i\neq 1} \mu_i A_i.$$

This implies that

$$A_1 X_1 = \sum_{i\neq 1} \mu_i A_i X_1. \tag{4.2.4}$$

It follows from (4.2.3) and (4.2.4) that

$$\begin{aligned}\textstyle\sum_{i\neq 1} A_i W_i &= A_1 X_1 + \textstyle\sum_{i\neq 1,2} A_i X_i \\ &= \textstyle\sum_{i\neq 1} \mu_i A_i X_1 + \textstyle\sum_{i\neq 1,2} A_i X_i \\ &= A_2 \mu_2 X_1 + \textstyle\sum_{i\neq 1,2} A_i (X_i + \mu_i X_1).\end{aligned}$$

Therefore, we have

$$A_2(W_2 - \mu_2 X_1) + \sum_{i\neq 1,2} A_i[W_i - (X_i + \mu_i X_1)] = \mathbf{0}.$$

Since the vectors $A_2, \cdots, A_{n+1}$ are linearly independent, we have $W_2 - \mu_2 X_1 = \mathbf{0}$, i.e.,$W_2 = \mu_2 X_1$. Because $W_2 \succ 0$ and $X_1 \succ \mathbf{0}$, this implies that $\mu_2 > 0$.

Conversely, suppose that the linear combination coefficients of a non-basis column of $A$ from basis columns of $A$ have at least one positive coefficient. Because system (4.2.2) has a feasible solution, it follows from Lemma 4.2.6 that it must have a basic feasible solution. Assume that $W$ is the basic feasible solution with its first row being zero vector. Then we have

$$\sum_{i\neq 1} A_i W_i = B.$$

By assumption, $A_1 = \sum_{i\neq 1} \mu_i A_i$ with at least one parameter, say, $\mu_p > 0$. Therefore there exists a unique index $q$ such that

$$\frac{W_q}{\mu_q} = \min_{\prec}\{\frac{W_h}{\mu_h} \mid \mu_h > 0\}$$

where $\min_{\prec}$ means the minimum in the lexicographical order over all row vectors $\frac{W_h}{\mu_h}$ for which $\mu_h > 0$. Clearly, $q \neq 1$. This leads to the following equation

$$\begin{aligned}
&\tfrac{1}{\mu_q} A_1 W_q + \textstyle\sum_{i\neq 1,q} A_i(W_i - \tfrac{\mu_i}{\mu_q} W_q) \\
&= \textstyle\sum_{i\neq 1} \tfrac{\mu_i}{\mu_q} A_i W_q + \sum_{i\neq 1,q} A_i W_i - \sum_{i\neq 1,q} \tfrac{\mu_i}{\mu_q} A_i W_q \\
&= \textstyle\sum_{i\neq 1} A_i W_i + \sum_{i\neq 1,q} (\tfrac{\mu_i}{\mu_q} A_i W_q - \tfrac{\mu_i}{\mu_q} A_i W_q) \\
&= B.
\end{aligned}$$

Note that $W_1 = \mathbf{0}$, $W_i \succ \mathbf{0}$ for $i \neq 1$. We have $\frac{W_q}{\mu_q} \succ \mathbf{0}$ and $W_i - \frac{\mu_i}{\mu_q} W_q \succ \mathbf{0}$ for all $i \neq q$. Let

$$\begin{aligned}
X_1 &= \tfrac{W_q}{\mu_q}, \\
X_q &= \mathbf{0}, \\
X_i &= W_i - \tfrac{\mu_i}{\mu_q} W_q, \quad (i \neq 1,\, q).
\end{aligned}$$

So $X$ is a basic feasible solution. By Lemma 4.2.6 the system (4.2.2) has exactly two basic feasible solutions. □

Note that the proof of the above theorem also tells us how to move from one basic feasible solution to another one.

**Corollary 4.2.8** *If all entries of the first row of $A$ are ones, then system* (4.2.2) *has a solution if and only if it has two basic feasible solutions.*

A natural way to use Theorem 4.2.7 is to replace $B$ in system (4.2.2) by the identity matrix $\mathbf{I}$ and to make all entries of the first row of $A$ be ones.

## 4.3. Merrill's algorithm

We wish to compute a fixed point of a u.s.c. point-to-set mapping $\psi$. Let us assume that $\psi$ satisfies the following condition (see Merrill [1972]).

**Definition 4.3.1 (Merrill's Condition)** *Let $\psi : \mathbb{R}^n \mapsto \mathbb{R}^n$ be a u.s.c. point-to-set mapping with nonempty, convex and compact values. There exist $w \in \mathbb{R}^n$, $\mu > 0$ and $\delta > 0$ such that for all $x \notin \bar{B}(w, \mu)$, $f(x) \in \psi(x)$ and $z \in \bar{B}(x, \delta)$,*

$$(f(x) - x)^\top (w - z) > 0.$$

In the end we will be able to find a fixed point of $\psi$ by Merrill's algorithm if $\psi$ satisfies Merrill's condition.

Let $\mathcal{T}$ be a triangulation of $\mathbb{R}^n \times [0,1]$ with restriction $\mathcal{T}^0 \subseteq \mathbb{R}^n \times \{0,1\}$. Such a triangulation is called *two-layered.* We first specify an element of $\mathbb{R}^n \times [0,1]$ by $y = (x_1, \cdots, x_n, t)^\top$ with $x \in \mathbb{R}^n$ and $t \in [0,1]$. Define a projection $p : \mathbf{R}^n \times [0,1] \mapsto \mathbf{R}^n$ by $p(x,t) = x$. Suppose that $\sigma =< y^1, y^n, \cdots, y^{n+2} >\in \mathcal{T}$. We define the diameter of the projection of $\sigma$ by

$$diam_p(\sigma) = \max\{||p(y^i) - p(y^j)|| \mid i, j \in I_{n+2}\}.$$

Moreover, the mesh size of the projection of $\mathcal{T}$ is defined by

$$mesh_p(\mathcal{T}) = \sup\{diam_p(\sigma) \mid \sigma \in \mathcal{T}\}.$$

For $i = 0$ or $1$, if $\sigma =< y^1, \cdots, y^{n+1} >\in \mathcal{T}^n$ and $\sigma \subseteq \mathbf{R}^n \times \{i\}$, then $\sigma_p =< p(y^1), \cdots, p(y^{n+1}) >$ is a simplex in $\mathbf{R}^n$. The collection of all such simplices $\sigma_p$ is denoted by $\mathcal{T}_i$.

Now we specify a two-layered triangulation for $\mathbf{R}^n \times [0,1]$. Recall the $K_1$-triangulation given in Chapter 1. Let $\tilde{K}_1$ denote the restriction of $K_1$ of $\mathbf{R}^{n+1}$ on $\mathbf{R}^n \times [0,1]$. Clearly, $\tilde{K}_1$ is two-layered. Furthermore, for any given $\epsilon > 0$, we define an $(n+1)\times(n+1)$ matrix by $P = [\epsilon e(1), \cdots, \epsilon e(n), e(n+1)]$, where $e(i)$ is the $i$-th unit vector of $\mathbf{R}^{n+1}$ for all $i$. Define

$$\tilde{K}_1(\epsilon) = \{< Py^1, \cdots, Py^{n+2} > \mid < y^1, \cdots, y^{n+2} >\in \tilde{K}_1\}.$$

Then $\tilde{K}_1(\epsilon)$ is also a two-layered triangulation of $\mathbf{R}^n \times [0,1]$. Moreover, it is true that

$$(\tilde{K}_1(\epsilon))_0 = (\tilde{K}_1(\epsilon))_1 = \epsilon K_1.$$

This leads to

$$mesh_p(\tilde{K}_1(\epsilon)) = \epsilon\sqrt{n}.$$

Given $\epsilon > 0$ and $\tilde{K}_1(\epsilon)$, we take an arbitrary point $(c,0)$ in $\mathbf{R}^n \times [0,1]$ such that $(c,0)$ lies in the interior of some simplex $\tau^0 \in (\tilde{K}_1(\epsilon))_0$. For ease of notation, in the sequel we simply use $\mathcal{T}$ to denote $\tilde{K}_1(\epsilon)$. Consider the following point-to-set homotopy mapping $\phi : \mathbf{R}^n \times [0,1] \mapsto \mathbf{R}^n$ defined by

$$\phi(x,t) = (1-t)\{c\} + t\psi(x).$$

A point $y$ is called a *fixed point* of $\phi$ if $p(y) \in \phi(y)$. Clearly, $(c,0)$ is the only fixed point of $\phi$ in $\mathbf{R}^n \times \{0\}$ and any fixed point $y$ of $\phi$ on $\mathbf{R}^n \times \{1\}$ projects a fixed point of $\psi$, i.e., $p(y) \in \psi(p(y))$. Merrill's algorithm will move from the point $(c,0)$ to a fixed point of $\psi$ if Merrill's condition is satisfied.

Define a labeling rule $l : \mathbf{R}^n \times [0,1] \mapsto \mathbf{R}^n$ by $l(y) = f(y) - p(y)$ where $f(y)$ is an element in $\phi(y)$. Given a simplex $\sigma =< y^1, \cdots, y^{n+1} >\in \mathcal{T}^n$, we call the following $(n+1) \times (n+1)$ matrix

$$L(\sigma) = \begin{bmatrix} 1 & \cdots & 1 \\ l(y^1) & \cdots & l(y^{n+1}) \end{bmatrix}$$

the *label matrix* of $\sigma$. $\sigma$ is said to be *complete* if the following system

$$L(\sigma)X = \mathbf{I}, \ X \succ \mathbf{0}$$

has a solution, where $\mathbf{I}$ is the identity matrix of rank $n+1$. Let $\mu$ be the first column of $X$ when $X$ exists. Clearly, $\sigma$ is completely labelled if it is complete. Further, if $\sigma$ is complete and lies on $\mathbb{R}^n \times \{1\}$, the point $y^* = \sum_{i=1}^{n+1} \mu_i y^i$ is a fixed point of the piecewise linear approximation of $\phi$ with respect to $\mathcal{T}$ and hence $p(y^*)$ is an approximate fixed point of $\psi$.

**Definition 4.3.2** *An $(n+1)$-simplex of $\mathcal{T}$ is a node if it has at least one complete facet. Two nodes are adjacent and connected by an arc or edge if they share a common complete facet. The degree of a node $\sigma$ is the number of nodes adjacent to $\sigma$, denoted by $deg(\sigma)$. Let $V$ be the set of nodes and let $A$ be the set of arcs. In this way we obtain a graph $G = (V, A)$.*

Recall that $\tau^0$ is the $n$-simplex lying on $\mathbb{R}^n \times \{0\}$ and containing $(c, 0)$ in its interior. Let $\sigma^0$ be the unique $(n+1)$-simplex of $\mathcal{T}$ having $\tau^0$ as its facet. Then we have

**Lemma 4.3.3** *The $n$-simplex $\tau^0$ is the only complete simplex on $\mathbb{R}^n \times \{0\}$.*

Proof: Note that $l(x, 0) = c - x$ and $(c, 0) \in int(\tau^0)$. Let $y^1, \cdots, y^{n+1}$ be the vertices of $\tau^0$. Note that $y^i = (x^i, 0)$ and $p(y^i) = x^i$ for all $i$. Then there is a unique vector $(b_1, \cdots, b_{n+1})$ such that $c = \sum_{i=1}^{n+1} b_i x^i$ and $\sum_{i=1}^{n+1} b_i = 1$, $b_i > 0$ for all $i \in I_{n+1}$. Since $x^1, \cdots, x^{n+1}$ are affinely independent, $c - x^1, \cdots, c - x^{n+1}$ must be also affinely independent. This implies that the matrix

$$L(\tau^0) = \begin{bmatrix} 1 & \cdots & 1 \\ c - x^1 & \cdots & c - x^{n+1} \end{bmatrix}$$

is non-singular. So $L(\tau^0)w = e(1)$ has a unique solution $w$. This solution $w$ is in fact $(b_1, \cdots, b_{n+1})^{\mathsf{T}}$. Since $L(\tau^0)$ is non-singular, $L(\tau^0)X = \mathbf{I}$ must also have a unique solution $W$. Notice that the first column of $W$ is $w$. Thus $W \succ \mathbf{0}$ and $\tau^0$ is complete by definition.

Now suppose that $\tau = < z^1, \cdots, z^{n+1} >$ is an arbitrary complete simplex of $\mathcal{T}^n$ lying on $\mathbb{R}^n \times \{0\}$. It follows that $L(\tau)X = \mathbf{I}$, $X \succ \mathbf{0}$ has a solution $X$. Let $(\mu_1, \cdots, \mu_{n+1})$ be the first column of $X$. Then we have

$$\sum_{i=1}^{n+1} \mu_i l(z^i) = \mathbf{0}, \ \sum_{i=1}^{n+1} \mu_i = 1, \ \mu_i \geq 0, \ i \in I_{n+1}.$$

By noticing that $z^1, \cdots, z^{n+1}$ are on $\mathbb{R}^n \times \{0\}$, we have

$$c = \sum_{i=1}^{n+1} \mu_i p(z^i).$$

So $(c,0) \in \tau$. However by hypothesis $c \in int(\tau^0)$. This implies that $\tau = \tau^0$. Hence $\tau^0$ is the only complete simplex on $\mathbb{R}^n \times \{0\}$. □

**Lemma 4.3.4** *Given the graph $G = (V, A)$ defined as above, for each node $\sigma$ of $G$, we have:*
*(i) if $\sigma$ has a complete facet lying on $\mathbb{R}^n \times \{0\}$ or $\mathbb{R}^n \times \{1\}$, then $deg(\sigma) = 1$;*
*(ii) in all other cases, $deg(\sigma) = 2$.*

Proof: (i) Let $\sigma = < y^1, \cdots, y^{n+1} >$ be the node with a complete facet $\tau = < y^1, \cdots, y^{n+1} >$ lying on $\mathbb{R}^n \times \{0\}$. Then the system

$$L(\tau)W = \mathbf{I},\ W \succ \mathbf{0}$$

has a solution $X$. Let

$$L(\sigma) = \begin{bmatrix} 1 & \cdots & 1 \\ l(y^1) & \cdots & l(y^{n+2}) \end{bmatrix}.$$

Consider the following system

$$L(\sigma)X = \mathbf{I},\ X \succeq \mathbf{0}. \tag{4.3.5}$$

Let $Y$ be the $(n+2) \times (n+1)$ matrix whose first $n+1$ rows are the first $n+1$ rows of $X$ and last row is $(0, \cdots, 0)$. Clearly, $Y$ is a basic feasible solution of system (4.3.5). Then by Corollary 4.2.8 we know that system (4.3.5) has exactly two basic feasible solutions. This means that $\sigma$ has exactly two complete facets. One is $\tau$ and the other is, say, $\rho$. By Theorem 1.4.6, there is exactly another $(n+1)$-simplex $\sigma^1$ of $\mathcal{T}$ having $\rho$ as its facet. Clearly $\sigma^1$ is a node and adjacent to $\sigma$. Thus $\deg(\sigma) = 1$.

(ii) Let $\sigma = < y^1, \cdots, y^{n+2} >$ be a node. Then $\sigma$ has a complete facet $\tau$. By Corollary 4.2.8 $\sigma$ has exactly two complete facets. One is $\tau$ and the other is, say, $\rho$. By assumption neither $\tau$ nor $\rho$ lies on $\mathbb{R}^n \times \{0\}$ or $\mathbb{R}^n \times \{1\}$. So by Theorem 1.4.6 each of $\tau$ and $\rho$ is a facet of exactly another $(n+1)$-simplex of $\mathcal{T}$. This leads to $deg(\sigma) = 2$. □

By using Theorem 1.4.14, Lemmas 4.3.3, 4.3.4, we have

**Theorem 4.3.5** *For the graph $G = (V, A)$ defined as above, each connected component of $G$ has one of the following five forms:*
*(i) a simple circuit;*
*(ii) a finite simple path whose two end nodes all have a complete facet lying on $\mathbb{R}^n \times \{1\}$;*
*(iii) an infinite simple path starting with an $(n+1)$-simplex which has a complete facet lying on $\mathbb{R}^n \times \{1\}$;*
*(iv) a finite simple path which starts with the $(n+1)$-simplex $\sigma^0$ and ends with another $(n+1)$-simplex having a complete facet on $\mathbb{R}^n \times \{1\}$;*
*(v) an infinite simple path which starts with the $(n+1)$-simplex $\sigma^0$.*

So in order to ensure the path beginning with $\sigma^0$ to be finite, we have to impose some conditions on $\psi$. Merrill's condition is a sufficient condition for this.

**Theorem 4.3.6** *If $mesh_p(\mathcal{T}) < \delta$ and Merrill's condition holds, the simple path beginning with $\sigma^0$ is finite.*

Proof: We will prove a stronger statement that there is only a finite number of nodes. Let $\eta = \delta + \max\{\mu, ||w - c||\}$. Define $D = \bar{B}(w, \eta) \times [0, 1]$. Clearly $D$ is a compact and convex subset of $\mathbb{R}^n \times [0, 1]$. This means that there is only a finite number of simplices in $D$. We will show that no nodes lie outside of $D$. Let $\tau$ be an $n$-simplex having a vertex $(z, t)$ outside $D$. That is, $||z - w|| > \eta$. Let $s = w - z$. Let $y$ be any vertex of $\tau$. We have to consider the following two cases.

Case (1). If $y = (x, 1)$, we have $l(y) = f(x) - x$ with $f(x) \in \psi(x)$. Since $z \notin \bar{B}(w, \mu + \delta)$ and $z \in \bar{B}(x, \delta)$, we have $x \notin \bar{B}(w, \mu)$. It follows from Merrill's condition that

$$s^\top l(y) = (w - z)^\top (f(x) - x) > 0.$$

Case (2). If $y = (x, 0)$, we have $l(y) = c - x$. Since $mesh_p(\mathcal{T}) < \delta$ and $||z - x|| < \delta$, we have

$$\begin{aligned} s^\top l(y) &= (w - z)^\top (c - x) \\ &= (w - z)^\top (w - z) + (w - z)^\top (c - w) + (w - z)^\top (z - x) \\ &> \eta^2 - \eta(||c - w|| + ||z - x||) \\ &> \eta^2 - \eta(||c - w|| + \delta) \\ &\geq \eta^2 - \eta^2 = 0. \end{aligned}$$

It follows from the above discussions that

$$s^\top l(y^i) > 0 \text{ for all } i \in I_{n+1}. \tag{4.3.6}$$

Now suppose to the contrary that $\tau$ is complete, then the system

$$L(\tau)W = \mathbf{0}, W \succ \mathbf{0}$$

has a solution $W$. Let $(\mu_1, \cdots, \mu_{n+1})^\top$ be the first column of $W$. We have

$$\sum_{i=1}^{n+1} \mu_i l(y^i) = \mathbf{0}, \ \sum_{i=1}^{n+1} \mu_i = 1, \ \mu_i \geq 0, \ i = 1, \cdots, n+1.$$

This implies that

$$\sum_{i=1}^{n+1} \mu_i s^\top l(y^i) = 0.$$

On the other hand it follows from ( 4.3.6) that

$$\sum_{i=1}^{n+1} \mu_i s^\top l(y^i) > 0.$$

This is a contradiction. □

Beginning with $\sigma^0$, Merrill's algorithm traces the unique finite simple path of adjacent simplices (i.e., nodes) and ends up with a node having a complete facet $\tau$ on $\mathbb{R}^n \times \{1\}$ provided that Merrill's condition is satisfied and the mesh size is small enough. $\tau$ yields an approximate fixed point of $\psi$. Now we give the formal steps of this method.

Merrill's Algorithm on $\mathbb{R}^n$

**Step (0)** Set $mesh_p(\mathcal{T}) < \delta$. Let $\tau^0$ be the unique $n$-simplex of $\mathcal{T}^n$ containing $(c, 0)$ in its interior. Let $\sigma^0$ be the unique $(n+1)$-simplex in $\mathcal{T}$ which has $\tau^0$ as its facet. Let $y^+$ be the vertex of $\sigma^0$ that is not a vertex of $\tau^0$. Set $k = 0$.

**Step (1)** Compute $W = L^{-1}(\tau^k)$. Let $W_i$ denote the $i$-th row of $W$. Compute $l(y^+)$ and let $q = (1, l^\top(y^+))^\top$. Let $v = (v_1, \cdots, v_{n+1})^\top = Mq$. Compute

$$\frac{W_s}{v_s} = \min_{\prec}\{\frac{W_h}{v_h} \mid v_h > 0\}.$$

Notice that $s$ is unique by Theorem 4.2.7. The column $(1, l^\top(y^s))^\top$ is eliminated. Let $\tau^{k+1}$ be the facet of $\sigma^k$ opposite $y^s$. If $\tau^{k+1}$ lies on $\mathbb{R}^n \times \{1\}$, this facet yields an approximate fixed point of $\psi$ and stop. (For example, let $z^1, \cdots, z^{n+1}$ be the vertices of $\tau^{k+1}$. We can take the point $z^* = \sum_{i=1}^{n+1} \frac{z^i}{n+1}$ as an approximate fixed point of $\psi$). If $\tau^{k+1}$ does not lie on $\mathbb{R}^n \times \{1\}$, go to Step (2).

**Step (2)** Find a simplex $\sigma^{k+1}$ sharing the facet $\tau^{k+1}$ with $\sigma^k$, and let $y^+$ be the vertex of $\sigma^{k+1}$ not being a vertex of $\tau^{k+1}$. Set $k = k+1$ and return to Step (1).

One can easily prove that if Merrill's condition is satisfied, an approximate fixed point of any desired accuracy can be found within a finite number of steps. Moreover, one can also immediately give a constructive proof for the following existence theorem of Merrill [1972].

**Theorem 4.3.7** *If a point-to-set mapping $\psi : \mathbb{R}^n \mapsto \mathbb{R}^n$ satisfies Merrill's condition, there exists at least one fixed point $x^* \in \mathbb{R}^n$ for $\psi$, i.e., $x^* \in \psi(x^*)$.*

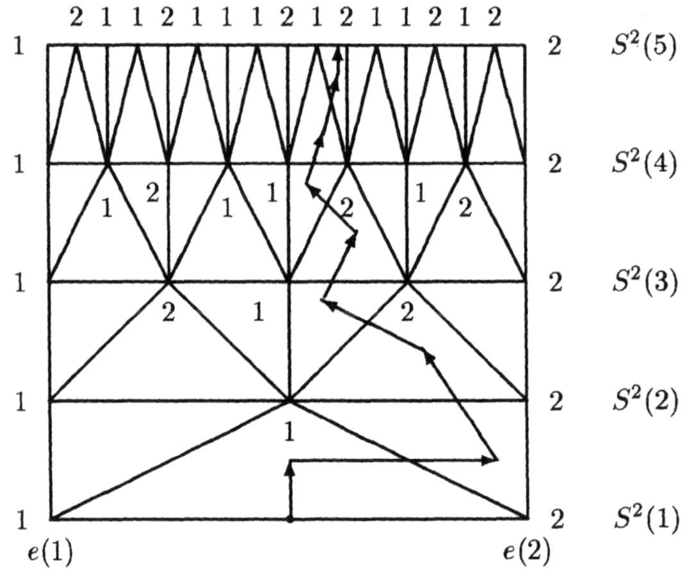


*Figure 4.1.* Illustration of the homotopy algorithm for $S^n$.

## 4.4. Eaves' algorithm on $S^n \times [1, \infty)$

In this and coming sections we present the homotopy algorithms of Eaves [1972] and Eaves and Saigal [1972] on $S^n \times [1, \infty)$ and $\mathbb{R}^n \times [1, \infty)$, respectively. Let us first give a simple example to motivate the reader. Consider the problem of finding a fixed point of a continuous function $f : S^2 \mapsto S^2$. We imbed $S^2$ in the set $S^2 \times [1, \infty)$ and triangulate each set $S^2 \times \{k\}$ with mesh size of $2^{-k+1}$ for $k \in \mathbb{N}$. Label $l(e(1), 1) = 1$ and $l(e(2), 1) = 2$. For $(x, k) \in S^2 \times [2, \infty)$, let $l(x, k) = l^2(x)$ where $l^2$ is the standard labeling rule given in the previous chapter. We can illustrate this idea in Figure 4.1. Clearly, as $k$ goes to infinity, the sequence of complete labelled simplices traced by the algorithm converges to a fixed point of $f$ since the mesh size of simplices on $S^2 \times \{k\}$ converges to zero. This is the basic idea underlying simplicial homotopy algorithm. Compared with Merrill's restart algorithm, this is indeed a natural homotopy method since the function is continuously deformed, whereas in Merrill's restart algorithm the deformation is periodically changing (i.e., the parameter $c$ is changing) for each restart.

Let $D$ be an $m$-dimensional convex subset of $\mathbb{R}^n$ and let $\delta_1 > \delta_2 > \cdots$ be a sequence of positive numbers with limit zero. For each $k \in \mathbb{N}$, let $D(k)$ denote the set $D \times \{k\}$. It is convenient to say that $D(k)$ is level $k$ of the set $D \times [1, \infty)$. A triangulation $\mathcal{T}$ of the set $D \times [1, \infty)$ with $(m+1)$-simplices is called a *continuously refining (CR) triangulation* of $D \times [1, \infty)$ if it satisfies

(a) $\mathcal{T}^0 \subseteq D \times \mathbb{N}$;

(b) $\mathcal{T}(k) = \{\tau \in \mathcal{T}^m \mid \tau \subseteq D(k)\}$ is a triangulation of $D(k)$ for each $k \in \mathbb{N}$;

(c) for each $k \in \mathbb{N}$, the mesh size of $\mathcal{T}(k)$ is equal to $\delta_k$.

Such triangulation was first introduced by Eaves [1972] with a restriction that

$$\frac{\delta_{k+1}}{\delta_k} = \frac{1}{2}.$$

Now we are going to describe Eaves' algorithm for computing a fixed point on the unit simplex $S^n$. Given a continuous function $f : S^n \mapsto S^n$, we want to find a fixed point of $f$. We first construct a CR triangulation $\mathcal{T}$ of $S^n \times [1,\infty)$ such that $\mathcal{T}(1)$ consists of exactly one $(n-1)$-simplex, i.e., $S^n(1)$. Then define a labeling rule $l : \mathcal{T}^0 \mapsto I_n$ as follows. For each $i \in I_n$, $l((e(i),1)) = i$. For $k > 1$,

$$l((x,k)) = \min\{j \in I_n \mid f_j(x) - x_j = \min_h (f_h(x) - x_h) \text{ and } x_j > 0\}.$$

Since $f$ is a continuous function and the labeling $l$ satisfies the hypothesis of Sperner lemma, there must exist a completely labeled simplex $\sigma_k$ in each level $S^n(k)$ and a c.l.simplex yields an approximate fixed point of $f$. Clearly, as $k$ increases, the accuracy of the approximation is also improved.

**Definition 4.4.1** *An $(n-1)$-simplex $\tau$ of $\mathcal{T}$ is completely labelled (c.l.) if the $n$ vertices of $\tau$ carry all labels from 1 up to $n$. An $n$-simplex of $\mathcal{T}$ is a node if it has at least one c.l. facet. Two nodes are adjacent and connected by an arc or edge if they share a common complete facet. The degree of a node $\sigma$ is the number of nodes adjacent to $\sigma$, denoted by $deg(\sigma)$. Let $V$ be the set of nodes and let $A$ be the set of arcs. In this way we obtain a graph $G = (V, A)$.*

Let $\sigma^1$ be the unique $n$-simplex of $\mathcal{T}$ having the c.l.simplex $S^n(1)$ as its facet. Clearly, $\sigma^1$ is a node. One can easily prove

**Lemma 4.4.2** *Let $G = (V, A)$ be the graph defined above and let $v$ be a node of $G$. Then,*
*(i) $deg(\sigma_1) = 1$;*
*(ii) In all other cases, $deg(v) = 2$.*

**Theorem 4.4.3** *For the graph $G = (V, A)$ defined as above, each connected component of $G$ has one of the following two forms:*
*(i) a simple circuit;*
*(ii) an infinite simple path which starts with the $n$-simplex $\sigma^1$.*

**Theorem 4.4.4** *For any $L \in \mathbb{N}$, the simple path starting with $\sigma^1$ will lead to a c.l. $(n-1)$-simplex of $\mathcal{T}(L)$ within a finite number of steps.*

Proof: Consider the set $M = S^n \times [1, L]$. Due to the properness of the labeling rule, there is no c.l. simplex of $\mathcal{T}^{n-1}$ lying on $bd(S^n) \times [1, L]$. Because $M$ is compact, the number of nodes is finite. □

Now we can give the formal steps of Eaves' algorithm.

The Homotopy Algorithm of Eaves

**Step (0)** Set $\tau^0 = S^n(1)$. Let $y^+$ be the vertex of $\sigma^1$ not being a vertex of $\tau^0$. Set $t = 1$.

**Step (1)** Compute $l(y^+)$. Let $y^-$ be the vertex of $\tau^{t-1}$ with label $l(y^+)$.

**Step (2)** Let $\tau^t$ be the facet of $\sigma^t$ opposite $y^-$. If $\tau^t$ is an element of $\mathcal{T}(k)$ for some $k$ and $k$ is sufficiently big, then $\tau^t$ gives a satisfactory approximation solution and stop. Otherwise, let $\sigma^{t+1}$ be a simplex of $\mathcal{T}$ sharing the facet $\tau^t$ with $\sigma^t$, and let $y^+$ be the vertex of $\sigma^{t+1}$ not being a vertex of $\tau^t$. Set $t = t + 1$. Return to Step (1).

Now we are ready to present a concrete $CR$ triangulation of $S^n \times [1, \infty)$. Instead of giving the original triangulation of Eaves [1972], we present a more general and sophisticated triangulation first introduced by van der Laan and Talman [1980] and Shamir [1980]. It has a striking property: namely, the mesh size ratio $\delta_{k+1}/\delta_k$ between any two successive levels $D(k+1)$ and $D(k)$ can be made arbitrary. This is in contrast to all $CR$ triangulations existing before 1980 which have only one possible ratio, namely, $\frac{1}{2}$. This new CR triangulation makes it possible for homotopy algorithms to obtain quadratic convergence. See Todd [1976a, p. 101] and van der Laan [1980b, pp. 157-161] for detailed explanation. Here we follow the approach of van der Laan and Talman [1980] and van der Laan [1980b]. To triangulate $S^n \times [1, \infty)$, we choose an arbitrary sequence of increasing positive integers $d_1, d_2, \cdots$, such that $d_1 = 1$, and $d_{m+1}$ is a multiple $k_m$ of $d_m$ for all $m \in \mathbb{N}$. We will associate the mesh size of the triangulation of each level $S^n(m)$ with $d_m^{-1}$. It is sufficient to describe only the triangulation of $S^n \times [m, m+1]$. By combining the triangulations of $S^n \times [m, m+1]$ for all pairs $[m, m+1]$, we obtain the triangulation of $S^n \times [1, \infty)$. For each $m \in \mathbb{N}$, let $G(m)$ be the $K_2(d_m)$-triangulation of $S^n$ with mesh size $1/d_m$. Then $G(m) \times \{m\}$ is a triangulation $\mathcal{T}(m)$ of $S^n(m)$. Recall that $K_2(d_m)$ is the collection of all $(n-1)$-simplices $\sigma(y^1, \pi(I_n))$ with vertices $y^1, \cdots, y^n$ in $S^n$ such that

(a) each component of $y^1$ is a non-negative multiple of $d_m^{-1}$;

(b) $\pi(I_n) = (\pi(1), \cdots, \pi(n))$ is a permutation of the elements in $I_n$;

(c) $y^{i+1} = y^i + q(\pi(i))/d_m$, $i = 1, \cdots, n-1$

where $q(j) = e(j+1) - e(j)$, $j \in I_{n-1}$, and $q(n) = e(1) - e(n)$. Here $e(i)$ is the $i$-th unit vector of $\mathbf{R}^n$, $i \in I_n$. Clearly, $y^1 = y^n + q(\pi(n))/d_m$. In such a way, each simplex of $G(m)$ may have $n$ representations, since each vertex of the simplex can be chosen as $y^1$. From now on in the rest of the section we will represent each simplex in a unique way as follows. For a given vertex $y$, let $\alpha(y) = (\alpha_1(y), \cdots, \alpha_{n-1}(y))$ be the vector with components

$$\alpha_h(y) = (1 - \sum_{l=1}^{h} y_h)d_m, \ h \in I_{n-1}.$$

Obviously, each $\alpha_h(y)$ is an integer. Now define a function $\chi : G^0(m) \mapsto I_n$ by

$$\chi(y) = 1 + (\sum_{h=1}^{n-1} \alpha_h(y)) \bmod n.$$

**Lemma 4.4.5** *Let $\sigma(y^1, \pi(I_n))$ be an $(n-1)$-simplex of $G(m)$. Then for each $i \in I_n$, there exists an index $j_i \in I_n$ such that $\chi(y^{j_i}) = i$. Furthermore, $\{\chi(y^i) \mid i \in I_n\} = I_n$.*

Proof: It suffices to show that for each $j \in I_{n-1}$,

$$\chi(y^{j+1}) = \chi(y^j) \bmod n + 1. \tag{4.4.7}$$

For some $j \in I_n$, let $\chi(y^j) = i$. In case $\pi(j) \neq n$, then $y_h^{j+1} = y_h^j$ for $h \neq \pi(j), \pi(j)+1$, $y_h^{j+1} = y_h^j - 1/d_m$ for $h = \pi(j)$, and $y_h^{j+1} = y_h^j + 1/d_m$ for $h = \pi(j)+1$. Hence $\alpha_h(y^{j+1}) = \alpha_h(y^j)$ for $h \neq \pi(j)$ and $\alpha_h(y^{j+1}) = \alpha_h(y^j) + 1$ for $h = \pi(j)$. This leads to $\sum_{h=1}^{n-1} \alpha_h(y^{j+1}) = 1 + \sum_{h=1}^{n-1} \alpha_h(y^j)$. So (4.4.7) holds. In case $\pi(j) = n$, then $y_h^{j+1} = y_h^j$ for $h = 1, n$, $y_n^{j+1} = y_n^j - 1/d_m$, and $y_1^{j+1} = y_1^j + 1/d_m$. Hence $\alpha_h(y^{j+1}) = \alpha_h(y^j) - 1$ for $h \in I_{n-1}$ and $\sum_{h=1}^{n-1} \alpha_h(y^{j+1}) = \sum_{h=1}^{n-1} \alpha_h(y^j) + 1 - n$. So again (4.4.7) holds. □

It follows from this lemma that if $\chi(y^1) = 1$, then $\chi(y^j) = j$ for all $j \in I_n$. In the following we will always assume that any simplex of $G(m)$ will be represented in such a way that $\chi(y^1) = 1$. We call it $\chi$-representation of a simplex. Suppose that $\sigma^1(y^1, \pi(I_n))$ is a $\chi$-representation. Let $\sigma^2(\bar{y}^1, \rho(I_n))$ be the adjacent simplex obtained from $\sigma^1$ by replacing the vertex $y^i$ of $\sigma$. Then $\sigma^2$ is also a $\chi$-representation if $y^i$ is replaced according to Table 4.1.

To triangulate $S^n \times [m, m+1]$ for any $m \in \mathbf{N}$, we take a particular simplex $\sigma^m(u^1, \beta(I_n))$ from $G(m)$ and call $\sigma^m$ the *starting simplex* on level $S^n(m)$. (If $m = 1$, clearly $\sigma^m = S^n$.) Then we choose non-negative integers $\lambda_1^m, \cdots, \lambda_n^m$ such that $\sum_{h=1}^{n} \lambda_h^m = k_m = d_{m+1}/d_m$. The point

$$c(\sigma^m) = \sum_{h=1}^{n} \lambda_h^m u^i / k_m$$

| | $y^1$ becomes | $\pi(I_n)$ becomes |
|---|---|---|
| $i=1$ | $y^1+q(\pi(1))/d_m$ | $(\pi(2),\cdots,\pi(n),\pi(1))$ |
| $1<i\leq n$ | $y^1$ | $(\pi(1),\cdots,\pi(i-2),\pi(i),\pi(i-1),\pi(i+1),\cdots,\pi(n))$ |

TABLE 4.1. $i$ is the index of the vertex to be replaced.

is called the *central point* of $\sigma^m$. It will appear that the parameters $\lambda_1^m,\cdots,\lambda_n^m$ determine the structure of the triangulation of $S^n\times[m,m+1]$.

Now let $\tau(y^1,\pi(I_n))$ be an arbitrary simplex of $G(m)$. The point

$$c(\tau)=\sum_{h=1}^{n}\lambda_h^m y^i/k_m$$

is called the *central point* of $\tau$. The triangulation of $S^n\times[m,m+1]$ will be such that the $n$-simplex which is the convex hull of $\tau\times\{m\}$ and $c(\tau)\times\{m+1\}$, is a simplex of this triangulation. In particular, the $n$-simplex which is the convex hull of $\sigma^m\times\{m\}$ and $c(\sigma^m)\times\{m+1\}$, is also a simplex of this triangulation, since $c(\tau)=c(\sigma^m)$ and $\tau(y^1,\pi(I_n))=\sigma(u^1,\beta(I_n))$. To triangulate $S^n\times[m,m+1]$ we first triangulate $\tau(y^1,\pi(I_n))\times[m,m+1]$ and then demonstrate that the union of the triangulations over all $(n-1)$-simplices $\tau(y^1,\pi(I_n))$ of $G(m)$ is a triangulation of $S^n\times[m,m+1]$. To triangulate the set $\tau(y^1,\pi(I_n))\times[m,m+1]$ we define for any proper subset $T$ of $I_n$ the set of vertices $A^\tau(T)$ of the triangulation $G(m+1)$ in $\tau(y^1,\pi(I_n))$ by

$$A^\tau(T)=\{y\in\tau \mid y=c(\tau)+\textstyle\sum_{h\in T}\mu_h q(\pi(h))/d_{m+1},\ \text{for positive integers } \mu_h, h\in T\}.$$

Note that $A^\tau(\emptyset)=\{c(\tau)\}$. Now a triangulation of $\tau\times[m,m+1]$ is obtained by connecting all the vertices $x$ of $G(m+1)$ in $A^\tau(T)$ on level $S^n(m+1)$ with the vertices $y^i$, $i\in I_n\setminus T$, on level $S^n(m)$. In particular, the convex hull of $c(\tau)\times\{m+1\}$ and $y^i\times\{m\}$ for all $i\in I_n$, is a simplex of the triangulation. Now we are ready to prove the following theorem of van der Laan and Talman [1980].

**Theorem 4.4.6** *The union of the triangulations of $\tau(y^1,\pi(I_n))\times[m,m+1]$ over all $(n-1)$-simplices $\tau(y^1,\pi(I_n))$ of $G(m)$ is a triangulation of $S^n\times[m,m+1]$.*

Proof: It is clear that the triangulation of $\tau\times[m,m+1]$ is well-defined for each simplex $\tau$ of $G(m)$. Let $\tau^1(y^1,\beta(I_n))$, $\tau^2(z^1,\gamma(I_n))$ be two adjacent $(n-1)$-simplices of $G(m)$ and $x$ a vertex of $G(m+1)$ in $\tau^1\cap\tau^2$. Then it is sufficient to prove that, if in the triangulation of $\tau^1 \times [m, m+1]$, $x$ is connected with a vertex $y$ of $G(m)$ in $\tau^1 \cap \tau^2$, $x$ is also connected with $y$ in the triangulation of $\tau^2 \times [m, m+1]$. Since $\tau^1$ and $\tau^2$ are adjacent, there exists a unique index $j$ such that $y^h = z^h$ for $h \neq j$ and $\beta(h) = \gamma(h)$ for $h \neq j-1$, $j$, with the convention that $j-1 = n$ if $j = 1$. Furthermore, $\beta(j-1) = \gamma(j)$ and $\beta(j) = \gamma(j-1)$. So all common vertices of $\tau^1$ and $\tau^2$ have the same index, namely, if for some $i$, $v = y^i$ we have $v = z^i$ for the same index $i$.

So we only need to show that $x \in A^{\tau^1}(T)$ implies $x \in A^{\tau^2}(T)$ for a vertex $x$ of $G(m+1)$ in the common facet of $\tau^1$ and $\tau^2$. Since $x$ is a vertex of $G(m+1)$ in the facet of $\tau^1$ opposite to $y^j$, there exist unique integers $\theta_h$ for $h \neq j-1$, $j$ such that

$$x = y^{j+1} + \sum_{h \neq j-1,j} \theta_h q(\beta(h))/d_{m+1} \tag{4.4.8}$$

with the convention that $y^{j+1} = y^1$ if $j = n$. If $x \in A^{\tau^1}(T)$ there also exist unique integers $\mu_h$ with $\mu_h > 0$ for $h \in T$ and $\mu_h = 0$ for $h \notin T$, such that

$$x = c(\tau^1) + \sum_{h=1}^{n} \mu_h q(\beta(h))/d_{m+1}. \tag{4.4.9}$$

Equivalently,

$$\begin{aligned}
x &= \textstyle\sum_{h=1}^n \lambda_h^m y^h/k_m + \mu_{j-1} q(\beta(j-1))/d_{m+1} + \mu_j q(\beta(j))/d_{m+1} \\
&\quad + \textstyle\sum_{h \neq j-1,j} \mu_h q(\beta(h))/d_{m+1} \\
&= \textstyle\sum_{h=1}^j \lambda_h^m \{y^{j+1} + \sum_{s=j+1}^n q(\beta(s))/d_m + \sum_{s=1}^{h-1} q(\beta(s))/d_m\}/k_m \\
&\quad + \textstyle\sum_{h=j+1}^n \lambda_h^m \{y^{j+1} + \sum_{s=j+1}^{h-1} q(\beta(s))/d_m\}/k_m + \mu_{j-1} q(\beta(j-1))/d_{m+1} \\
&\quad + \mu_j q(\beta(j))/d_{m+1} + \textstyle\sum_{h \neq j-1,j} \mu_h q(\beta(h))/d_{m+1} \\
&= y^{j+1} + (\textstyle\sum_{h=1}^j \lambda_h^m)(\sum_{s=j+1}^n q(\beta(s))/d_{m+1}) + \sum_{s=1}^{j-2} \lambda_j^m q(\beta(s))/d_{m+1} \\
&\quad + \lambda_j^m q(\beta(j-1))/d_{m+1} + \textstyle\sum_{h=1}^{j-1} \sum_{s=1}^{h-1} \lambda_h^m q(\beta(s))/d_{m+1} \\
&\quad + \textstyle\sum_{h=j+1}^n \sum_{s=j+1}^{h-1} \lambda_h^m q(\beta(s))/d_{m+1} + \mu_{j-1} q(\beta(j-1))/d_{m+1} \\
&\quad + \mu_j (-\textstyle\sum_{s \neq j} q(\beta(s))/d_{m+1} + \sum_{h \neq j-1,j} \mu_h q(\beta(h))/d_{m+1}.
\end{aligned}$$

Defining $\bar{\theta}_i$ to be the coefficient of $q(\beta(i))$ for $i \neq j-1$, $j$, we obtain that

$$x = y^{j+1} + \sum_{i \neq j-1,j} \bar{\theta}_i q(\beta(i))/d_{m+1} + (\lambda_j^m + \mu_{j-1} - \mu_j) q(\beta(j-1))/d_{m+1}. \tag{4.4.10}$$

Comparing (4.4.10) with (4.4.8), we have that $\bar{\theta}_i = \theta_i$ for $i \neq j-1$, $j$ and

$$\lambda_j^m + \mu_{j-1} - \mu_j = 0. \tag{4.4.11}$$

On the other hand it follows from (4.4.9) that

$$\begin{aligned}
x &= \textstyle\sum_{h \neq j} \lambda_h^m y^h/k_m + \lambda_j^m (y^{j-1} + q(\beta(j-1))/d_m)/k_m \\
&\quad + \mu_{j-1} q(\beta(j-1))/d_{m+1} + \mu_j q(\beta(j))/d_{m+1} + \textstyle\sum_{h \neq j-1,j} \mu_h q(\beta(h))/d_{m+1}.
\end{aligned}$$

Therefore, because $z^i = y^i$ for $i \neq j$, $\gamma(j-1) = \beta(j)$ and $\gamma(j) = \beta(j-1)$, we obtain that

$$\begin{aligned} x &= \textstyle\sum_{h \neq j} \lambda_h^m z^h / k_m + \lambda_j^m (z^{j-1} + q(\gamma(j))/d_m)/k_m \\ &\quad + \mu_{j-1} q(\gamma(j))/d_{m+1} + \mu_j q(\gamma(j-1))/d_{m+1} + \textstyle\sum_{h \neq j-1, j} \mu_h q(\gamma(h))/d_{m+1} \\ &\quad + \lambda_j^m q(\gamma(j-1))/d_{m+1} - \lambda_j^m q(\gamma(j-1))/d_{m+1}. \end{aligned}$$

From (4.4.11) it follows that $\mu_j - \lambda_j^m = \mu_{j-1}$ and $\mu_{j-1} + \lambda_j^m = \mu_j$. Consequently,

$$x = c(\tau^2) + \sum_{h=1}^{n} \mu_h q(\gamma(h))/d_{m+1}$$

with $\mu_h$ the same coefficients as in (4.4.9). Hence $x \in A^{\tau^2}(T)$. □

Following from the above proof, we have

**Corollary 4.4.7** *If a vertex $x$ of $G(m+1)$ belongs to two adjacent simplices $\tau^1(y^1, \beta(I_n))$ and $\tau^2(z^1, \gamma(I_n))$ of $G(m)$ and for some $T \subseteq I_n$,*

$$x = c(\tau^1) + \sum_{h \in T} \mu_h q(\beta(h))/d_{m+1},$$

*then for the same $\mu_h$*

$$x = c(\tau^2) + \sum_{h \in T} \mu_h q(\gamma(h))/d_{m+1}.$$

Now we illustrate the above theorem by an example. Take $n = 3$, $d_1 = 1$, $d_2 = 3$, $d_3 = 12$, $\lambda_1^1 = 1$, $\lambda_2^1 = 1$, $\lambda_3^1 = 1$, $\lambda_1^2 = 1$, $\lambda_2^2 = 2$ and $\lambda_3^2 = 1$. So, $k_1 = 3$ and $k_2 = 4$. In the Figure 4.2 the triangulations $G(1)$ and $G(2)$ of $S^3$ are given, where $\sigma^1 = S^3$, $c(\sigma^1) = (1/3, 1/3, 1/3)^\top$. The vertices of $G(2)$ in $A^{\sigma^1}(T)$ are connected with $e(i)$ for $i \notin T$. In the Figure 4.3 the triangulations $G(2)$ and $G(3)$ are depicted, where $\sigma^2 = \sigma(u^1, \beta(I_3))$ with $u^1 = (1/3, 1/3, 1/3)^\top$ and $\beta(I_3) = (1,3,2)$, and $c(\sigma^2) = (2/12, 7/12, 3/12)^\top$. Notice that the vertices of $G(2)$ are drawn heavily. For any simplex $\tau$ of $G(2)$ the areas $A^\tau(T)$ are indicated by the heavy lines. Observe that those heavy lines are linked in a consistent way. Take for example, $\tau^1 = \tau(y^1, \beta)$ with $y^1 = (1/3, 1/3, 1/3)^\top$, $\beta = (2,1,3)$, and $\tau^2 = \tau(z^1, \gamma)$ with $z^1 = (0,0,1)^\top$, $\gamma = (3,1,2)$. Clearly, $\tau^1$ and $\tau^2$ are adjacent. The region $A^{\tau^1}(\{1,2\}) = A_1$ adjoins to the region $A^{\tau^2}(\{1,2\}) = B_1$ in harmony. It is also true for $A^{\tau^1}(\{1\})$ and $A^{\tau^2}(\{1\})$.

## 4.5. Eaves-Saigal's algorithm on $\mathbb{R}^n \times [1, \infty)$

In this section we describe the algorithm of Eaves and Saigal [1972]. Instead of presenting their original $CR$ triangulation of $\mathbb{R}^n \times [1, \infty)$, we give a

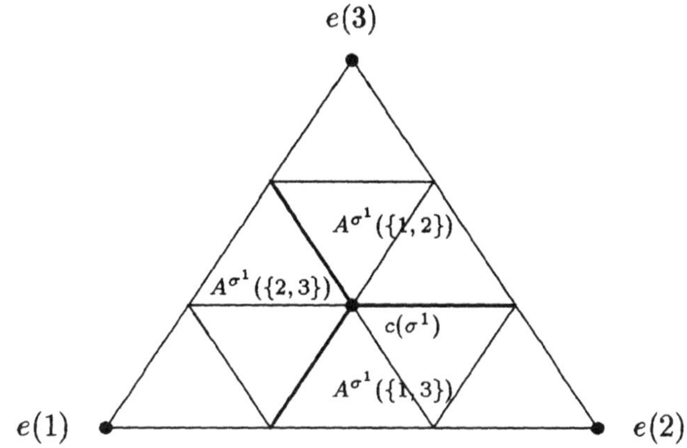


*Figure 4.2.* Illustration of $G(1)$ and $G(2)$ of $S^3$.

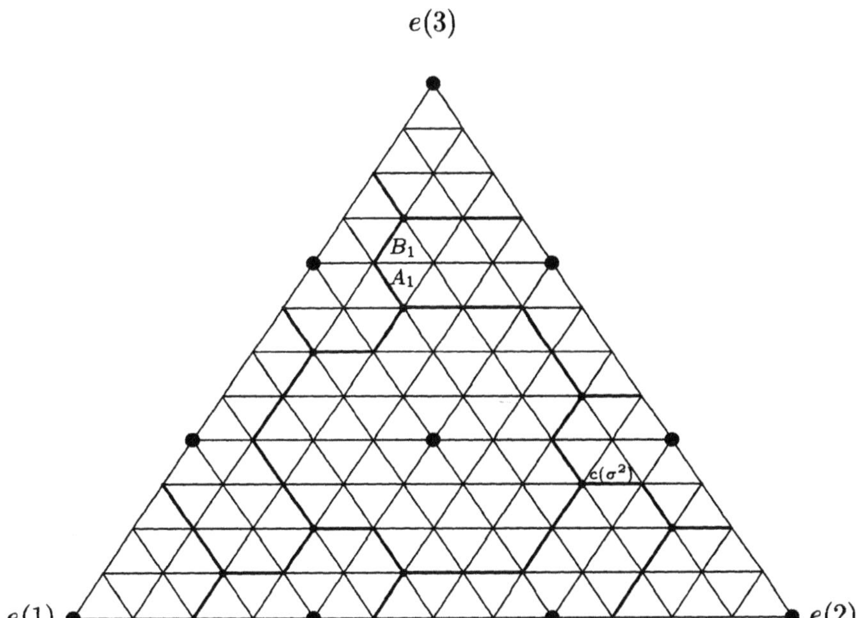


*Figure 4.3.* Illustration of $G(2)$ and $G(3)$ of $S^3$.

generalization of the $CR$ triangulation introduced in the previous section. Let $d_1, d_2, \cdots$, be a sequence of increasing positive integers such that $d_{i+1} = k_i d_i$, $i \in \mathbb{N}$, with $k_i$ an arbitrary integer bigger than one. On each level

$\mathbb{R}^n(m)$, $\mathbb{R}^n$ is triangulated by the $K_1(d_m)$-triangulation. Recall that the $K_1(d_m)$-triangulation of $\mathbb{R}^n$ with grid size $1/d_m$ is the collection of all $n$-simplices $\sigma(y^1, \pi(I_{n+1}))$ with vertices $y^1, \cdots, y^{n+1}$ in $\mathbb{R}^n$ such that

(a) each component of $y^1$ is a multiple of $1/d_m$;

(b) $\pi(I_{n+1}) = (\pi(1), \cdots, \pi(n+1))$ is a permutation of the elements in $I_{n+1}$;

(c) $y^{i+1} = y^i + u(\pi(i))/d_m$, $i \in I_n$

where $u(i) = e(i)$, $i \in I_n$, and $q(n+1) = -\sum_{h=1}^{n} e(i)$. As in Section 4.4 we can describe each $n$-simplex in the $\chi$-representation. Given any vertex $y$ of $K_1(d_m)$, we define

$$\alpha(y) = d_m y,$$

and

$$\chi(y) = 1 + (\sum_{h=1}^{n} \alpha_h(y)) \bmod (n+1).$$

Now the triangulation of $\mathbb{R}^n \times [m, m+1]$ for any $m \in \mathbb{N}$, is constructed in the same way as described in Section 4.4 for $S^n \times [m, m+1]$. So, let $\sigma^m(y^1, \pi(I_{n+1}))$ be the starting simplex of $K_1(d_m)$ and let $\lambda_1^m, \cdots, \lambda_{n+1}^m$ be arbitrarily chosen non-negative integers such that $\sum_{h=1}^{n+1} \lambda_h^m = k_m$. The point

$$c(\sigma^m) = \sum_{h=1}^{n+1} \lambda_h^m y^h / k_m$$

is the central point of $\sigma^m$. To subdivide the set $\tau \times [m, m+1]$ for each $\chi$-represented simplex $\tau(v^1, \beta(I_{n+1})$ of $K_1(d_m)$, we again define the set $A^\tau(T)$, with $T$ a proper subset of $I_{n+1}$, and connect all vertices $x$ of $K_1(d_{m+1})$ in $A^\tau(T)$ on level $\mathbb{R}^n(m+1)$ with the vertices $v^i$ of $\tau$ on level $\mathbb{R}^n(m)$. The union of the triangulations of $\tau \times [m, m+1]$ for all $n$-simplices $\tau$ of $K_1(d_m)$, is a simplicial subdivision of $\mathbb{R}^n \times [m, m+1]$. Combining these triangulations for all pairs $[m, m+1]$ with $m \in \mathbb{N}$, we immediately obtain a $CR$-triangulation of $\mathbb{R}^n \times [1, \infty)$. We denote this triangulation by $\mathcal{T}$.

Suppose that we want to compute a fixed point of a point-to-set mapping $\psi : \mathbb{R}^n \mapsto \mathbb{R}^n$. Let $(c, 1)$ be a point contained in the interior of an $n$-simplex $\tau^1$ on level $\mathbb{R}^n(1)$. Define a vector labeling rule $l : \mathbb{R}^n \times [1, \infty) \mapsto \mathbb{R}^n$ by

$$l(y) = \begin{cases} c - x & \text{if } y = (x, 1) \\ f(x) - x & \text{if } y = (x, t) \end{cases}$$

where $f(x) \in \psi(x)$ and $t > 1$. Given a simplex $\tau = \langle y^1, \cdots, y^{n+1} \rangle \in \mathcal{T}^n$, we call the following $(n+1) \times (n+1)$ matrix

$$L(\tau) = \begin{bmatrix} 1 & \cdots & 1 \\ l(y^1) & \cdots & l(y^{n+1}) \end{bmatrix}$$

the *label matrix* of $\tau$. $\tau$ is said to *complete* if the following system

$$L(\tau)X = \mathbf{I}, \ X \succ \mathbf{0}$$

has a solution, where $\mathbf{I}$ is the identity matrix of rank $n+1$. Let $\mu$ be the first column of $X$ when $X$ exists. Clearly, $\tau$ is completely labelled if it is complete. Further, if $\tau$ is complete and lies on $\mathbf{R}^n(m)$ for $m > 1$, the point $y^* = \sum_{i=1}^{n+1} \mu_i y^i$ is a fixed point of the piecewise linear approximation of $\phi$ with respect to $\mathcal{T}$ and hence the projection $p(y^*)$ of $y^*$ on $\mathbf{R}^n$ is an approximate fixed point of $\psi$. Analogous with the results of Merrill's algorithm we have the following parallel results.

**Definition 4.5.1** *An $(n+1)$-simplex of $\mathcal{T}$ is a node if it has at least one complete facet. Two nodes are adjacent and connected by an arc or edge if they share a common complete facet. The degree of a node $\sigma$ is the number of nodes adjacent to $\sigma$, denoted by $deg(\sigma)$. Let $V$ be the set of nodes and let $A$ be the set of arcs. In this way we obtain a graph $G = (V, A)$.*

Let $\sigma^1$ be the unique $(n+1)$-simplex of $\mathcal{T}$ having $\tau^1$ as its facet. Then we have

**Lemma 4.5.2** *The $n$-simplex $\tau^1$ is the only complete simplex on $\mathbf{R}^n(1)$.*

**Lemma 4.5.3** *Given the graph $G = (V, A)$ defined as above, for each node $\sigma$ of $G$, we have:*
*(i) if $\sigma$ has a complete facet lying on $\mathbf{R}^n \times \{1\}$, then $deg(\sigma) = 1$;*
*(ii) in all other cases, $deg(\sigma) = 2$.*

By using Theorem 1.4.14, Lemmas 4.5.2, 4.5.3, we have

**Theorem 4.5.4** *For the graph $G = (V, A)$ defined as above, each connected component of $G$ has one of the following two forms:*
*(i) a simple circuit;*
*(ii) an infinite simple path which starts with the $(n+1)$-simplex $\sigma^1$.*

In order to ensure the path beginning with $\sigma^1$ to be confined in a set $D \times [1,\infty)$ where $D$ is a compact subset of $\mathbf{R}^n$, we have to impose some conditions on $\psi$. Merrill's condition is a sufficient condition for this.

**Theorem 4.5.5** *If $mesh(\mathcal{T}(1)) < \delta$ and Merrill's condition holds, the simple path beginning with $\sigma^1$ is confined in a set $D \times [1,\infty)$ for some compact subset $D$ of $\mathbf{R}^n$.*

Now we can give the steps of the simplicial homotopy algorithm of Eaves and Saigal [1972].

The Algorithm of Eaves and Saigal on $\mathbf{R}^n \times [1,\infty)$

**Step (0)** Set $mesh(\mathcal{T}(1)) < \delta$. Let $\tau^1$ be the unique $n$-simplex of $\mathcal{T}^n$ containing $(c,1)$ in its interior. Let $\sigma^1$ be the unique $(n+1)$-simplex in $\mathcal{T}$ which has $\tau^1$ as its facet. Let $y^+$ be the vertex of $\sigma^1$ that is not a vertex of $\tau^1$. Set $k = 1$.

**Step (1)** Compute $W = L^{-1}(\tau^k)$. Let $W_i$ denote the $i$-th row of $W$. Compute $l(y^+)$ and let $q = (1, l^\top(y^+))^\top$. Let $v = (v_1, \cdots, v_{n+1})^\top = Mq$. Compute

$$\frac{W_s}{v_s} = \min_{\prec}\{\frac{W_h}{v_h} \mid v_h > 0\}.$$

Notice that $s$ is unique by Theorem 4.2.7. The column $(1, l^\top(y^s))^\top$ is eliminated. Let $\tau^{k+1}$ be the facet of $\sigma^k$ opposite $y^s$. If $\tau^{k+1}$ lies on $\mathbb{R}^n \times \{m\}$ for some $m > 1$, this facet yields an approximate fixed point of $\psi$. When the accuracy is satisfactory, stop. Otherwise, go to Step (2).

**Step (2)** Find a simplex $\sigma^{k+1}$ sharing the facet $\tau^{k+1}$ with $\sigma^k$, and let $y^+$ be the vertex of $\sigma^{k+1}$ not being a vertex of $\tau^{k+1}$. Set $k = k + 1$ and return to Step (1).

We wrap up this chapter by giving a few notes. The $CR$ triangulations of Eaves [1972] and Eaves and Saigal [1972] can be obtained by specifying the parameters $d_{i+1} = 2d_i$ for all $i \in \mathbb{N}$ and parameters $\lambda_i^m$ of the triangulations given in this and previous sections. See also Todd [1976a, 1977] and Kojima [1978]. More recently, Kojima and Yamamoto [1982], Eaves [1984], Broadie and Eaves [1987], Doup and Talman [1987c] and Dang [1991] have developed a series of $CR$ triangulations with arbitrary refining factors. Furthermore, homotopy algorithms are intensively studied in Eaves [1984]. Finally we point out that the great potentiality of homotopy algorithms is their capability of computing several (many, or all) solutions of certain nonlinear equations. See e.g., Kojima and Mizuno [1983].

CHAPTER 5

# Variable Dimension Restart Algorithms

## 5.1. Introduction

In the previous chapter we discussed two efficient methods that are the restart method of Merrill and the homotopy method of Eaves (and Saigal). In both methods we need to imbed an $n$-dimensional problem into $(n+1)$-dimensional space. Moreover, a path of $(n+1)$-simplices is generated. In this chapter we will introduce a more efficient algorithm which was developed by van der Laan and Talman [1979]. This algorithm is called the variable dimension restart algorithm. It does not need an extra dimension. More precisely, this algorithm starts with any point in the region of interest (say, $S^{n+1}$ or $\mathbb{R}^n$), generates a path of adjacent $t$-simplices of varying dimension and terminates with a completely labelled simplex yielding an approximation of a fixed point. Notice that $t$ can vary between 0 and $n$. Because the starting point can be arbitrarily chosen in the region of interest, this method is called the variable dimension restart algorithm. So when the accuracy of the current approximate solution is not satisfactory, the algorithm can be restarted at this approximate solution with a finer triangulation in the hope that within a small number of steps a better approximate solution is found. In particular, when the algorithm starts with the point $e(1)$ of the unit simplex, it operates in the same way as Kuhn's variable dimension algorithm given in Chapter 3 does.

This chapter is organized as follows. In Section 2 we give a full description of the variable dimension restart algorithm on the unit simplex $S^n$. In Section 3 we discuss this algorithm on the unbounded region $\mathbb{R}^n$ and give a simple convergent condition for the algorithm. By the limit argument we obtain an existence theorem from which Brouwer's fixed point theorem can be easily derived. Finally in Section 4 we conclude the chapter by indicating some related references.

## 5.2. Van der Laan-Talman's algorithm on $S^n$

Given a continuous function $f$ from the unit simplex $S^n$ into itself, our goal is to approximate a fixed point of $f$. Let $S^n$ be triangulated and every vertex in $S^n$ be labelled by a proper labeling rule. In this section, we will assume that $S^n$ is triangulated by $K_2(m)$ given in Definition 3.2.8 and every vertex is labelled by $l^2$ given in Definition 3.2.6. Now we discuss how the variable dimension restart algorithm of van der Laan and Talman [1979] searches for a completely labelled simplex in $S^n$. First we need to extend the $(n \times (n-1))$-matrix $Q$ given in Section 3.2 to the $n \times n$-matrix $Q$ given below:

$$Q = \begin{bmatrix} -1 & 0 & \cdots & 0 & 1 \\ 1 & -1 & \cdots & 0 & 0 \\ \vdots & \vdots & & \vdots & \vdots \\ 0 & 0 & \cdots & -1 & 0 \\ 0 & 0 & \cdots & 1 & -1 \end{bmatrix}.$$

Let $q(1)$, $\cdots$, $q(n)$ be the columns of $Q$. In other words, we can define $q(j) = e(j+1) - e(j)$ for $j \in I_n$ by specifying $j+1 = 1$ when $j = n$.

**Definition 5.2.1** *Let $v$ be an arbitrary point in $K_2^0(m)$. For $T \subset I_n$, the set $A(v,T)$ is given by*

$$A(v,T) = \{\, x \in S^n \mid x = v + \sum_{j \in T} \lambda_j q(j),\ \lambda_j \geq 0,\ j \in T \,\}.$$

The point $v$ will be the starting point of the algorithm. Note that $A(v,\emptyset) = \{\, v \,\}$ and that if $S \subset T \subset I_n$, then $A(v,S) \subseteq A(v,T)$. If $v$ is an interior point of $S^n$, then the dimension of $A(v,T)$ is equal to $|T|$, i.e., $dim(A(v,T)) = |T|$. Furthermore, by recalling $S^n(k)$ given Section 3.5, we see $A(e(1),\emptyset) = S^n(1)$ and $A(e(1), I_k) = S^n(k+1)$ for each $k \in I_{n-1}$. It will become clear that Kuhn's variable dimension algorithm is a particular case of van der Laan and Talman's algorithm. The sets $A(v,T)$, $T \subset I_n$, are shown in Figure 5.1 for $n = 3$, $m = 12$ and $v = (1/3, 1/3, 1/3)^\top$.

The following results can be easily derived.

**Lemma 5.2.2** *For $\emptyset \subset T \subset I_n$, the boundary of the set $A(v,T)$ can be expressed as*

$$bd(A(v,T)) = (\cup_{j \in T} A(v, T \setminus \{\, j \,\})) \cup (\cup_{j \in T}(A(v,T) \cap S_j^n)).$$

**Lemma 5.2.3** *Let $T$ be a proper subset of $I_n$. When the $K_2(m)$-triangulation with grid size $m^{-1}$ is restricted to $A(v,T)$, then it induces a triangulation of $A(v,T)$. Moreover, the induced triangulation of $A(v,T)$ is the collection of $t$-simplices $\sigma(y^1, \pi(T))$ in $A(v,T)$ with vertices $y^1$, $\cdots$, $y^{t+1}$ such that*

*(a) $y^1 \in A(v,T)$ is an element of $K_2^0(m)$;*

*(b) $\pi = (\pi(1), \cdots, \pi(t))$ is a permutation of the elements of the set $T$;*

*(c) $y^{i+1} = y^i + q(\pi(i))/m$, $i = 1$, $\cdots$, $t$.*

Now Definitions 3.5.2 and 3.5.3 can be generalized in the following ways.

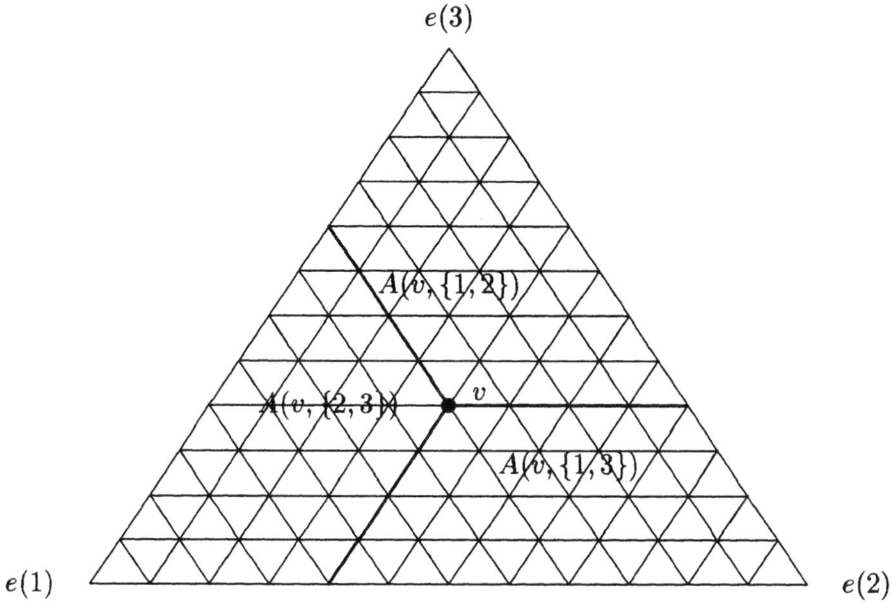


*Figure 5.1.* The sets $A(v,T)$ of $S^n$ for $n = 3$ and $m = 12$.

**Definition 5.2.4** *For $\emptyset \subset T \subseteq I_n$, a $(t-1)$-simplex $\sigma$ is $T$-complete if its vertices bear all labels of the set $T$ with $t = |T|$.*

Observe that $T = I_n$ if $t = n$ and that any zero-dimensional simplex $\{ w \}$ is $\{ l(w) \}$-complete. The following observation is easy but important.

**Lemma 5.2.5** *Let $l : K_2^0(m) \mapsto I_n$ be a proper labeling rule. Then for any $\emptyset \subset T \subset I_n$ with $t = |T|$, there is no $T$-complete $(t-1)$-simplex in $A(v,T) \bigcap bd(S^n)$.*

Proof: Clearly, $A(v,T) \bigcap bd(S^n) \subseteq \bigcup_{i \in T} S_i^n$. It follows from the labeling function $l$ that $x_i = 0$ implies $l(x) \neq i$ for any vertex $x \in K_2^0(m)$. Hence $A(v,T) \bigcap bd(S^n)$ does not contain any $T$-complete $(t-1)$-simplex. □

**Definition 5.2.6** *A simplex $\sigma \in K_2^+(m)$ is a node if $\sigma = \{v\}$, or $\sigma$ is a $t$-simplex in $A(v,T)$ and has a $T$-complete facet for some $T$ with $\emptyset \subset T \subset I_n$. Two nodes $\sigma_1$, $\sigma_2$ are adjacent if they lie in the same set $A(v,T)$ and if either they share a common $T$-complete facet or one is a $T$-complete facet of the other. The degree of a node $\sigma$ is the number of nodes adjacent to $\sigma$, denoted by $deg(\sigma)$. Let $V$ be the set of nodes and let $A$ be the set of arcs. In this way a graph $G = (V, A)$ is obtained.*

From the above definitions it is easy to verify that every c.l. simplex is a node. Now we have the following result which generalizes Lemma 3.5.4.

**Lemma 5.2.7** *Let $G = (V, A)$ be the graph defined above and let $\sigma$ be a node of $G$.*
*(i) If $\sigma = \{v\}$, then $deg(\sigma) = 1$;*
*(ii) If $\sigma$ is a completely labeled simplex of $K_2(m)$, then $deg(\sigma) = 1$;*
*(iii) In all other cases, $deg(\sigma) = 2$.*

Proof: (i) If $\sigma = \{v\}$, then $\sigma_1 =< v, y^+ >$ is a 1-dimensional simplex, where $y^+ = v + q(l(v))/m$. Let $T^0 = \{l(v)\}$. Then $\sigma$ is a $T^0$-complete facet of $\sigma_1$ in the set $A(v, T^0)$. Hence $\sigma_1$ is also a node. Moreover, $\sigma_1$ is the only node adjacent to $\sigma$. So, $deg(\{v\}) = 1$.

(ii) If $\sigma$ is a completely labeled simplex of $K_2(m)$, it implies that $\sigma$ belongs to $A(v, T)$ for some $T \subset I_n$ with $|T| = n - 1$. Let $\tau$ be the unique $T$-complete facet of $\sigma$. First, we show that $\tau$ does not lie in the boundary of $S^n$. Suppose to the contrary that $\tau \subseteq bd(S^n)$. It would mean that $\tau \subseteq S^n_h$ for some $h \in T$. However, it contradicts Lemma 5.2.5. Next, we consider the case in which $\tau \subseteq \bigcup_{h\in T} A(v, T\backslash\{h\})$. Thus, $\tau \subseteq A(v, T\backslash\{h\})$ for some $h \in T$. In this case $\tau$ is a simplex in $A(v, T\backslash\{h\})$ with a $(T\backslash\{h\})$-complete facet. So, $\tau$ is a node. Moreover, $\tau$ is $T$-complete. Since $\tau$ is in the boundary of $A(v, T)$, $\tau$ is the only node adjacent to $\sigma$, i.e., $deg(\sigma) = 1$. Finally, if $\tau \not\subseteq bd(A(v, T))$, it follows from (ii) of Theorem 1.4.6 that there exists exactly one simplex $\sigma'$ of $K_2(m)$ sharing the common facet $\tau$ with $\sigma$. It is clear that $\sigma'$ is the only node adjacent to $\sigma$. Hence $deg(\sigma) = 1$.

(iii) Let $\sigma$ in $A(v, T)$ be a $t$-simplex, where $t = |T|$, with vertices $y^1$, $\cdots$, $y^{t+1}$. Assume that $\sigma$ is a node. In case the set $\{l(y^i) \mid i = 1, \cdots, t+1\}$ is equal to $T$, $\sigma$ has exactly two $T$-complete facets $\tau_1$, $\tau_2$. Consider $\tau_1$ for example. Suppose that $\tau_1 \subseteq \bigcup_{h\in T} A(v, T\backslash\{h\})$. It means that $|T| \geq 1$ and $\tau_1 \subseteq A(v, T\backslash\{h\})$ for some $h \in T$. If $T \setminus \{h\} = \emptyset$, then $\tau_1 = \{v\}$. Otherwise, $\tau_1$ clearly has a $(T\backslash\{h\})$-complete facet. Hence in both cases $\tau_1$ is a node adjacent to $\sigma$. In case $\tau_1 \not\subseteq bd(A(v, T))$, it follows from (ii) of Theorem 1.4.6 that there exists exactly one simplex $\sigma'$ in $A(v, T)$ sharing the common facet $\tau_1$ with $\sigma$. Again, $\sigma'$ is a node adjacent to $\sigma$. Notice that both $\tau_1$ and $\tau_2$ do not lie in the boundary of $S^n$. Thus, $deg(\sigma) = 2$. In case $\sigma$ has a vertex $y^+$ whose label, say $j$, does not belong to $T$, we have $t < n - 1$. Let $\tau$ be the only $T$-complete facet of $\sigma$. Again, $\tau$ does not lie in the boundary of $S^n$. In case $\tau \subseteq \bigcup_{h\in T} A(v, T\backslash\{h\})$, then $\tau$ is a node adjacent to $\sigma$. Next, in case $\tau \not\subseteq bd(A(v, T))$, then there exists exactly one simplex $\sigma'$ in $A(v, T)$ sharing the common facet $\tau$ with $\sigma$ and $\sigma'$ is a node adjacent to $\sigma$. On the other hand, let $\bar{\sigma}$ be the unique $(t+1)$-simplex in $A(v, T \bigcup \{j\})$ with $\sigma$ as its facet. It is clear that $\bar{\sigma}$ is a node adjacent to $\sigma$. There are no other nodes adjacent to $\sigma$. Hence, $deg(\sigma) = 2$. □

By noticing the fact that the number of simplices in $S^n$ is finite, we immediately have the following result.

**Theorem 5.2.8** *For the graph $G = (V, A)$ defined as above, each connected component of $G$ has one of the following three forms:*
*(i) a finite simple circuit;*
*(ii) a finite simple path whose end nodes are c.l. $(n-1)$-simplices;*
*(iii) a finite simple path starts with the node $\{v\}$ and ends with a c.l. $(n-1)$-simplex in $S^n$.*

The next result is now straightforward by applying Theorem 1.4.14.

**Corollary 5.2.9** *Let $m \in \mathbb{N}$ be given. Then there is a finite simple path which starts with $\{v\}$ and ends with a completely labeled simplex in $K_2(m)$.*

Now we can describe the algorithm of van der Laan and Talman in detail. In the description the $n$-vector $R$ stands for the distance vector between the starting point $v$ and the vertex $y^1$ of the last generated simplex $\sigma(y^1, \cdots, y^{t+1})$, i.e., $y^1 = v + \sum_{h=1}^{n} R_h q(h)/m$.

The Algorithm of van der Laan and Talman on $S^n$

**Step (0)** Set $t = 0$, $y^1 = v$, $T = \emptyset$, $\sigma = \sigma(y^1, \pi(T))$ and $R = \mathbf{0}$. Set $y^+ = y^1$.

**Step (1)** Compute $l(y^+)$. If $l(y^+)$ is not an element of $T$, go to Step (3). Otherwise, there exists exactly one vertex $y^s \neq y^+$ with $l(y^s) = l(y^+)$.

**Step (2)** If $s = t+1$ and $R_{\pi(t)} = 0$, go to Step (4). Otherwise, $\sigma(y^1, \pi(T))$ and $R$ are adapted according to Table 5.1 by replacing $y^s$. Return to Step (1) with $y^+$ equal to the new vertex of $\sigma$.

**Step (3)** If $t = n-1$, a completely labeled simplex is found and the algorithm stops. If $t < n-1$, a $(T \bigcup \{ l(y^+) \})$-complete simplex is found and the dimension is increased. Set $T = T \bigcup \{ l(y^+) \}$ and $\pi(T) = (\pi(1), \cdots, \pi(t), l(y^+))$. Then $\sigma = \sigma(y^1, \pi(T))$, $t = t+1$ and $y^+ = y^{t+1}$. Return to Step (1).

**Step (4)** Decrease the dimension by eliminating $y^s$. Set $T = T \backslash \{ \pi(t) \}$, $\pi(T) = (\pi(1), \cdots, \pi(t-1))$, $\sigma = \sigma(y^1, \pi(T))$, and $t = t-1$. Return to Step (2) with $y^s$ the vertex of $\sigma$ with label equal to the deleted element of $T$.

The algorithm is demonstrated in Figure 5.2 for $S^n$ with $n = 3$, $m = 12$ and $v = (1/3, 1/3, 1/3)^\top$.

## 5.3. Van der Laan-Talman's algorithm on $\mathbb{R}^n$

Now we consider the problem of finding a fixed point of a continuous function $f$ from $\mathbb{R}^n$ into $\mathbb{R}^n$. We will use $K_1$-triangulation given in Chapter 1

| | $y^1$ becomes | $\pi(T)$ becomes | $R$ becomes |
|---|---|---|---|
| $s = 1$ | $y^1 + q(\pi(1))/m$ | $(\pi(2), \cdots, \pi(t), \pi(1))$ | $R + e(\pi(1))$ |
| $1 < s \leq t$ | $y^1$ | $(\pi(1), \cdots, \pi(s-2), \pi(s), \pi(s-1), \pi(s+1), \cdots, \pi(t))$ | $R$ |
| $s = t + 1$ | $y^1 - q(\pi(t))/m$ | $(\pi(t), \pi(1), \cdots, \pi(t-1))$ | $R - e(\pi(t))$ |

TABLE 5.1. $s$ is the index of the vertex to be replaced.

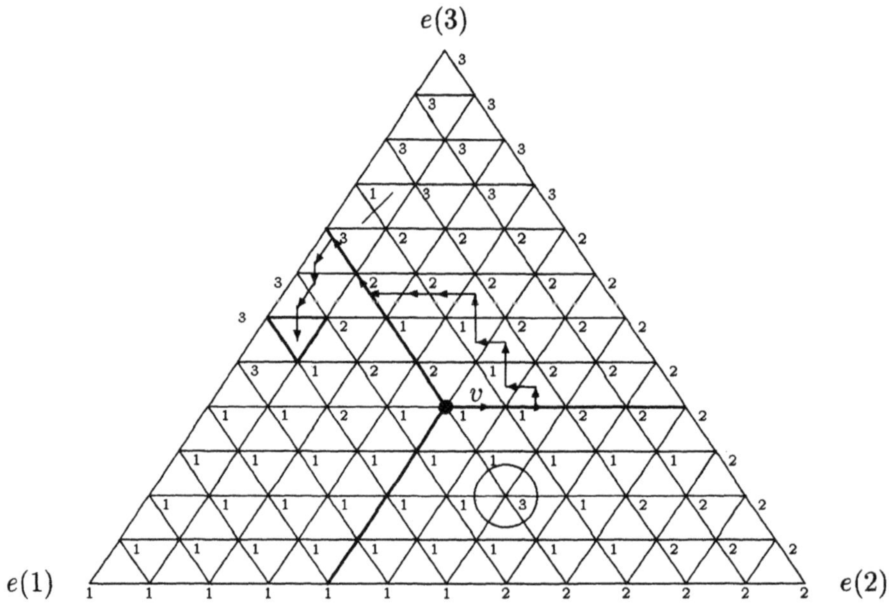


*Figure 5.2.* Illustration of van der Laan-Talman's algorithm on $S^n$.

to underlie the algorithm of van der Laan and Talman [1979,1980b,1981a], van der Laan [1980b] and Talman [1980]. First we need to redefine $K_1$ such that $K_1$ can have any desirable mesh size.

**Definition 5.3.1** *Given a positive integer $m$, the $K_1(m)$-triangulation of $\mathbb{R}^n$ with grid size $m^{-1}$ is the collection of all $n$-simplices $\sigma(y^1, \pi)$ with vertices $y^1$, $\cdots$, $y^{n+1}$ in $\mathbb{R}^n$ such that*
*(a) each component of $y^1$ is a multiple of $m^{-1}$;*
*(b) $\pi = (\pi(1), \cdots, \pi(n))$ is a permutation of the elements in $I_n$;*
*(c) $y^{i+1} = y^i + m^{-1}u(\pi(i))$, $i \in I_n$ where $u(i) = e(i)$, $i \in I_n$.*

It is easy to see that $mesh(K_1(m)) = \sqrt{n}/m$. So $mesh(K_1(m))$ will converge to zero as $m$ goes to infinity. For the sake of description, we define

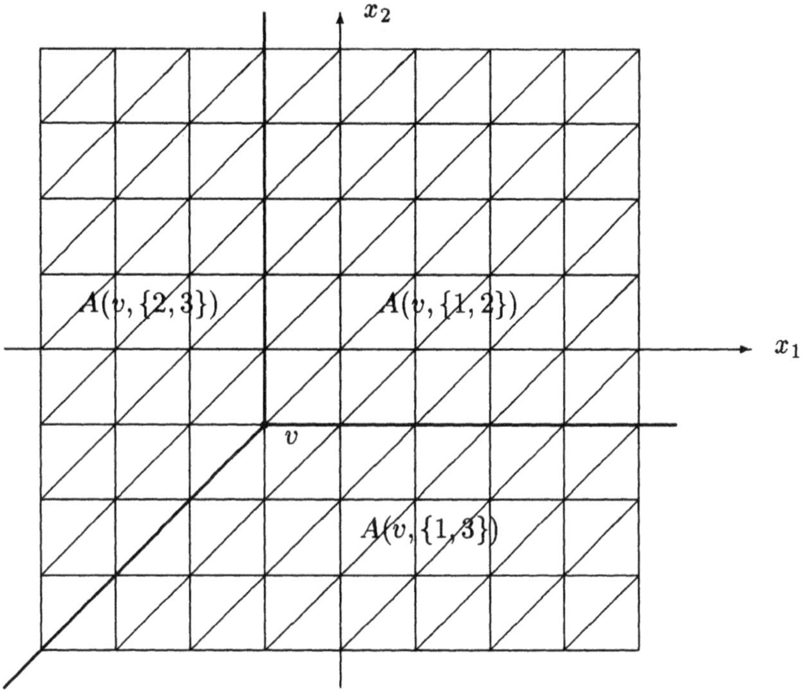


*Figure 5.3.* The sets $A(v,T)$ of $\mathbb{R}^n$ for $n = 2$.

$u(n+1) = -e$, i.e., $u(n+1) = -\sum_{i=1}^{n} e(i)$. Now we can adapt all definitions given in the previous section.

**Definition 5.3.2** *Let $v$ be an arbitrary point in $K_1^0(m)$. For $T \subset I_{n+1}$, the set $A(v,T)$ is given by*

$$A(v,T) = \{\, x \in \mathbb{R}^n \mid x = v + \sum_{j \in T} \lambda_j u(j),\ \lambda_j \geq 0,\, j \in T \,\}.$$

The point $v$ will be the starting point of the algorithm. Note that $A(v, \emptyset) = \{\, v \,\}$ and that if $S \subset T \subset I_{n+1}$, then $A(v,S) \subset A(v,T)$. The dimension of $A(v,T)$ is equal to $|T|$, i.e., $dim(A(v,T)) = |T|$. The sets $A(v,T)$, $T \subset I_{n+1}$, are shown in Figure 5.3 for $n = 2$, $m = 1$ and $v = (-1,-1)^\top$.

The following results can be easily derived.

**Lemma 5.3.3** *For $\emptyset \subset T \subset I_{n+1}$, the boundary of the set $A(v,T)$ can be expressed as*

$$bd(A(v,T)) = \cup_{j \in T} A(v, T \setminus \{\, j \,\}).$$

**Lemma 5.3.4** *Let $T$ be a proper subset of $I_{n+1}$. When the $K_1(m)$-triangulation with grid size $m^{-1}$ is restricted to $A(v,T)$, then it induces*

*a triangulation of $A(v,T)$. Moreover, the induced triangulation of $A(v,T)$ is the collection of $t$-simplices $\sigma(y^1, \pi(T))$ in $A(v,T)$ with vertices $y^1, \cdots, y^{t+1}$ such that*
*(a) $y^1 \in A(v,T)$ is an element of $K_1^0(m)$;*
*(b) $\pi = (\pi(1), \cdots, \pi(t))$ is a permutation of the elements of the set $T$;*
*(c) $y^{i+1} = y^i + u(\pi(i))/m$, $i = 1, \cdots, t$.*

**Definition 5.3.5** *Let $f : \mathbf{R}^n \mapsto \mathbf{R}^n$ be a function and let $\mathcal{T}$ be a triangulation of $\mathbf{R}^n$. The labeling rule $l : \mathcal{T}^0 \mapsto I_{n+1}$ is given by*

$$l(y) = \min\{\{n+1\} \cup \{i \in I_n \mid f_i(y) > y_i\}\}.$$

**Definition 5.3.6** *For $\emptyset \subset T \subseteq I_{n+1}$, a $(t-1)$-simplex $\sigma$ is $T$-complete if its vertices bear all labels of the set $T$ with $t = |T|$. An $n$-simplex is a completely labelled (c.l.) simplex if it is $I_{n+1}$-complete.*

Observe that $T = I_{n+1}$ if $t = n+1$ and that any zero-dimensional simplex $\{w\}$ is $\{l(w)\}$-complete. The following theorem says that a completely labelled simplex yields a good approximation of a fixed point of a continuous function.

**Theorem 5.3.7** *Let $f : \mathbf{R}^n \mapsto \mathbf{R}^n$ be a continuous function, let $\mathcal{T}$ be a triangulation of $\mathbf{R}^n$ and let $\epsilon > 0$, $0 < h < \frac{\epsilon}{2\sqrt{n}}$ be such that $mesh(\mathcal{T}) \leq h$, and for any $x, y \in \mathbf{R}^n$, $\max_i |x_i - y_i| \leq h$ implies*

$$\max_i |f_i(x) - f_i(y)| \leq \frac{\epsilon}{2\sqrt{n}}.$$

*Given the labeling rule $l$ and a completely labelled simplex $\sigma$, then we have*

$$||x - f(x)|| \leq \epsilon$$

*for all $x \in \sigma$.*

Proof: We may assume that $\sigma =< y^1, \cdots, y^{n+1} >$ and $l(y^i) = i$ for all $i \in I_{n+1}$. Now it follows from the labeling rule $l$ that

$$f_i(y^i) > y_i^i$$

and

$$f_i(y^{n+1}) \leq y_i^{n+1}$$

for all $i \in I_n$. This leads to

$$\begin{aligned} x_i - f_i(x) &= (f_i(y^i) - f_i(x)) + (y_i^i - f_i(y^i)) + (x_i - y_i^i) \\ &\leq \tfrac{\epsilon}{2\sqrt{n}} + 0 + h \\ &\leq \tfrac{\epsilon}{\sqrt{n}} \end{aligned}$$

and

$$\begin{aligned} x_i - f_i(x) &= (f_i(y^{n+1}) - f_i(x)) + (y_i^{n+1} - f_i(y^{n+1})) + (x_i - y_i^{n+1}) \\ &\geq -\tfrac{\epsilon}{2\sqrt{n}} + 0 - h \\ &\geq -\tfrac{\epsilon}{\sqrt{n}} \end{aligned}$$

for all $i \in I_n$. Consequently, we have

$$\max_i |x_i - f_i(x)| \leq \frac{\epsilon}{\sqrt{n}}.$$

Now it is easy to see that

$$||x - f(x)|| \leq \epsilon$$

for all $x \in \sigma$. □

**Definition 5.3.8** *A simplex $\sigma \in K_1^+(m)$ is a node if $\sigma = \{v\}$, or $\sigma$ is a t-simplex in $A(v,T)$ and has a $T$-complete facet for some $\emptyset \subset T \subset I_n$. Two nodes $\sigma_1$, $\sigma_2$ are adjacent if they lie in the same set $A(v,T)$ and if either they share a common $T$-complete facet or one is a $T$-complete facet of the other. The degree of a node $\sigma$ is the number of nodes adjacent to $\sigma$, denoted by $deg(\sigma)$. Let $V$ be the set of nodes and let $A$ be the set of arcs. In this way a graph $G = (V, A)$ is obtained.*

From the above definitions it is easy to verify that every c.l. simplex is a node. Arguing by analogy of Lemma 5.2.7 leads to the following result.

**Lemma 5.3.9** *Let $G = (V, A)$ be the graph defined above and let $\sigma$ be a node of $G$.*
*(i) If $\sigma = \{v\}$, then $deg(\sigma) = 1$;*
*(ii) If $\sigma$ is a completely labeled simplex of $K_1(m)$, then $deg(\sigma) = 1$;*
*(iii) In all other cases, $deg(\sigma) = 2$.*

**Theorem 5.3.10** *For the graph $G = (V, A)$ defined as above, each connected component of $G$ has one of the following five forms:*
*(i) a simple circuit;*
*(ii) a finite simple path whose end nodes are both c.l. $n$-simplices;*
*(iii) an infinite simple path which starts with a c.l. $n$-simplex;*
*(iv) an infinite simple path which starts with the node $\{v\}$;*
*(v) a finite simple path which starts with the node $\{v\}$ and ends with a c.l. $n$-simplex.*

In order to find a c.l. $n$-simplex, we need to exclude Case (iv). To do so, we impose an extra condition on the function $f$.

**Theorem 5.3.11** *Let $f : \mathbb{R}^n \mapsto \mathbb{R}^n$ be a continuous and bounded function and let $K_1(m)$ be a triangulation of $\mathbb{R}^n$. Then there exists a finite simple path which starts with $\{v\}$ and ends with a c.l. $n$-simplex $\sigma$. Moreover, $\sigma$ yields a good approximation of a fixed point of $f$.*

Proof: Since $f$ and $v$ are bounded, there exists a constant $M > 0$ such that $-M < f_i(x) < M$ for all $i \in I_n$ and for every $x \in \mathbb{R}^n$ and $-M < v_i < M$ for all $i \in I_n$. Define an $n$-dimensional cube by

$$C(M) = \{x \in \mathbb{R}^n \mid -(M+1) \leq x_i \leq M+1,\ i \in I_n\}.$$

We will prove that the algorithm cannot go outside $C(M)$. In other words, there are no nodes outside $C(M)$. Suppose to the contrary that there is a node $\sigma$ outside $C(M)$. This implies that there exists a nonempty proper subset $T$ of $I_{n+1}$ such that $\sigma \subseteq A(v,T) \setminus C(M)$ has a $T$-complete facet $\tau$. Let $y^1, \cdots, y^{t+1}$ be the vertices of $\sigma$, where $t = |T|$. We have to consider the following two cases.

(1). In case $n+1 \notin T$, there exists an index $i \in T$ such that $y_i^1 > M+1, \cdots, y_i^{t+1} > M+1$. It follows that $f_i(y^1) < M < M+1 < y_i^1, \cdots, f_i(y^{t+1}) < M < M+1 < y_i^{t+1}$. This implies that $l(y^j) \neq i$ for any $j \in I_{t+1}$. Hence $\sigma$ is not a node by definition.

(2). In case $n+1 \in T$, there are two possibilities. (2.1): there exists an index $i \in T$ with $i \in I_n$ such that $y_i^1 > M+1, \cdots, y_i^{t+1} > M+1$. This is the same as Case (1) above. (2.2): there is an index $i \in I_n$ but $i \notin T$ such that $y_i^1 < -(M+1), \cdots, y_i^{t+1} < -(M+1)$. It follows that $y_i^1 < -(M+1) < -M < f_i(y^1), \cdots, y_i^{t+1} < -(M+1) < -M < f_i(y^{t+1})$. This implies that $l(y^j) \neq n+1$ for any $j \in I_{t+1}$. Hence $\sigma$ is not a node by definition.

It follows from Theorem 1.4.14 that since the number of simplices in $C(M)$ is finite, the path starting with $\{v\}$ must be simple and finite.

The last part of the theorem is obvious, since $f$ is continuous. □

Now we are ready to establish the following existence theorem.

**Theorem 5.3.12** *Let $f : \mathbb{R}^n \mapsto \mathbb{R}^n$ be a continuous and bounded function. Then $f$ has a fixed point.*

Proof: We will constructively prove this result by using the above algorithm. Let $\{K_1(2^m) \mid m \in \mathbb{Z}_+\}$ be a sequence of triangulations of $\mathbb{R}^n$. Clearly, the mesh size will converge to zero as $m$ goes to infinity. With respect to $K_1(2^m)$, let $y^m$ be the vertex of the c.l. simplex $\sigma^m$ found by the above algorithm with $l(y^m) = 1$. The existence of $\sigma^m$ is guaranteed by Theorem 5.3.11. Moreover, let $v = y^m$ be the starting point of the algorithm for $K_1(2^{m+1})$. Notice that $y^m$ is also a vertex of $K_1(2^{m+1})$. For $K_1(1)$, we just take an arbitrary vertex as the starting point $v$. Clearly the sequence $\{y^m \mid m \in \mathbb{Z}_+\}$ has a convergent subsequence. For simplicity we may assume it is $\{y^m \mid m \in \mathbb{Z}_+\}$ itself. By Theorem 5.3.7 it is readily seen that the cluster point of this sequence is a fixed point of $f$. □

As a consequence of the above result, we can derive Brouwer's fixed point theorem. But first let us consider a more general result. Let $B$ be a subset of $\mathbb{R}^n$ with $dim(B) = n$. Let $h$ be an interior point of $B$. For each point $z \in bd(B)$, let $r(z, h)$ denote the set of points in $\mathbb{R}^n \setminus B$ which are the points on the straight line starting with $h$ and going through $z$. A set $B$ is called *star-shaped* if there exists a point $h \in int(B)$ such that $z, y \in bd(B)$ with $y \neq z$ imply $r(z, h) \cap r(y, h) = \emptyset$. Clearly convex sets are star-shaped.

**Theorem 5.3.13** *Let $B$ be an $n$-dimensional star-shaped compact subset of $\mathbb{R}^n$ and let $g : B \mapsto B$ be a continuous function. Then $g$ has a fixed point in $B$.*

Proof: Let $h$ and $r$ be defined as above. Then we extend $g$ to $f$. Define the function $f : \mathbb{R}^n \mapsto \mathbb{R}^n$ by

$$f(x) = \begin{cases} g(x), & x \in B; \\ g(z), & x \in r(z, h) \text{ and } z \in bd(B). \end{cases}$$

It is easy to verify that $f$ is continuous and bounded. By Theorem 5.3.12 there exists a point $x^* \in \mathbb{R}^n$ such that $f(x^*) = x^*$. Since $f(\mathbb{R}^n) \subseteq B$, this means that $x^* \in B$. By definition we have $g(x^*) = f(x^*)$. Thus, $g(x^*) = x^*$. □

Clearly, Brouwer theorem is also proved. We point out here that different conditions can also guarantee the convergence of the above algorithm. With respect to this point, the interested reader should refer to van der Laan [1980b]. Now we give the detailed steps of the algorithm of van der Laan and Talman on $\mathbb{R}^n$. In the description the $(n+1)$-vector $R$ stands for the distance vector between the starting point $v$ and the vertex $y^1$ of the last generated simplex $\sigma(y^1, \cdots, y^{t+1})$, i.e., $y^1 = v + \sum_{h=1}^{n+1} R_h u(h)/m$. Let $E(i)$ be the $i$-th unit vector of $\mathbb{R}^{n+1}$, for $i \in I_{n+1}$.

The Algorithm of van der Laan and Talman on $\mathbb{R}^n$

**Step (0)** Set $t = 0$, $y^1 = v$, $T = \emptyset$, $\sigma = \sigma(y^1, \pi(T))$ and $R = \mathbf{0}$. Set $y^+ = y^1$.

**Step (1)** Compute $l(y^+)$. If $l(y^+)$ is not an element of $T$, go to Step (3). Otherwise, there exists exactly one vertex $y^s \neq y^+$ with $l(y^s) = l(y^+)$.

**Step (2)** If $s = t+1$ and $R_{\pi(t)} = 0$, go to Step (4). Otherwise, $\sigma(y^1, \pi(T))$ and $R$ are adapted according to Table 5.2 by replacing $y^s$. Return to Step (1) with $y^+$ equal to the new vertex of $\sigma$.

**Step (3)** If $t = n$, a completely labeled simplex is found and the algorithm stops. If $t < n$, a $(T\bigcup\{\, l(y^+)\,\})$-complete simplex is found and the dimension is increased. Set $T = T\bigcup\{\, l(y^+)\,\}$ and $\pi(T) = (\pi(1),\cdots,\pi(t), l(y^+)\,)$. Then $\sigma = \sigma(y^1, \pi(T))$, $t = t+1$ and $y^+ = y^{t+1}$. Return to Step (1).

**Step (4)** Decrease the dimension by eliminating $y^s$. Set $T = T\backslash\{\,\pi(t)\,\}$, $\pi(T) = (\pi(1),\cdots,\pi(t-1))$, $\sigma = \sigma(y^1, \pi(T))$, and $t = t-1$. Return to Step (2) with $y^s$ the vertex of $\sigma$ with label equal to the deleted element of $T$.

| | $y^1$ becomes | $\pi(T)$ becomes | $R$ becomes |
|---|---|---|---|
| $s=1$ | $y^1 + u(\pi(1))/m$ | $(\pi(2),\cdots,\pi(t),\pi(1))$ | $R + E(\pi(1))$ |
| $1 < s \leq t$ | $y^1$ | $(\pi(1),\cdots,\pi(s-2),\pi(s),\pi(s-1),\pi(s+1),\cdots,\pi(t))$ | $R$ |
| $s=t+1$ | $y^1 - u(\pi(t))/m$ | $(\pi(t),\pi(1),\cdots,\pi(t-1))$ | $R - E(\pi(t))$ |

TABLE 5.2. $s$ is the index of the vertex to be replaced.

## 5.4. Concluding remarks

After van der Laan and Talman [1979] developed the first variable dimension restart algorithm, several new algorithms have been proposed. These algorithms can be classified according to the number of rays along which the algorithms may leave the starting point. From this definition, the algorithm introduced in Sections 5.2 and 5.3 is called the $(n+1)$-ray algorithm. On $\mathbf{R}^n$ there are, for example, the $2n$-ray algorithm by van der Laan and Talman [1981a], the $2^n$-ray algorithm by Wright [1981], the $(3^n-1)$-ray algorithm by Kojima and Yamamoto [1984], and the 2-ray algorithm by Saigal [1983] and Yamamoto [1983]. According to Kojima and Yamamoto [1984] and van der Laan and Seelen [1984] the $(n+1)$-ray algorithm, the $2^n$-ray algorithm and the $(3^n-1)$-ray algorithm are very efficient. However, the performance of the 2-ray algorithm is quite poor. On the unit simplex $S^{n+1}$, besides the $(n+1)$-ray algorithm, there is the $(2^{n+1}-2)$-ray algorithm by Doup, van der Laan and Talman [1987]. Both algorithms are quite efficient; see Doup [1988]. The $(2^{n+1}-2)$-ray algorithm has a number of interesting economic and game theoretic properties (see van der Laan and Talman [1987a], Kamiya and Talman [1991] and Herings [1997]). On $S^{n+1}\times\mathbf{R}^m_+$, the $(n+m+1)$-ray algorithm was introduced by Hofkes [1990] , the $(n+1)2^m$-ray algorithm by Yamamoto and Yang [1993], and the $(2^{n+m+1}-2)$-ray algorithm by Talman, Yamamoto and Yang [1995]. In such cases, the algorithms are designed to compute equilibria in economy models with production. On the simplotope there are several algorithms (see e.g., van der Laan and Talman [1982], Doup and Talman [1987b], and Talman and Yang [1994]). These algorithms are proposed to compute a Nash equilibrium and its refinements. Kremers and Talman [1994] proposed a variable dimension algorithm for linear complementarity problem. Furthermore, several unifying approaches for these algorithms were given by Kojima and Yamamoto [1982], Freund [1984a], Yamamoto [1984a], and van der Laan and Talman [1986]. However, there is always some exception case to which the unifying approaches do not apply. It is worth pointing out that the great potentiality of variable dimension restart algorithms is their flexibility of working directly on various sets.

In this book we do not discuss the index (or orientation) theory of simplices. In this theory each completely labelled simplex is associated with an index from the set $\{-1, +1\}$. This index is closely related to the Brouwer degree of a mapping. The interested reader is referred to Shapley [1974], Eaves and Scarf [1976], Todd [1976c], Charnes, Garcia and Lemke [1977], Prüfer and Siegberg [1981], van der Laan and Talman [1981b], van der Laan [1980b, 1984], and Yamamoto [1984b].

There are a number of excellent books and proceedings on fixed point methods and applications. Scarf [1973] and Todd [1976a] are classic, though the latter one contains some inaccuracies. The reader should also consult Hu and Robinson [1973], Karamardian [1977], Peitgen and Walther [1979], Forster [1980], van der Laan [1980b], Robinson [1980], Talman [1980], Allgower, Glashoff and Peitgen [1981], Garcia and Zangwill [1981], Eaves [1984], Wang [1986a], Talman and van der Laan [1987], Doup [1988], Allgower and Georg [1990], Shoven and Whalley [1992], and van den Elzen [1993].

CHAPTER 6

# An Integer Labeling Algorithm for Solving a Class of Integer Programming Problems

## 6.1. Introduction

In this chapter we consider the following problem. Given an arbitrary simplex $P$, for example, the convex hull of $m+1$ $(0 \leq m \leq n)$ affinely independent vectors of $\mathbb{R}^n$, the problem is that of determining whether or not $P$ contains an integer point. This problem is not quite as innocent as it appears. In fact we first prove that it is intractable in the sense that it is $NP$-complete. We then develop an algorithm to solve the problem. The algorithm is based on both a specific integer labeling rule and the $K_1$-triangulation of $\mathbb{R}^n$. The main feature of the algorithm can be described as follows. We first subdivide $\mathbb{R}^n$ into $n$-dimensional simplices such that all integer points of $\mathbb{R}^n$ are the vertices of the triangulation, and then assigns an integer to each integer point of $\mathbb{R}^n$ according to the labeling rule. Starting from an arbitrarily chosen integer point, the algorithm generates a sequence of adjacent simplices of varying dimension and terminates with either the YES or (exclusively) NO answer within a finite number of steps. In the YES case, the algorithm finds an integer point in $P$. The NO answer shows that there is no integer point in $P$. The algorithm is derived from the algorithm of van der Laan and Talman [1979] (see Chapter 5) although their algorithm was introduced to compute a fixed point of a continuous function.

Our work was motivated by the works of Scarf [1981a-b] and of Dang and van Maaren [1993]. Scarf's method is based on primitive sets and has so far only been characterized for problems of low dimension (see White, Caplin and Van der Heyden [1985]). For some recent developments related to Scarf's method the interested reader should consult Scarf [1985, 1986], Scarf and Shallcross [1993], Bárány, Howe and Scarf [1994] and Bárány, Scarf and Shallcross [1998]. In some sense our algorithm can be seen as an easy realization of Scarf's method in the setting of triangulations. Our algorithm is developed for simplices in a standard form. We prove in a constructive manner that any simplex can be transformed to an simplex in the standard form using a unimodular matrix. This idea of transformation was inspired by Scarf [1981b] for $n = 2$ and White [1983] for $n > 2$. The method of Dang and van Maaren [1993] attempts to solve the same problem stated above, is based on the $J_1$-triangulation and is not involved with the above unimodular transformation. Their method is derived from the 2-ray algorithm of Yamamoto [1983] and does not guarantee to give a right answer in the sense that it may not find a solution even if the problem has one. Finally we point out that the integer programming problems are well studied; see Schrijver [1986] and Nemhauser and Wolsey [1988]. The reader is also suggested to refer to Lenstra [1983] and Lovász and Scarf [1992] for polynomial algorithms on integer linear programming problems when the dimension is fixed in advance.

This chapter is organized as follows. In Section 2 we formally present the problem and prove it is NP-complete. Some basic properties of the problem are derived. In Section 3 the labeling rule and the algorithm are introduced. The algorithm is proposed for simplices in a standard form. In Section 4 we demonstrate how to transform an arbitrary simplex into a simplex in the standard form using a unimodular matrix. In Section 5 we discuss a practical implementation of the algorithm, report a number of tested problems and conclude with some remarks. This chapter is based on Yang [1994,1995] and Li et al [1998].

## 6.2. The problem: an NP-complete problem

The problem we consider is that of testing the integer feasibility of an $m$-dimensional $(0 \le m \le n)$ simplex $P$ given by

$$P = \{x \in \mathbb{R}^n \mid a^{i\top} x \le b_i, \text{ for all } i \in I_{n+1}\},$$

where $a^{i\top} = (a_{i,1}, \cdots, a_{i,n})$ is the $i$-th row of an $(n+1) \times n$ matrix $A$ for $i = 1, \cdots, n+1$, and $b = (b_1, \cdots, b_{n+1})^\top$ is a vector of $\mathbb{R}^{n+1}$. Throughout the chapter it is assumed that $a^1, \cdots, a^{n+1}$, and $b$ are integer vectors, and that the following assumption also holds.

**Assumption 6.2.1** *The origin of $\mathbb{R}^n$ is contained in the interior of the convex hull of the vectors $a^1, \cdots, a^{n+1}$. In other words, there exists a unique vector $\lambda \in \mathbb{R}^{n+1}_{++}$ such that $\sum_{i \in N} \lambda_i a^i = 0$ and $\sum_{i \in N} \lambda_i = 1$.*

This assumption implies that any $n$ vectors among $a^1, \cdots, a^{n+1}$ are linearly independent. Notice that when $P$ is a full-dimensional simplex in $\mathbb{R}^n$, the assumption is fulfilled.

Let $N$ denote the set $I_{n+1}$ and let $N_{-i}$ denote the set $N$ without the index $i$ for any $i \in N$.

It is well-known that the integer feasibility problem of arbitrary polytopes is *NP-complete.* We can even prove that the integer feasibility problem of arbitrary simplices is also $NP$-complete. To our knowledge, this result is not known in the literature. For those who are not familiar with the theory of complexity, we recommend them to an excellent book of Papadimitriou and Steiglitz [1982].

**Lemma 6.2.2** *The integer feasibility problem of arbitrary simplices is $NP$-complete.*

Proof: Clearly, the problem is in $NP$. It is well-known that the integer knapsack problem (see Papadimitriou and Steiglitz [1982]) is an $NP$-complete problem. The integer knapsack problem is stated as follows: Given $n+2$ arbitrary positive integers $c_1, \cdots, c_{n+1}$, and $k$, are there any non-negative integers $y_1, \cdots, y_{n+1}$ such that $\sum_{i=1}^{n+1} c_i y_i = k$? By using the polynomial-time algorithm of Kannan and Bachem [1979] we can find an $(n+1) \times (n+1)$ unimodular matrix $U$ such that $(c_1, c_2, \cdots, c_{n+1})U = (h, 0, \cdots, 0)$, $(c_1, c_2, \cdots, c_{n+1})/h$ are integer vectors and $h > 0$. Moreover, the set $Q = \{y \in Z^{n+1} \mid \sum_{i=1}^{n+1} c_i y_i = k\}$ is nonempty if and only if $k/h$ is an integer. When the set $Q$ is nonempty, the set $W = \{y \in \mathbf{Z}_+^{n+1} \mid \sum_{i=1}^{n+1} c_i y_i = k\}$ is nonempty if and only if the set

$$\begin{aligned} V &= \{y \in \mathbf{Z}^{n+1} \mid -Uy \leq 0,\, y_1 = k/h\} \\ &= \{y \in \mathbf{Z}^{n+1} \mid y = (k/h, x),\, \bar{A}x \leq \bar{b},\, x \in \mathbf{Z}^n\}, \end{aligned}$$

is nonempty. Let $S = \{x \in \mathbb{R}^n \mid \bar{A}x \leq \bar{b}\}$. It is readily seen that $W$ is nonempty if and only if the set $S$ contains an integer point. Notice that $S$ is a full-dimensional simplex in $\mathbb{R}^n$. This means that any instance $P_1$ of the integer knapsack problem can be polynomially transformed into an instance $P_2$ of the integer feasibility problem of full-dimensional simplices such that $P_1$ has a solution if and only if $P_2$ has a solution. Hence the problem at hand is $NP$-complete. □

Our algorithm to be presented will be applied to a simplex $P$ in some standard form. With respect to an $(n+1) \times n$ matrix $A$, we introduce the following assumptions:

**Assumption 6.2.3**
(*a*) $a_{n+1,j} \leq 0$ *for all* $j \in I_n$;
(*b*) $a_{i,i} > 0$ *and* $a_{i,j} \leq 0$ *for all* $i \in I_n$ *and all* $j \in I_n$ *with* $j \neq i$;
(*c*) $|a_{i,j}| < a_{i,i}$ *for all* $i \in I_n$ *and all* $j \in I_n$ *with* $j \neq i$;
(*d*) $\sum_{j \neq i} |a_{i,j}| < a_{i,i}$ *for all* $i \in I_n$.

A simplex $P$ is said to be in the weak standard form if its associated matrix $A$ satisfies (*a*) and (*b*). A simplex $P$ is said to be in the standard form if it is in the weak standard form and (*c*) is also satisfied by $A$. Furthermore, a simplex $P$ is said to be in the strong standard form if it is in the standard form and (*d*) is also satisfied by $A$. Observe that the standard form bears many similarities with the classical Hermite normal form (see e.g., Nemhauser and Wolsey [1988]). In Section 4 we shall show that any simplex $P$ can be represented in the standard form. Our algorithm will be applied to any simplex in any standard form. In practice, the efficiency of the algorithm will be improved as the form becomes stronger.

For a given simplex $P$ with matrix $A$, let $B$ be the matrix of $A$ without the last row and let $d$ be the vector of $b$ without the last component. Let $x^R = B^{-1}d$ and $x^I = (\lfloor x_1^R \rfloor, \cdots, \lfloor x_n^R \rfloor)$. Define

$$\Omega = \{x \in \mathbb{R}^n \mid {a^i}^\top x \leq b_i, \quad \text{for all } i \in I_n\},$$

$$\Psi = \{x \in \mathbb{R}^n \mid {a^i}^\top x > b_i, \quad \text{for all } i \in I_n\}.$$

The above notation will be used throughout the chapter. An integer point $\bar{y} \in \Omega$ is the greatest integer point if $\bar{y} \geq x$ for any $x \in \mathbb{Z}^n \cap \Omega$. The following results will be used for the convergence proof of the algorithm.

**Lemma 6.2.4** *Let a simplex $P$ be given in the weak standard form. For any two integer points $x^1$ and $x^2$ in $\Omega$, the integer point*

$$\bar{x} = (\max\{x_1^1, x_1^2\}, \cdots, \max\{x_n^1, x_n^2\})^\top.$$

*is in $\Omega$. Moreover, $\Omega$ has a unique greatest integer point.*

Proof: Since integer points $x^1$ and $x^2$ in $\Omega$, it implies that for $h = 1, 2$, it holds that

$$\sum_{j=1}^{n} a_{i,j} x_j^h \leq b_i, \; i = 1, \cdots, n.$$

Since $a_{i,j} \leq 0$ for $j \neq i$, it is easy to see that

$$\begin{aligned} a_{i,i} x_i^h &\leq b_i - \textstyle\sum_{j \neq i} a_{i,j} x_j^h \\ &\leq b_i - \textstyle\sum_{j \neq i} a_{i,j} \max\{x_j^1, x_j^2\} \\ &= b_i - \textstyle\sum_{j \neq i} a_{i,j} \bar{x}_j \end{aligned}$$

for all $i \in I_n$. This means that

$$a_{i,i} \bar{x}_i \leq b_i - \sum_{j \neq i} a_{i,j} \bar{x}_j$$

for all $i \in I_n$. Hence

$$\sum_{j=1}^{n} a_{i,j} \bar{x}_j \leq b_i$$

for all $i \in I_n$, i.e., $\bar{x} \in \Omega$. □

The following theorem gives an apparently new characterization of square Leontief matrices which generalizes a result of Dantzig [1955] (see also Gale [1960]). For other characterizations, we refer to Theorem 4 of Veinott [1968]. The following proof was inspired by that work.

**Theorem 6.2.5** *Let a simplex $P$ be given in the weak standard form. Then for any $w \in \mathbb{Z}^n$ there exist two integer points $\beta \in \Omega$ and $\gamma \in \Psi$ such that $\beta << w << \gamma$.*

Proof: By the Assumption 6.2.1, it is clear that $B^{-1}$ exists. We first prove that $B^{-1}$ is non-negative. Let $D = [a^1, a^2, \cdots, a^n]$. By the Assumptions 6.2.1 and 6.2.3(a) there exists a unique vector $x^0 \in \mathbb{R}^n_{++}$ such that $Dx^0 = -a^{n+1} > 0$. Next we show that for any $x \in \mathbb{R}^n$ $Dx \geq 0$ implies $x \geq 0$. We can decompose $D$ such that $D = \alpha I - F$ where $\alpha \in \mathbb{Z}_{++}$, $I$ is the identity matrix and $F$ is a non-negative matrix. We need to consider the following two cases.

Case (1). Assume that $-a^{n+1} >> 0$. Now suppose on the contrary that there exists $x^1 \in \mathbb{R}^n$ such that $Dx^1 \geq 0$ but $x^1 \not\geq 0$. Then some component of $x^1$ is negative. Let $\mu = max_{i \in I_n}\{-\frac{x_i^1}{x_i^0}\}$, say $\mu = -\frac{x_1^1}{x_1^0}$. Clearly, $\mu > 0$. Define $x^2 = x^1 + \mu x^0$. Note that $x^2 \geq 0$ and $x_1^2 = 0$. But on the other hand, we have

$$\alpha x^2 = \alpha I(x^0 + \mu x^1) >> Fx^0 + \mu F x^1 \geq 0,$$

yielding a contradiction with $x_1^2 = 0$. Now we show that $D^{-1}$ is non-negative. Let $y^i$ be the $i$-th column of $D^{-1}$. Then $Dy^i = e(i)$. By the above assertion we know $y^i \geq 0$. So $D^{-1}$ is non-negative.

Case (2). Suppose that some component of $a^{n+1}$ is zero. Let $I = \{i \in I_n \mid a_{n+1,i} = 0\}$ and let $J = I_n \backslash I$. Clearly, $|I| < n$ by the Assumption 6.2.1. In the Case (1) above, $I$ is empty. Suppose that for each $k \in I$ there exists $l \in J$ such that $a_{l,k} < 0$. We shall prove that this can be reduced to Case (1). Choose $k \in I$ and $l \in J$ with $a_{l,k} < 0$. Then let $x_l^1 = x_l^0 - \epsilon$ for some $\epsilon > 0$, and $x_i^1 = x_i^0$ for $i \neq l$. If $\epsilon$ is small enough, then $x^1 >> 0$ and

$$\begin{array}{lll} \sum_{h \in I_n} a_{h,i} x_h^1 & \geq 0, & \text{for all } \; i \in I \setminus \{k\} \\ \sum_{h \in I_n} a_{h,i} x_h^1 & > 0, & \text{for all } \; i \in J \cup \{k\}. \end{array}$$

So let $I$, $J$ and $x^0$ become $I \setminus \{k\}$, $J \cup \{k\}$ and $x^1$ respectively. Repeat the above process. We can reduce $I$ to be empty. Then this case is reduced to Case (1) above. We are done. Now assume that $a_{j,i} = 0$ for all $i \in I$ and all $j \in J$. We shall prove that this is impossible. We first consider a simple case in which $I = I_k$ for $k = |I|$. Then $J = I_n \setminus I_k$. Let $y^1 = (x_1^0, \cdots, x_k^0)^\top$ and $y^2 = (x_{k+1}^0, \cdots, x_n^0)^\top$. Define

$$X = \begin{bmatrix} a_{1,1} & \cdots & a_{k,1} \\ \vdots & \vdots & \vdots \\ a_{1,k} & \cdots & a_{k,k} \end{bmatrix}, \quad Y = \begin{bmatrix} a_{k+1,k+1} & \cdots & a_{n,k+1} \\ \vdots & \vdots & \vdots \\ a_{k+1,n} & \cdots & a_{n,n} \end{bmatrix},$$

$$Z = \begin{bmatrix} a_{1,k+1} & \cdots & a_{k,k+1} \\ \vdots & \vdots & \vdots \\ a_{1,n} & \cdots & a_{k,n} \end{bmatrix}.$$

Since $y^2 >> 0$, $Yy^2 >> 0$, the off-diagonal elements of $Y$ are all non-positive and the diagonal elements of $Y$ are all positive, it immediately follows from the Case (1) above that $Y^{-1} > 0$. Note that $Z \leq 0$. Let $z = (y^{1\top}, (-Y^{-1}Zy^1)^\top)^\top$. Clearly, $z > 0$ and $Dz = 0$. Then we have that $D(x^0 + \lambda z) = -a^{n+1}$ for all $\lambda \in R$. This contradicts the fact that $x^0$ is the unique solution of $Dx = -a^{n+1}$. The general case can be shown as follows. It is not difficult to find two permutation matrices $Q_1$ and $Q_2$ such that

$$Q_1 D Q_2 = \begin{bmatrix} X & 0 \\ Z & Y \end{bmatrix},$$

where $X$ is an $|I| \times |I|$ matrix, $Y$ is an $|J| \times |J|$ matrix, and the diagonal elements of $X$ and $Y$ are all positive. Now the rest of the proof goes alone the same line as before.

Since $B = D^\top$ and $B^{-1} = (D^{-1})^\top$, $B^{-1}$ is also non-negative. It is easy to choose a big enough $\gamma \in \mathbf{Z}^n$ and a big enough $y \in \mathbf{R}^n$ such that $\gamma >> w$, $\gamma \geq B^{-1}y$ and $y >> d$. It is clear that $\gamma \in \Psi$. Similarly, we can find $\beta$ as required in the theorem. □

The above proof implies

**Corollary 6.2.6** *Suppose that an $(n+1) \times n$ matrix $A$ satisfies Assumptions 6.2.1 and 6.2.3 (a) and (b). Then for any $y \in Z^n_+$ there exists $x \in \mathbf{Z}^n_+$ such that $a^{i\top}x \geq y_i$ for all $i \in I_n$. Moreover, $B^{-1}$ is non-negative.*

## 6.3. The integer labeling algorithm

In this section the algorithm will be proposed which can be applied to any simplex in any of the three standard forms.

**(1) The labeling rule and the algorithm:** Now we introduce the following labeling rule.

**Labeling Rule I:** Let a labeling function $L : \mathbf{Z}^n \mapsto N \cup \{0\}$ be given as follows. To $x \in \mathbf{Z}^n$ the label $L(x) = i$ is assigned if $i$ is the smallest index for which

$$a^{i\top}x - b_i = \max\{a^{h\top}x - b_h \mid a^{h\top}x - b_h > 0, \quad \text{for some } h \in N\}.$$

If $a^{h\top}x - b_h \leq 0$ for all $h \in N$, then the label $L(x) = 0$ is assigned to $x$.

Notice that if there exists a point $x \in \mathbf{Z}^n$ satisfying $L(x) = 0$, then $P$ contains at least one integer point. Let $T$ be the $K_1$-triangulation of $\mathbf{R}^n$ as described in Chapter 1. This triangulation is such that the collection of the vertices of simplices in $\mathcal{T}$ is the set of all integer points of $\mathbb{R}^n$. Given an $n$-simplex $\sigma(x^1, \cdots, x^{n+1})$ in $\mathcal{T}$, let

$$L(\sigma) = \{L(x^1), \cdots, L(x^{n+1})\}.$$

An $n$-simplex $\sigma$ is called *a completely labelled (c.l.) simplex* if $|L(\sigma)| = n+1$. Specifically, an $n$-simplex $\sigma$ is called *a completely labelled simplex of type I* if $L(\sigma) = \{0\} \cup N_{-i}$ for some index $i \in N$. An $n$-simplex $\sigma$ is called *a completely labelled simplex of type II* if $L(\sigma) = N$. Observe that a completely labelled simplex of type $I$ has a vertex with label 0, that is an integer point of $P$.

Define the set $\{q(1), \cdots, q(n+1)\}$ of $n+1$ vectors of $\mathbb{R}^n$ by

$$q(i) = -e(i), \; i = 1, \cdots, n, \quad \text{and} \quad q(n+1) = \sum_{i=1}^{n} e(i).$$

Let $v$ be an arbitrary integer point of $\mathbb{R}^n$. The point $v$ will be the starting point of the algorithm. Define for $T$ being a proper subset of $N$ the regions $A(T)$ by

$$A(T) = \{x \in R^n \mid x = v + \sum_{j \in T} \lambda_j q(j), \; \lambda_j \geq 0, \quad \text{for all } j \in T\},$$

with $A(\emptyset) = \{v\}$. Notice that the dimension of $A(T)$ equals $t$ with $t = |T|$, and that $bd(A(T)) = \cup_{h \in T} A(T \setminus \{h\})$. The $K_1$-triangulation subdivides any set $A(T)$ into $t$-simplices $\sigma(x^1, \pi(T))$ with vertices $x^1, \cdots, x^{t+1}$, where $x^1$ is a vertex in $A(T)$, $\pi(T) = (\pi(1), \cdots, \pi(t))$ is a permutation of the elements of the set $T$, and $x^{i+1} = x^i + q(\pi(i))$, $i = 1, \cdots, t$. For a proper subset $T$ of $N$ a $(t-1)$-simplex $\sigma(x^1, \cdots, x^t)$, $1 \leq t \leq n$, is called T-complete if the $t$ vertices of $\sigma$ carry all labels of the set $T$. Note that every vertex $y$ as a zero-dimensional simplex $\{y\}$ is $\{L(y)\}$-complete in case $L(y) \neq 0$.

Now the algorithm generates a sequence of adjacent $t$-simplices in $A(T)$ having $T$-complete common facets. Formally the steps of the integer labeling algorithm are described as follows. In the description the $(n+1)$-vector $R$ stands for the distance vector between the starting point $v$ and the vertex $x^1$ of the last generated simplex $\sigma(x^1, \cdots, x^{t+1})$, i.e., $x^1 = v + \sum_{h=1}^{n+1} R_h q(h)$. $\delta$ denotes the number of steps. In Table 6.1, $E(i)$ denotes the $i$-th unit vector of $\mathbb{R}^{n+1}$, for $i \in I_{n+1}$.

### The Integer Labeling (IL) Algorithm

**Step (0)** Set $t = 0$, $x^1 = v$, $T = \emptyset$, $\pi(T) = \emptyset$, $\sigma = < x^1 >$, $\bar{x} = x^1$, $R_i = 0$, $i \in N$, and $\delta = 1$.

| | $x^1$ becomes | $\pi(T)$ becomes | $R$ becomes |
|---|---|---|---|
| $s = 1$ | $x^1 + q(\pi(1))$ | $(\pi(2), \ldots, \pi(t), \pi(1))$ | $R + E(\pi(1))$ |
| $1 < s < t+1$ | $x^1$ | $(\pi(1), \ldots, \pi(s-2), \pi(s), \pi(s-1), \pi(s+1), \ldots, \pi(t))$ | $R$ |
| $s = t+1$ | $x^1 - q(\pi(t))$ | $(\pi(t), \pi(1), \ldots, \pi(t-1))$ | $R - E(\pi(t))$ |

TABLE 6.1. Pivot rules if the vertex $x^s$ of is to be replaced.

**Step (1)** Calculate $l(\bar{x})$ and set $L = L(\bar{x})$. If $L = 0$, an integer point is found and the algorithm terminates. If $L$ is not an element of $T$, go to Step (3). Otherwise $L = L(x^s)$ for exactly one vertex $x^s \neq \bar{x}$ of $\sigma$.

**Step (2)** If $s = t + 1$ and $R_{\pi(t)} = 0$, go to Step (4). Otherwise $\sigma$ and $R$ are adapted according to Table 6.1 by replacing $x^s$. Set $\delta = \delta + 1$. Return to Step (1) with $\bar{x}$ equal to the new vertex of $\sigma$.

**Step (3)** If $t = n$, a completely labelled simplex of type $II$ is found and the algorithm terminates. Otherwise, a $(T \bigcup \{ L \})$-complete simplex is found and $T$ becomes $T \bigcup \{ L \}$, $\pi(T)$ becomes $(\pi(1), \cdots, \pi(t), L)$, $\sigma$ becomes $\sigma(x^1, \pi(T))$, and $t$ becomes $t + 1$. Set $\delta = \delta + 1$. Return to Step (1) with $\bar{x}$ equal to $x^{t+1}$.

**Step (4)** Let, for some $k$, $k \leq t$, $x^k$ be the vertex of $\sigma$ with label $\pi(t)$. Then $T$ becomes $T \backslash \{ \pi(t) \}$, $\pi(T)$ becomes $(\pi(1), \cdots, \pi(t-1))$, $\sigma$ becomes $\sigma(x^1, \pi(T))$, $t$ becomes $t - 1$, and return to Step (2) with $s = k$ and $\delta = \delta + 1$.

Without loss of generality we may assume that the algorithm is initiated at an infeasible integer point $v$, i.e., $v \notin P$. Notice that every simplex $\sigma(x^1, \pi(T))$ generated by the algorithm lies in $A(T)$ and is a $t$-simplex of the simplicial subdivision of $A(T)$ induced by the $K_1$-triangulation of $\mathbb{R}^n$. For any $\beta, \gamma \in \mathbf{Z}^n$ with $\beta << \gamma$, define

$$C^n(\beta, \gamma) = \{ x \in \mathbb{R}^n \mid \beta \leq x \leq \gamma \}.$$

One can easily obtain the following result.

**Lemma 6.3.1** *Choose $\beta, \gamma \in \mathbf{Z}^n$ with $\beta << v << \gamma$. Then for any nonempty proper subset $T$ of $N$ with $t = |T|$, the $K_1$-triangulation also induces a triangulation of $A(T) \cap C^n(\beta, \gamma)$. For any $(t-1)$-simplex $\tau \subseteq A(T) \cap C^n(\beta, \gamma)$, then either $\tau \subseteq bd(A(T) \cap C^n(\beta, \gamma))$ and is a facet of exactly one simplex in $A(T) \cap C^n(\beta, \gamma)$, or $\tau \not\subseteq bd(A(T) \cap C^n(\beta, \gamma))$ and is a common facet of exactly two simplices in $A(T) \cap C^n(\beta, \gamma)$.*

**Definition 6.3.2** *A simplex $\sigma$ is a node if it satisfies one of the following conditions:*
*(a) $\sigma = \{v\}$ if $L(v) \neq 0$;*
*(b) $\sigma$ is a $t$-simplex in $A(T)$ for some proper subset $T$ of $N$ with $t = |T| \geq 1$ and at least one facet of $\sigma$ is $T$-complete.*

**Definition 6.3.3** *Two nodes $\sigma^1$ and $\sigma^2$ are adjacent and therefore connected by an arc $e = \{\sigma^1, \sigma^2\}$ if both $\sigma^1$ and $\sigma^2$ are in $A(T)$ for some proper subset $T$ of $N$, $t = |T|$, and one of the following cases occurs:*
*(a) $\sigma^1$ and $\sigma^2$ are both $t$-simplices and share a common $T$-complete facet;*
*(b) either $\sigma^1$ is a $T$-complete facet of $\sigma^2$ and $\sigma^2$ is a $t$-simplex or $\sigma^2$ is a $T$-complete facet of $\sigma^1$ and $\sigma^1$ is a $t$-simplex.*

*Moreover, the degree of a node $\sigma$ is equal to the number of nodes adjacent to $\sigma$, denoted by $deg(\sigma)$.*

From the above two definitions we obtain a graph $G = (V, A)$ where $V$ is the set of nodes and $A$ is the set of arcs.

**Lemma 6.3.4** *Let $\sigma$ be a node in the graph $G = (V, A)$. Then*
*(i) $deg(\sigma) = 1$ when $\sigma = \{v\}$;*
*(ii) $deg(\sigma) = 1$ when $\sigma$ is a completely labelled simplex of type $II$ or $\sigma$ has a vertex labelled with 0;*
*(iii) $deg(\sigma) = 2$ in all other cases.*
*Moreover, there exists a unique simple path starting with $\sigma^0 = \{v\}$.*

Proof: (i) If $\sigma = \{v\}$, then $\sigma^1 = \{v, y\}$ is a 1-dimensional simplex, where $y = v + q(L(v))$. Let $T^0 = \{L(v)\}$. Then $\sigma$ is a $T^0$-complete facet of $\sigma^1$ in the set $A(T^0)$. Hence $\sigma^1$ is also a node. Moreover, $\sigma^1$ is the only node adjacent to $\sigma$. So $deg(\{v\}) = 1$.

(ii) We first consider the case in which the node $\sigma$ is a completely labelled simplex of type $II$. This implies that $\sigma$ lies in a set $A(T)$ for some $T \subset N$ with $|T| = n$. Let $\tau$ be the unique $T$-complete facet of $\sigma$. First we consider the case in which $\tau \subseteq \cup_{i \in T} A(T \setminus \{i\})$. Thus $\tau \subseteq A(T \setminus \{h\})$ for some $h \in T$. In this case $\tau$ is a simplex in $A(T \setminus \{h\})$ with a $(T \setminus \{h\})$-complete facet. So, $\tau$ is a node. Since $\tau$ is on the boundary of $A(T)$, $\tau$ is the only node adjacent to $\sigma$, i.e., $deg(\sigma) = 1$. Next if $\tau$ does not lie on the boundary of $A(T)$, it follows from Lemma 6.3.1 that there exists exactly one simplex $\sigma^1$ in $A(T)$ sharing the common facet $\tau$ with $\sigma$. It is clear that $\sigma^1$ is the only node adjacent to $\sigma$. Hence $deg(\sigma) = 1$. Now we consider the case in which $\sigma$ is a node with a vertex labelled by zero. Then $\sigma$ lies in a set $A(T)$ for some $T \subset N$ with $|T| \leq n$. Moreover, $\sigma$ has a unique $T$-complete facet $\tau$. First we consider the case in which $\tau \subseteq \cup_{i \in T} A(T \setminus \{i\})$. Thus $\tau \subseteq A(T \setminus \{h\})$ for some $h \in T$. In this case $\tau$ is a simplex in $A(T \setminus \{h\})$ with a $(T \setminus \{h\})$-complete facet. So, $\tau$ is a node. Since $\tau$ is on the boundary of $A(T)$, $\tau$ is the only node adjacent to $\sigma$, i.e., $deg(\sigma) = 1$. Next if $\tau$ does not lie on the boundary of $A(T)$, it follows from Lemma 6.3.1 that there exists exactly one simplex $\sigma^1$ in $A(T)$ sharing the common facet $\tau$ with $\sigma$. It is clear that $\sigma^1$ is the only node adjacent to $\sigma$. Again $deg(\sigma) = 1$.

(iii) Let a node $\sigma$ in $A(T)$ be a $t$-dimensional simplex, where $t = |T|$, with vertices $x^1, \cdots, x^{t+1}$. First consider the case in which $\{L(x^i) \mid i \in I_{t+1}\}$ is equal to $T$. In this case $\sigma$ has two $T$-complete facets $\tau_1$ and $\tau_2$. Consider $\tau_1$ for example. Suppose that $\tau_1 \subseteq \cup_{i \in T} A(T \setminus \{i\})$. This means that $|T| \geq 1$ and $\tau_1 \subseteq A(T \setminus \{h\})$ for some $h \in T$. When $|T| = 1$, $\tau_1 = \{v\}$. Hence $\tau_1$ is a node adjacent to $\sigma$. When $|T| > 1$, $\tau_1$ clearly has a $(T \setminus \{h\})$-complete facet. Hence $\tau_1$ is a node adjacent to $\sigma$. In the case $\tau_1$ does not lie in the boundary of $A(T)$, it follows from Lemma 6.3.1 that there exists exactly one simplex $\sigma^1$ in $A(T)$ sharing the common facet $\tau_1$ with $\sigma$. Evidently, $\sigma^1$ is a node adjacent to $\sigma$. The same discussion holds for $\tau_2$. Thus, $deg(\sigma) = 2$. Now we consider the case in which $\sigma$ has a vertex $x^+$ whose label, say $j$, does not belong to $T$. Clearly, $|T| < n - 1$. Let $\tau$ be the only $T$-complete facet of $\sigma$. First suppose that $\tau \subseteq \cup_{i \in T} A(T \setminus \{i\})$. Then $\tau \subseteq A(T \setminus \{h\})$ for some $h \in T$. $\tau$ is a node adjacent to $\sigma$. Next, in the case $\tau$ does not lie on the boundary of $A(T)$, then by Lemma 6.3.1 there exists exactly one simplex $\sigma^1$ in $A(T)$ sharing the common facet $\tau$ with $\sigma$. So $\sigma^1$ is a node adjacent to $\sigma$. On the other hand, let $\sigma^2$ be the unique $(t+1)$-dimensional simplex in $A(T \cup \{j\})$ with $\sigma$ as its facet. Clearly $\sigma^2$ is a node adjacent to $\sigma$. There are no other nodes adjacent to $\sigma$. Hence $deg(\sigma) = 2$. □

Lemma 6.3.4 implies that the simple path starting with $\sigma^0 = \{v\}$ either leads to another node $\sigma^l$ of degree one or goes to infinity. In the former case the algorithm terminates with an integer point in $P$ or a completely labelled simplex of type $II$. The coming result shows that the path will never leave the set $C^n(\beta, \gamma)$ if $\beta$ and $\gamma$ are integer points such that $\beta \in \Omega$, $\gamma \in \Psi$ and $\beta << v << \gamma$. The proof is similar to that of Theorem 5.3.11.

**Lemma 6.3.5** *Let $P$ be a simplex in the weak standard form. For any proper subset $T$ of $N$, there is no $T$-complete $(t-1)$-simplex in $A(T) \cap bd(C^n(\beta, \gamma))$ if the two integer vectors $\beta$ and $\gamma$ are chosen such that $\beta \in \Omega$ and $\gamma \in \Psi$ with $\beta << v << \gamma$.*

Proof: By Theorem 6.2.5 there exists two integer vectors $\beta$ and $\gamma$ such that $\beta \in \Omega$ and $\gamma \in \Psi$ with $\beta << v << \gamma$. For each $i \in I_n$, define

$$\begin{aligned} D(\gamma_i) &= \{x \in \mathbf{Z}^n \mid x_i = \gamma_i\} \\ D(\beta_i) &= \{x \in \mathbf{Z}^n \mid x_i = \beta_i\}. \end{aligned}$$

We need to consider the following two cases:

Case (1) Given $i \in I_n$ and a point $x \in D(\beta_i)$, we have that $x_i = \beta_i$ and for $h \neq i$, $x_h = \beta_h + \alpha_h$ with $\alpha_h \geq 0$. Since $\beta^\top a^i - b_i \leq 0$, and

$\sum_{h\neq i} a_{i,h}\alpha_h \le 0$ by the Assumption 6.2.3(b), we have

$$\begin{array}{rcl} x^\top a^i - b_i & = & a_{i,i}x_i + \sum_{h\neq i} a_{i,h}x_h - b_i \\ & = & a_{i,i}\beta_i + \sum_{h\neq i} a_{i,h}(\beta_h + \alpha_h) - b_i \\ & = & \beta^\top a^i - b_i + \sum_{h\neq i} a_{i,h}\alpha_h \\ & \le & 0. \end{array}$$

This implies that $L(x) \neq i$. So there is no label $i$ in $D(\beta_i)$ for any $i \in I_n$.

Case (2) Given $i \in I_n$ and a point $x \in D(\gamma_i)$, we have that $x_i = \gamma_i$ and for $h \neq i$, $x_h = \gamma_h - \alpha_h$ with $\alpha_h \ge 0$. Since $\beta \in \Psi$, it is easy to see that $\beta^\top a^{n+1} - b_{n+1} < 0$. Since $\sum_{h\neq i} a_{n+1,h}\alpha_h \le 0$ by the Assumption 6.2.3(a), we have

$$\begin{array}{rcl} x^\top a^{n+1} - b_{n+1} & = & \gamma^\top a^{n+1} - b_{n+1} + \sum_{h\neq i} a_{n+1,h}\alpha_h \\ & < & 0. \end{array}$$

This implies that $L(x) \neq n+1$. So there is no $n+1$ label in $D(\gamma_i)$ for any $i \in I_n$.

In order to obtain the desired results, we need to address the following two cases.

Case (a). For any nonempty proper subset $T$ of $N$ with $n+1 \notin T$ and $t = |T|$, then we have that $A(T) \cap bd(C^n(\beta,\gamma)) = \cup_{i\in T} A(T) \cap D(\beta_i)$. If a $(t-1)$-dimensional simplex $\sigma$ lies on $A(T) \cap bd(C^n(\beta,\gamma))$, there exists an index $i \in T$ such that $\sigma \subseteq A(T) \cap D(\beta_i)$. Since $\sigma$ has no label $i$ by Case (1) above, $\sigma$ is not $T$-complete.

Case (b). For any nonempty proper subset $T$ of $N$ with $n+1 \in T$ and $t = |T|$, then we have that

$$A(T) \cap bd(C^n(\beta,\gamma)) = (\cup_{i\in T\setminus\{n+1\}} A(T) \cap D(\beta_i)) \cup (\cup_{i\in I_n} A(T) \cap D(\gamma_i)).$$

If a $(t-1)$-dimensional simplex $\sigma$ lies on $A(T) \cap bd(C^n(\beta,\gamma))$, there exists either an index $i \in T \setminus \{n+1\}$ such that $\sigma \subseteq A(T) \cap D(\beta_i)$, or an index $i \in I_n$ such that $\sigma \subseteq A(T) \cap D(\gamma_i)$. In the former case $\sigma$ has no label $i \in T \setminus \{n+1\}$ by Case (1) above. In the latter case $\sigma$ has no label $n+1$ by Case (2) above. Hence in either case $\sigma$ is not $T$-complete. □

This result is shown in Figure 6.1 by the following example.

Example 1: The simplex is given by

$$P = \{\, x \in R^2 \mid a^{i\top} x \le b_i,\ i = 1, \cdots, 3 \,\},$$

where $a^1 = (2,-1)^\top$, $a^2 = (-1,3)^\top$, and $a^3 = (-1,-1)^\top$, $b_1 = 1$, $b_2 = -1$, and $b_3 = 1$. Let $v = (0,1)^\top$. Choose $\beta = (-3,-3)^\top$ and $\gamma = (3,2)^\top$. There are no $T$-complete simplices on $A(T) \cap bd(C^2(\beta,\gamma))$ for any $T \subset I_3$.

Summarizing the above discussion, we come to the global convergence result of the algorithm.

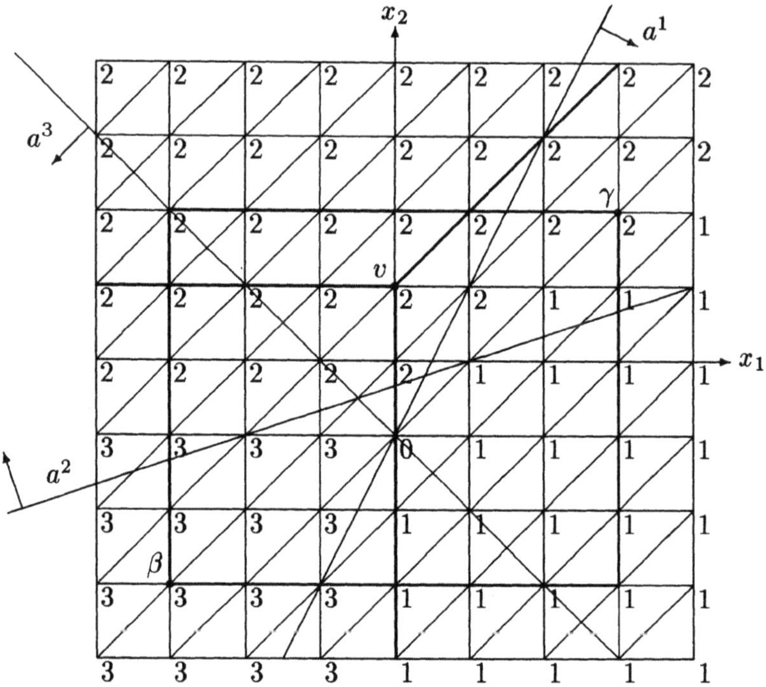

*Figure 6.1.* There are no $T$-complete simplices on $A(T) \cap bd(C^n(\beta, \gamma))$.

**Theorem 6.3.6** *Let a simplex $P$ be given in the weak standard form. Then the algorithm will terminate with either an integer point in $P$ or a completely labelled simplex of type $II$, within a finite number of steps. Moreover, the algorithm only operates on $C^n(\beta, \gamma)$.*

Proof: Suppose that the simple path starting with $\{v\}$ goes out the set $C^n(\beta, \gamma)$. Then there exists a node $\sigma$ in $C^n(\beta, \gamma)$ with $deg(\sigma) = 2$ for some $T \subset N$ such that a $T$-complete facet $\tau$ lies on $A(T) \cap bd(C^n(\beta, \gamma))$. This contradicts Lemma 6.3.5. So the algorithm will never leave the set $C^n(\beta, \gamma)$. Notice that the nodes in $C^n(\beta, \gamma)$ are finite since the number of simplices in $C^n(\beta, \gamma)$ is finite. It follows from Lemma 6.3.4 and Theorem 1.4.14 that the path will never visit a node twice. This completes the proof. □

Although the above theorem guarantees the finite convergence of the algorithm, it may well be the case that the algorithm only finds a completely labelled simplex of type $II$ even if the simplex $P$ does contain integer points. The next basic theorem states that the algorithm will always find the unique greatest integer point $\bar{y}$ in $\Omega$ if we restrict the starting point $v$ in $\Psi$. Hence if the label of $\bar{y}$ is zero, $\bar{y}$ must be an integer point in $P$. If the label of $\bar{y}$ is $n+1$, then the simplex $P$ does not contain any integer point at all.

**Theorem 6.3.7 (Test Theorem)** *Let a simplex $P$ be given in the weak standard form. Starting with any integer point $v$ in $\Psi$, the algorithm terminates with either the greatest integer point in $P$ or a completely labelled simplex of type $II$ indicating that there is no integer point in $P$, within a finite number of iterations.*

Proof: According to Lemma 6.2.4, $\Omega$ has a unique greatest integer point, say $\bar{y}$. It is clear that $v >> \bar{y}$. In fact $\bar{y} << x$ for all $x \in \Psi$. By Theorem 6.2.5 there exists an integer point $\gamma \in \Psi$ such that $\gamma >> v$. Let $\beta = \bar{y}$. Consider the set $C^n(\beta, \gamma)$. Then all the conditions of Lemma 6.3.5 are satisfied. By Theorem 6.3.6 the algorithm will either lead to an integer point in $P$ or terminates with a completely labelled simplex of type $II$. Since the operation of the algorithm is confined to $C^n(\beta, \gamma)$, and the point $\bar{y}$ is the unique point which either has the label zero or the label $n+1$, then the algorithm must terminate with a node of degree one with one vertex being $\bar{y}$. When the label of $\bar{y}$ is zero, then $\bar{y}$ is an integer point in $P$. When the label of $\bar{y}$ is $n+1$, the simplex $P$ contains no integer point. This completes the proof. □

In the third subsection we shall discuss how to get an integer point in $\Psi$. By slightly modifying the proof of Theorem 6.3.7, we have the following corollary which extends Theorem 6.3.7.

**Corollary 6.3.8** *Let a simplex $P$ be given in the weak standard form. Starting with any point $v \in \mathbf{Z}^n$ with $v \geq x^I$, the algorithm terminates with either the greatest integer point in $P$ or a completely labelled simplex of type $II$ indicating that there is no integer point in $P$, within a finite number of iterations.*

Proof: First note that $x^I \geq \bar{y}$ where $\bar{y}$ is the greatest integer point in $\Omega$. By Theorem 6.2.5 there exists an integer point $\gamma \in \Psi$ with $\gamma >> v$. Let $\beta = \bar{y}$. Consider the set $C^n(\beta, \gamma)$. Let $S = \{i \in I_n \mid v_i = \bar{y}_i\}$. If $S$ is empty, it is reduced to Theorem 6.3.7. So we consider the case in which $S$ is nonempty. For each $i \in I_n$, define

$$\begin{aligned} D(\gamma_i) &= \{x \in \mathbf{Z}^n \mid x_i = \gamma_i\} \\ D(\beta_i) &= \{x \in \mathbf{Z}^n \mid x_i = \beta_i\}. \end{aligned}$$

Analogous to Lemma 6.3.4, we have the following results: (1) there is no label $i$ in $D(\beta_i)$ for any $i \in I_n$; (2) there is no $n+1$ label in $D(\gamma_i)$ for any $i \in I_n$.

In order to obtain the desired results, we need to address the following three cases.

Case (a). For any nonempty proper subset $T$ of $N$ with $n+1 \notin T$, $T \subseteq I_n \backslash S$ and $t = |T|$, then we have that $A(T) \cap bd(C^n(\beta, \gamma)) = \cup_{i \in T} A(T) \cap D(\beta_i)$. If a $(t-1)$-dimensional simplex $\sigma$ lies on $A(T) \cap bd(C^n(\beta, \gamma))$, there exists an index $i \in T$ such that $\sigma \subseteq A(T) \cap D(\beta_i)$. Since $\sigma$ has no label $i$ by the Case (1) above, $\sigma$ is not $T$-complete.

Case (b). For any nonempty proper subset $T$ of $N$ with $n+1 \in T$, $T \subseteq N \setminus S$ and $t = |T|$, then we have that

$$A(T) \cap bd(C^n(\beta,\gamma)) = (\cup_{i \in T\setminus\{n+1\}} A(T) \cap D(\beta_i)) \cup (\cup_{i \in I_n} A(T) \cap D(\gamma_i)).$$

If a $(t-1)$-dimensional simplex $\sigma$ lies on $A(T) \cap bd(C^n(\beta,\gamma))$, there exists either an index $i \in T \setminus \{n+1\}$ such that $\sigma \subseteq A(T) \cap D(\beta_i)$, or an index $i \in I_n$ such that $\sigma \subset A(T) \cap D(\gamma_i)$. In the former case $\sigma$ has no label $i \in T \setminus \{n+1\}$ by Case (1) above. In the latter case $\sigma$ has no label $n+1$ by Case (2) above. Hence in either case $\sigma$ is not $T$-complete.

Case (c). For any $i \in S$, there is no label $i$ in $D(\beta_i)$ by the Case (1). Since $L(x) \neq i$ for any $x \in Z^n \cap D(\beta_i)$, by Step (3) of the algorithm a point $y = x + q(i) = x - e(i)$ lying outside $C^n(\beta,\gamma)$ cannot occur.

It follows from Cases (a), (b) and (c) that the algorithm will never leave the set $C^n(\beta,\gamma)$. Now the rest of the proof goes alone the same line as in Theorem 6.3.7. □

The next result can be somehow viewed as a generalization of Sperner lemma.

**Theorem 6.3.9** *Let a simplex $P$ be given in the weak standard form. Then there exists at least one completely labelled simplex.*

Proof: In case the simplex $P$ contains no integer points, the result follows directly from Theorem 6.3.6. If the simplex $P$ contains integer points, then we label all integer points in $P$ by $n+1$. It follows from Theorem 6.3.6 that there exists a completely labelled simplex of type $II$, say $\sigma^1$. Let $x$ be the vertex of $\sigma^1$ with label $n+1$. If $x$ is an integer point in $P$, then restore the true label of $x$, i.e., $L(x) = 0$. Hence in this case $P$ also has a completely labelled simplex. □

From Theorem 6.3.7 and Corollary 6.3.8 we can easily see that the algorithm in fact gives a way for solving the integer linear programming problem

$$\begin{array}{ll} \min & a^{(n+1)\top} x, \\ \text{s.t.} & a^{i\top} x \le b_i,\ i = 1,\cdots,n, \\ & x \in \mathbf{Z}^n, \end{array}$$

under the assumption that the origin of $\mathbf{R}^n$ is contained in the interior of the convex hull of the vectors $a^1, \cdots, a^{n+1}$. Let us illustrate the algorithm by two examples. Recall in Example 1 the simplex is given by

$$P = \{x \in \mathbf{R}^2 \mid a^{i\top} x \le b_i,\ i = 1,\cdots,3\},$$

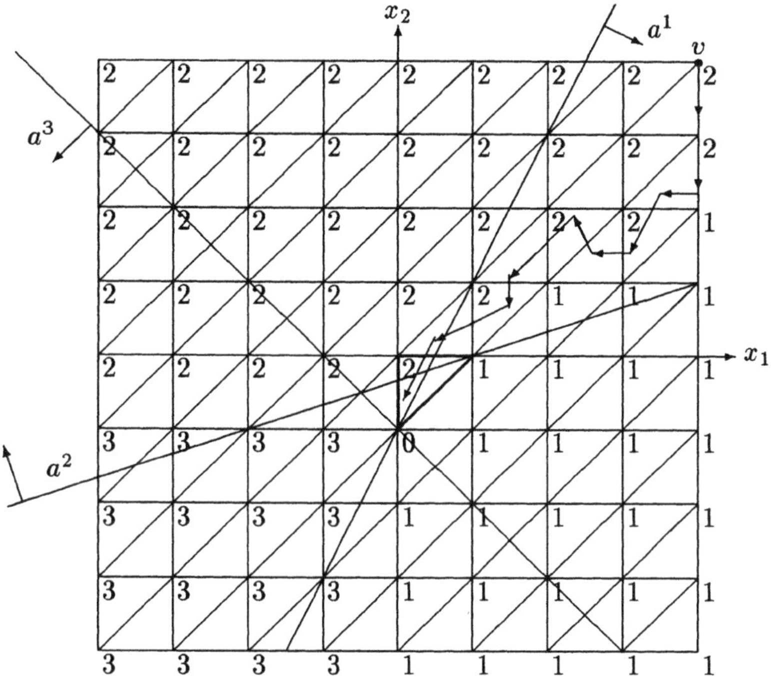


*Figure 6.2.* The IL algorithm finds an integer point of P.

where $a^1 = (2,-1)^\top$, $a^2 = (-1,3)^\top$, and $a^3 = (-1,-1)^\top$, $b_1 = 1$, $b_2 = -1$, and $b_3 = 1$. The path generated by the algorithm from $v = (4,4)^\top$ leads to the integer point $(0,-1)^\top$ in $P$ and are shown in Figure 6.2. The point $\bar{y} = (0,-1)^\top$ is found.

Example 2: The simplex is given by

$$P = \{\, x \in \mathbb{R}^2 \mid a^{i\top} x \leq b_i,\ i = 1,\cdots,3 \,\},$$

where $a^1 = (5,-1)^\top$, $a^2 = (0,1)^\top$, and $a^3 = (-3,0)^\top$, $b_1 = 1$, $b_2 = 2$, and $b_3 = -1$. The path generated by the algorithm from $v = (4,4)^\top$ leads to the unique completely labelled simplex of type $II$ and is demonstrated in Figure 6.3. The point $\bar{y} = (0,2)^\top$ with label 3 is found.

**(2) Special cases:** Now we are going to discuss simplices in the strong standard from. This case has been discussed in Scarf [1986]. In this case it can be shown that primitive sets coincide with simplices. So the results here are very close to those of Scarf. We point out that simplices in the strong standard form are very restrictive. Nevertheless, simplices in this form have very nice properties as the next theorem indicates. The geometric context of the theorem can be easily understood in two dimension.

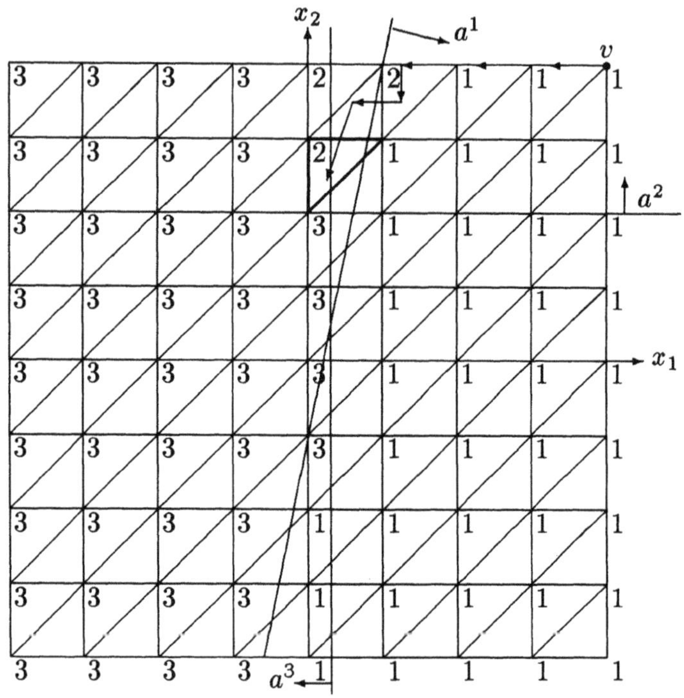

*Figure 6.3.* The IL algorithm finds the unique c.l. simplex.

**Theorem 6.3.10 (Exclusion Theorem)** *Let a simplex $P$ be given in the strong standard form. Then the Labeling Rule I precludes the possibility of the coexistence of a completely labelled simplex of type I and a completely labelled simplex of type II. In other words, $P$ contains an integer point if and only if there are no completely labelled simplices of type II.*

Proof: We only need to consider the case in which $P$ contains an integer point, say $x^0$, i.e., $Ax^0 \leq b$. Let us suppose to the contrary that there is a completely labelled simplex of type $II$, say $\sigma(x^1, \pi)$ with vertices $x^1$, $\cdots$, $x^{n+1}$, where $\pi = (\pi(1), \cdots, \pi(n+1))$ is a permutation of the $n+1$ elements of $N$, and

$$x^{i+1} = x^i + q(\pi(i)),\ i = 1, \cdots, n;$$
$$x^1 = x^{n+1} + q(\pi(n+1)).$$

Now it is easy to see that there exist non-negative integers $k_1^1$, $\cdots$, $k_{n+1}^1$ such that

$$x^1 = x^0 + \sum_{i \in N} k_i^1 q(i),$$

and

$$\min_{h \in N} k_h^1 = 0.$$

Let

$$l = argmin\{\, \pi^{-1}(h) \mid k_h^1 = \max_{j \in N} k_j^1 \,\}.$$

Then there exist non-negative integers $k_1^i$, $\cdots$, $k_{n+1}^i$ such that

$$x^i = x^0 + \sum_{j \in N} k_j^i q(j)$$

for $i \in N_{-1}$. Notice that

$$k_l^i = \max_{j \in N} k_j^i$$

for any $i \in N$.

The following cases need to be addressed:

(1). If $1 \leq l \leq n$, for any $i \in N$ it holds

$$\begin{array}{rcl} a^{l\top} x^i - b_l & = & a^{l\top} x^0 - b_l + \sum_{h=1}^{n+1} k_h^i a^{l\top} q(h) \\ & \leq & -k_l^i a_{l,l} - \sum_{h=1,h\neq l}^{n} k_h^i a_{l,h} + k_{n+1}^i \sum_{h=1}^{n} a_{l,h} \\ & \leq & -k_l^i a_{l,l} - \sum_{h=1,h\neq l}^{n} k_l^i a_{l,h} + k_{n+1}^i \sum_{h=1}^{n} a_{l,h} \\ & \leq & (k_{n+1}^i - k_l^i) \sum_{h=1}^{n} a_{l,h} \\ & \leq & 0. \end{array}$$

This implies that $L(x^i) \neq l$ for $i \in N$.

(2). If $l = n+1$, for any $i \in N$ it holds

$$\begin{array}{rcl} a^{(n+1)\top} x^i - b_{n+1} & = & a^{(n+1)\top} x^0 - b_{n+1} + \sum_{h=1}^{n+1} k_h^i a^{(n+1)\top} q(h) \\ & \leq & \sum_{h=1}^{n+1} k_h^i a^{(n+1)\top} q(h) \\ & \leq & -\sum_{h=1}^{n} k_h^i a_{n+1,h} + k_{n+1}^i \sum_{h=1}^{n} a_{n+1,h} \\ & \leq & -\sum_{h=1}^{n} k_{n+1}^i a_{n+1,h} + k_{n+1}^i \sum_{h=1}^{n} a_{n+1,h} \\ & \leq & (k_{n+1}^i - k_{n+1}^i) \sum_{h=1}^{n} a_{n+1,h} \\ & = & 0. \end{array}$$

This implies that $L(x^i) \neq n+1$ for all $i \in N$. We conclude from (1) and (2) with a contradiction. The proof is complete. □

The above result implies that if a simplex $P$ is given in the strong standard form, then the starting point $v$ can be chosen arbitrarily. Moreover, it is easy to derive the following result.

**Corollary 6.3.11** *Let a simplex $P$ be given in the strong standard form If $P$ contains an integer point, then the Labeling Rule results in at least $n+1$ completely labelled simplices of type I and no completely labelled simplex of type II.*

Proof: Consider the simplest case in which $P$ contains a single integer point, say $w$. The general case can be shown in a similar way. Choose $w$ to be the starting point since it is allowed. Take an arbitrary index $k$ from $N$ and set $L(w) = k$ artificially. Then the algorithm will terminate with a completely labelled simplex, say $\sigma_k$, within a finite number of steps. If $w$ is a vertex of $\sigma_k$, then restore the true label of $w$, i.e., $L(w) = 0$. Hence $\sigma_k$ is of type $I$. Otherwise, it must be of type $II$. However, this can not happen according to Theorem 6.3.10. Clearly, for every $k \in N$ a different simplex of type $I$ is obtained. □

We remark that in general the Exclusion Theorem does not hold for the problems in the standard form.

**(3) How to get an integer point in $\Psi$:** In order to obtain an integer point in $\Psi$, we need to modify some definitions in Part 1 of this section. First we introduce the following labeling rule.

**Labeling Rule II:** Let a labeling function $L : \mathbf{Z}^n \mapsto N$ be given as follows.

To $x \in \mathbf{Z}^n$ the label $L(x) = i$ is assigned if $i$ is the smallest index for which

$$a^{i\top}x - b_i = \min\{a^{h\top}x - b_h \mid a^{h\top}x - b_h \leq 0, \quad \text{for some} \quad h \in I_n\}.$$

If $a^{h\top}x - b_h > 0$ for all $h \in I_n$, then the label $L(x) = n + 1$ is assigned to $x$.

Notice that if there exists a point $x \in \mathbf{Z}^n$ satisfying $L(x) = n + 1$, then $x$ is in the set $\Psi$. We will apply the algorithm in Part 1 of this section to find a completely labelled simplex of type $II$. In order to do so, we need to replace the set $\{q(1), \cdots, q(n+1)\}$ by the set $\{\bar{q}(1), \cdots, \bar{q}(n+1)\}$ where $\bar{q}(i) = e(i)$ for all $i \in I_n$ and $\bar{q}(n+1) = -\sum_{i=1}^{n} e(i)$. Accordingly, the sets $A(T)$ also change by using $\bar{q}(i)$ instead of $q(i)$. Now we prove the following result.

**Lemma 6.3.12** *Let a simplex $P$ be given in the weak standard form. For any proper subset $T$ of $N$, there is no $T$-complete $(t-1)$-simplex in $A(T) \cap bd(C^n(\beta, \gamma))$ with respect to the labeling rule $II$ if the two integer vectors $\beta$ and $\gamma$ are chosen such that $\beta \in \Omega$ and $\gamma \in \Psi$ with $\beta << v << \gamma$.*

Proof: By Theorem 6.2.5 there exists two integer vectors $\beta$ and $\gamma$ such that $\beta \in \Omega$ and $\gamma \in \Psi$ with $\beta << v << \gamma$. For each $i \in I_n$, define

$$\begin{aligned} D(\gamma_i) &= \{x \in \mathbf{Z}^n \mid x_i = \gamma_i\} \\ D(\beta_i) &= \{x \in \mathbf{Z}^n \mid x_i = \beta_i\}. \end{aligned}$$

We need to consider the following two cases:

Case (1) Given $i \in I_n$ and a point $x \in D(\gamma_i)$, we have that $x_i = \gamma_i$ and for $h \neq i$, $x_h = \gamma_h - \alpha_h$ with $\alpha_h \geq 0$. Since $a^{i\top}\gamma - b_i > 0$, and $\sum_{h \neq i} a_{i,h}\alpha_h \leq 0$ by the Assumption 6.2.3 (b), we have

$$\begin{aligned} a^{i\top}x - b_i &= a_{i,i}x_i + \textstyle\sum_{h \neq i} a_{i,h}x_h - b_i \\ &= a_{i,i}\gamma_i + \textstyle\sum_{h \neq i} a_{i,h}(\gamma_h - \alpha_h) - b_i \\ &= a^{i\top}\gamma - b_i - \textstyle\sum_{h \neq i} a_{i,h}\alpha_h \\ &> 0. \end{aligned}$$

This implies that $L(x) \neq i$. So there is no label $i$ in $D(\gamma_i)$ for any $i \in I_n$.
Case (2) Given $i \in I_n$ and a point $x \in D(\beta_i)$, we have that $x_i = \gamma_i$ and for $h \neq i$, $x_h = \beta_h + \alpha_h$ with $\alpha_h \geq 0$. Since $\gamma \in \Omega$, it is easy to see that $a^{i\top}\beta - b_i \leq 0$. Moreover, $\sum_{h \neq i} a_{i,h}\alpha_h \leq 0$ by the Assumption 6.2.3 (b). Now we have

$$\begin{aligned} a^{i\top}x - b_i &= a_{i,i}x_i + \textstyle\sum_{h \neq i} a_{i,h}x_h - b_i \\ &= a_{i,i}\beta_i + \textstyle\sum_{h \neq i} a_{i,h}(\beta_h + \alpha_h) - b_i \\ &= a^{i\top}\beta - b_i + \textstyle\sum_{h \neq i} a_{i,h}\alpha_h \\ &\leq 0. \end{aligned}$$

This implies that $L(x) \neq n+1$. So there is no $n+1$ label in $D(\beta_i)$ for any $i \in I_n$.

In order to obtain the desired results, we need to address the following two cases.
Case (a). For any nonempty proper subset $T$ of $N$ with $n+1 \notin T$ and $t = |T|$, then we have that $A(T) \cap bd(C^n(\beta, \gamma)) = \cup_{i \in T} A(T) \cap D(\gamma_i)$. If a $(t-1)$-dimensional simplex $\sigma$ lies on $A(T) \cap bd(C^n(\beta, \gamma))$, there exists an index $i \in T$ such that $\sigma \subseteq A(T) \cap D(\gamma_i)$. Since $\sigma$ has no label $i$ by Case (1) above, $\sigma$ is not $T$-complete.
Case (b). For any nonempty proper subset $T$ of $N$ with $n+1 \in T$ and $t = |T|$, then we have that

$$A(T) \cap bd(C^n(\beta, \gamma)) = (\cup_{i \in T \setminus \{n+1\}} A(T) \cap D(\gamma_i)) \cup (\cup_{i \in I_n} A(T) \cap D(\beta_i)).$$

If a $(t-1)$-dimensional simplex $\sigma$ lies on $A(T) \cap bd(C^n(\beta, \gamma))$, there exists either an index $i \in T \setminus \{n+1\}$ such that $\sigma \subseteq A(T) \cap D(\gamma_i)$, or an index $i \in I_n$ such that $\sigma \subseteq A(T) \cap D(\beta_i)$. In the former case $\sigma$ has no label $i \in T \setminus \{n+1\}$ by the Case (1) above. In the latter case $\sigma$ has no label $n+1$ by Case (2) above. Hence in either case $\sigma$ is not $T$-complete. □

Following the same line of Theorem 6.3.6, we have the following global convergence result of the algorithm.

**Theorem 6.3.13** *Let a simplex $P$ be given in the weak standard form. By using the labeling rule II, the algorithm will terminate with a completely*

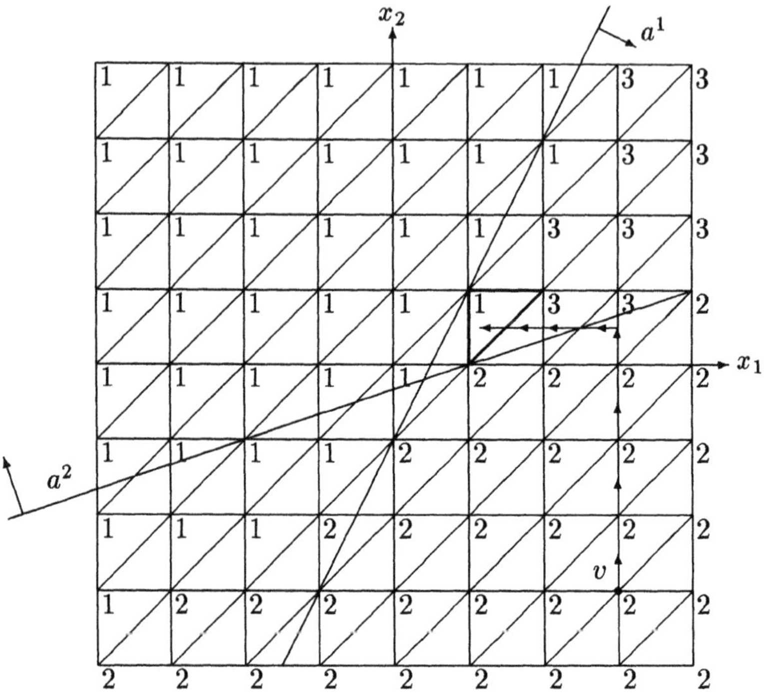

*Figure 6.4.* The IL algorithm finds integer points in $\Psi$.

*labelled simplex of type II, within a finite number of steps. Moreover, the algorithm only operates on* $C^n(\beta, \gamma)$.

The above theorem shows that starting with any point $v \in \mathbf{Z}^n$, the algorithm always finds an integer point in $\Psi$, because the vertex with label $n+1$ of a completely labelled simplex of type *II* is an integer point of $\Psi$. Let us illustrate the algorithm by Example 1 in Figure 6.4. Starting with $v = (3, -3)^\top$, the algorithm finds a completely labelled simplex whose vertices are all integer vectors in $\Psi$.

## 6.4. Unimodular transformation

In order to conform to the standard form, let us come back to the original problem. We are given an $m$-dimensional simplex

$$P = \{x \in \mathbf{R}^n \mid Ax \leq b\},$$

where $a^{i\top} = (a_{i,1}, \cdots, a_{i,n})$ is the $i$-th row of the matrix $A$ for $i \in N$, and $b = (b_1, \cdots, b_{n+1})^\top$.

A linear transformation $U$ is called *a unimodular transformation* if $U$ is bijective on $\mathbf{Z}^n$. $U$ is *unimodular* if and only if the entries of $U$ are integer and the determinant of $U$ is equal to 1 or $-1$. Let $\bar{A}$ be the product $AU$. If $U$ is unimodular, it is readily seen that given an integer point $x$ and $y = Ux$, $Ax \leq b$ if and only if $\bar{A}x \leq b$. Let $\mathbf{I}$ be the identity matrix. Clearly, $\mathbf{I}$ is a unimodular matrix. In order to get a useful unimodular matrix, we need to study some other examples and their effect when they postmultiply a matrix $A$ and premultiply a vector $x$:

(i) Interchange: $U$ is equal to $\mathbf{I}$ except that the $k$-th column of $U$ is $e(l)$ and the $l$-th column of $U$ is $e(k)$. This transformation switches columns $k$ and $l$ of $A$ and switches the $k$-th and $l$-th components of $x$.

(ii) Reversal of sign: $U$ is equal to $\mathbf{I}$ except that the $k$-th column of $U$ is equal to $-e(k)$. $U$ changes the sign of the entries of the $k$-th column of $A$ and of the $k$-th component of $x$.

(iii) Addition: $U$ is equal to $\mathbf{I}$ except that the $(k, l)$-th entry of $U$, $k \neq l$, is equal to one. This transformation replaces the $l$-th column of $A$ by the sum of columns $k$ and $l$ and replaces the $k$-th component of $x$ by the sum of components $k$ and $l$.

The following result can be found in Newman [1972].

**Theorem 6.4.1** *Every unimodular matrix can be expressed as a finite product of unimodular matrices of type (i), (ii) and (iii).*

The next basic result says that every simplex can be brought into the standard form by a unimodular transformation. We point out that this theorem was first discovered by Pnueli [1968] and independently rediscovered by the current author. Pnueli's result was formulated in a different form and proved in a somehow different way. White [1983] also independently discovered a weaker version of this theorem. More precisely, White proved by induction that every simplex $P$ can be transformed into the weak standard form.

**Theorem 6.4.2 (Transformation Theorem)** *For any given m-dimensional simplex*

$$P = \{x \in \mathbb{R}^n \mid Ax \leq b\},$$

*there exists an $n \times n$ unimodular matrix $U$ such that $\bar{A} = AU$ has the standard form.*

Proof: Notice that the origin of $\mathbb{R}^n$ is in the interior of the convex hull of the vectors $a^1, \cdots, a^{n+1}$. This implies that there are $n+1$ strictly positive convex combination coefficients $\lambda_1, \cdots, \lambda_{n+1}$ such that

$$\sum_{i=1}^{n+1} \lambda_i a^i = \mathbf{0}. \tag{6.4.1}$$

Moreover, the corresponding convex combination coefficients remain unchanged under any unimodular transformation of $A$. The subsequent proof consists of four phases.

Phase $I$: We first implement the procedure for $A$:

Step (0) Set $i = n + 1$.
Step (1) Set $j = i - 1$. If $a_{i,j} = 0$, choose $a_{i,l} \neq 0$ for some $1 \leq l \leq j - 1$ and switch column $j$ with column $l$. Set $k = j - 1$.
Step (2) If $a_{i,k} = 0$, do nothing. Otherwise use the Euclidean algorithm to find the greatest common divisor of $a_{i,j}$ and $a_{i,k}$, denoted by $r = g.c.d(a_{i,j}, a_{i,k})$, and $p$, $q$ relatively prime such that $pa_{i,j} + qa_{i,k} = r$. Set $A' = AD$, where $D$ is the identity matrix $\mathbf{I}$ in all but columns $k$ and $j$. In column $k$, we have $d_{j,k} = -a_{i,k}/r$, $d_{k,k} = a_{i,j}/r$, and $d_{s,k} = 0$ otherwise. In column $j$, we have $d_{j,j} = p$, $d_{k,j} = q$, and $d_{s,j} = 0$ otherwise. If $k > 1$, return to Step (2) with $k = k - 1$. Otherwise check whether $a_{i,j} \leq 0$. If not, reverse the sign of column $j$. Set $i = i - 1$ and $A = A'$. Go to Step (3).
Step (3) If $i = 1$, stop. Otherwise go to Step (1).

Let us consider Step (2) where $A' = AD$ with entries $a'_{s,t}$. Note that

$$detD = pa_{i,j}/r + qa_{i,k}/r = 1.$$

Hence $D$ is a unimodular matrix. Moreover, for $s \in N$

$$a'_{s,t} = a_{s,t}, \quad t \neq j, k$$

$$a'_{s,j} = pa_{s,j} + qa_{s,k}$$

and

$$a'_{s,k} = -a_{i,k}a_{s,j}/r + a_{i,j}a_{s,k}/r.$$

In particular, $a'_{i,j} = r$ and $a'_{i,k} = 0$. It is clear that all other operations are also unimodular transformations. Therefore, after a finite number of steps, the procedure brings $A$ into the form:

$$(a_{i,j}) = \begin{bmatrix} + & ? & \cdots & ? & ? \\ - & ? & \cdots & ? & ? \\ 0 & - & \cdots & ? & ? \\ \vdots & \vdots & & \vdots & \vdots \\ 0 & 0 & \cdots & - & ? \\ 0 & 0 & \cdots & 0 & - \end{bmatrix} \tag{6.4.2}$$

where " + " stands for a positive entry, and " − " for a zero or negative entry. Notice that in the above matrix entries " − " can not be zero.

For $k = 2, \cdots, n+1$, we denote the submatrix of $A$ obtained by taking rows 1 through $k$ and columns 1 through $k-1$ of $A$ by $A_k$, i.e.,

$$A_k = \begin{bmatrix} + & ? & \cdots & ? & ? \\ - & ? & \cdots & ? & ? \\ 0 & - & \cdots & ? & ? \\ \vdots & \vdots & & \vdots & \vdots \\ 0 & 0 & \cdots & - & ? \\ 0 & 0 & \cdots & 0 & - \end{bmatrix} \tag{6.4.3}$$

being a $k \times (k-1)$ matrix.

Phase $II$: Suppose that for some $k$, $2 \leq k \leq n$, we have unimodularly transformed $A_k$ into $A'_k$ satisfying the conditions of the standard form. This implies that $A_{k+1}$ is brought into the form:

$$\bar{A}_{k+1} = \begin{bmatrix} + & - & \cdots & - & ? \\ - & + & \cdots & - & ? \\ \vdots & \vdots & & \vdots & \vdots \\ - & - & \cdots & + & ? \\ - & - & \cdots & - & ? \\ 0 & 0 & \cdots & 0 & - \end{bmatrix}$$

of which the $k \times (k-1)$ upper left submatrix is $A'_k$ and the last column is the last column of $A_{k+1}$. Let $a'_i$ denote the $i$-th row of $A'_k$ for $i = 1, \cdots, k$. Note that the zero row vector of $\mathbb{R}^{k-1}$ is in the interior of the convex hull of all rows of $A'_k$. More precisely, the corresponding convex combination coefficients are $\lambda'_i = \lambda_i / \sum_{h=1}^{k} \lambda_h$ for $i = 1, \cdots, k$. So all the conditions of Corollary 6.2.6 are satisfied by $A'_k$. Thus there exists a positive integer combination of columns one through $k-1$ of $\bar{A}_{k+1}$ for which the last element is zero, the $k$-th element is strictly negative and the other elements are strictly positive. These positive integer combination coefficients are solutions of the following system of linear equations:

$$\begin{aligned} a'_1 x &\geq 1 \\ &\vdots \\ a'_{k-1} x &\geq 1 \\ x &\in \mathbf{Z}_{+}^{k-1}. \end{aligned} \tag{6.4.4}$$

A solution of system (6.4.4) can be found by the algorithm in Part 3 of Section 6.3 within a finite number of steps from any starting point $v$ in $\mathbf{Z}^{k-1}$. More precisely, we adopt the following labeling rule:

To $y \in \mathbf{Z}^{k-1}$ we assign the label $L(y) = i$ if $i$ is the smallest index for which

$$a_i' y - 1 = \min\{\, a_h' y - 1 \mid a_h' y - 1 \le 0, \text{ for some } h \in I_{k-1} \,\}.$$

If $a_h' y - 1 > 0$ for all $h = 1, \cdots, k-1$, then the label $L(y) = k$ is assigned to $y$.

Furthermore, we define

$$q(i) = E(i), \quad i \in \{1, 2, \cdots, k-1\}$$

where $E(i)$ is the $i$-th unit vector of $\mathbf{R}^{k-1}$.

By subtracting a large positive integer multiple of the above combination from the last column of $\bar{A}_{k+1}$, we can therefore transform $\bar{A}_{k+1}$ (hence $A_{k+1}$) into the form:

$$\hat{A}_{k+1} = (\hat{a}_{i,j}) = \begin{bmatrix} + & - & \cdots & - & - \\ - & + & \cdots & - & - \\ \vdots & \vdots & & \vdots & \vdots \\ - & - & \cdots & + & - \\ - & - & \cdots & - & + \\ 0 & 0 & \cdots & 0 & - \end{bmatrix}. \tag{6.4.5}$$

Phase $III$: Next we shall give a procedure to transform the matrix $\hat{A}_{k+1}$ to the standard form. The procedure is described as follows:

Step (a) If there are indices $i$ and $j$ $(i \neq j)$ for some $1 \le i, j \le k$ with $\hat{a}_{i,i} \le |\hat{a}_{i,j}|$, we can find a positive integer $c$ and an integer $d \in \{0, 1, \cdots, \hat{a}_{i,i} - 1\}$ such that $|\hat{a}_{i,j}| = c\hat{a}_{i,i} + d$, where $c$ is the lower integer part of $|\hat{a}_{i,j}|/\hat{a}_{i,i}$, and then add $c$ multiple of column $i$ to column $j$.

Step (b) Repeat Step (a) until there are no indices $i$ and $j$ $(i \neq j)$ for $1 \le i, j \le k$ with $\hat{a}_{i,i} \le |\hat{a}_{i,j}|$.

It is obvious that the above operation is a unimodular transformation. We still have to demonstrate that the procedure is feasible. Let $k = n$. For $k \le n$, the proof is the same. Recall that the origin of $R^n$ is in the interior of the convex hull of the vectors $a^1, \cdots, a^{n+1}$. This implies that there are $n+1$ strictly positive convex combination coefficients $\lambda_1, \cdots, \lambda_{n+1}$ such that

$$\sum_{i=1}^{n+1} \lambda_i a^i = \mathbf{0}. \tag{6.4.6}$$

We will show that $\hat{a}_{i,i} \le |\hat{a}_{i,j}|$ implies $|\hat{a}_{j,i}| < \hat{a}_{j,j}$. System ( 6.4.6) means that

$$\sum_{i=1}^{n+1} \lambda_i \hat{a}_{i,h} = 0, \; h = 1, \cdots, n.$$

It is easily derived that

$$\lambda_i \hat{a}_{i,i} \geq \lambda_j |\hat{a}_{j,i}|,$$

and

$$\lambda_i |\hat{a}_{i,j}| \leq \lambda_j \hat{a}_{j,j}.$$

Notice that at least one of the above inequalities holds with strict inequality, say, $\lambda_i \hat{a}_{i,i} > \lambda_j |\hat{a}_{j,i}|$. Moreover, it holds that $\lambda_i \hat{a}_{i,i} \leq \lambda_i |\hat{a}_{i,j}|$. All of this together implies that

$$\lambda_j |\hat{a}_{j,i}| < \lambda_i \hat{a}_{i,i} \leq \lambda_i |\hat{a}_{i,j}| \leq \lambda_j \hat{a}_{j,j}.$$

It follows that $|\hat{a}_{j,i}| < \hat{a}_{j,j}$.

We are now ready to prove that the new generated column $j$, denoted by $(b_{1,j}, \cdots, b_{n+1,j})^{\top}$, preserves the same pattern as before. Note that

$$b_{h,j} = \hat{a}_{h,j} + c\hat{a}_{h,i},\ h \in N.$$

It is readily seen that

$$b_{i,j} = -d \leq 0,\ |b_{i,j}| < \hat{a}_{i,i},$$

and

$$b_{h,j} \leq 0, \text{ for all } \ h \neq i,\ j.$$

Observe that

$$\begin{aligned} \lambda_i \hat{a}_{i,i} &\leq \lambda_i |\hat{a}_{i,j}| \\ &= \lambda_i (c\hat{a}_{i,i} + d) \\ &\leq \lambda_j \hat{a}_{j,j} \\ &= \lambda_j (b_{j,j} + c|\hat{a}_{j,i}|). \end{aligned}$$

It follows that

$$\begin{aligned} \lambda_j b_{j,j} &\geq c\lambda_i \hat{a}_{i,i} + d\lambda_i - c\lambda_j |\hat{a}_{j,i}| \\ &= c(\lambda_i \hat{a}_{i,i} - \lambda_j |\hat{a}_{j,i}|) + d\lambda_i \\ &> d\lambda_i \\ &\geq 0. \end{aligned}$$

Hence we have that $b_{j,j} > 0$. On the other hand, it is easy to see that the procedure will terminate within a finite number of steps since each entry $\hat{a}_{i,j}$ is finite. So, the above procedure will bring $\hat{A}_{k+1}$ to the standard form, denoted by $A'_{k+1}$, within a finite number of steps.

Phase *IV*:

Step (0) Implement Phase *I* and obtain $A$ as in ( 6.4.2). Set $k = 2$ and

$A'_2 = A_2$ as in ( 6.4.3).
Step (1) Implement Phase $II$ for $A'_k$ and obtain $\hat{A}_{k+1}$ as in ( 6.4.5). Implement Phase $III$ for $\hat{A}_{k+1}$ and obtain $A'_{k+1}$. Set $k = k + 1$.
Step (2) If $k = n + 1$, set $\bar{A} = A'_{n+1}$ and stop. Otherwise, go to Step (1).

It is obvious that $A_2$ has the standard form and Step (1) always brings $A_{k+1}$ to the standard form $A'_{k+1}$ within a finite number of steps.

Hence the procedure produces the matrix $\bar{A} = AU$ satisfying the desired conditions in the theorem where $U$ is a unimodular matrix. □

We shall conclude this section with one example to illustrate the proof.

**Example 6.4.3** *We are given a simplex*

$$P = \{x \in \mathbb{R}^4 \mid Ax \leq e\},$$

*where $A$ is given by*

$$A = \begin{bmatrix} -7 & 3 & 10 & -6 \\ -8 & -2 & 1 & -5 \\ 9 & -6 & -7 & 6 \\ -1 & 6 & -3 & 4 \\ 7 & -1 & -1 & 1 \end{bmatrix}.$$

Then

$$U_1 = \begin{bmatrix} -5 & 1 & 0 & 0 \\ -31 & 6 & 0 & 0 \\ 165 & -31 & -1 & 0 \\ 169 & -32 & -1 & -1 \end{bmatrix}$$

brings $A$ to

$$F_1 = \begin{bmatrix} 578 & -107 & -4 & 6 \\ -578 & 109 & 4 & 5 \\ 0 & -2 & 1 & -6 \\ 0 & 0 & -1 & -4 \\ 0 & 0 & 0 & -1 \end{bmatrix}.$$

Then

$$U_2 = \begin{bmatrix} 0 & 1 & 0 & 0 \\ -1 & 4 & 0 & 0 \\ 10 & -11 & -1 & 0 \\ 9 & -14 & -1 & -1 \end{bmatrix}$$

brings $A$ to

$$F_2 = \begin{bmatrix} 43 & -21 & -4 & 6 \\ -33 & 43 & 4 & 5 \\ -10 & -22 & 1 & -6 \\ 0 & 0 & -1 & -4 \\ 0 & 0 & 0 & -1 \end{bmatrix}.$$

Then

$$U_3 = \begin{bmatrix} 0 & 1 & -1 & 0 \\ -1 & 4 & -3 & 0 \\ 10 & -11 & 0 & 0 \\ 9 & -14 & 4 & -1 \end{bmatrix}$$

brings $A$ to

$$F_3 = \begin{bmatrix} 43 & -21 & -26 & 6 \\ -33 & 43 & -6 & 5 \\ -10 & -22 & 33 & -6 \\ 0 & 0 & -1 & -4 \\ 0 & 0 & 0 & -1 \end{bmatrix}.$$

We now have to solve the following system of inequalities:

$$\begin{array}{rcl} 43x_1 - 21x_2 - 26x_3 & \geq & 1 \\ -33x_1 + 43x_2 - 6x_3 & \geq & 1 \\ -10x_1 - 22x_2 + 33x_3 & \geq & 1 \end{array} \tag{6.4.7}$$

$$x \in \mathbf{Z}_+^3.$$

By setting the starting point $v = (-10, 5, 10)^\top$, the algorithm in the previous section finds a solution $x = (23, 21, 21)^\top$ of system (6.4.7) within 65 steps. Then

$$U_4 = \begin{bmatrix} 0 & 1 & -1 & 0 \\ -1 & 4 & -3 & 6 \\ 10 & -11 & 0 & 3 \\ 9 & -14 & 4 & 8 \end{bmatrix}$$

brings $A$ to

$$F_4 = \begin{bmatrix} 43 & -21 & -26 & 0 \\ -33 & 43 & -6 & -49 \\ -10 & -22 & 33 & -9 \\ 0 & 0 & -1 & 59 \\ 0 & 0 & 0 & -1 \end{bmatrix}.$$

Finally,

$$U = \begin{bmatrix} 0 & 1 & -1 & 1 \\ -1 & 4 & -3 & 10 \\ 10 & -11 & 0 & -8 \\ 9 & -14 & 4 & -6 \end{bmatrix}$$

brings $A$ to

$$\bar{A} = \begin{bmatrix} 43 & -21 & -26 & -21 \\ -33 & 43 & -6 & -6 \\ -10 & -22 & 33 & -31 \\ 0 & 0 & -1 & 59 \\ 0 & 0 & 0 & -1 \end{bmatrix}.$$

## 6.5. Numerical results

We have programmed in Pascal and C, and implemented the integer labeling algorithm. Let $P = \{x \in \mathbb{R}^n \mid Ax \leq b\}$ be a simplex in the standard form. Our goal is to test whether $P$ contains an integer point or not. Recall that the Fibonacci numbers are defined by $f(k+2) = f(k+1) + f(k)$, $k = 1, 2, \cdots$, with $f(1) = f(2) = 1$. We will use the Fibonacci numbers later. In the following, "yes" means $P$ contains an integer point and "no" means $P$ contains no integer point. In Table 6.2 where $b_i = f(i) + c$, $i \in N$, we give several instances of Example 6.5.1 for $n = 10$. In Table 6.3 where $b_i = i(-1)^i n + k$, $i = 1, \cdots, n$, and $b_{n+1} = -c$, $k$ and $c$ are positive integers, we give several instances of Example 6.5.1 for $n = 10$. In Table 6.4 where $b_i = i(-1)^i n + k$, $i = 1, \cdots, n$, and $b_{n+1} = -c$, $k$ and $c$ are positive integers, we give several instances of Example 6.5.2 for $n = 10$.

**Example 6.5.1** *The coefficients of* $A = (a_{i,j})$ *are given by*

$$\begin{aligned} &a_{i,j} = -n(n+2-i) + j - 1, \ j \neq i, i, j = 1, \cdots, n, \\ &a_{n+1,j} = -j, \ j = 1, \cdots, n, \\ &a_{i,i} = \textstyle\sum_{j \in N, j \neq i} |a_{j,i}|, \ i = 1, \cdots, n. \end{aligned}$$

**Example 6.5.2** *The coefficients of* $A = (a_{i,j})$ *are given by*

$$\begin{aligned} &a_{i,j} = -(n - i + 1), \ i \neq j, \ i, j = 1, \cdots, n, \\ &a_{n+1,j} = -(n - j + 1), \ j = 1, \cdots, n, \\ &a_{i,i} = n(n+1)/2, \ i = 1, \cdots, n. \end{aligned}$$

| $c$ | the greatest integer point or no integral solution | steps |
|---|---|---|
| $-10$ | no integral solution | 1457 |
| $-2$ | no integral solution | 1682 |
| 10 | $(0,0,0,0,0,0,0,0,0,0)^{\mathsf{T}}$ | 1450 |
| 50 | $(0,0,0,0,0,0,0,0,0,0)^{\mathsf{T}}$ | 1450 |
| 250 | $(31,29,26,23,21,18,15,12,9,6)^{\mathsf{T}}$ | 2444 |
| 500 | $(136,125,113,102,90,78,65,52,39,25)^{\mathsf{T}}$ | 2799 |
| 900 | $(294,270,246,221,195,168,141,113,84,54)^{\mathsf{T}}$ | 2189 |
| 1000 | $(343,316,287,258,228,197,165,132,98,63)^{\mathsf{T}}$ | 2382 |
| 2500 | $(906,833,758,680,601,519,435,348,258,166)^{\mathsf{T}}$ | 2421 |
| 5000 | $(1869,1718,1563,1404,1240,1071,897,717,532,342)^{\mathsf{T}}$ | 2752 |
| 7500 | $(2842,2613,2377,2135,1885,1628,1364,1091,810,520)^{\mathsf{T}}$ | 1843 |
| 10000 | $(3811,3504,3188,2863,2528,2184,1829,1463,1086,697)^{\mathsf{T}}$ | 2269 |

TABLE 6.2. Tested instances of Example 6.5.1 by the IL algorithm for $n = 10$.

| $k$, $c$ | the greatest point or no integral solution | steps |
|---|---|---|
| 190, 3 | $(3,3,2,2,2,2,1,1,1,1)^{\mathsf{T}}$ | 3041 |
| 189, 3 | no integral solution | 3093 |
| 300, 1404 | $(54,50,45,41,36,31,26,21,15,10)^{\mathsf{T}}$ | 2396 |
| 300, 1405 | no integral solution | 2396 |
| 400, 2211 | $(85,78,71,64,56,49,41,33,24,16)^{\mathsf{T}}$ | 2190 |
| 400, 2212 | no integral solution | 2190 |
| 500, 3095 | $(119,110,100,90,79,68,57,46,34,22)^{\mathsf{T}}$ | 2967 |
| 500, 3096 | no integral solution | 2967 |
| 600, 4655 | $(179,165,150,135,119,103,86,69,51,33)^{\mathsf{T}}$ | 2631 |
| 600, 4656 | no integral solution | 2631 |

TABLE 6.3. Tested instances of Example 6.5.1 by the IL algorithm for $n = 10$.

**Example 6.5.3** *The coefficients of* $A = (a_{i,j})$ *are given by*

$$\begin{aligned} a_{i,j} &= -1, \quad i \neq j,\ i = 1, \cdots, n+1;\ j = 1, \cdots, n, \\ a_{i,i} &= n+1, \quad i = 1, \cdots, n. \end{aligned}$$

| k, c | the greatest point or no integral solution | steps |
|---|---|---|
| 400, 3847 | $(94, 87, 78, 71, 62, 55, 44, 37, 25, 19)^\top$ | 198 |
| 400, 3848 | no integral solution | 198 |
| 500, 4848 | $(118, 109, 99, 90, 78, 69, 56, 47, 32, 23)^\top$ | 184 |
| 500, 4849 | no integral solution | 184 |
| 600, 5899 | $(144, 133, 120, 109, 95, 84, 68, 57, 39, 28)^\top$ | 171 |
| 600, 5900 | no integral solution | 171 |
| 700, 6811 | $(166, 153, 139, 126, 110, 97, 79, 65, 46, 32)^\top$ | 202 |
| 700, 6812 | no integral solution | 202 |

TABLE 6.4. Tested instances of Example 6.5.2 by the IL algorithm for $n = 10$.

| $n$, $k$, $c$ | the greatest integer point or no integer point | steps |
|---|---|---|
| 10, 10, 1000 | the greatest integer point | 191 |
| 20, 10, 1000 | the greatest integer point | 1566 |
| 30, 10, 1000 | the greatest integer point | 6043 |
| 40, 10 , 1000 | the greatest integer point | 14652 |
| 50, 10, 1000 | the greatest integer point | 26603 |
| 60, 10, 1000 | the greatest integer point | 51184 |
| 70, 10, 1000 | no integer point in $P$ | 56331 |
| 40, 10, 500 | the greatest integer point | 13921 |
| 45, 10, 500 | no integer point in $P$ | 14912 |

TABLE 6.5. Tested instances of Example 6.5.2 by the IL algorithm.

In Table 6.5 where $b_i = c$ for all $i \in I_n$ and $b_{n+1} = -k$, we give several instances of Example 6.5.2. In Table 6.6 where $b_i = i(-1)^i n + k$, $i = 1, \cdots,$ $n$, and $b_{n+1} = -c$, $k$ and $c$ are positive integers, we give several instances of Example 6.5.3. We remark that this example is relatively easier than the first two examples since it satisfies the conditions of the strong standard form.

The above preliminary results indicate that the integer labeling algorithm seems to work very well. Since the operation of the algorithm consists

| $n$, $k$, $c$ | steps of the IL algorithm | yes or no |
|---|---|---|
| 10, 10, 100 | 103 | no |
| 20, 100, 100 | 122 | yes |
| 30, 200, 500 | 316 | yes |
| 40, 500, 1500 | 424 | yes |
| 50, 1000, 4000 | 533 | yes |
| 60, 2000, 5000 | 388 | yes |
| 70, 4000, 9000 | 212 | yes |
| 80, 5000, 8000 | 224 | yes |
| 90, 6000, 7000 | 267 | yes |
| 100, 600, 7000 | 6228 | yes |
| 110, 400, 7000 | 8792 | yes |

TABLE 6.6. Tested instances of Example 6.5.3 by the IL algorithm.

only of addition and multiplication of integers, the algorithm is therefore computationally very simple and stable. Besides the above examples, we have tested various other instances. For details, see Li et al [1998]. In this paper we report on a number of tested problems with up to 500 variables. Here we point out that the unimodular transformation of a simplex into a simplex in the standard form could be very time-consuming. We have not programmed and implemented the unimodular transformation.

Now we conclude this chapter with some remarks. At first sight it may be somehow surprising that simplicial algorithms have such immediate applications in integer programming. It is our understanding that these applications are quite natural, since integer programming problems are just the most extreme cases of highly nonlinear problems. It would be of great interest to extend this approach to the integer feasibility problem of general polytopes. Moreover, the theoretic complexity of the above algorithm is also worth investigating.

# CHAPTER 7

## Refinement and Stability of Stationary Points

### 7.1. Introduction

The purpose of this chapter is twofold. On the one hand, a new solution to the stationary point problem or the variational inequality problem on the unit simplex will be introduced. This new solution concept is called *a robust stationary point.* It is a refinement of the concept of stationary point and was essentially motivated from economic equilibrium problems, non-cooperative games, biological and engineering problems. We recommend the interested reader to Arrow and Hurwicz [1958], Arrow, Block and Hurwicz [1959], Wu and Jiang [1962], Selten [1975], Myerson [1978], Kreps and Wilson [1982], and van Damme [1987] for the various motivations. Mathematically speaking, a continuous function from the unit simplex $S^n$ into $\mathbb{R}^n$ may have multiple stationary points and some of them are undesirable from a point of view of stability. So it is important to eliminate those undesirable stationary points. One way of achieving this goal is to refine the concept of a stationary point. It will be shown that every continuous function on the unit simplex has at least one stationary point, although a stationary point need not be robust. It will also be shown that when applying this refined concept to game-theoretic or economic equilibrium problems, it is very meaningful and intuitive. On the other hand, we shall spend a fairly large portion of time on the computation of robust stationary points. To do so, a new simplicial algorithm will be developed. This algorithm is called *an adaptive simplicial algorithm.*

As we have seen in Chapter 1, the stationary point problem is equivalent to the fixed point problem. To compute a fixed point or a stationary point on the unit simplex, several algorithms have been proposed (see Scarf [1967b, 1973], Kuhn [1968, 1969], Eaves [1972], Kuhn and MacKinnon [1975], Kellogg, Li and Yorke [1976], Smale [1976], van der Laan and Talman [1979], Doup and Talman [1987a], Doup, van der Laan and Talman [1987]). See Chapters 3, 4 and 5. However, no existing algorithms can guarantee to find a robust stationary point. This motivates us to design a device which is always able to find a robust stationary point. This device is the adaptive simplicial algorithm.

The adaptive simplicial algorithm can be roughly described as follows. Starting from an arbitrarily chosen interior point $v$ in $S^n$, the algorithm generates a piecewise linear path of points in $S^n$. This path is traced by alternating linear programming pivot steps to follow a linear piece of the path and replacement steps in a simplicial subdivision of the interior of $S^n$ to move from one simplex to the next one. Within a finite number of function evaluations and linear programming pivot steps the algorithm finds an approximate robust stationary point of the function of any a priori chosen accuracy. The path generated by the algorithm corresponds to a sequence of $\theta$-robust stationary points of the piecewise linear approximation of the function with respect to the underlying simplicial subdivision, where $0 < \theta \leq 1$. This simplicial subdivision differs from other simplicial subdivisions of $S^n$. We call it the *P-triangulation.* When the path generated by the algorithm approaches the boundary of $S^n$, $\theta$ goes to zero and the mesh size of the triangulation along the path automatically converges to zero. This makes the algorithm different from all other simplicial algorithms. Generally speaking, the algorithm is a blend of a simplicial restart algorithm and a homotopy algorithm and maintains the basic properties of both. This can be seen as follows. If the algorithm converges to a solution on the boundary of $S^n$, it shares the property with a homotopy algorithm that the variable $\theta$ can be considered as a homotopy parameter (see Chapter 4), in the sense that when $\theta$ tends to zero, the mesh size of the triangulation also tends to zero. However it should be emphasized that the algorithm does not need an extra dimension which is required by homotopy algorithms. If, however, the algorithm converges to a solution in the interior of $S^n$, it behaves exactly as a variable dimension restart algorithm does (see Chapter 5).

Although it may not be apparent from the arguments of this chapter, the algorithm is inspired by the procedure proposed by Yamamoto [1993] for the determination of a proper Nash equilibrium of a non-cooperative $n$-person game. The algorithm can be seen as a constructive combinatorial analog of that procedure in case the starting point of the algorithm is chosen to be the barycenter of $S^n$.

This chapter consists of five sections. In Section 2 we introduce the definition of a robust stationary point and prove the existence of a robust stationary point for a continuous function on the unit simplex. In Section 3 we specify the $P$-triangulation of the unit simplex. In Section 4 we give a detailed description of the algorithm. In Section 5 we give several numerical examples and discuss how to solve the degeneracy problem. In Section 6 we generalize the concept of robust stationary point to general polytopes. Sections 1-5 are based on Yang [1996] and Section 6 is drawn from van der Laan, Talman and Yang [1996].

## 7.2. The concept of robust stationary point

In this section we first give the definition of a robust stationary point and then show the nonemptiness of the set of robust stationary points of a continuous function on the unit simplex. Let a function $f : S^n \mapsto \mathbb{R}^n$ be given.

**Definition 7.2.1** *For given $\theta > 0$ a point $x \in S^n$ is a $\theta$-robust stationary point of $f$ if*
*(a) $x$ is an interior point of $S^n$;*
*(b) $x_k \leq \theta x_l$ if $f_k(x) < f_l(x)$, for any $k, l \in I_n$.*

**Definition 7.2.2** *A point $x^* \in S^n$ is a robust stationary point of $f$ on $S^n$ if there exist sequences $\{\, \theta_t \mid t \in \mathbb{N} \,\}$ of positive numbers and $\{\, x(\theta_t) \mid t \in \mathbb{N} \,\}$ of $\theta_t$-robust stationary points $x(\theta_t)$ of $f$ such that*

$$\lim_{t \to \infty} \theta_t = 0 \ \text{ and } \ \lim_{t \to \infty} x(\theta_t) = x^*.$$

We remark that if a stationary point $x^*$ of $f$ lies in the interior of $S^n$, then $x^*$ must be a robust stationary point of $f$ with equal values of the components. Examples given in Section 7.5 will demonstrate that the concept of a robust stationary point is indeed a refinement of that of a stationary point.

**Lemma 7.2.3** *Let $f : S^n \mapsto \mathbb{R}^n$ be a continuous function. If $x^* \in S^n$ is a robust stationary point of $f$, then $x^*$ is also a stationary point of $f$.*

Proof: We only need to consider two cases. If $x^*$ lies in the interior of $S^n$, it implies that $f_i(x^*) = f_j(x^*)$ for $i, j \in I_n$. Hence we have

$$(x^* - x)^\top f(x^*) = \sum_{i=1}^{n} (x_i^* - x_i) f_i(x^*) = 0$$

for any $x \in S^n$. It means that $x^*$ is a stationary point of $f$. On the other hand, if $x^*$ is on the boundary of $S^n$, there exists a proper subset $J$ of $I_n$ such that $x_j^* = 0$ for all $j \in J$. It follows from Definitions 7.2.1 and 7.2.2 that $f_k(x^*) = f_l(x^*)$ for $k, l \in I_n \backslash J$ and $f_k(x^*) \geq f_j(x^*)$ for $k \in I_n \backslash J$ and $j \in J$. Now for given $l \in I_n \backslash J$, we have

$$(x^* - x)^\top f(x^*) = \sum_{i \in I_n \backslash J} (x_i^* - x_i) f_i(x^*) - \sum_{j \in J} x_j f_j(x^*) \geq \sum_{i=1}^{n} (x_i^* - x_i) f_l(x^*) = 0$$

for any $x \in S^n$. It implies that $x^*$ is a stationary point of $f$. □

A fundamental question arises here. Does a continuous function $f$ from $S^n$ to $\mathbb{R}^n$ have a robust stationary point? The following theorem provides an affirmative answer.

**Theorem 7.2.4** *Let $f : S^n \mapsto \mathbb{R}^n$ be a continuous function. Then $f$ has at least one robust stationary point in $S^n$.*

Proof: We first show that there exists at least one $\theta$-robust stationary point, for any $\theta$, $0 < \theta < 1$. Given such a $\theta$, let $\delta = \frac{1}{n}\theta^{n-1}$ and define

$$S(\theta) = \{ x \in S^n \mid x_i \geq \delta, \ i = 1, ..., n \}.$$

It is clear that $S(\theta)$ is a nonempty, convex, compact subset of $S^n$. We further define a correspondence $F : S(\theta) \mapsto S(\theta)$ by

$$F(x) = \{ y \in S(\theta) \mid \text{ if } f_i(x) < f_j(x) \text{ then } y_i \leq \theta y_j \text{ for any } i, j \}.$$

$F(x)$ is obviously a closed and convex set for every $x \in S(\theta)$. Given $x \in S(\theta)$ and $i \in I_n$, let $\Delta(i)$ be the number of $j$'s such that $f_i(x) < f_j(x)$ and let

$$y_i^* = \theta^{\Delta(i)} / \sum_{l=1}^{n} \theta^{\Delta(l)}.$$

Then $y_i^* \geq \delta$ for $i = 1, \cdots, n$. Hence $y^* \in F(x)$ and therefore $F(x)$ is nonempty. Moreover the continuity of $f$ guarantees that $F$ is upper semi-continuous. Thus $F$ satisfies all conditions of Kakutani's fixed point theorem and so there exists a point $x(\theta) \in S(\theta)$ such that $x(\theta) \in F(x(\theta))$. It is easily seen that $x(\theta)$ is a $\theta$-robust stationary point of $f$.

So, for every $0 < \theta < 1$, $f$ has a $\theta$-robust stationary point $x(\theta)$. Now let us take a sequence $\{ \theta_t \mid t \in \mathbb{N} \}$ of numbers between 0 and 1 converging to zero and a sequence of $\theta_t$-robust stationary points of $f$. Since $S^n$ is a compact set, there exists a subsequence converging to a cluster point $x^* \in S^n$. Clearly, $x^*$ is a robust stationary point of $f$. □

## 7.3. The $P$-triangulation of the unit simplex

In this section a new simplicial subdivision is introduced which will underlie the adaptive simplicial algorithm. This simplicial subdivision is called *$P$-triangulation* of $S^n$. Let $v$ be a point in the interior of $S^n$. The point $v$ will be the starting point of the algorithm. We rearrange the components of $v$ in decreasing order to obtain a vector $p = (p_1, \cdots, p_n)^\top \in S^n$ by

$$\begin{aligned} p_i &= v_{j_i}, \text{ for } i \in I_n, \\ p_l &\geq p_m, \text{ for } l \leq m, \text{ and } l, m \in I_n, \end{aligned}$$

where $(j_1, j_2, \ldots, j_n)$ is a permutation of $(1, 2, \cdots, n)$. For $t \in (0, 1]$, let

$$p_i(t) = p_i t^{i-1} / \sum_{j \in I_n} p_j t^{j-1}, \text{ for } i \in I_n,$$

and define

$$p_i(0) = \lim_{t \to 0^+} p_i(t) = \begin{cases} 1, & \text{for } i = 1, \\ 0, & \text{for } i \neq 1. \end{cases}$$

It is readily seen that $p_1(t) \geq p_2(t) \geq \cdots \geq p_n(t)$ for $t \in [0,1]$.

**Definition 7.3.1** *For $t \in [0,1]$, the set $A(t)$ is given by*

$$A(t) = \{\, x \in \mathbb{R}^n \mid \sum_{i \in I_n} x_i = 1, \quad \sum_{j \in J} x_j \leq \sum_{j=1}^{k} p_j(t) \text{ for any } J \subset I_n \text{ with } k = |J| \,\}.$$

It is easily seen that $A(0) = S^n$, and that if $v$ is the barycenter of $S^n$, then $A(1) = \{\, v \,\}$. More generally for every $t \in [0,1]$ we have that $v \in A(t)$ and $v$ is a vertex of $A(1)$. Moreover $A(t)$ is a polytope for every $t \in [0,1]$.

For $J \subset I_n$ and $t \in [0,1]$ , we define $b(J)$ and $c_J(t)$ by

$$\begin{aligned} b(J) &= \textstyle\sum_{j \in J} e(j), \\ c_J(t) &= \textstyle\sum_{j=1}^{l} p_j(t) \quad \text{with } l = |J|. \end{aligned}$$

Let $\mathcal{K} = \{\, K = (K_1, K_2, \cdots, K_m) \mid \emptyset \subset K_1 \subset \cdots \subset K_m \subset I_n \,\}$. We say that $K \in \mathcal{K}$ conforms to $J \in \mathcal{K}$, if it holds that every component of $K$ is also a component of $J$. For $K \in \mathcal{K}$ and a positive integer $l$, let

$$F(l, K) = \{\, x \in A(2^{-l}) \mid b^\top(K_i)x = c_{K_i}(2^{-l}) \text{ for every } i \in I_m \,\}.$$

Then $F(l, K)$ is a face of $A(2^{-l})$ with dimension equal to $n - 1 - m$. For $K \in \mathcal{K}$, let

$$F(0,1;K) = \{\, x \mid x = \nu v + (1-\nu)z, \text{ for some } z \in F(1,K), \text{ and some } \nu \in [0,1] \,\}$$

and for $l \in \mathbb{N}$

$$F(l, l+1; K) = \{\, x \mid x = \nu y + (1-\nu)z, \text{ for some } y \in F(l,K), \text{ some } z \in F(l+1,K), \text{ and some } \nu \in [0,1] \,\}.$$

For $K \in \mathcal{K}$ and $l \in \mathbb{N}$, we denote the union of $F(i-1, i; K)$ over $i = 1, 2, \cdots, l$ by $F(l, K)$. We remark that $F(l, K)$ is not necessarily a convex set. For $l \in \mathbb{Z}_+$, we denote the union of $F(l, l+1; K)$ over all $K \in \mathcal{K}$ by $F(l, l+1)$. For $K \in \mathcal{K}$, we denote the union of $F(l, l+1; K)$ over all $l = 0, 1, \cdots$ by $F(K)$. Notice that the dimension of $F(K)$ is equal to $n - m$, and that the union of $F(K)$ over all $K \in \mathcal{K}$ is the interior of $S^n$.

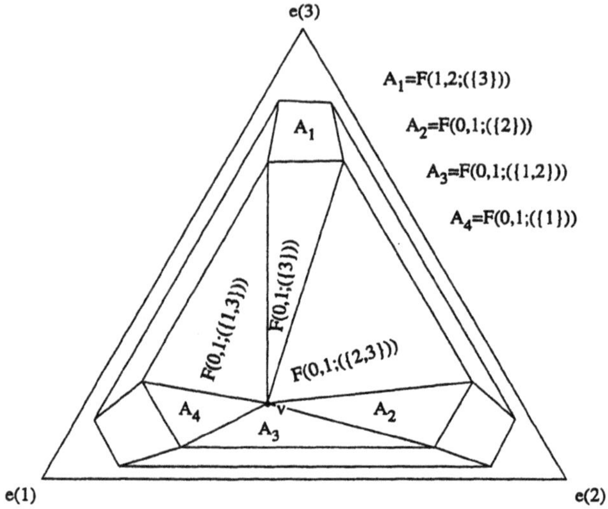


*Figure 7.1.* The subdivision of $int(S^n)$ for $n = 3$.

Figure 7.1 shows the subdivision of $S^n$ for $n = 3$ and $v = (1/2, 1/3, 1/6)^\top$. A simplicial subdivision underlying the algorithm must be such that every set $F(l, l+1; K)$ is subdivided into $t$-dimensional simplices. Such a triangulation can be described as follows. For $K \in \mathcal{K}$, we denote $v(0, K) = v$ and for $l \in \mathbb{N}$, let $v(l, K)$ be an interior point (e.g., the barycenter) of $F(l, K)$. For $K \in \mathcal{K}$, if $K$ consists of $n-1$ components, then $F(l, K)$ is a vertex of $A(2^{-l})$. For $l \in \mathbb{N}$ and $K \in \mathcal{K}$, let $F(l, K(n-1))$ be a vertex of $F(l, K)$, i.e., $K(n-1)$ has $n-1$ components and $K$ conforms to $K(n-1)$. Moreover let $(J_1, J_2, \cdots, J_t) = \gamma(K, K(n-1))$ be a conformation of $K$ and $K(n-1)$, where $t = n - m$, i.e., $J_1 = K(n-1)$, $J_l \in \mathcal{K}$ for $l = 2, \cdots, t-1$, $J_t = K$, $J_l$ conforms to $J_{l-1}$ and has one component less than $J_{l-1}$ for $l = 2, \cdots, t$. For given $l \in \mathbb{Z}_+$, $K \in \mathcal{K}$ and $\gamma(K, K(n-1))$, the subset $F(l, l+1; K, \gamma(K, K(n-1)))$ of $F(l, l+1; K)$ is defined to be the convex hull of $v(l, J_1)$, $v(l, J_2)$, $\cdots$, $v(l, J_t)$, $v(l+1, J_1)$, $v(l+1, J_2)$, $\cdots$, and $v(l+1, J_t)$, so

$$\begin{aligned} F(l, l+1; I, \gamma(K, K(n-1))) = \{\, & x \in S^n \mid x = v(l, K(n-1)) + \alpha q_0 \\ & + \textstyle\sum_{j=1}^{t-1} \alpha_j q_j(\alpha),\ 0 \le \alpha \le 1, \\ & \text{and } 0 \le \alpha_{t-1} \le \cdots \le \alpha_1 \le 1 \,\}, \end{aligned}$$

where $q_0 = (v(l+1, J_1) - v(l, J_1))$, and for $j = 1, \cdots, t-1$, $0 \le \alpha \le 1$,

$$q_j(\alpha) = \alpha(v(l+1, J_{j+1}) - v(l+1, J_j)) + (1-\alpha)(v(l, J_{j+1}) - v(l, J_j)).$$

The dimension of $F(l,l+1;K,\gamma(K,K(n-1)))$ is equal to $t$ and $F(l,l+1;K)$ is the union of $F(l,l+1;K,\gamma(K,K(n-1)))$ over all conformations $\gamma(K,K(n-1))$ and over all index sets $K(n-1)$ conformed by $K$.

**Definition 7.3.2** *Let $d$ be an arbitrary positive integer. For given $l \in \mathbb{Z}_+$, $K \in \mathcal{K}$ and $\gamma(K,K(n-1))$, the set $G^d(l,l+1;K,\gamma(K,K(n-1)))$ is the collection of $t$-simplices $\sigma(a,\pi)$ with vertices $y^1, \cdots, y^{t+1}$ in $F(l,l+1;K,\gamma(K,K(n-1)))$ such that*
*(a) $y^1 = v(l,K(n-1)) + a(0)d^{-1}q_0 + \sum_{j=1}^{t-1} a(j)q_j(a(0)/d)/(a(0)+ld)$ where $a=(a(0),a(1),\cdots,a(n-2))^\top$ is a vector of integers such that $0 \le a(0) \le d-1$ and $a(n-2) = \cdots = a(t) = 0 \le a(t-1) \le \cdots \le a(2) \le a(1) \le a(0)+ld$;*
*(b) $\pi = (\pi(1),\cdots,\pi(t))$ is a permutation of $(0,1,\cdots,t-1)$ such that $s < s'$ if for some $q \in \{0,1,\cdots,t-2\}$ it holds that $\pi(s) = q$, $\pi(s') = q+1$, $a(q) = a(q+1)$ in case $q \ge 1$, and $a(0)+ld = a(1)$ in case $q = 0$;*
*(c) Let $i$ be such that $\pi(i) = 0$. Then*

$$\begin{aligned}
y^{j+1} &= y^j + q_{\pi(j)}(a(0)/d)/(a(0)+ld),\ j = 1,\cdots,i-1,\\
y^{i+1} &= v(l,K(n-1)) + (a(0)+1)d^{-1}q_0\\
&\quad + \sum_{j=1}^{t-1} a(j)q_j((a(0)+1)/d)/(a(0)+1+ld)\\
&\quad + \sum_{j=1}^{i-1} q_{\pi(j)}((a(0)+1)/d)/(a(0)+1+ld),\\
y^{j+1} &= y^j + q_{\pi(j)}((a(0)+1)/d)/(a(0)+1+ld),\ i < j \le t.
\end{aligned}$$

The set $G^d(l,l+1;K,\gamma(K,K(n-1)))$ is a simplicial subdivision of $F(l,l+1;K,\gamma(K,K(n-1)))$ with grid size $d^{-1}$. Moreover, the union $G^d(l,l+1;K)$ of $G^d(l,l+1;K,\gamma(K,K(n-1)))$ over all conformations $\gamma(K,K(n-1))$ and all $K(n-1)$ conformed by $K$ is a simplicial subdivision of $F(l,l+1;K)$, and the union $G^d(l,l+1)$ of $G^d(l,l+1;K)$ over all sets $K \in \mathcal{K}$ induces a triangulation of $F(l,l+1)$. Taking the union $G^d(l)$ of $G^d(j,j+1)$ over $j = 0, 1, \cdots, l-1$, we obtain a simplicial subdivision of $A(2^{-l})$ with grid size $d^{-1}$. The union of $G^d(l)$ over all $l \in \mathbb{Z}_+$ is a simplicial subdivision of the interior of $S^n$ and is called *P-triangulation* of $S^n$ with grid size $d^{-1}$. Observe that for $K \in \mathcal{K}$ and $l \in \mathbb{N}$ the union $G^d(l,K)$ of $G^d(i-1,i;K)$ over $i = 1,2,\cdots,l$, is a simplicial subdivision of the set $F(K,l)$, and for $K \in \mathcal{K}$ the union $G^d(K)$ of $G^d(l,l+1;K)$ over $l = 0,1,\cdots$, is a simplicial subdivision of the set $F(K)$. The $P$-triangulation of $S^n$ for $n=3$, $d=2$ and $v = (1/3,1/3,1/3)^\top$ is illustrated in Figure 7.2.

For given $l \in \mathbb{Z}_+$ the mesh size of $G^d(l,l+1)$ is equal to

$$\delta_{l,d} = \sup\{\, diam(\sigma) \mid \sigma \in G^d(l,l+1)\,\}.$$

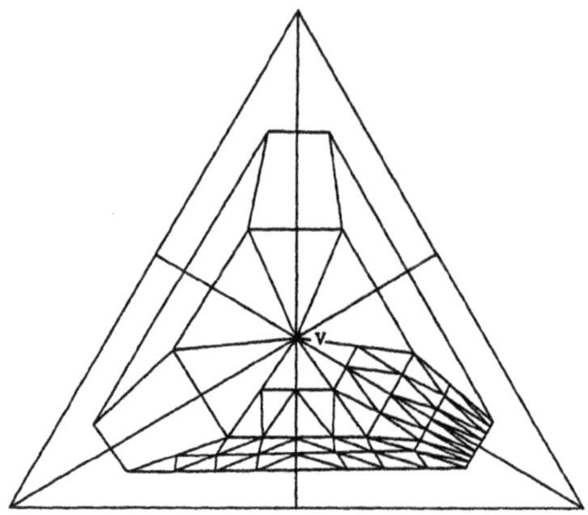


*Figure 7.2.* Illustration of $P$-triangulation of $S^n$.

Now we have the following property.

**Lemma 7.3.3** *Let $d$ be a given positive integer. For the $P$-triangulation of $S^n$ with grid size $d^{-1}$, it holds that*

$$\lim_{l \to \infty} \delta_{l,d} = 0.$$

## 7.4. The adaptive simplicial algorithm

In this section we discuss how to operate the adaptive simplicial algorithm in the $P$-triangulation of $S^n$ to approximate a robust stationary point of a continuous function on $S^n$. Starting at the point $v$, the algorithm will generate a sequence of adjacent simplices of the $P$-triangulation in the set $F(K)$ having $K$-complete common facets, for varying $K \in \mathcal{K}$.

**Definition 7.4.1** *Let the function $f : S^n \mapsto \mathbb{R}^n$ be given. For given $K = (K_1, \cdots, K_m) \in \mathcal{K}$ and $s = t-1$ or $t$, where $t = n - m$, an $s$-simplex $\sigma$ with vertices $y^1, \cdots, y^{s+1}$ is $K$-complete if the system of linear equations*

$$\sum_{i=1}^{s+1} \lambda_i \begin{pmatrix} f(y^i) \\ 1 \end{pmatrix} - \sum_{j=1}^{m} \mu_j \begin{pmatrix} b(K_j) \\ 0 \end{pmatrix} - \beta \begin{pmatrix} e \\ 0 \end{pmatrix} = \begin{pmatrix} \mathbf{0} \\ 1 \end{pmatrix} \tag{7.4.1}$$

*has a solution $\lambda_i^* \geq 0$, $i = 1, \cdots, s+1$, $\mu_j^* \geq 0$, $j = 1, \cdots, m$, and $\beta^* \in \mathbb{R}$.*

A solution $\lambda_i^*$, $i = 1, \cdots, s+1$, $\mu_j^*$, $j = 1, \cdots, m$, and $\beta^*$ will be denoted by $(\lambda^*, \mu^*, \beta^*)$. For $s = t-1$ we assume that the system (7.4.1) has a unique solution with $\lambda_i^* > 0$, $i = 1, \cdots, t$, and $\mu_j^* > 0$, $j = 1, \cdots, m$, and that for $s = t$ at most one variable of $(\lambda^*, \mu^*)$ is equal to zero (*nondegeneracy assumption*). We remark that this assumption can be dropped if we use the lexicographic pivoting method in linear programming to solve system (7.4.1) (see Section 5).

The algorithm starts by leaving the point $v$ in one out of $n!$ directions. This direction is uniquely determined by $f(v)$. Because of the nondegeneracy assumption, all components of the vector $f(v)$ are different. Let $(i_1, \cdots, i_n)$ be the permutation of $(1, \cdots, n)$ such that $f_{i_1}(v) > \cdots > f_{i_n}(v)$. Then the 0-dimensional simplex $\{v\}$ is $K^0$-complete with $K^0 = (K_1^0, \cdots, K_{n-1}^0)$ where $K_j^0 = \{i_1, \cdots, i_j\}$ for $j = 1, \cdots, n-1$. Moreover, $\{v\}$ is a facet of a unique 1-simplex $\sigma^0$ in $F(K^0)$, where $\sigma^0 = \sigma(a, \pi)$ with $a = \mathbf{0}$ and $\pi = (0)$. Since for given $K \in \mathcal{K}$ a $K$-complete $t$-simplex has at most two $K$-complete facets and a facet of a $t$-simplex in $F(K)$ is a facet of at most one other $t$-simplex in $F(K)$, we obtain that the $K$-complete $t$-simplices $\sigma(a, \pi)$ in $F(K)$, determine sequences of adjacent $t$-simplices in $F(K)$ with $K$-complete common facets. As described below, the sequences of the $K$-complete $t$-simplices in $F(K)$ can be uniquely linked for varying $K \in \mathcal{K}$ to obtain sequences of adjacent simplices of varying dimension. Under the nondegeneracy assumption, one of these sequences starts with $\sigma^0$ in $F(K^0)$ and is followed by the algorithm. Thus, starting at the point $v$, the algorithm generates a unique sequence of $K$-complete adjacent $t$-simplices in $F(K)$ of varying dimension. In this way within a finite number of steps either the algorithm reaches a point $\bar{x}$ in an $(n-1)$-dimensional simplex for which $\bar{f}_i(\bar{x}) = \bar{f}_j(\bar{x})$ for every $i$ and $j \in I_n$, where $\bar{f}$ is the piecewise linear approximation of $f$ with respect to the $P$-triangulation, or for every $l \in \mathbb{N}$ the algorithm finds a $K$-complete $(t-1)$-simplex in $F(l, K)$ for some $K \in \mathcal{K}$. Suppose the latter case holds, then we have the following result.

**Lemma 7.4.2** *For $l \in \mathbb{N}$ and $K \in \mathcal{K}$, let $\sigma$ with vertices $y^1, \cdots, y^t$ be a $K$-complete $(t-1)$-simplex of the $P$-triangulation lying in $F(l, K)$. Let $(\lambda^*, \mu^*, \beta^*)$ be the corresponding unique solution of system (7.4.1). Then $x = \sum_{i=1}^t \lambda_i^* y^i$ is a $2^{-l}$-robust stationary point of the P.L. approximation $\bar{f}$ of $f$ with respect to the $P$-triangulation. Moreover, $x$ is a stationary point of $\bar{f}$ on $A(2^{-l})$.*

Proof: Since $K = (K_1, K_2, \cdots, K_m) \in \mathcal{K}$, there exist $l_1 < l_2 < \cdots < l_m$ such that

$$\begin{aligned} K_1 &= \{ i_1, \cdots, i_{l_1} \}, \\ K_2 &= \{ i_1, \cdots, i_{l_1}, i_{l_1+1}, \cdots, i_{l_2} \}, \end{aligned}$$

$$\begin{array}{rcl} & \vdots & \\ K_m & = & \{ i_1, \cdots, i_{l_m} \}, \\ I_n \backslash K_m & = & \{ i_{l_m+1}, \cdots, i_n \}. \end{array}$$

Then it follows from equation (7.4.1) that

$$\begin{array}{rcl} \bar{f}_{i_1}(x) & = & \cdots = \bar{f}_{i_{l_1}}(x) = \mu_1^* + \cdots + \mu_m^* + \beta^* \\ & > & \bar{f}_{i_{l_1}+1}(x) = \cdots = \bar{f}_{i_{l_2}}(x) = \mu_2^* + \cdots + \mu_m^* + \beta^* \\ & \vdots & \\ & > & \bar{f}_{i_{l_{m-1}+1}}(x) = \cdots = \bar{f}_{i_{l_m}}(x) = \mu_m^* + \beta^* \\ & > & \bar{f}_{i_{l_m+1}}(x) = \cdots = \bar{f}_{i_n}(x) = \beta^*, \end{array}$$

where $\mu_i^* > 0$ for $i = 1, \cdots, m$. Now it is not difficult to check that

$$x_i \leq 2^{-l} x_j \text{ whenever } \bar{f}_i(x) < \bar{f}_j(x).$$

It means that $x$ is a $2^{-l}$-robust stationary point of the P.L. approximation $\bar{f}$ of $f$ with respect to the $P$-triangulation.

Moreover, for each face $F(l, K)$, $K \in \mathcal{K}$, let $F^*(K)$ be the set of all $n$-dimensional vectors $y$ such that every point of $F(l, K)$ is a solution of the linear programming problem

$$\begin{array}{c} \max\ y^\top \hat{x}, \\ \text{s.t. } \hat{x} \in A(2^{-l}). \end{array}$$

Then the stationary point problem for $\bar{f}$ on $A(2^{-l})$ is the problem of finding a point $x^*$ in $A(2^{-l})$ such that $\bar{f}(x^*) \in F^*(K)$ for a minimum face $F(l, K)$ of $A(2^{-l})$ containing $x^*$. Duality theorem implies that $F^*(K) = \{y \mid y = \sum_{i=1}^m \mu_i b(K_i) + \beta e,\ \mu_i \geq 0 \text{ for } i = 1, \cdots, m, \text{ and } \beta \in \mathbb{R}\}$. It follows from equation (7.4.1) that $\bar{f}(x) \in F^*(K)$. Hence $x$ is a stationary point of $\bar{f}$ on $A(2^{-l})$. □

In order to extend the domain of the P.L. approximation $\bar{f}$ of $f$, we recall from Lemma 7.3.3 that for a given positive integer $d$, the mesh size $\delta_{l,d}$ converges to zero as $l$ goes to infinity. We can therefore take $\bar{f}(x)$ to be $f(x)$ if $x$ lies on the boundary of $S^n$, since $f$ is a continuous function. Hence $\bar{f}$ is also a continuous function from $S^n$ into $\mathbb{R}^n$.

For each $t \in [0, 1/2]$, let $V(t)$ denote the set of stationary points of $f$ on $A(t)$. We summarize the above discussions by the following observations:

(P1) For each $t \in [0, 1/2]$, the set $V(t)$ is a nonempty closed set.

(P2) For each $t \in (0, 1/2]$, $x \in A(t)$ is a $t$-robust stationary point of f on $S^n$ if and only if $x$ belongs to the set $V(t)$.

(P3) For each $t \in [0,1/2]$, if $x \in V(t)$ lies in the interior of $A(t)$, then all the components of $f(x)$ are the same.

The next lemma shows that a $2^{-l}$-robust stationary point of $\bar{f}$ on $S^n$ is an approximate $2^{-l}$-robust stationary point of $f$ on $S^n$.

**Lemma 7.4.3** *Let $l \in \mathbb{Z}_+$ and $d \in \mathbb{N}$ and let $\eta_{l,d} = \sup\{ diam(f(\sigma)) \mid \sigma \in G^d(l-1,l) \}$. Let $x$ be a $2^{-l}$-robust stationary point of the P.L. approximation $\bar{f}$ of $f$ with respect to the P-triangulation, such that $x \in F(l, K^l)$ for some $K^l \in \mathcal{K}$. Then $f(x)$ lies in the $\eta_{l,d}$-neighborhood of $F^*(K^l)$, i.e., there is a $y \in F^*(K^l)$ such that $||y - f(x)|| \leq \eta_{l,d}$.*

Proof: Let $y^1, \cdots, y^t$ be the vertices of a $(t-1)$-simplex of $G^d(l-1,l)$ in $F(l, K^l)$ containing $x$. Then $\bar{f}(x) = \sum_{j=1}^t \lambda_j^* f(y^j)$ lies in $F^*(K^l)$, where $\lambda_1^*, \cdots, \lambda_t^*$ are convex combination coefficients such that $x = \sum_{j=1}^t \lambda_j^* y^j$. Clearly

$$\begin{aligned} ||\bar{f}(x) - f(x)|| &= ||\sum_{j=1}^{t} \lambda_j^* f(y^j) - f(x)|| \\ &= ||\sum_{j=1}^{t} \lambda_j^* (f(y^j) - f(x))|| \\ &\leq \sum_{j=1}^{t} \lambda_j^* ||f(y^j) - f(x)|| \\ &\leq \eta_{l,d}. \end{aligned}$$

It also holds that

$$||\bar{f}(z) - f(z)|| \leq \eta_{l,d}, \quad \text{for any } z \in F(l-1,l).$$

□

Define a point-to-set mapping $U : [0,1] \mapsto S^n$ by

$$U(t) = \begin{cases} A(t), & \text{for } t \in [0,1/2], \\ 2(1-t)A(2^{-1}) + (2t-1)\{v\}, & \text{for } t \in [1/2,1], \end{cases}$$

Denote the set of stationary points of $f$ on $U(t)$ by $Y(t)$ for $t \in [0,1]$. Observe that $Y(t) = V(t)$ for $t \in [0,1/2]$, and that $U(t)$ is contained in $A(2^{-1})$ for $t \in [1/2,1]$. As $t$ decreases from 1 to 0, $U(t)$ first expands from the starting point $v$ to the set $A(2^{-1})$ and then to the whole set $S^n$. It is easy to show that the mapping $U$ is a continuous point-to-set mapping with nonempty, convex and compact values.

Define a function $g : S^n \times [0,1] \mapsto S^n$ by

$$g(x,t) = \arg\min\{\, \|x + f(x) - y\|^2 \mid y \in U(t) \,\}.$$

It follows from Corollary 1.6.13 that $g$ is continuous. Moreover, it is easy to see that $x = g(x,t)$ if and only if $x \in Y(t)$ for each $t \in [0,1]$. By applying Theorem 1.6.12, we have the following result.

**Theorem 7.4.4** *Let $f : S^n \mapsto \mathbf{R}^n$ be a continuous function and let $U : [0,1] \mapsto S^n$ be the point-to-set mapping as defined above. Then there exists a connected set of stationary points of $f$ on $U(t)$, $0 \le t \le 1$, linking $v$ and a robust stationary point of $f$ on $S^n$.*

Let us now consider the case in which the algorithm converges to a boundary point of $S^n$. Since $S^n$ is compact and $f$ is continuous on $S^n$, the error $\eta_{l,d}$ in Lemma 7.4.3 tends to zero as $d$ is fixed and $\delta_{l,d}$ goes to zero when $l$ goes to infinity. Let $x^l$ be a $2^{-l}$-robust stationary point of $\bar{f}$. We can consider $x^l$ as an approximate $2^{-l}$-robust stationary point of $f$. Then the algorithm generates a sequence $\{\, x^l \mid l \in \mathbf{N} \,\}$ of approximate $2^{-l}$-robust stationary points of $f$ which therefore has a cluster point $x^*$. For simplicity of notation we can assume that this sequence itself converges to $x^*$. We are now ready to state the following theorem.

**Theorem 7.4.5** *Suppose that for a given positive integer d the vector $x^l$ is an approximate $2^{-l}$-robust stationary point generated by the algorithm, for every $l \in \mathbf{N}$, i.e., for each $l \in \mathbf{N}$, $x^l \in F(l, K^l)$ with $K^l \in \mathcal{K}$ is a $2^{-l}$-robust stationary point of $\bar{f}$. Then any cluster point $x^*$ of the sequence $\{\, x^l \mid l \in \mathbf{N} \,\}$ is a robust stationary point of $f$ on $S^n$.*

Proof: By definition, $x^*$ is a robust stationary point of $\bar{f}$ on $S^n$. Notice that $x^*$ lies on the boundary of $S^n$. We shall demonstrate that for any given $\epsilon > 0$, there exists a positive integer $M$, such that for $l \in \mathbf{N}$ with $l > M$, there is a $2^{-l}$-robust stationary point $y^l \in A(2^{-l})$ of $f$ on $S^n$ which is in the $\epsilon$-neighborhood of $x^l$.

For $(x,t) \in S^n \times [0,1]$, let

$$H(x,t) = x - g(x,t),$$

where $g$ is defined as before. This is a continuous homotopy defined on $S^n \times [0,1]$ between $H(x,1) = x - v$ and $H(x,0) = x - g(x,0)$. Now set

$$H^{-1}(\mathbf{0}) = \{\, (x,t) \in S^n \times [0,1] \mid H(x,t) = \mathbf{0} \,\}.$$

Let $W(t)$ denote the set of stationary points of $\bar{f}$ on $U(t)$ for $t \in [0,1]$. Similar to $H$, we can construct a continuous function $G$ with respect to $\bar{f}$, i.e.,

$$G : S^n \times [0,1] \mapsto \mathbf{R}^n,$$

such that

$$W(t) = \{ x \in S^n \mid G(x,t) = 0 \}, t \in [0,1].$$

Set

$$G^{-1}(\mathbf{0}) = \{ (x,t) \in S^n \times [0,1] \mid G(x,t) = \mathbf{0} \}.$$

For each $l \in \mathbf{N}$, let

$$\xi^l = (x^l, 2^{-l}).$$

It is clear that $\lim_{l\to\infty} \xi^l = \xi^* = (x^*, 0)$, since $\lim_{l\to\infty} x^l = x^*$. We define

$$N(\epsilon) = \{ (x,t) \in S^n \times [0,1] \mid ||(x,t)-(z,s)|| < \epsilon \text{ for some } (z,s) \in H^{-1}(0) \}.$$

Clearly, it holds that

$$||H(\psi)|| > 0 \text{ for any } \psi \in S^n \times [0,1] \backslash N(\epsilon).$$

Observe that $N(\epsilon)$ is open, so the set $S^n \times [0,1] \backslash N(\epsilon)$ is compact. From the compactness we derive that the minimum can be attained. Thus, for some $\nu > 0$

$$\min\{ ||H(\psi)|| \mid \psi \in S^n \times [0,1] \backslash N(\epsilon) \} > \nu.$$

Hence, if $\psi \in S^n \times [0,1]$ satisfies

$$||H(\psi)|| \leq \nu, \tag{7.4.2}$$

then $\psi$ must be in $N(\epsilon)$. Because $H$ is uniformly continuous on $S^n \times [0,1]$ and $\bar{f}$ is the P.L. approximation of $f$, it follows that

$$||H(\psi) - G(\psi)|| < \nu \tag{7.4.3}$$

for any $\psi = (x,t) \in S^n \times [0,1]$ under the condition that the diameter of the simplices in which $x$ lies is small enough, say, smaller than $\Delta > 0$.

Lemma 7.3.3 states that given a positive integer $d$, as $l$ goes to infinity, the mesh size $\delta_{l,d}$ converges to zero. This implies that there exists a positive integer $M$ such that for every $l \in \mathbf{N}$ with $l > M$, it holds that

$$\delta_{l,d} < \Delta.$$

Since for any $l \in \mathbf{N}$ with $l > M$, $\xi^l \in G^{-1}(0)$, i.e., $G(\xi^l) = 0$, it follows from (7.4.3) that

$$||H(\xi^l)|| < \nu.$$

By (7.4.2) $\xi^l$ must be in $N(\epsilon)$. This implies that for any $l \in \mathbf{N}$ with $l > M$, there is $\psi^l \in H^{-1}(0)$ which is in the $\epsilon$-neighborhood of $\xi^l$. Without loss of generality we may assume that $\psi^l = (y^l, 2^{-l})$. This is what we claimed.

Now let us take a sequence of positive numbers $\{ \epsilon_k \mid k \in \mathbb{N} \}$ with limit zero. Then for any $k$, there exists a $y^{l_k}$ being a $2^{-l_k}$-robust stationary point of $f$ such that $||x^{l_k} - y^{l_k}|| \leq \epsilon_k$. Since $\lim_{k\to\infty} x^{l_k} = x^*$ and $\lim_{k\to\infty} \epsilon_k = 0$, it immediately follows that

$$\lim_{k\to\infty} y^{l_k} = x^*.$$

Hence $x^*$ is a robust stationary point of $f$ on $S^n$. □

In case the algorithm terminates with an $(n-1)$-dimensional simplex $\sigma$ with vertices $y^1, \cdots, y^n$, then $\bar{x} = \sum_{i=1}^{n} \lambda_i^* y^i$ is a robust stationary point of $\bar{f}$. If the accuracy of approximation is not satisfactory, the algorithm can be restarted at the point $\bar{x}$ with a smaller grid size $d^{-1}$ to find a better approximate robust stationary point hopefully within a small number of steps. Suppose that the algorithm generates a sequence $\{ \bar{x}^h \mid h \in \mathbb{N} \}$, where $\bar{x}^h$ is the robust stationary point of $\bar{f}$ corresponding to the grid size $d_h^{-1}$ for an increasing sequence of positive integers $\{ d_h \mid h \in \mathbb{N} \}$. It is readily seen that for every $l \in \mathbb{Z}_+$, the mesh size $\delta_{l,d_h}$ tends to zero as $h$ goes to infinity. Therefore the sequence $\{ \bar{x}^h \mid h \in \mathbb{N} \}$ has a subsequence converging to a cluster point $x^*$. Clearly, $x^*$ is a robust stationary point of $f$ on $S^n$.

As described above, starting at the point $v$, the algorithm generates a unique sequence of adjacent $t$-simplices $\sigma(a, \pi)$ in $F(K)$ for varying $K \in \mathcal{K}$ of varying dimension $t = n - m$. When, with respect to some $\sigma(a, \pi)$ with vertices $y^1, \cdots, y^{t+1}$ in some $G^d(l, l+1; K, \gamma(K, K(n-1)))$ for some $l \in \mathbb{Z}_+$ and $\gamma(K, K(n-1))$, the variable $\lambda_q$, for some $q$, $1 \leq q \leq t+1$, becomes zero through a linear programming pivot step in (7.4.1), then the replacement step determines the unique $t$-simplex $\bar{\sigma}(\bar{a}, \bar{\pi})$ in $F(l, l+1; K, \gamma(K, K(n-1)))$ sharing with $\sigma$ the common facet $\tau$ opposite vertex $y^q$ unless this facet lies on the boundary of $F(l, l+1; K, \gamma(K, K(n-1)))$. If $\tau$ does not lie on the boundary of the set $F(l, l+1; K, \gamma(K, K(n-1)))$, then $\bar{\sigma}(\bar{a}, \bar{\pi})$ can be obtained from $a$ and $\pi$ as given in Table 7.1, where $E(j-1)$ is the $j$-th unit vector in $\mathbb{R}^{n-1}$, $j = 1, \cdots, n-1$.

| | $\bar{\pi}$ | $\bar{a}$ |
|---|---|---|
| $q = 1$ | $(\pi(2), \ldots, \pi(t), \pi(1))$ | $a + E(\pi(1))$ |
| $1 < q < t+1$ | $(\pi(1), \ldots, \pi(q-2), \pi(q), \pi(q-1), \pi(q+1), \ldots, \pi(t))$ | $a$ |
| $q = t+1$ | $(\pi(t), \pi(1), \ldots, \pi(t-1))$ | $a - E(\pi(t))$ |

TABLE 7.1. Parameters of $\bar{\sigma}$ if the vertex $y^q$ of is to be replaced.

The algorithm continues with $\bar{\sigma}$ by making a linear programming (LP) pivot step in (7.4.1) with $(f^\top(\bar{y}),1)^\top$, where $\bar{y}$ is the vertex of $\bar{\sigma}$ opposite the facet $\tau$. In case a facet $\tau$ of a simplex in $G^d(l,l+1;K,\gamma(K,K(n-1)))$ is not a facet of another simplex in $G^d(l,l+1;K,\gamma(K,K(n-1)))$, then $\tau$ lies on the boundary of $F(l,l+1;K,\gamma(K,K(n-1)))$. According to Definition 7.3.2 we have the following lemma.

**Lemma 7.4.6** *Let $\sigma(a,\pi)$ be a $t$-simplex in $F(l,l+1;K,\gamma(K,K(n-1)))$. The facet $\tau$ of $\sigma$ opposite the vertex $y^q$, $1 \le q \le t+1$, lies on the boundary of this set if and only if one of the following cases occurs:*
*(i) $1 < q < t+1$, $\pi(q) = h+1$, $\pi(q-1) = h$ for some $h \in \{0,1,\cdots,t-2\}$, and $a(h) = a(h+1)$ in case $h \ge 1$, and $a(0)+ld = a(1)$ in case $h = 0$;*
*(ii) $q = t+1$, $\pi(t) = t-1$, and $a(t-1) = 0$;*
*(iii) $q = 1$, $\pi(1) = 0$, and $a(0) = d-1$;*
*(iv) $q = t+1$, $\pi(t) = 0$, and $a(0) = 0$.*

Suppose the algorithm generates the simplex $\sigma(a,\pi)$ as given in Lemma 7.4.6 and $\lambda_q$ becomes zero after making an LP pivot step in (7.4.1). Then the facet $\tau$ of $\sigma$ opposite to the vertex $y^q$ is $K$-complete. In case $(iii)$ the facet $\tau$ lies in the face $F(l+1,K)$ of $A(2^{-l-1})$ and the algorithm reaches a $2^{-l-1}$-robust stationary point $\bar{x} = \sum_{i=2}^{t+1} \lambda_i^* y^i$ of $\bar{f}$ lying in $F(l+1,K)$. If $l$ is large enough, then $\bar{x}$ is an approximate robust stationary point of $f$. Otherwise, the algorithm proceeds with $\bar{\sigma}$ by making an LP pivot step in (7.4.1) with $(f^\top(\bar{y}),1)^\top$, where $\bar{y}$ is the vertex of $\bar{\sigma}$ opposite the facet $\tau$ and $\bar{\sigma}$ in $F(l+1,l+2;K,\gamma(K,K(n-1)))$ is obtained according to Table 7.1.

In case $(iv)$ the facet $\tau$ lies in the face $F(l,K)$ of $A(2^{-l})$ and the algorithm continues with $\bar{\sigma}$ by making an LP pivot step in (7.4.1) with $(f^\top(\bar{y}),1)^\top$, where $\bar{y}$ is the vertex of $\bar{\sigma}$ opposite the facet $\tau$ and $\bar{\sigma}$ in $F(l-1,l;K,\gamma(K,K(n-1)))$ is also obtained from Table 7.1.

In case $(i)$ and if $h \ge 1$, the facet $\tau$ is a facet of the $t$-simplex $\bar{\sigma} = \sigma(a,\pi)$ in $F(l,l+1;K)$ lying in the subset $F(l,l+1;K,\bar{\gamma}(K,K(n-1)))$ with

$$\bar{\gamma}(K,K(n-1)) = (J_1,\cdots,J_h,\bar{J}_{h+1},J_{h+2},\cdots,J_t),$$

where $\bar{J}_{h+1} \in \mathcal{K}$, $\bar{J}_{h+1} \ne J_{h+1}$, is uniquely determined by the properties that $\bar{J}_{h+1}$ conforms to $J_h$, has one component less than $J_h$, and is conformed by $J_{h+2}$. In case $(i)$ and if $h = 0$, then $\tau$ is a facet of the $t$-simplex $\bar{\sigma} = \sigma(a,\pi)$ in $F(l,l+1;K,\bar{\gamma}(K,\bar{K}(n-1)))$ with $\bar{K}(n-1)$ and $\bar{\gamma}$ defined as follows. Let $J_1 = K(n-1) = (K_1,\cdots,K_{n-1})$. In case $J_2 = (K_1,\cdots,K_{n-2})$, we have $\bar{K}(n-1) = (K_1,\cdots,K_{n-2},\bar{K}_{n-1})$ with $\bar{K}_{n-1} = K_{n-2} \bigcup (I_n \backslash K_{n-1})$. In case $J_2 = (K_2,\cdots,K_{n-1})$, let $\bar{K}(n-1) = (\bar{K}_1,K_2,\cdots,K_{n-1})$ with $\bar{K}_1 = K_2 \backslash K_1$. Finally if $J_2 = (K_1,\cdots,K_s,K_{s+2},\cdots,K_{n-1})$ for some $s \in \{1,\cdots,n-3\}$, we have $\bar{K}(n-1) = (K_1,\cdots,K_s,\bar{K}_{s+1},K_{s+2},\cdots,K_{n-1})$ with $\bar{K}_{s+1} = K_s \bigcup (K_{s+2} \backslash K_{s+1})$. Then $\bar{\gamma}(K,\bar{K}(n-1)) = (\bar{K}(n-1),J_2,\cdots,J_t)$.

In both subcases of case (*i*) the algorithm continues with making a pivot step in (7.4.1) with $(f^\top(\bar{y}), 1)^\top$, where $\bar{y}$ is the vertex of the new $t$-simplex $\bar{\sigma}$ opposite the facet $\tau$.

In case (*ii*) the facet lies in the set $F(l, l+1; J_{t-1})$ of $F(K)$. More precisely, $\tau$ is the $(t-1)$-simplex $\sigma(a, \bar{\pi})$ in $F(l, l+1; \bar{K}, \bar{\gamma}(\bar{K}, K(n-1)))$, where $\bar{K} = J_{t-1}$, $\bar{\gamma}(\bar{K}, K(n-1)) = (J_1, \cdots, J_{t-1})$, and $\bar{\pi} = (\pi(1), \cdots, \pi(t-1))$. The algorithm now proceeds with making a pivot step in (7.4.1) with $(-b^\top(K_s), 0)^\top$, where $K_s$ is the unique component of $J_{t-1}$ but not of $J_t$.

Finally, if through an LP pivot step in (7.4.1), the variable $\mu_s$ becomes 0 for some $s \in \{ 1, \cdots, m \}$, then the algorithm terminates with the approximate robust stationary point $\bar{x} = \sum_i \lambda_i^* y^i$ of $f$ and restarts at the point $\bar{x}$ with a smaller grid size when the accuracy is not satisfactory in case $m = 1$. In case $m \geq 2$, the simplex $\sigma(a, \pi)$ is a facet of a unique $(t+1)$-simplex $\sigma$ in $F(\bar{K})$ with $\bar{K} = (K_1, \cdots, K_{s-1}, K_{s+1}, \cdots, K_m)$. More precisely, $\bar{\sigma} = \sigma(a, \bar{\pi})$ lies in $F(l, l+1; \bar{K}, \bar{\gamma}(\bar{K}, K(n-1)))$, where $\bar{\gamma}(\bar{K}, K(n-1)) = (\gamma, \bar{K})$, and $\bar{\pi} = (\pi(1), \cdots, \pi(t), t)$. The algorithm continues by making a pivot step in (7.4.1) with $(f^\top(\bar{y}), 1)^\top$, where $\bar{y}$ is the vertex of $\bar{\sigma}$ opposite the facet $\sigma$. This concludes the description of how the algorithm works in the $P$-triangulation of $S^n$.

## 7.5. Examples and resolution of degeneracy

In this section we give some examples to show the applications of the concept of a robust stationary point and the power of the algorithm as well.

**Example 7.5.1 Walrasian Equilibrium Problem** *There are two commodities in a pure exchange economy. The excess demand function is given by* $z(p) = (p_1 p_2^2(1 - p_1^2), -p_1^2 p_2(1 - p_1^2))^\top$ *for* $p \in S^2$. *There are two Walrasian equilibria, being stationary points of* $z$: $\bar{p} = (1, 0)^\top$, $\hat{p} = (0, 1)^\top$. *Only* $\bar{p}$ *is a robust stationary point which is more sensible than* $\hat{p}$ *in economic terms. We give some explanation. The basic idea of the Walrasian tatonnement process is as follows. Suppose that an economy is in disequilibrium. Then the Walrasian auctioneer would increase the price of a commodity if the excess demand of that commodity is positive, and decrease the price of a commodity if the excess demand of that commodity is negative. A sensible equilibrium should be stable against some data perturbation. In this example, suppose that the economy is slightly perturbed away from the equilibrium* $(0, 1)^\top$. *Then the Walrasian tatonnement process would lead to the equilibrium* $(1, 0)^\top$.

**Example 7.5.2 Stationary Point Problem** *The function is defined by* $f(x) = (x_1 + x_2, x_2 + x_3, x_3 + x_1)^\top$ *for* $x \in S^3$. *The set of stationary points* *is*

$$\{\,(1/3,1/3,1/3)^\top,(1,0,0)^\top,(0,1,0)^\top,(0,0,1)^\top\,\}.$$

*However, $f$ just has one robust stationary point:* $(1/3,1/3,1/3)^\top$.

**Example 7.5.3 Nash Equilibrium Problem** *The marginal expected payoff function is given by $f(x)=(x_1-9x_3,-7x_3,-9x_1-7x_2-7x_3)^\top$ for $x\in S^3$. This example is due to Myerson [1978] (see Example 2.3.3). There are three Nash equilibria, being stationary points of $f$:* $(1,0,0)^\top$, $(0,1,0)^\top$ *and* $(0,0,1)^\top$. *The function $f$ just has one robust stationary point, being the unique proper Nash equilibrium:* $(1,0,0)^\top$.

Now we briefly discuss how to resolve the degeneracy problem for system ( 7.4.1). For a $(t-1)$-simplex $\tau(y^1,\cdots,y^t)$ and an element $K=(K_1,\cdots,K_m)\in\mathcal{K}$ with $t=n-m$, we define a matrix with respect to $\tau$ and $K$ by

$$M_{K,\tau}=\begin{bmatrix} 1 & \cdots 1 & 0 & \cdots & 0 & 0 \\ f(y^1) & \cdots f(y^t) & -b(K_1) & \cdots & -b(K_m) & -e \end{bmatrix}.$$

We say that $\tau$ is almost $K$-complete if $M_{K,\tau}^{-1}$ exists and is semilexicopositive. Now we have the following lemma.

**Lemma 7.5.4** *For the starting point $v$, let $(i_1,\cdots,i_n)$ be a permutation of $(1,\cdots,n)$ such that $f_{i_1}(v)\geq f_{i_2}(v)\geq\cdots\geq f_{i_n}(v)$ and $f_{i_s}(v)=f_{i_t}(v)$ implies both $s<t$ and $i_s>i_t$. Set $K^0=(K_1^0,\cdots,K_{n-1}^0)$ with $K_j^0=\{\,i_1,\cdots,i_j\,\}$ for $j\in I_{n-1}$. Then $<v>$ is an almost $K^0$-complete simplex in $F(K^0)$ but is not almost $K$-complete in $F(K)$ for any other $K\in\mathcal{K}$.*

For a $t$-simplex $\sigma(y^1,\cdots,y^{t+1})$ and an element $K=(K_1,\cdots,K_m)\in\mathcal{K}$ with $t=n-m$, we define a matrix with respect to $\sigma$ and $K$ by

$$M_{K,\sigma}=\begin{bmatrix} 1 & \cdots 1 & 0 & \cdots & 0 & 0 \\ f(y^1) & \cdots f(y^{t+1}) & -b(K_1) & \cdots & -b(K_m) & -e \end{bmatrix}.$$

Now we can replace system ( 7.4.1) by the following system of equations

$$\begin{array}{rcl} M_{K,\sigma}X & = & \mathbf{I} \\ X & \succeq_s & \mathbf{0}. \end{array}$$

The rest of the algorithm is the same as in the preceding section. We shall demonstrate the algorithm with Example 7.5.3 in a pedagogic manner. We will see that this example is degenerate. The algorithm starts at the point $v=(1/3,1/3,1/3)^\top$. The steps in the course of the algorithm are calculated below.

Step 1: $v = (1/3, 1/3, 1/3)^\top$, $f(v) = (-8/3, -7/3, -23/3)^\top$, $\tau^0 =< v >$, and $K^0 = (\{2\}, \{2, 1\})$. Set $M = M_{K^0,\tau^0}$. We have

$$M = \begin{bmatrix} 1 & 0 & 0 & 0 \\ -8/3 & 0 & -1 & -1 \\ -7/3 & -1 & -1 & -1 \\ -23/3 & 0 & 0 & -1 \end{bmatrix}, \quad M^{-1} = \begin{bmatrix} 1 & 0 & 0 & 0 \\ 1/3 & 1 & -1 & 0 \\ 15/3 & -1 & 0 & 1 \\ -23/3 & 0 & 0 & -1 \end{bmatrix}.$$

Step 2: $x^1 = (13/42, 19/42, 10/42)^\top$, $f(x^1) = (-77/42, -70/42, -320/42)^\top$, $M^{-1}(1, f^\top(x^1))^\top = (1, 1/6, -33/42, -1/21)^\top$. $v$ is replaced by $x^1$. Set $\tau^1 =< x^1 >$ and $K^1 = K^0$. Let $M = M_{K^1,\tau^1}$. We have

$$M = \begin{bmatrix} 1 & 0 & 0 & 0 \\ -77/42 & 0 & -1 & -1 \\ -70/42 & -1 & -1 & -1 \\ -320/42 & 0 & 0 & -1 \end{bmatrix}, \quad M^{-1} = \begin{bmatrix} 1 & 0 & 0 & 0 \\ 1/6 & 1 & -1 & 0 \\ 81/14 & -1 & 0 & 1 \\ -160/21 & 0 & 0 & -1 \end{bmatrix}.$$

Step 3: $x^2 = (2/7, 4/7, 1/7)^\top$, $f(x^2) = (-1, -1, -53/7)^\top$, $M^{-1}(1, f^\top(x^2))^\top = (1, 1/6, -11/14, -1/21)^\top$. (Notice that the problem is now becoming degenerate.) $x^1$ is replaced by $x^2$. Set $\tau^2 =< x^2 >$ and $K^2 = K^0$. Let $M = M_{K^2,\tau^2}$. We have

$$M = \begin{bmatrix} 1 & 0 & 0 & 0 \\ -1 & 0 & -1 & -1 \\ -1 & -1 & -1 & -1 \\ -53/7 & 0 & 0 & -1 \end{bmatrix}, \quad M^{-1} = \begin{bmatrix} 1 & 0 & 0 & 0 \\ 0 & 1 & -1 & 0 \\ 46/7 & -1 & 0 & 1 \\ -53/7 & 0 & 0 & -1 \end{bmatrix}.$$

Step 4: $x^3 = (5/21, 14/21, 2/21)^\top$, $f(x^3) = (-13/21, -14/21, -157/21)^\top$, $M^{-1}(1, f^\top(x^3))^\top = (1, 1/21, -2/7, -2/21)^\top$. $(0, -b^\top(\{2\}))^\top$ is replaced by $(1, f^\top(x^3))^\top$. Increase the dimension. Set $\tau^3 =< x^2, x^3 >$ and $K^3 = (\{2, 1\})$. Let $M = M_{K^3,\tau^3}$. We have

$$M = \begin{bmatrix} 1 & 1 & 0 & 0 \\ -1 & -13/21 & -1 & -1 \\ -1 & -14/21 & -1 & -1 \\ -53/7 & -157/21 & 0 & -1 \end{bmatrix}, \quad M^{-1} = \begin{bmatrix} 1 & -21 & 21 & 0 \\ 0 & 21 & -21 & 0 \\ 46/7 & 5 & -6 & 1 \\ -53/7 & 2 & -2 & -1 \end{bmatrix}.$$

Step 5: $x^4 = (13/42, 25/42, 4/42)^\top$, $f(x^4) = (-23/42, -28/42, -320/42)^\top$, $M^{-1}(1, f^\top(x^4))^\top = (-1/2, 5/2, 3/14, 2/7)^\top$. $x^3$ is replaced by $x^4$. Set $\tau^4 =< x^2, x^4 >$ and $K^4 = K^3$. Let $M = M_{K^4,\tau^4}$. We have

$$M = \begin{bmatrix} 1 & 1 & 0 & 0 \\ -1 & -23/42 & -1 & -1 \\ -1 & -28/42 & -1 & -1 \\ -53/7 & -320/42 & 0 & -1 \end{bmatrix}, \quad M^{-1} = \begin{bmatrix} 1 & -42/5 & 42/5 & 0 \\ 0 & 42/5 & -42/5 & 0 \\ 46/7 & 16/5 & -21/5 & 1 \\ -53/7 & -2/5 & 2/5 & -1 \end{bmatrix}.$$

Step 6: $x^5 = (5/14, 7/14, 2/14)^\top$, $f(x^5) = (-13/14, -1, -54/7)^\top$,

$$M^{-1}(1, f^\top(x^5))^\top = (2/5, 3/5, 18/35, 16/35)^\top.$$

$x^4$ is replaced by $x^5$. Set $\tau^5 =< x^2, x^5 >$ and $K^5 = K^3$. Let $M = M_{K^5,\tau^5}$. We have

$$M = \begin{bmatrix} 1 & 1 & 0 & 0 \\ -1 & -13/14 & -1 & -1 \\ -1 & -1 & -1 & -1 \\ -53/7 & -54/7 & 0 & -1 \end{bmatrix}, \ M^{-1} = \begin{bmatrix} 1 & -14 & 14 & 0 \\ 0 & 14 & -14 & 0 \\ 46/7 & 2 & -3 & 1 \\ -53/7 & -2 & 2 & -1 \end{bmatrix}.$$

Step 7: $x^6 = (13/42, 19/42, 10/42)^\top$, $f(x^6) = (-77/42, -70/42, -320/42)^\top$, $M^{-1}(1, f^\top(x^6))^\top = (10/3, -7/3, 2/7, 8/21)^\top$. Note that $x^6 = x^1$. $x^2$ is replaced by $x^6$. Set $\tau^6 =< x^6, x^5 >$ and $K^6 = K^3$. Let $M = M_{K^6,\tau^6}$. We have

$$M = \begin{bmatrix} 1 & 1 & 0 & 0 \\ -77/42 & -13/14 & -1 & -1 \\ -70/42 & -1 & -1 & -1 \\ -320/42 & -54/7 & 0 & -1 \end{bmatrix}, \ M^{-1} = \begin{bmatrix} 3/10 & -21/5 & 21/5 & 0 \\ 7/10 & 21/5 & -21/5 & 0 \\ 227/35 & 16/5 & -21/5 & 1 \\ -269/35 & -2/5 & 2/5 & -1 \end{bmatrix}.$$

Step 8: $x^7 = (16/42, 16/42, 10/42)^\top$, $f(x^7) = (-37/21, -35/21, -163/21)^\top$, $M^{-1}(1, f^\top(x^7))^\top = (7/10, 3/10, 3/35, -1/35)^\top$. $x^6$ is replaced by $x^7$. Set $\tau^7 =< x^7, x^5 >$ and $K^7 = K^3$. Let $M = M_{K^7,\tau^7}$. We have

$$M = \begin{bmatrix} 1 & 1 & 0 & 0 \\ -37/21 & -13/14 & -1 & -1 \\ -35/21 & -1 & -1 & -1 \\ -163/21 & -54/7 & 0 & -1 \end{bmatrix}, \ M^{-1} = \begin{bmatrix} 3/7 & -6 & 6 & 0 \\ 4/7 & 6 & -6 & 0 \\ 316/49 & 26/7 & -33/7 & 1 \\ -379/49 & -2/7 & -2/7 & -1 \end{bmatrix}.$$

Step 9: $x^8 = (3/7, 3/7, 1/7)^\top$, $f(x^8) = (-6/7, -1, -55/7)^\top$,

$$M^{-1}(1, f^\top(x^8))^\top = (-3/7, 4/7, 6/49, 32/49)^\top.$$

$x^5$ is replaced by $x^8$. Set $\tau^8 =< x^7, x^8 >$ and $K^8 = K^3$. Let $M = M_{K^8,\tau^8}$. We have

$$M = \begin{bmatrix} 1 & 1 & 0 & 0 \\ -37/21 & -6/7 & -1 & -1 \\ -35/21 & -1 & -1 & -1 \\ -163/21 & -55/7 & 0 & -1 \end{bmatrix}, \ M^{-1} = \begin{bmatrix} 3/5 & -21/5 & 21/5 & 0 \\ 2/5 & 21/5 & -21/5 & 0 \\ 224/35 & 16/5 & -21/5 & 1 \\ -273/35 & -2/5 & 2/5 & -1 \end{bmatrix}.$$

Step 10: $x^9 = (7/14, 5/14, 2/14)^\top$, $f(x^9) = (-11/14, -1, -56/7)^\top$,

$$M^{-1}(1, f^\top(x^9))^\top = (-3/10, 13/10, 3/35, 4/35)^\top.$$

$x^8$ is replaced by $x^9$. Set $\tau^9 =< x^7, x^9 >$ and $K^9 = K^3$. Let $M = M_{K^9,\tau^9}$. We have

$$M = \begin{bmatrix} 1 & 1 & 0 & 0 \\ -37/21 & -11/14 & -1 & -1 \\ -35/21 & -1 & -1 & -1 \\ -163/21 & -56/7 & 0 & -1 \end{bmatrix}, \ M^{-1} = \begin{bmatrix} 9/13 & -42/13 & 42/13 & 0 \\ 4/13 & 42/13 & -42/13 & 0 \\ 580/91 & 38/13 & -51/13 & 1 \\ -713/91 & -10/13 & 10/13 & -1 \end{bmatrix}.$$

Step 11: $x^{10} = (19/42, 13/42, 10/42)^\top$, $f(x^{10}) = (-71/42, -70/42, -332/42)^\top$, $M^{-1}(1, f^\top(x^{10}))^\top = (10/13, 3/13, -94/91, 8/91)^\top$. $x^7$ is replaced by $x^{10}$. Set $\tau^{10} =< x^{10}, x^9 >$ and $K^{10} = K^3$. Let $M = M_{K^{10},\tau^{10}}$. We have

$$M = \begin{bmatrix} 1 & 1 & 0 & 0 \\ -71/42 & -11/14 & -1 & -1 \\ -70/42 & -1 & -1 & -1 \\ -332/42 & -56/7 & 0 & -1 \end{bmatrix}, \; M^{-1} = \begin{bmatrix} 9/10 & -21/5 & 21/5 & 0 \\ 1/10 & 21/5 & -21/5 & 0 \\ 221/35 & 16/5 & -21/5 & 1 \\ -277/35 & -2/5 & 2/5 & -1 \end{bmatrix}.$$

Step 12: $x^{11} = (4/7, 2/7, 1/7)^\top$, $f(x^{11}) = (-5/7, -1, -57/7)^\top$,

$$M^{-1}(1, f^\top(x^{11}))^\top = (10/13, 3/13, -94/91, 8/91)^\top.$$

$x^9$ is replaced by $x^{11}$. Set $\tau^{11} =< x^{10}, x^{11} >$ and $K^{11} = K^3$. Let $M = M_{K^{11},\tau^{11}}$. We have

$$M = \begin{bmatrix} 1 & 1 & 0 & 0 \\ -71/42 & -5/7 & -1 & -1 \\ -70/42 & -1 & -1 & -1 \\ -332/42 & -57/7 & 0 & -1 \end{bmatrix}, \; M^{-1} = \begin{bmatrix} 12/13 & -42/13 & 42/13 & 0 \\ 1/13 & 42/13 & -42/13 & 0 \\ 82/13 & 38/13 & -51/13 & 1 \\ -103/13 & -10/13 & 10/13 & -1 \end{bmatrix}.$$

Step 13: Decrease the dimension. Set $K^{12} = (K_1^{12}, K_2^{12}) = (\{1\}, \{1, 2\})$. $M^{-1}(0, -b^\top(K_1^{12}))^\top = (42/13, -42/13, -38/13, 10/13)^\top$. $(1, f^\top(x^{10}))^\top$ is replaced by $(0, -b^\top(K_1^{12}))^\top$. Set $\tau^{12} =< x^{11} >$. Let $M = M_{K^{12},\tau^{12}}$. We have

$$M = \begin{bmatrix} 1 & 0 & 0 & 0 \\ -5/7 & -1 & -1 & -1 \\ -1 & 0 & -1 & -1 \\ -57/7 & 0 & 0 & -1 \end{bmatrix}, \; M^{-1} = \begin{bmatrix} 1 & 0 & 0 & 0 \\ 2/7 & -1 & 1 & 0 \\ 50/7 & 0 & -1 & 1 \\ -57/7 & 0 & 0 & -1 \end{bmatrix}.$$

Step 14: $x^{12} = (14/21, 5/21, 2/21)^\top$, $f(x^{12}) = (-4/21, -14/21, -175/21)^\top$, $M^{-1}(1, f^\top(x^{12}))^\top = (1, -4/21, -11/21, 4/21)^\top$. $x^{11}$ is replaced by $x^{12}$. Set $\tau^{13} =< x^{12} >$ and $K^{13} = K^{12}$. Let $M = M_{K^{13},\tau^{13}}$. We have

$$M = \begin{bmatrix} 1 & 0 & 0 & 0 \\ -4/21 & -1 & -1 & -1 \\ -14/21 & 0 & -1 & -1 \\ -175/21 & 0 & 0 & -1 \end{bmatrix}, \; M^{-1} = \begin{bmatrix} 1 & 0 & 0 & 0 \\ 10/21 & -1 & 1 & 0 \\ 161/21 & 0 & -1 & 1 \\ -175/21 & 0 & 0 & -1 \end{bmatrix}.$$

Step 15: $x^{13} = (16/21, 4/21, 1/21)^\top$, $f(x^{13}) = (7/21, -7/21, -179/21)^\top$, $M^{-1}(1, f^\top(x^{12}))^\top = (1, -4/21, -11/21, 4/21)^\top$. $x^{12}$ is replaced by $x^{13}$. Set $\tau^{14} =< x^{13} >$ and $K^{14} = K^{12}$. Let $M = M_{K^{14},\tau^{14}}$. We have

$$M = \begin{bmatrix} 1 & 0 & 0 & 0 \\ 7/21 & -1 & -1 & -1 \\ -7/21 & 0 & -1 & -1 \\ -179/21 & 0 & 0 & -1 \end{bmatrix}, \; M^{-1} = \begin{bmatrix} 1 & 0 & 0 & 0 \\ 2/3 & -1 & 1 & 0 \\ 172/21 & 0 & -1 & 1 \\ -179/21 & 0 & 0 & -1 \end{bmatrix}.$$

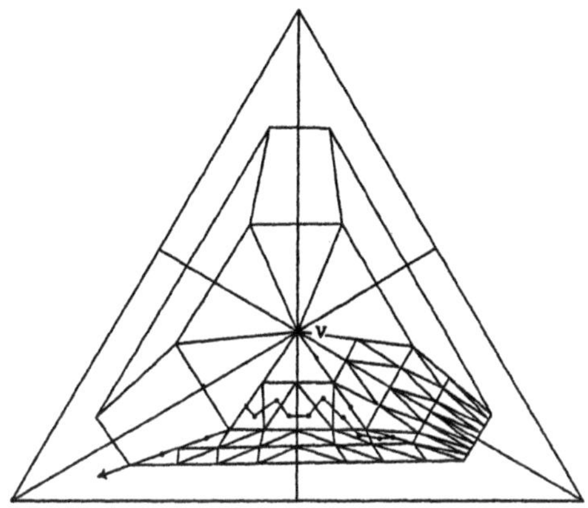


*Figure 7.3.* Illustration of the adaptive algorithm on $S^3$.

It is easy to confirm that at the $13+2(k-1)th$ step for $k \in \mathbb{N}$, we get an approximate robust stationary point given by

$$y^k = (\frac{1}{1+2^{-k}+2^{-2k}}, \frac{2^{-k}}{1+2^{-k}+2^{-2k}}, \frac{2^{-2k}}{1+2^{-k}+2^{-2k}})^\top.$$

For each $k \in \mathbb{N}$, $y^k$ is in fact a $2^{-k}$-robust stationary point of $f$. The path followed by the algorithm is illustrated in Figure 7.3. The algorithm converges to the robust stationary point $(1,0,0)^\top$.

## 7.6. Extension to general polytopes

The concept of robust stationary point is extended in Talman and Yang [1994] to the simplotope. In this paper a simplicial algorithm is also developed to compute a proper Nash equilibrium of a finite-person game in normal form. More recently, van der Laan, Talman and Yang [1996] have considerably generalized this concept to general polytopes and have also introduced several interesting applications in game theory and general equilibrium theory. Moreover, they proposed an algorithm to compute the refined stationary point. Here we can only give a short description of this generalization by focusing on a full-dimensional polytope P in $\mathbb{R}^n$.

Consider an arbitrary full-dimensional polytope

$$\mathrm{P} = \{x \in \mathbb{R}^n \mid {a^i}^\top x \leq b_i, \text{ for all } i \in I_m\}.$$

It is assumed that P is simple and that no constraints are redundant.

For each subset $I$ of $I_m$, define

$$F(I) = \{x \in \mathrm{P} \mid a^{i\top} x = b_i \text{ for all } i \in I\}.$$

Then $F(I)$ is a face of P unless it is empty. Note that $F(\emptyset) = \mathrm{P}$. Let

$$\mathcal{I} = \{I \subseteq I_m \mid F(I) \text{ is a nonempty face of } \mathrm{P}\}.$$

Let a function $f : \mathrm{P} \mapsto \mathbb{R}^n$ be given.

**Definition 7.6.1** *For given $\theta > 0$ a point $x \in \mathrm{P}$ is a $\theta$-robust stationary point of $f$ if*

*(a) $x$ is an interior point of* P;

*(b) for some $I \in \mathcal{I}$, $f(x) = \sum_{h \in I_m} \mu_h a^h$ with $\mu_h \geq 0$ for all $h \in I$ and $\mu_h = 0$ for all $h \in I_m \setminus I$, if $\mu_k < \mu_l$, then $b_l - a^{l\top} x \leq \theta(b_k - a^{k\top} x)$.*

**Definition 7.6.2** *A point $x^* \in \mathrm{P}$ is a robust stationary point of $f$ on* P *if there exist sequences $\{\theta_l \mid l \in \mathsf{N}\}$ of positive numbers and $\{x(\theta_l) \mid l \in \mathsf{N}\}$ of $\theta_l$-robust stationary points $x(\theta_l)$ of $f$ such that*

$$\lim_{l \to \infty} \theta_l = 0 \quad \textit{and} \quad \lim_{l \to \infty} x(\theta_l) = x^*.$$

Now we are going to give a geometric interpretation of a robust stationary point. For a stationary point $x^* \in F(I)$ for some $I \in \mathcal{I}$, we have that $\mu_i > 0$ implies $b_i - a^{i\top} x^* = 0$ for any $i$. One might think of this relation as the complementary slackness condition as in the linear programming theory. In order for $x^*$ to be a robust stationary point, it is reasonable to require that the higher $\mu_i$ is, the stronger this equality should be. That is to say, in a neighborhood of $x^*$ for any $\theta > 0$ there should be a point $x(\theta)$ such that if $\mu_i(\theta) > \mu_j(\theta)$, then $x(\theta)$ must be $\theta$ times closer to $F(\{i\})$ than to $F(\{j\})$, i.e., $b_i - a^{i\top} x(\theta) \leq \theta(b_j - a^{j\top} x(\theta))$.

The next two examples demonstrate that the concept of a robust stationary point is indeed a proper refinement of that of a stationary point.

**Example 7.6.3** *Take $n = 2$ and $a^1 = -a^3 = (1,0)^\top$, $a^2 = -a^4 = (0,1)^\top$, $b_1 = b_2 = 1$ and $b_3 = b_4 = 0$. Then $\mathrm{P} = C^2 = \{x \in \mathbb{R}^2 \mid x_1 \leq 1, x_2 \leq 1, -x_1 \leq 0, \text{ and } -x_2 \leq 0\}$. Let a continuous function $f : \mathrm{P} \mapsto \mathbb{R}^2$ be given by*

$$f(x) = (x_1 - 1, x_2 - 1)^\top.$$

It is not difficult to show that for a continuous function $f : C^2 \mapsto \mathbb{R}^2$, $x^*$ is a stationary point of $f$ if and only if it holds

$$\begin{array}{lll} f_i(x^*) \leq 0 & \text{if} & x_i^* = 0, \\ f_i(x^*) = 0 & \text{if} & 0 < x_i^* < 1, \\ f_i(x^*) \geq 0 & \text{if} & x_i^* = 1. \end{array}$$

The set of stationary points of the function is equal to:

$$\{(0,0)^\top,\ (1,1)^\top,\ (1,0)^\top,\ (0,1)^\top\}.$$

However, only $(0,0)^\top$ is a robust stationary point of $f$. To show this, let for $\theta \in (0,1)$

$$x_1(\theta) = \theta,\ x_2(\theta) = \theta^2.$$

Clearly, $x(\theta)$ lies in the interior of P for each $\theta \in (0,1)$. Note that

$$f(x(\theta)) = \mu_3 a^3 + \mu_4 a^4 = \mu_3(-1,0)^\top + \mu_4(0,-1)^\top,$$

with $\mu_3 > 0$ and $\mu_4 > 0$. In fact we have $\mu_4 = 1-\theta^2 > \mu_3 = 1-\theta$. It also holds

$$b_4 - a^{4\top} x(\theta) = \theta^2 \le \theta(b_3 - a^{3\top} x(\theta)) = \theta^2.$$

Hence for each $\theta \in (0,1)$, $x(\theta)$ is a $\theta$-robust stationary point. It is easy to see that

$$\lim_{\theta \to 0^+} x(\theta) = (0,0)^\top.$$

It is also easy to check why all other stationary points are not robust stationary points. We leave it to the reader.

**Example 7.6.4** *Let $f : C^2 \mapsto \mathbb{R}^2$ be given by*

$$f(x) = (x_2(1-x_1)^2(x_2-1), x_1(1-x_2)^2(x_1-1))^\top,$$

*where $C^2$ is as given in Example 7.6.3.*

The set of stationary points of this function is equal to:

$$\{(0,x)^\top \mid 0 \le x \le 1\} \cup \{(1,x)^\top \mid 0 \le x \le 1\}$$
$$\cup \{(x,1)^\top \mid 0 \le x \le 1\} \cup \{(x,0)^\top \mid 0 \le x \le 1\}.$$

It is remarkable that this function has only one stationary point $(0,0)^\top$ being a robust stationary point. We will prove it. For $\theta \in (0,1)$, let

$$x(\theta) = (\theta, \theta^2)^\top.$$

Clearly, $x(\theta)$ is an interior point of $C^2$. For each $\theta \in (0,1)$ we have

$$f(x(\theta)) = \mu_3 a^3 + \mu_4 a^4 = \mu_3(-1,0)^\top + \mu_4(0,-1)^\top,$$

with $\mu_3 = \theta^2(1-\theta)^2(1-\theta^2) > 0$ and $\mu_4 = \theta(1-\theta^2)^2(1-\theta) > 0$. Since $\frac{\mu_4}{\mu_3} = \frac{(1+\theta)}{\theta} > 1$, $\mu_4 > \mu_3$. It also holds

$$b_4 - a^{4\top} x(\theta) = \theta^2 \le \theta(b_3 - a^{3\top} x(\theta)) = \theta^2.$$

Hence $x(\theta)$ is a $\theta$-robust stationary point. Since

$$\lim_{\theta \to \theta+} x(\theta) = (0,0)^\top,$$

the point $(0,0)^\top$ is a robust stationary point by definition. It is left to the reader to check that any other stationary point is not a robust stationary point.

CHAPTER 8

# The Computation of a Continuum of Zero Points of a Point-to-Set Mapping

## 8.1. Introduction

So far in the literature most numerical methods are designed to search for a single fixed point or zero point of the underlying function. As a matter of fact, most of the existing fixed point theorems only guarantee the existence of a single fixed point. As far as we are aware of, Browder theorem is the only result on the existence of a connected set of fixed points up to now. Recall in Chapter 1 this theorem is stated as follows. Let $X$ be a nonempty, compact and convex subset of $\mathbb{R}^n$ and let $\phi : X \times [0,1] \mapsto X$ be a nonempty-valued, convex-valued and compact-valued u.s.c. point-to-set mapping. Then there exists a connected subset $C$ of $X \times [0,1]$ such that $x \in \phi(x,t)$ for every $(x,t) \in C$, $C \cap X \times \{0\} \neq \emptyset$ and $C \cap X \times \{1\} \neq \emptyset$. Note that for each fixed $t \in [0,1]$, there exists a fixed point of $\phi(\cdot,t)$ by Kakutani theorem. Although our intuition tells us there must exist a connected set of fixed points from level zero to level one since $t$ is a free variable, it is by no means easy to prove this. Amann [1972] presented a theorem for the existence of three fixed points. This theorem says: Let $D_i = \{x \in \mathbb{R}^n \mid u^i \ll x \ll v^i\}$ for $i \in I_2$ and $D = \{x \in \mathbb{R}^n \mid u^1 \ll x \ll v^2\}$ where $u^1 \ll v^1$, $u^2 \ll v^2$, $u^1 \leq u^2$, $v^1 \leq v^2$, and $cl(D_1) \cap cl(D_2) = \emptyset$. Let $f : cl(D) \mapsto \mathbb{R}^n$ be a function satisfying that (a) $u \leq v$ implies $f(u) \leq f(v)$; and (b) $u^i \ll f(u^i)$, $f(v^i) \ll v^i$ for all $i \in I_2$. Then there exists at least one fixed point in each set $D_1$, $D_2$, and $D \backslash (cl(D_1) \cup cl(D_2))$. However, neither Browder nor Amann gives constructive proofs for their results.

In this chapter we investigate the existence problem of a connected set of zero points or fixed points for general upper semi-continuous correspondences mapping from the $n$-dimensional unit cube $U^n$ into $\mathbb{R}^n$. It will be shown that an upper semi-continuous correspondence satisfying certain mild conditions has a connected set of zero points linking the origin $\mathbf{0}$ and the point $e$ of $U^n$. This result is so powerful that Browder theorem and Kakutani theorem can be shown as special cases. We will also present the first algorithm of this kind for computing the connected set of zero points. Hence the above theorem will be demonstrated in a constructive manner. The algorithm can be roughly described as follows. Starting with the point $\mathbf{0}$ (or $e$), the algorithm generates a piecewise linear path of points in $U^n$ that terminates at $e$ (or $\mathbf{0}$) within a finite number of steps. It will be shown that each point on the path yields an approximate zero point. The finer the triangulation is, the more accurate each approximate solution will be. Using the limit argument we will be able to show the existence of a connected set of zero points linking the two points $\mathbf{0}$ and $e$. In addition, here we have to tackle some specific degeneracy problems, although in most literature it is assumed that degeneracy does not occur for simplicial algorithms. An important application of the algorithm introduced here concerns the problem of computing a connected set of constrained equilibria in the economic model with price rigidities described in Section 2.5. As it is shown, there are two trivial constrained equilibria in this economic model. When applied to this economic model, the algorithm starts with one trivial constrained equilibrium, generates a set of approximate constrained equilibria and terminates with the other trivial constrained equilibrium within a finite number of steps. As a consequence of our general results we prove that there exists a connected set of constrained equilibria linking both trivial constrained equilibria. Thus, this algorithm allows us to find all kinds of constrained equilibria. Related problems have been considered by, e.g., van der Laan [1982], Cornielje and van der Laan [1986], and Herings [1995]. In van der Laan [1982], a simplicial algorithm with an integer labeling rule is proposed to compute an unemployment (or supply-constrained) equilibrium of a continuous excess demand function. In particular, Herings [1995] has intensively studied the structure of the set of constrained equilibria and has many interesting results in this respect.

This chapter consists of six sections. In Section 2 we formulate the problem. This problem is very general so that the equilibrium problem with price rigidities described in Section 2.5 can be put in this framework as a special case. In Section 3 we present the algorithm in detail and prove its convergence. In Section 4 we analyze the accuracy of the approximate solutions obtained by the algorithm and show that there exists a connected set of zero points containing two trivial zero points. With respect to the economic model in the presence of price rigidities described in Section 2.5, it is shown that there exists a connected set of constrained equilibria linking the two trivial constrained equilibria. An illustration of the algorithm by Example 2.5.8 can be found in Section 5. Some remarks and open questions are described in Section 6. There we will also show how Browder theorem and Kakutani theorem can be derived as special cases of our general results. This chapter except Section 6 is based on Herings, Talman and Yang [1996].

## 8.2. The problem and an economic model with price rigidities

We consider the problem of computing a set of zero points for any correspondence $\zeta$ from the unit cube $U^n$ into $\mathbb{R}^n$ satisfying the following assumption.

**Assumption 8.2.1** *The correspondence $\zeta : U^n \mapsto \mathbb{R}^n$ satisfies:*
*(a) $\zeta$ is a u.s.c. correspondence with nonempty, compact, and convex values;*
*(b) for each $q \in U^n$, there exists $z \in \zeta(q)$ such that for every $j \in I_n$ with $q_j = 0$ it holds that $z_j \geq 0$, and for every $j \in I_n$ with $q_j = 1$ it holds that $z_j \leq 0$;*
*(c) for each $q \in U^n$ and for each $z \in \zeta(q)$, there exists $p(q,z) \in \mathbb{R}^n_{++}$ such that $z^\top p(q,z) = 0$.*

It is clear that the correspondence $\zeta$ has two trivial zero points: $\mathbf{0}$ and $e$, i.e., $\mathbf{0} \in \zeta(\mathbf{0})$ and $\mathbf{0} \in \zeta(e)$. In the subsequent sections we will show that there exists a connected set of zero points for the correspondence $\zeta$ linking the two trivial zero points. In effect, an algorithm will be introduced to approximate the connected set of zero points. Therefore, the existence of a connected set of zero points for any correspondence satisfying Assumption 8.2.1 (see Theorem 8.4.3) will be demonstrated in a constructive manner. The following example shows that a function satisfying Assumption 8.2.1 indeed has a connected set of zero points.

**Example 8.2.2** *Let a function $f : U^2 \mapsto \mathbb{R}^2$ be defined by*

$$\begin{aligned} f_1(x_1, x_2) &= x_2^m - x_1^n, \\ f_2(x_1, x_2) &= x_1^n - x_2^m, \end{aligned}$$

*where $m$ and $n$ are any given positive integers. It is clear that $f$ is continuous on $U^2$. So Assumption 8.2.1 (a) is met. It is also easy to check that Assumptions 8.2.1 (b) and (c) are satisfied. In fact, the following subset $C$ of $U^2$*

$$C = \{(x, x^{\frac{n}{m}})^\top \mid x \in [0,1]\}$$

*is a connected set of zero points of $f$ linking $(0,0)^\top$ and $(1,1)^\top$.*

Now we translate the economic model with price rigidities introduced in Section 2.5 into the framework described in Assumption 8.2.1. Notice that by Theorem 2.5.6 the excess demand correspondence of the economy with price rigidities satisfies Assumption 8.2.1. More precisely, Assumption 8.2.1 (a) is the same as Property (i) of Theorem 2.5.6, Assumption 8.2.1 (b) is weaker than Properties (ii) and (iii) of Theorem 2.5.6, and Assumption 8.2.1 (c) is weaker than Property (iv) of Theorem 2.5.6. Hence as a by-product, it

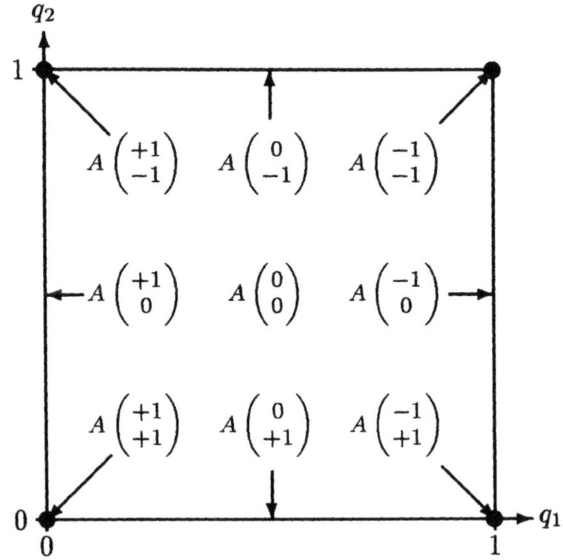


*Figure 8.1.* The subsets $A(s)$ of $U^n$ for $n = 2$.

will be shown that there is a connected set of constrained equilibria linking the two trivial constrained equilibria in the economy with price rigidities.

## 8.3. An algorithm for computing a continuum of zero points

Let a correspondence $\zeta : U^n \mapsto \mathbf{R}^n$ satisfy Assumption 8.2.1. In this section a simplicial algorithm will be developed to approximate and to prove the existence of a connected set of zero points of the correspondence $\zeta$ with the two trivial zero points.

For a sign vector $s \in \{-1, 0, +1\}^n$, define the subset $A(s)$ of $U^n$ by

$$A(s) = \{x \in U^n \mid x_j = 0 \text{ if } s_j = +1;\ x_j = 1 \text{ if } s_j = -1\}.$$

It is easy to see that the dimension of $A(s)$ is equal to $|I^0(s)|$. Notice that the set $A(\mathbf{0})$ equals the set $U^n$. All $3^n$ possible sets $A(s)$ are illustrated for $n = 2$ in Figure 8.1.

Let $\mathcal{T}$ be a triangulation of the unit cube $U^n$. According to Theorem 1.4.6, for any sign vector $s$ the collection of $t$-faces of the simplices in $\mathcal{T}$ lying in the set $A(s)$ induces a triangulation of $A(s)$ with $t = |I^0(s)|$. So, each facet $\tau$ of a $t$-simplex $\sigma \subseteq A(s)$ either lies on the boundary of $A(s)$ and is only a facet of $\sigma$ or is a facet of exactly one other $t$-simplex in $A(s)$. For computational purposes, we can choose the $K_1$-triangulation for $U^n$ having

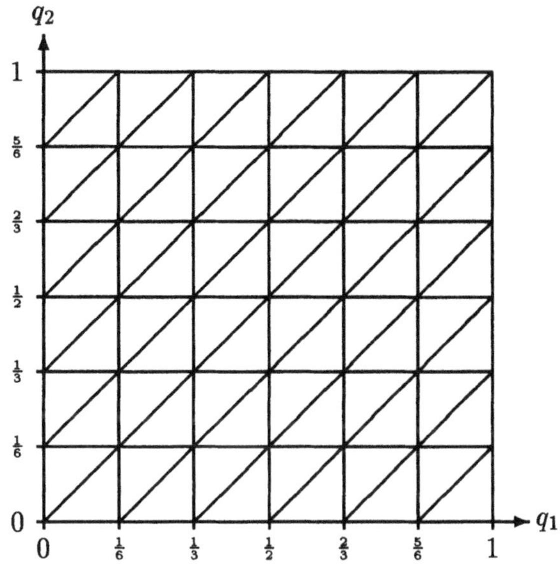


*Figure 8.2.* The $K_1$-triangulation of $U^n$ for $n = 2$.

an arbitrarily chosen mesh size. For $n = 2$, the $K_1$-triangulation of $U^n$ is shown in Figure 8.2 with grid size $\frac{1}{6}$.

Let $Z : U^n \mapsto \mathbb{R}^n$ be any piecewise linear approximation of $\zeta$ with respect to the triangulation $\mathcal{T}$ of $U^n$. In the sequel $Z$ will be chosen such that the following boundary condition is satisfied.

**Assumption 8.3.1** *The piecewise linear approximation $Z : U^n \mapsto \mathbb{R}^n$ of $\zeta$ with respect to $\mathcal{T}$ is such that for each vertex $x$ in the triangulation $\mathcal{T}$ it holds that $Z_j(x) \geq 0$ if $x_j = 0$, and $Z_j(x) \leq 0$ if $x_j = 1$.*

Notice that it is always possible to choose a piecewise linear approximation $Z$ of $\zeta$ with respect to $\mathcal{T}$ satisfying Assumption 8.3.1 if $\zeta$ satisfies Assumption 8.2.1.

To approximate and to prove the existence of a connected set of zero points of $\zeta \in U^n$ containing $\mathbf{0}$ and $e$ we constructively prove that for any piecewise linear approximation $Z$ of $\zeta$ satisfying Assumption 8.3.1, there exists a piecewise linear path $f : [0, 1] \mapsto U^n$ of points in $U^n$ connecting $\mathbf{0}$ and $e$, and having some specific properties. Let $\widetilde{C}$ denote the image of $[0, 1]$ by $f$, so $\widetilde{C} = f([0, 1])$. The piecewise linear path will be constructed such that for every point $q \in \widetilde{C}$ there is a number $\beta \in \mathbb{R}$ satisfying

$$\begin{aligned} 0 \leq Z_j(q) \leq \beta, &\quad \text{if } q_j = 0, \\ Z_j(q) = \beta, &\quad \text{if } 0 < q_j < 1, \\ 0 \geq Z_j(q) \geq \beta, &\quad \text{if } q_j = 1. \end{aligned} \tag{8.3.1}$$

As will be shown in the next section, if mesh($\mathcal{T}$) is small enough, then $\beta$ will be arbitrarily close to zero for every $q \in \widetilde{C}$, i.e., every point $q$ on the path is an approximate zero point of $\zeta$ when the mesh size of the underlying triangulation $\mathcal{T}$ is taken small enough.

Let $\sigma(x^1, \ldots, x^{t+1})$ be a $t$-simplex of $\mathcal{T}^+$ in $U^n$ and let $s$ be a sign vector with $|I^0(s)| = t$. Let $i_1 < \cdots < i_{n-t}$ be such that $I^-(s) \cup I^+(s) = \{i_1, \ldots, i_{n-t}\}$. Consider solutions $(\lambda_1, \ldots, \lambda_{t+1}, \mu_1, \ldots, \mu_{n-t}, \beta)^\top \in \mathbb{R}^{n+2}$ of the following system of equations:

$$\sum_{j=1}^{t+1} \lambda_j \begin{pmatrix} 1 \\ Z(x^j) \end{pmatrix} + \sum_{j=1}^{n-t} \mu_j \begin{pmatrix} 0 \\ s_{i_j} e(i_j) \end{pmatrix} + \beta \begin{pmatrix} 0 \\ -1^n \end{pmatrix} = \begin{pmatrix} 1 \\ 0^n \end{pmatrix}. \tag{8.3.2}$$

If $\lambda_j \geq 0$ for all $j \in I_{t+1}$, and $\mu_j \geq 0$ for all $j \in I_{n-t}$, then $(\lambda_1, \ldots, \lambda_{t+1}, \mu_1, \ldots, \mu_{n-t}, \beta)$ is called an admissible solution to (8.3.2). Notice that $\beta$ is unrestricted and that $\sum_{j=1}^{t+1} \lambda_j = 1$. For an admissible solution to (8.3.2) corresponding to a simplex $\sigma(x^1, \ldots, x^{t+1})$ in $A(s)$ it holds that the point $q$ satisfies (8.3.1), where $q = \sum_{j=1}^{t+1} \lambda_j x^j \in \sigma$. Notice that the linear system (8.3.2) has $n+1$ equations and $n+2$ variables and hence there is one degree of freedom. The piecewise linear path $\widetilde{C}$ connecting $\mathbf{0}$ and $e$ will be obtained by an algorithm that generates a sequence of adjacent simplices of varying dimension such that with respect to each simplex of this sequence there exists a set of admissible solutions. Each linear piece of the path $\widetilde{C}$ within some simplex of $\mathcal{T}$ will be traced by making a pivot step in system (8.3.2) with one of the variables. Since $\beta$ is an unconstrained variable, its corresponding column is not subject to leaving the system.

An admissible solution to (8.3.2) is said to be degenerate if at least two of the variables $\lambda_j$, $j \in I_{t+1}$, and $\mu_j$, $j \in I_{n-t}$, equal zero. For any piecewise linear approximation $Z$ satisfying Assumption 8.3.1 of a correspondence $\zeta$ satisfying Assumption 8.2.1 with respect to any triangulation $\mathcal{T}$ of $U^n$ it holds that $Z(\mathbf{0}) = \mathbf{0}$ and $Z(e) = \mathbf{0}$. Hence $\lambda_1 = 1$, $\mu_j = 0$, for all $j \in I_n$, and $\beta = 0$ is a degenerate admissible solution to (8.3.2) corresponding to the 0-simplex $\sigma(\mathbf{0})$ and any sign vector $s$ with $|I^0(s)| = 0$. A similar degenerate admissible solution exists corresponding to the 0-simplex $\sigma(e)$. So the usual nondegeneracy assumption stating that there exist no degenerate admissible solutions makes no sense for the problem under consideration. To cope with the degeneracy problem, we will adopt lexicographic pivot rules from linear programming theory. Recall this in Chapter 4. A related approach is used in Wright [1981], where an algorithm is presented to compute a zero point of a correspondence $\phi : \mathbb{R}^n \mapsto \mathbb{R}^n$.

For a $(t-1)$-simplex $\tau(x^1, \ldots, x^t)$ in $U^n$ and a sign vector $s$ with $t = |I^0(s)|$, the $(n+1) \times (n+1)$ matrix $A_{s,\tau}$ is defined by

$$A_{s,\tau} = \begin{bmatrix} 1 & \cdots & 1 & 0 & \cdots & 0 & 0 \\ Z(x^1) & \cdots & Z(x^t) & s_{i_1} e(i_1) & \cdots & s_{i_{n-t}} e(i_{n-t}) & -1^n \end{bmatrix}.$$

**Definition 8.3.2** *Let $\tau$ be a $(t-1)$-simplex of $U^n$ and let $s$ be a sign vector with $|I^0(s)| = t$. Then $\tau$ is s-complete if $A_{s,\tau}^{-1}$ exists and is semilexicopositive.*

When $\tau$ is $s$-complete, then the first column of $A_{s,\tau}^{-1}$ corresponds to an admissible solution to (8.3.2) for a $t$-simplex $\sigma$ being the convex hull of $\tau$ and some vertex $x^{t+1} \in U^n$, and where $\lambda_{t+1} = 0$.

We will now describe an algorithm that generates a finite sequence of simplices starting with the simplex $< \mathbf{0} >$ and terminating with the simplex $< e >$ satisfying that for every $(t-1)$-simplex $\tau$ in the sequence there exists a sign vector $s$ with $t = |I^0(s)|$ such that $\tau$ is an $s$-complete simplex in $A(s)$. Moreover, any two successive simplices in the sequence either are facets of the same simplex or one is a facet of the other.

**Lemma 8.3.3** *Let the sign vector $s \in \{-1, 0, +1\}^n$ be such that $s_j = +1$ for each $j \in I_{n-1}$ and $s_n = 0$. Then the 0-simplex $\tau(\mathbf{0})$ is an s-complete simplex in $A(s)$ and is not an $\tilde{s}$-complete simplex in $A(\tilde{s})$ for any other sign vector $\tilde{s} \in \mathbb{R}^n$.*

Proof: Suppose that $\tau(\mathbf{0})$ is an $\tilde{s}$-complete simplex in $A(\tilde{s})$. Assumption 8.2.1 (c) and Assumption 8.3.1 imply that $Z(\mathbf{0}) = \mathbf{0}$. Since $\tau$ is 0-dimensional and $\tau$ lies in $A(\tilde{s})$, it has to hold that $|I^0(\tilde{s})| = 1$ and $I^-(\tilde{s}) = \emptyset$. Then $A_{\tilde{s},\tau}$ is given by

$$A_{\tilde{s},\tau} = \begin{bmatrix} 1 & 0 & \cdots & 0 & 0 \\ 0^n & e(i_1) & \cdots & e(i_{n-1}) & -1^n \end{bmatrix}$$

where $i_1 < i_2 < \cdots < i_{n-1}$. Define $i_0 = 0$ and $i_n = n+1$. Let $j$ be the unique element in the set $I^0(\tilde{s})$ and let $k \in I_n$ be such that $i_{k-1} < j < i_k$. Then

$$A_{\tilde{s},\tau}^{-1} = \begin{bmatrix} 1 & 0 & \cdots & 0 & 0 & 0 & \cdots & 0 \\ 0^n & e(i_1) & \cdots & e(i_{k-1}) & -1^n & e(i_k - 1) & \cdots & e(i_{n-1} - 1) \end{bmatrix}.$$

It is clear that this matrix is semilexicopositive if and only if $k = n$. □

**Lemma 8.3.4** *Let the sign vector $s \in \{=1, 0, +1\}^n$ be such that $s_1 = 0$ and $s_j = -1$ for each $j \in I_n \setminus \{1\}$. Then the 0-simplex $\tau(e)$ is an $s$-complete simplex in $A(s)$ and is not an $\tilde{s}$-complete simplex in $A(\tilde{s})$ for any other sign vector $\tilde{s}$.*

Proof: Suppose that $\tau(e)$ is an $\tilde{s}$-complete simplex in $A(\tilde{s})$. Assumption 8.2.1 (c) and Assumption 8.3.1 imply that $Z(e) = \mathbf{0}$. Since $\tau$ is 0-dimensional and $\tau$ lies in $A(\tilde{s})$, it holds that $|I^0(\tilde{s})| = 1$ and $I^+(\tilde{s}) = \emptyset$. Hence $A_{\tilde{s},\tau}$ is given by

$$A_{\tilde{s},\tau} = \begin{bmatrix} 1 & 0 & \cdots & 0 & 0 \\ 0^n & -e(i_1) & & -e(i_{n-1}) & -1^n \end{bmatrix}$$

where $i_1 < i_2 < \cdots < i_{n-1}$. Define $i_0 = 0$ and $i_n = n+1$. Let $j$ be the unique element in the set $I^0(\tilde{s})$ and let $k \in I_n$ be such that $i_{k-1} < j < i_k$. Then it is easily verified that

$$A_{\tilde{s},\tau}^{-1} = \begin{bmatrix} 1 & 0 & \cdots & 0 & 0 & 0 & \cdots & 0 \\ & & & & 1^{n-1} & & & \\ 0^n & -e(i_1) & & -e(i_{k-1}) & -1 & -e(i_k - 1) & & -e(i_{n-1} - 1) \end{bmatrix}.$$

Clearly, this matrix is semilexicopositive if and only if $k = 1$. □

The following lemma is well-known in linear programming theory and can be easily verified by showing that $\tilde{A}\tilde{A}^{-1} = \mathbf{I}$ with $\tilde{A}$ as in the lemma. We will use the following notation in this and next chapters. Let $A$ be a $k^1 \times k^2$ matrix. Then, for $j \in I_{k^1}$, $A_{j\cdot}$ denotes row $j$ of the matrix $A$, and for $j \in I_{k^2}$, $A_{\cdot j}$ denotes column $j$ of the matrix $A$.

**Lemma 8.3.5** *Let $A = (A_{\cdot 1}, \ldots, A_{\cdot n+1})$ be a non-singular $(n+1) \times (n+1)$ matrix and let $x \in \mathbb{R}^{n+1}$. Let $k \in I_{n+1}$ and let $\tilde{A} = (A_{\cdot 1}, \ldots, A_{\cdot k-1}, x, A_{\cdot k+1}, \ldots, A_{\cdot n+1})$. Then either $(A^{-1}x)_k = 0$ and $\tilde{A}$ is singular, or $(A^{-1}x)_k \neq 0$ and the following matrix*

$$\tilde{A}^{-1} = \begin{bmatrix} (A^{-1})_{1\cdot} - \frac{(A^{-1}x)_1}{(A^{-1}x)_k}(A^{-1})_{k\cdot} \\ \vdots \\ (A^{-1})_{k-1\cdot} - \frac{(A^{-1}x)_{k-1}}{(A^{-1}x)_k}(A^{-1})_{k\cdot} \\ \frac{1}{(A^{-1}x)_k}(A^{-1})_{k\cdot} \\ (A^{-1})_{k+1\cdot} - \frac{(A^{-1}x)_{k+1}}{(A^{-1}x)_k}(A^{-1})_{k\cdot} \\ \vdots \\ (A^{-1})_{n+1\cdot} - \frac{(A^{-1}x)_{n+1}}{(A^{-1}x)_k}(A^{-1})_{k\cdot} \end{bmatrix}$$

*exists.*

The next lemma describes the cases that may occur when a $t$-simplex $\sigma$ in $A(s)$ has at least one $s$-complete facet $\tau$ and a lexicographic pivot step to $A_{s,\tau}$ is performed in the case the column in (8.3.2) to be added corresponds to the vertex of $\sigma$ opposite $\tau$.

**Lemma 8.3.6** *Let $\sigma$ be a $t$-simplex of $A(s)$ where $s$ is a sign vector with $t$ zero components. If $\sigma$ has an $s$-complete facet $\tau$, then exactly one of the following cases holds:*
*(i) The simplex $\sigma$ is an $\tilde{s}$-complete simplex in $A(\tilde{s})$ for precisely one sign vector $\tilde{s}$ with $|I^0(\tilde{s})| = t+1$ and no facets of $\sigma$ other than $\tau$ are $s$-complete;*
*(ii) The simplex $\sigma$ has exactly one other $s$-complete facet $\tilde{\tau}$ and $\sigma$ is not an $\tilde{s}$-complete simplex in $A(\tilde{s})$ for any sign vector $\tilde{s}$.*

Proof: Let $x^{t+1}$ be the vertex of $\sigma$ not contained in $\tau$, and let $y = A_{s,\tau}^{-1}(1, Z(x^{t+1})^\top)^\top$. Since $A_{s,\tau}A_{s,\tau}^{-1} = \mathbf{I}$, it holds that $\sum_{i=1}^{t}(A_{s,\tau}^{-1})_{i,1} = 1$ and $\sum_{i=1}^{t}(A_{s,\tau}^{-1})_{i,j} = 0$ for any $j \in I_{n+1} \setminus \{1\}$. Suppose that the first $n$ components of $y$ are non-positive. Then

$$0 \geq (e^{t^\top}, 0^{n-t+1^\top})y = (e^{t^\top}, 0^{n-t+1^\top})A_{s,\tau}^{-1}(1, Z(x^{t+1})^\top)^\top = (1, 0^{n^\top})(1, Z(x^{t+1})^\top)^\top = 1.$$

This is a contradiction. So it is possible to choose $k \in I_n$ such that $\frac{1}{y_k}(A_{s,\tau}^{-1})_{k\cdot}$ is minimal according to the lexicographic ordering over all row vectors $\frac{1}{y_j}(A_{s,\tau}^{-1})_{j\cdot}$ for which $y_j > 0$ and $j \in I_n$. It is clear that $k$ is uniquely determined, since otherwise $A_{s,\tau}^{-1}$ would be singular. Now it holds that either $k \in I_n \setminus I_t$ or $k \in I_t$.

If $k \in I_n \setminus I_t$, then let $\tilde{s}_{i_{k-t}} = 0$ and $\tilde{s}_j = s_j$ for $j \in I_n \setminus \{i_{k-t}\}$. Clearly $\sigma \subseteq A(\tilde{s})$ and $|I^0(\tilde{s})| - 1$ equals the dimension of $\sigma$. The matrix $A_{\tilde{s},\sigma}$ is obtained by deleting column $k$ of $A_{s,\tau}$ and adding the vector $(1, Z(x^{t+1})^\top)^\top$ between columns $t$ and $t+1$. Using Lemma 8.3.5, $A_{\tilde{s},\sigma}^{-1}$ exists and is semilexicopositive. So $\sigma$ is an $\tilde{s}$-complete $t$-simplex in $A(\tilde{s})$.

If $k \in I_t$, then let $\tilde{\tau}$ be the facet of $\sigma$ opposite the vertex $x^k$. Using Lemma 8.3.5 and the choice of $k$ it follows that $A_{s,\tilde{\tau}}^{-1}$ exists and is semilexicopositive. Hence $\tilde{\tau}$ is an $s$-complete simplex in $A(s)$.

It follows directly from Lemma 8.3.5 that if some other column is replaced, then the inverse of the new matrix is not semilexicopositive. □

The next lemma gives the cases that may occur when an $s$-complete facet $\tau$ of a $t$-simplex in $A(s)$ lies in a $(t-1)$-dimensional set $A(\tilde{s})$ and a lexicographic pivot step to $A_{s,\tau}$ is performed in the case the column in (8.3.2) to be added is equal to $(0, \tilde{s}_{\tilde{k}}e(\tilde{k})^\top)^\top$ for the unique element $\tilde{k}$ in the set $(I^-(\tilde{s}) \cup I^+(\tilde{s})) \setminus (I^-(s) \cup I^+(s))$.

**Lemma 8.3.7** *Let $\tau$ be an s-complete $(t-1)$-simplex of $A(s)$ where $s$ is a sign vector with $t$ zero components. Suppose that $\tau$ lies in $A(\tilde{s})$ where $\tilde{s}$ has $t-1$ zero components. Then exactly one of the following cases holds:*
*(i) The simplex $\tau$ is equal to $<\mathbf{0}>$ or equal to $<e>$;*
*(ii) The simplex $\tau$ is an $\bar{s}$-complete simplex in $A(\bar{s})$ for precisely one sign vector $\bar{s} \neq s$ with $|I^0(\bar{s})| = t$ and $\tau$ has no $\tilde{s}$-complete facets;*
*(iii) Precisely one facet of $\tau$ is $\tilde{s}$-complete and $\tau$ is not an $\bar{s}$-complete simplex in $A(\bar{s})$ for any sign vector $\bar{s} \neq s$.*

Proof: For some unique index $\tilde{k} \in I_n$, it must hold that $s_{\tilde{k}} = 0$ and $\tilde{s}_{\tilde{k}} \neq 0$. Let $y = A_{s,\tau}^{-1}(0, \tilde{s}_{\tilde{k}} e(\tilde{k})^\top)^\top$. Moreover, $\tilde{s}_h = s_h$ for $h \neq \tilde{k}$. Exactly one of the following three possibilities occurs: (i) $\tilde{s}_{\tilde{k}} = +1$ and $y_j \leq 0$ for all $j \in I_n$;(ii) $\tilde{s}_{\tilde{k}} = -1$ and $y_j \leq 0$, for all $j \in I_n$; (iii) there exists an index $j \in I_n$ such that $y_j > 0$. If $\tilde{s}_{\tilde{k}} = +1$ and $y_j \leq 0$ for all $j \in I_n$, then, since $A_{s,\tau} y = (0, \tilde{s}_{\tilde{k}} e(\tilde{k})^\top)^\top$, we have $y_1 = \cdots = y_t = 0$. So

$$1 = \sum_{j=1}^{n+1} (A_{s,\tau})_{\tilde{k}+1,j} y_j = \sum_{j=t+1}^{n+1} (A_{s,\tau})_{\tilde{k}+1,j} y_j = -y_{n+1},$$

where for the last equality it is used that $s_{\tilde{k}} = 0$. Hence $y_{n+1} = -1$, and consequently

$$\sum_{j=t+1}^{n} (A_{s,\tau})_{\cdot j} y_j = \sum_{j=1}^{n-t} \begin{pmatrix} 0 \\ s_{i_j} e(i_j) \end{pmatrix} y_{t+j} = \begin{pmatrix} 0 \\ e(\tilde{k}) \end{pmatrix} - \begin{pmatrix} 0 \\ e \end{pmatrix}.$$

Therefore $t = 1$ and $s_{i_j} = +1$ for all $j \in I_{n-t}$. So $\tilde{s} = +e$ and therefore $\tau = \{\mathbf{0}\}$.

If $\tilde{s}_{\tilde{k}} = -1$ and $y_j \leq 0$ for all $j \in I_n$, then again $y_1 = \cdots = y_t = 0$. So

$$-1 = \sum_{j=t+1}^{n+1} (A_{s,\tau})_{\tilde{k}+1,j} y_j = -y_{n+1},$$

and hence $y_{n+1} = 1$. Consequently

$$\sum_{j=t+1}^{n} (A_{s,\tau})_{\cdot j} y_j = \sum_{j=1}^{n-t} \begin{pmatrix} 0 \\ s_{i_j} e(i_j) \end{pmatrix} y_{t+j} = \begin{pmatrix} 0 \\ e \end{pmatrix} - \begin{pmatrix} 0 \\ e(\tilde{k}) \end{pmatrix}.$$

Therefore $t = 1$ and $s_{i_j} = -1$ for all $j \in I_{n-t}$. So $\tilde{s} = -e$ and therefore $\tau = \{e\}$.

If there exists some index $j \in I_n$ such that $y_j > 0$, then it is possible to choose $k \in I_n$ as in the proof of Lemma 8.3.6. Again, either $k \in I_n \setminus I_t$ or $k \in I_t$.

If $k \in I_n \setminus I_t$, then let $\bar{s}_{i_{k-t}} = 0$ and $\bar{s}_j = \tilde{s}_j$ for all $j \in I_n \setminus \{i_{k-t}\}$, and consider $A_{\bar{s},\tau}$. Using Lemma 8.3.5, the choice of $k$ guarantees that $A_{\bar{s},\tau}^{-1}$ is semilexicopositive and therefore $\tau$ is $\bar{s}$-complete in $A(\bar{s})$.

If $k \in I_t$, let $\tau'$ be the facet of $\tau$ opposite to the vertex $x^k$. By Lemma 8.3.5 and the choice of $k$, $A_{\tilde{s},\tau'}^{-1}$ is semilexicopositive and hence $\tau'$ is $\tilde{s}$-complete in $A(\tilde{s})$.

It follows directly from Lemma 8.3.5 that if some other column of $A_{s,\tau}$ is replaced, then the inverse of the new matrix is not semilexicopositive. □

In order to prove the convergence of the algorithm, we construct a graph $G = (V, A)$ as follows, where $V$ is the set of nodes and $A$ is the set of arcs. We say that a simplex $\sigma \in \mathcal{T}^+$ is a *node* if $\sigma$ is a $t$-simplex in $A(s)$ and has an $s$-complete facet $\tau$ for some sign vector $s \in \mathbb{R}^n$ with $t = |I^0(s)|$. Two nodes $\sigma_1$ and $\sigma_2$ are said to be *adjacent* (and are linked by an arc) if they lie in the same set $A(s)$ and if they share a common $s$-complete facet or one is an $s$-complete facet of the other. The degree of a node $\sigma$ is defined to be the number of nodes adjacent to $\sigma$, denoted by $\deg(\sigma)$. Let $\tau^0$ be the zero-dimensional simplex $< \mathbf{0} >$. As it is shown in Lemma 8.3.3, $\tau^0$ is $s^0$-complete for some unique sign vector $s^0 \in \mathbb{R}^n$. Let $\sigma^0$ be the unique simplex in $A(s^0)$ having $\tau^0$ as its facet. It is clear that $\sigma^0$ is a node and the degree of $\sigma^0$ is one. Similarly, as it is shown in Lemma 8.3.4, the simplex $< e >$ is $\bar{s}$-complete for some unique sign vector $\bar{s} \in \mathbb{R}^n$. Let $\tilde{\sigma}$ be the unique simplex in $A(\bar{s})$ having $< e >$ as its facet. It is clear that $\tilde{\sigma}$ is a node and the degree of $\tilde{\sigma}$ is one.

Using the lemmas above we easily obtain the following result.

**Lemma 8.3.8** *Let $G = (V, A)$ be the graph defined above and let $\sigma$ be a node of $G$.*
*(i) If $\sigma = \sigma^0$, then $deg(\sigma) = 1$;*
*(ii) If $\sigma = \tilde{\sigma}$, then $deg(\sigma) = 1$;*
*(iii) In all other cases, $deg(\sigma) = 2$.*

By applying Theorem 1.4.14, we are led to the following theorem.

**Theorem 8.3.9** *Let $\mathcal{T}$ be a triangulation of $U^n$, let $\zeta : U^n \mapsto \mathbb{R}^n$ satisfy Assumption 8.2.1, and let $Z : U^n \mapsto \mathbb{R}^n$ be a piecewise linear approximation of $\zeta$ with respect to $\mathcal{T}$ satisfying Assumption 8.3.1. Then there exists a unique finite sequence of simplices $\sigma^0, \ldots, \sigma^M$ such that $\sigma^M = \tilde{\sigma}$, whereas $\sigma^{k-1}$ and $\sigma^k$ are adjacent for each $k \in I_M$.*

Now we can give a detailed description of the steps of the algorithm.

### The HTY Algorithm

**Step (0)** Let $i = 0$, $t = 1$, $x^1 = \mathbf{0}$, $\tau^i = \tau(x^1)$, $s = (e^\top, 0)^\top$, and let $x^{t+1}$ be the unique vertex of the simplex in $A(s)$ containing $\tau^i$ as a facet opposite to it.

**Step (1)** Let $\sigma$ be equal to the convex hull of $x^{t+1}$ and $\tau^i$. Pivot $(0, Z(x^{t+1})^\top)^\top$ lexicographically into the linear system (8.3.2) corresponding to $A_{s,\tau^i}$, yielding as described in Lemma 8.3.6, a unique column $k \in I_n$ of $A_{s,\tau^i}$ which has to be replaced. If $k \in I_n \backslash I_t$, then go to Step 3 with $j' = i_{k-t}$.

**Step (2)** Set $i = i+1$ and let $\tau^i$ be the facet of $\sigma$ opposite the vertex $x^k$. If $\tau^i =< e >$, then the algorithm terminates. If $\tau^i$ lies in $A(\tilde{s})$ for some $\tilde{s}$ with $t-1$ zero components, then go to Step 4. Otherwise there is exactly one simplex $\tilde{\sigma}$ in $A(s)$ such that $\tilde{\sigma} \neq \sigma$ and $\tau^i$ is a facet of $\tilde{\sigma}$. Go to Step 1 with $x^{t+1}$ as the unique vertex in $\tilde{\sigma}$ opposite to the facet $\tau^i$.

**Step (3)** Define $\tilde{s}$ by $\tilde{s}_{j'} = 0$ and $\tilde{s}_j = s_j$ for each $j \in I_n \setminus \{j'\}$. There is a unique simplex $\tilde{\sigma}$ in $A(\tilde{s})$ having $\sigma$ as a facet. Set $i = i+1$, $t = t+1$, and go to Step 1 with $x^{t+1}$ as the unique vertex in $\tilde{\sigma}$ opposite to $\sigma$, $s = \tilde{s}$, and $\tau^i = \sigma$.

**Step (4)** Let $\sigma$ be equal to $\tau^i$. Pivot $(0, \tilde{s}_{\tilde{k}} e(\tilde{k})^\top)^\top$ lexicographically into the linear system (8.3.2) determined by $A_{s,\tau^i}$, where $\tilde{k} \in I_n$ is such that $s_{\tilde{k}} = 0$ and $\tilde{s}_{\tilde{k}} \neq 0$. By Lemma 8.3.7 there is a unique column $k \in I_n$ of $A_{s,\tau^i}$ which has to be replaced. If $k \in I_n \setminus I_t$, set $j' = i_{k-t}$, $s = \tilde{s}$, $t = t-1$, and $i = i-1$, and then go to Step (3). Otherwise set $s = \tilde{s}$ and $t = t-1$, and go to Step (2).

It is worthwhile to mention that we can also start the algorithm with the simplex $< e >$ and terminate with the simplex $< \mathbf{0} >$ . In the next theorem it is shown that the sequence of adjacent simplices generated by the algorithm yields a piecewise linear path of points in $U^n$ connecting $\mathbf{0}$ and $e$ such that every point $q$ on the path satisfies (8.3.1), i.e., for some $\beta \in \mathbb{R}$ it holds that $0 \leq Z_j(q) \leq \beta$ if $q_j = 0$, $Z_j(q) = \beta$ if $0 < q_j < 1$, and $0 \geq Z_j(q) \geq \beta$ if $q_j = 1$.

**Theorem 8.3.10** *Let $\mathcal{T}$ be a triangulation of $U^n$ and let the correspondence $\zeta : U^n \mapsto \mathbb{R}^n$ satisfy Assumption 8.2.1. Let $Z : U^n \mapsto \mathbb{R}^n$ be a piecewise linear approximation of $\zeta$ with respect to $\mathcal{T}$ satisfying Assumption 8.3.1. Then there exists a continuous, piecewise linear function $f : [0,1] \mapsto U^n$ satisfying $f(0) = \mathbf{0}$, $f(1) = e$, and for every $r \in [0,1]$ it* *holds that for some* $\beta \in \mathbb{R}$

$$\begin{array}{ll} 0 \leq Z_j(f(r)) \leq \beta, & \textit{if} \quad f_j(r) = 0, \\ Z_j(f(r)) = \beta, & \textit{if} \quad 0 < f_j(r) < 1, \\ 0 \geq Z_j(f(r)) \geq \beta & \textit{if} \quad f_j(r) = 1. \end{array}$$

Proof: Consider all different pairs of sign vectors and simplices $(s^0, \tau^0), \ldots, (s^{M'}, \tau^{M'})$ successively generated by the algorithm, satisfying that $\tau^k$ is an $s^k$-complete facet of a $t^k$-simplex in $A(s^k)$ for all $k \in \{0\} \cup I_{M'}$ with $t^k = |I^0(s^k)|$. Clearly, $A^{-1}_{s^k,\tau^k}$ is semilexicopositive for every $k \in \{0\} \cup I_{M'}$. Consequently $\sum_{j=1}^{n+1} (A_{s^k,\tau^k})_{\cdot j} (A^{-1}_{s^k,\tau^k})_{j,1} = (1, \mathbf{0}^\top)^\top$, or equivalently

$$\sum_{j=1}^{t^k} (A^{-1}_{s^k,\tau^k})_{j,1} \begin{pmatrix} 1 \\ Z(x^j) \end{pmatrix} + \sum_{j=1}^{n-t^k} (A^{-1}_{s^k,\tau^k})_{t^k+j,1} \begin{pmatrix} 0 \\ s^k_{i_j} e(i_j) \end{pmatrix} + (A^{-1}_{s^k,\tau^k})_{n+1,1} \begin{pmatrix} 0 \\ -e \end{pmatrix} = \begin{pmatrix} 1 \\ 0 \end{pmatrix}$$

where $\tau^k = \tau^k(x^1, \ldots, x^{t^k})$ and $(A^{-1}_{s^k,\tau^k})_{j,1} \geq 0$ for $j \in I_n$. Define $y^k = \sum_{j=1}^{t^k} (A^{-1}_{s^k,\tau^k})_{j,1} x^j$ for each $k \in \{0\} \cup I_{M'}$. Note that $y^0 = \mathbf{0}$ and $y^{M'} = e$. Define $f : [0,1] \mapsto U^n$ by

$$f(r) = (1 - M'r + \lfloor M'r \rfloor) y^{\lfloor M'r \rfloor} + (M'r - \lfloor M'r \rfloor) y^{\lfloor M'r \rfloor + 1}, \quad r \in [0,1]$$

where $y^{M'+1}$ is an arbitrary vector in $\mathbb{R}^n$. This function $f$ satisfies the requirements of the theorem. To verify this, remark that for every $k \in I_{M'}$, $A_{s^{k-1},\tau^{k-1}}$ and $A_{s^k,\tau^k}$ have $n$ columns in common. Let $B_{\bar{s}^{k-1},\sigma^{k-1}}$ be the $(n+1) \times (n+2)$ matrix which contains all the columns of $A_{s^{k-1},\tau^{k-1}}$ and $A_{s^k,\tau^k}$ for $k \in I_{M'}$. The matrix $B_{\bar{s}^{k-1},\sigma^{k-1}}$ yields the system as in (8.3.2) for a simplex $\sigma^{k-1}$ which is the convex hull of $\tau^{k-1}$ and $\tau^k$ and a sign vector $\bar{s}^{k-1}$ which is such that $I^-(\bar{s}^{k-1}) = I^-(s^{k-1}) \cup I^-(s^k)$ and $I^+(\bar{s}^{k-1}) = I^+(s^{k-1}) \cup I^+(s^k)$. It is easily verified that $\sigma^{k-1}$ is a simplex in $A(\bar{s}^{k-1})$. The first columns of both $A^{-1}_{s^{k-1},\tau^{k-1}}$ and $A^{-1}_{s^k,\tau^k}$, extended with a zero component, yield admissible solutions to the system in (8.3.2) induced by $B_{\bar{s}^{k-1},\sigma^{k-1}}$. Finally, when $\tilde{y}$, $\bar{y} \in \mathbb{R}^{n+2}$ both correspond to admissible solutions to the system $B_{\bar{s}^{k-1},\sigma^{k-1}} y = (1, \mathbf{0}^\top)^\top$, then $\lambda \tilde{y} + (1-\lambda)\bar{y}$ also corresponds to an admissible solution for every $\lambda \in [0,1]$. Since for every $k \in I_{M'}$ the simplex $\sigma^{k-1}$ lies in $A(\bar{s}^{k-1})$, it follows from (8.3.2) and Assumption 8.3.1 that for every $r \in [0,1]$ the point $q = f(r)$ satisfies (8.3.1). □

## 8.4. Existence of a continuum of zero points and accuracy analysis

In this section we will show that every correspondence $\zeta : U^n \mapsto \mathbb{R}^n$ satisfying Assumption 8.2.1 has a connected set of zero points containing the two trivial zero points.

**Theorem 8.4.1** *Let $\zeta : U^n \mapsto \mathbb{R}^n$ be a correspondence satisfying Assumption 8.2.1. For $r \in \mathbb{N}$, let $\mathcal{T}^r$ be a triangulation of $U^n$ with mesh size smaller than $\frac{1}{r}$, and let $Z^r : U^n \mapsto \mathbb{R}^n$ be a piecewise linear approximation of $\zeta$ with respect to $\mathcal{T}^r$ satisfying Assumption 8.3.1. Let $(q^r)_{r\in\mathbb{N}}$ be an arbitrary convergent sequence of points in $U^n$ with limit $q^*$ such that for any $r \in \mathbb{N}$ it holds that $q^r \in f^r([0,1])$, where $f^r$ corresponds to the function $f$ of Theorem 8.3.10 for the piecewise linear approximation $Z^r$. Then $\mathbf{0} \in \zeta(q^*)$.*

Proof: Let $\{(\lambda_1^r, \ldots, \lambda_{n+1}^r, x^{1^r}, \ldots, x^{n+1^r}, z^{1^r}, \ldots, z^{n+1^r}) \mid r \in \mathbb{N}\}$ be a sequence of points lying in the set

$$\prod_{j=1}^{n+1} [0,1] \times \prod_{j=1}^{n+1} U^n \times \prod_{j=1}^{n+1} \mathbb{R}^n$$

such that $\sum_{j=1}^{n+1} \lambda_j^r = 1$, $\sigma(x^{1^r}, \ldots, x^{n+1^r})$ is a simplex of $\mathcal{T}^r$, $z^{j^r} = Z^r(x^{j^r})$, and $q^r = \sum_{j=1}^{n+1} \lambda_j^r x^{j^r} \in f^r([0,1])$. Notice that it may happen that $\lambda_j^r = 0$ for some $j \in I_{n+1}$. By definition, $Z^r(q^r) = \sum_{j=1}^{n+1} \lambda_j^r z^{j^r}$. Define $z^r = Z^r(q^r)$. Then

$$z^r = \beta^r e - \sum_{j\in\{k\in I_n \mid q_k^r = 0\}} \mu_j^r e(j) + \sum_{j\in\{k\in I_n \mid q_k^r = 1\}} \mu_j^r e(j),$$

for some $\beta^r \in \mathbb{R}$ and some $\mu_j^r \geq 0$ for each $j \in \{k \in I_n \mid q_k^r = 0 \text{ or } q_k^r = 1\}$. Since $\cup_{q\in U^n}\zeta(q)$ is compact, the sequence given above remains in a compact set, and without loss of generality it can be assumed to converge to an element

$$(\lambda_1^*, \ldots, \lambda_{n+1}^*, x^{*1}, \ldots, x^{*n+1}, z^{*1}, \ldots, z^{*n+1}).$$

Define $z^* = \sum_{j=1}^{n+1} \lambda_j^* z^{*j}$. Clearly it holds that $z^r \mapsto z^*$. Since for every $r \in \mathbb{N}$ the mesh size of $\mathcal{T}^r$ is smaller than $\frac{1}{r}$, it holds for every $j \in I_{n+1}$ that $x^{*j} = q^*$. Using that $\zeta$ is upper semi-continuous and $\cup_{q\in U^n}\zeta(q)$ is bounded this implies that for every $j \in I_{n+1}$, $z^{*j} \in \zeta(q^*)$. Since $\zeta$ is convex-valued, $\sum_{j=1}^{n+1} \lambda_j^* = 1$, and $\lambda_j^* \geq 0$ for $j \in I_{n+1}$, it holds that $z^* \in \zeta(q^*)$.

If there is a subsequence $(q^{r^s})_{s\in\mathbb{N}}$ of $(q^r)_{r\in\mathbb{N}}$ such that for every $s \in \mathbb{N}$, $0 < q_j^{r^s} < 1$ for all $j \in I_n$, then $z^{r^s} = \beta^{r^s} e$. Since $z^* \in \zeta(q^*)$, there is some $p \in \mathbb{R}^n_{++}$ such that $p^\top z^* = 0$. Consequently $z^* = \mathbf{0}$. If there is not such a subsequence, then there exists a subsequence $(q^{r^s})_{s\in\mathbb{N}}$ of $(q^r)_{r\in\mathbb{N}}$ such that for some $k \in I_n$, $q_k^{r^s} = 0$ for all $s \in \mathbb{N}$, or for some $k \in I_n$, $q_k^{r^s} = 1$ for all $s \in \mathbb{N}$. In the first case, using that $Z^r$ satisfies Assumption 8.3.1, it holds that $0 \leq z_k^{r^s} \leq \beta^{r^s}$ and therefore $\mathbf{0} \leq z^{r^s}$ for all $s \in \mathbb{N}$. This implies that $z^* \geq \mathbf{0}$. In the second case, using again that $Z^r$ satisfies Assumption 8.3.1, it holds that $0 \geq z_k^{r^s} \geq \beta^{r^s}$ and therefore $\mathbf{0} \geq z^*$. In both cases the existence of a vector $p \in \mathbb{R}^n_{++}$ such that $p^\top z^* = 0$ implies that $z^* = \mathbf{0}$. Hence, $\mathbf{0} \in \zeta(q^*)$. □

From Theorem 8.4.1 the next result immediately follows.

**Corollary 8.4.2** *Let $\zeta : U^n \mapsto \mathbb{R}^n$ be a correspondence satisfying Assumption 8.2.1. For $r \in \mathbb{N}$, let $\mathcal{T}^r$ be a triangulation of $U^n$ with mesh size smaller than $\frac{1}{r}$, and let $Z^r : U^n \mapsto \mathbb{R}^n$ be a piecewise linear approximation of $\zeta$ with respect to $\mathcal{T}^r$, satisfying Assumption 8.3.1. Then for every $\varepsilon > 0$ there exists an $R \in \mathbb{N}$ such that for every $r \geq R$ it holds that $q^r \in f^r([0,1])$ implies $\|Z^r(q^r)\|_\infty < \varepsilon$.*

Proof: Suppose that a sequence $(q^r, Z^r(q^r))_{r\in\mathbb{N}}$ exists with $q^r \in f^r([0,1])$ and $\|Z^r(q^r)\|_\infty \geq \varepsilon$ for every $r \in \mathbb{N}$. Since $U^n$ is compact and $\cup_{q\in U^n}\zeta(q)$ is bounded, there exists a converging subsequence $(q^{r^s}, Z^{r^s}(q^{r^s}))_{s\in\mathbb{N}}$, with limit, say, $(q^*, z^*)$, where $\|z^*\|_\infty \geq \varepsilon > 0$. As in the proof of Theorem 8.4.1 it can be shown that $z^* = \mathbf{0}$, yielding a contradiction. □

Using Theorems 8.3.10 and 8.4.1 it will be shown that there exists a connected set $C$ such that $\mathbf{0}, e \in C$ and $\mathbf{0} \in \zeta(q)$ for all $q \in C$. For a nonempty, compact set $S \subset \mathbb{R}^n$, we define the continuous function $d_S : \mathbb{R}^n \mapsto \mathbb{R}$ by $d_S(x) = \min\{\|x - y\|_\infty \mid y \in S\}$.

**Theorem 8.4.3** *Let $\zeta : U^n \mapsto \mathbb{R}^n$ be a correspondence satisfying Assumption 8.2.1. Then there exists a connected set of points $C$ in $U^n$ such that $\mathbf{0}, e \in C$ and $\mathbf{0} \in \zeta(q)$ for every $q \in C$.*

Proof: Define $Q = \{q \in U^n \mid \mathbf{0} \in \zeta(q)\}$. Clearly, $\mathbf{0}, e \in Q$ and $Q$ is compact. Suppose the theorem is false. Then $e$ is not an element of the component set[1] of $Q$ containing $\mathbf{0}$. By Munkres [1975, p. 235], it holds for every compact set $X$ in some Euclidean space and every element $x \in X$, that the component set of $X$ containing $x$ equals the intersection of all sets containing $x$ which are both open and closed in $X$. Hence there exists a set $Q^0$ which is open and closed in $Q$, such that $\mathbf{0} \in Q^0$ and $e \notin Q^0$. Define $Q^1 = Q \setminus Q^0$. Then $Q^1$ is open and closed in $Q$, $\mathbf{0} \notin Q^1$, and $e \in Q^1$.

Since $Q$ is compact, $Q^0$ and $Q^1$ are disjoint, compact sets. Hence, there exists $\varepsilon > 0$ such that $\min\{\|q^0 - q^1\|_\infty \mid q^0 \in Q^0,\ q^1 \in Q^1\} \geq \varepsilon$. For every $r \in \mathbb{N}$, let $\mathcal{T}^r$ be a triangulation of $U^n$ with mesh size smaller than $\frac{1}{r}$, let $Z^r : U^n \mapsto \mathbb{R}^n$ be a piecewise linear approximation of $\zeta$ with respect to $\mathcal{T}^r$ satisfying Assumption 8.3.1, and let $f^r$ correspond to the function $f$ of Theorem 8.3.10 for the piecewise linear approximation $Z^r$. Define $g^r : [0,1] \mapsto \mathbb{R}$ by

$$g^r(t) = d_{Q^0}(f^r(t)) - d_{Q^1}(f^r(t)),\ t \in [0,1].$$

[1] Let $x$ be an element of a topological space $X$. The *component set* of $x$ in $X$ is defined as the union of all connected subsets of $X$ containing $x$.

Since $g^r$ is continuous, $g^r(0) \leq -\varepsilon$, and $g^r(1) \geq \varepsilon$, there exists a point $t^r \in [0,1]$ such that $g^r(t^r) = 0$. Hence $d_{Q^0}(f^r(t^r)) = d_{Q^1}(f^r(t^r)) = d_Q(f^r(t^r)) \geq \frac{1}{2}\varepsilon$. Without loss of generality it can be assumed that $(f^r(t^r))_{r\in\mathbb{N}}$ converges to a point $q^* \in U^n$. Hence,

$$d_Q(q^*) = d_Q(\lim_{r\mapsto\infty} f^r(t^r)) = \lim_{r\mapsto\infty} d_Q(f^r(t^r)) \geq \frac{1}{2}\varepsilon > 0.$$

However, by Theorem 8.4.1, $d_Q(q^*) = 0$, yielding a contradiction. □

Consequently we have also proved Theorem 2.5.7 in Chapter 2. For the remainder of this section, consider the case where $\zeta$ satisfying Assumption 8.2.1 is a continuous function, denoted by $z$. Then there exists a function $p : U^n \mapsto \mathbb{R}^n_{++}$, not necessarily continuous, satisfying $(p(q))^\top z(q) = 0$. We will now derive some properties of the solutions generated by the algorithm. Let $\varepsilon > 0$ be given and let $\delta > 0$ be such that for $\tilde{q}, \overline{q} \in U^n$, $\|\tilde{q} - \overline{q}\|_\infty < \delta$ implies $\|z(\tilde{q}) - z(\overline{q})\|_\infty < \varepsilon$. By the continuity of $z$ and the compactness of $U^n$ such a $\delta$ exists. Consider a triangulation $\mathcal{T}$ with mesh size less than $\delta$ and suppose $q$ lies in the image set $f([0,1])$ of the piecewise linear function $f$ given by Theorem 8.3.10. Then there exist numbers $\lambda_j \geq 0$ for all $j \in I_{n+1}$, $\mu_j \geq 0$ for all $j \in I_n$, and $\beta \in \mathbb{R}$ such that $\sum_{j=1}^{n+1} \lambda_j = 1$, $q = \sum_{j=1}^{n+1} \lambda_j x^j$ for vertices $x^j$ of a simplex in $\mathcal{T}$ containing $q$, and for every $j \in I_n$

$$\begin{array}{rcll} Z_j(q) & = & \beta - \mu_j, & \text{if } \; q_j = 0, \\ Z_j(q) & = & \beta, & \text{if } \; 0 < q_j < 1, \\ Z_j(q) & = & \beta + \mu_j, & \text{if } \; q_j = 1. \end{array}$$

Clearly $\|Z(q) - z(q)\|_\infty = \|\sum_{j=1}^{n+1} \lambda_j(z(x^j) - z(q))\|_\infty < \varepsilon$. Hence $Z_j(q) - \varepsilon < z_j(q) < Z_j(q) + \varepsilon$ for all $j \in I_n$. If for some $j \in I_n$, $q_j = 0$, then $\beta - \mu_j = Z_j(q) \geq 0$, so $\beta \geq \mu_j \geq 0$. If for some $j \in I_n$, $q_j = 1$, then $\beta + \mu_j = Z_j(q) \leq 0$, so $\beta \leq -\mu_j \leq 0$.

First consider the case where for some $j \in I_n$, $q_j = 0$, and for some $j \in I_n$, $q_j = 1$. Then $\beta = 0$ and $\mu_j = 0$ for any $j \in I_n$ satisfying $q_j = 0$ or $q_j = 1$. So $Z(q) = \mathbf{0}$ and therefore $-\varepsilon < z_j(q) < \varepsilon$ for all $j \in I_n$. Moreover, if $q_j = 0$, then $0 \leq z_j(q) < \varepsilon$, and if $q_j = 1$, then $-\varepsilon < z_j(q) \leq 0$.

Consider now the case where for every $j \in I_n$, $0 < q_j < 1$. So $Z(q) = \beta e$. Then

$$|\beta| \sum_{j=1}^{n} p_j(q) = |(p(q))^\top Z(q)| = |(p(q))^\top (Z(q) - z(q))| < \varepsilon \sum_{j=1}^{n} p_j(q).$$

So $|\beta| < \varepsilon$. Consequently $\beta - \varepsilon < z_j(q) < \beta + \varepsilon$ for all $j \in I_n$ with $|\beta| < \varepsilon$.

Next consider the case where for some $j \in I_n$, $q_j = 1$, and for every $j \in I_n$, $q_j > 0$. Since $\beta + \mu_j \leq 0$ if $q_j = 1$, it holds that $0 \leq \mu_j \leq -\beta$ when

$q_j = 1$. Moreover,

$$0 = (p(q))^\top z(q) < (\beta + \varepsilon) \sum_{j \in \{k \in I_n \mid 0 < q_k < 1\}} p_j(q)$$

implies $\beta + \varepsilon > 0$ and therefore $-\varepsilon < \beta \leq 0$. Hence

$$\begin{aligned} \beta - \varepsilon < z_j(q) \leq 0, &\quad \text{if} \quad q_j = 1, \\ \beta - \varepsilon < z_j(q) < \beta + \varepsilon, &\quad \text{if} \quad 0 < q_j < 1, \end{aligned}$$

with $-\varepsilon < \beta \leq 0$.

Finally, consider the case where for some $j \in I_n$, $q_j = 0$, and for every $j \in I_n$, $q_j < 1$. As in the previous paragraph, it can be shown that

$$\begin{aligned} 0 \leq z_j(q) < \beta + \varepsilon, &\quad \text{if} \quad q_j = 0, \\ \beta - \varepsilon < z_j(q) < \beta + \varepsilon, &\quad \text{if} \quad 0 < q_j < 1, \end{aligned}$$

with $0 \leq \beta < \varepsilon$.

All cases are summarized in the theorem below.

**Theorem 8.4.4** *Let $z : U^n \mapsto \mathbb{R}^n$ be a function satisfying Assumption 8.2.1. Let $\varepsilon > 0$ be given and choose $\delta > 0$ such that for all $\tilde{q}$, $\overline{q} \in U^n$, $\|\tilde{q} - \overline{q}\|_\infty < \delta$ implies $\|z(\tilde{q}) - z(\overline{q})\|_\infty < \varepsilon$. Let $\mathcal{T}$ be a triangulation with mesh size less than $\delta$ and let $q \in f([0,1])$ with $f$ as in Theorem 8.3.10. Then there is a $\beta \in \mathbb{R}$ satisfying $-\varepsilon < \beta < \varepsilon$, with $0 \leq \beta$ if $q_j = 0$ for some $j \in I_n$, and $\beta \leq 0$ if $q_j = 1$ for some $j \in I_n$, such that*

$$\begin{aligned} 0 \leq z_j(q) < \beta + \varepsilon, &\quad \textit{if} \quad q_j = 0, \\ \beta - \varepsilon < z_j(q) < \beta + \varepsilon, &\quad \textit{if} \quad 0 < q_j < 1, \\ \beta - \varepsilon < z_j(q) \leq 0, &\quad \textit{if} \quad q_j = 1. \end{aligned}$$

## 8.5. An illustration of the algorithm

In this section we apply the algorithm to compute the set of constrained equilibria in the economic model with price rigidities given in Example 2.5.8. Recall that in this example we have an exchange economy with two agents and two commodities. The set of admissible prices is given by $P = \{p \in \mathbb{R}^2_+ \mid \frac{1}{6} \leq p_1 \leq 2,\, p_2 = 1\}$. The initial endowments and the consumption sets of the two agents are given by $\omega^1 = (1,4)^\top$ and $\omega^2 = (2,1)^\top$, $X^1 = \mathbb{R}^2_+$ and $X^2 = \mathbb{R}^2_+$, respectively. The two agents have Cobb-Douglas utility functions given by

$$u^1(x_1, x_2) = (x_1)^{\frac{3}{4}} (x_2)^{\frac{1}{4}},\ x \in \mathbb{R}^2_+,$$

$$u^2(x_1, x_2) = (x_1)^{\frac{1}{4}}(x_2)^{\frac{3}{4}},\ x \in \mathbb{R}^2_+,$$

respectively. The rationing system $(l, L) : U^n \mapsto -\mathbb{R}^4_+ \times \mathbb{R}^4_+$ is given by

$$\begin{array}{rcl} l^1_1(q) = l^2_1(q) & = & -3\min\{1, 3q_1\},\ \ q \in U^2, \\ l^1_2(q) = l^2_2(q) & = & -5\min\{1, 3q_2\},\ \ q \in U^2, \\ L^1_1(q) = L^2_1(q) & = & 18\min\{1, 3 - 3q_1\},\ \ q \in U^2, \\ L^1_2(q) = L^2_2(q) & = & 5\min\{1, 3 - 3q_2\},\ \ q \in U^2. \end{array}$$

The pricing function $p : U^2 \mapsto \mathbb{R}^2_{++}$ is given by

$$\begin{array}{rcl} p_1(q) & = & \max\{\frac{1}{6}, \min\{\frac{11}{2}q_1 - \frac{5}{3}, 2\}\},\ \ q \in U^2, \\ p_2(q) & = & 1,\ \ q \in U^2. \end{array}$$

The excess demand function of agent 1 is given by

$$z^1(q) = \begin{cases} (90q_2, -15q_1)^\top, & 0 \le q_1 \le \frac{1}{3},\ 0 \le q_2 \le \frac{71}{360}, \\ (\frac{71}{4}, -\frac{71}{24})^\top, & 0 \le q_1 \le \frac{1}{3},\ \frac{71}{360} \le q_2 \le 1, \\ (\frac{90q_2}{33q_1 - 10}, -15q_2)^\top, & \frac{1}{3} \le q_1 \le \frac{2}{3},\ 0 \le q_2,\ 33q_1 + 360q_2 \le 82, \\ (\frac{82 - 33q_1}{132q_1 - 40}, \frac{33q_1 - 82}{24})^\top, & \frac{1}{3} \le q_1 \le \frac{2}{3},\ q_2 \le 1,\ 33q_1 + 360q_2 \ge 82, \\ (\frac{15q_2}{2}, -15q_2)^\top, & \frac{2}{3} \le q_1,\ 0 \le q_2 \le \frac{1}{6},\ 36q_1 + 5q_2 \le 36, \\ (\frac{5}{4}, -\frac{5}{2})^\top, & \frac{2}{3} \le q_1 \le \frac{211}{216},\ \frac{1}{6} \le q_2 \le 1, \\ (54 - 54q_1, 108q_1 - 108)^\top, & \frac{211}{216} \le q_1 \le 1,\ q_2 \le 1,\ 36q_1 + 5q_2 \ge 36. \end{cases}$$

The excess demand function of agent 2 is given by

$$z^2(q) = \begin{cases} (0,0)^\top, & 0 \le q_1 \le \frac{1}{3},\ 0 \le q_2 \le 1, \\ (\frac{33 - 99q_1}{66q_1 - 20}, \frac{33q_1 - 11}{4})^\top, & \frac{1}{3} \le q_1 \le \frac{2}{3},\ 0 \le q_2,\ 33q_1 + 60q_2 \le 71, \\ (\frac{90q_2 - 90}{33q_1 - 10}, 15 - 15q_2)^\top, & q_1 \le \frac{2}{3},\ q_2 \le 1,\ 33q_1 + 60q_2 \ge 71, \\ (-\frac{11}{8}, \frac{11}{4})^\top, & \frac{2}{3} \le q_1 \le 1,\ 0 \le q_2 \le \frac{49}{60}, \\ (\frac{15q_2 - 15}{2}, 15 - 15q_2)^\top, & \frac{2}{3} \le q_1 \le 1,\ \frac{49}{60} \le q_2 \le 1. \end{cases}$$

Now the total excess demand function, $z : U^2 \mapsto \mathbb{R}^2$, is given by the function $z = z^1 + z^2$. The zero points of $z$ can easily be determined analytically. The points

$$\begin{pmatrix} 0 \\ 0 \end{pmatrix}, \begin{pmatrix} \frac{1}{3} \\ 0 \end{pmatrix}, \begin{pmatrix} \frac{148}{231} \\ \frac{71}{420} \end{pmatrix}, \begin{pmatrix} \frac{148}{231} \\ \frac{349}{420} \end{pmatrix}, \begin{pmatrix} \frac{2}{3} \\ \frac{5}{6} \end{pmatrix}, \begin{pmatrix} \frac{211}{216} \\ \frac{5}{6} \end{pmatrix}, \begin{pmatrix} 1 \\ 1 \end{pmatrix}$$

and the convex combinations of any two successive points are all zero points of the function $z$.

The sequence of simplices generated by the algorithm is illustrated in Figure 8.3 where the grid size of the $K_1$-triangulation is equal to $\frac{1}{6}$. The

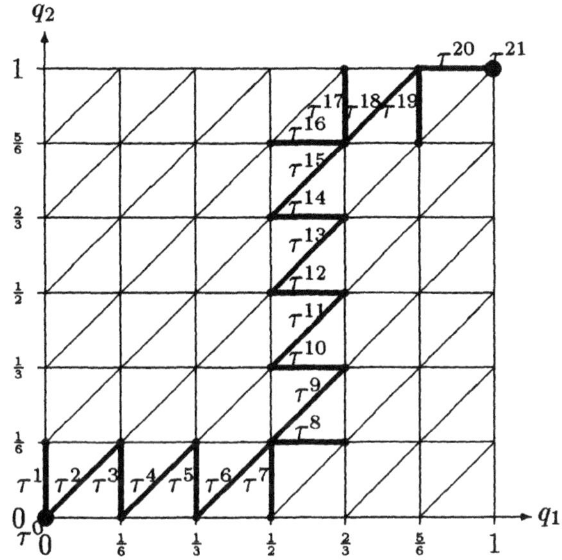


*Figure 8.3.* Illustration of the HTY algorithm for $n = 2$.

piecewise linear path $f([0,1])$ of points generated by the algorithm is given by the points

$$\begin{pmatrix} 0 \\ 0 \end{pmatrix}, \begin{pmatrix} \frac{1}{3} \\ 0 \end{pmatrix}, \begin{pmatrix} \frac{1}{2} \\ \frac{11}{120} \end{pmatrix}, \begin{pmatrix} \frac{113}{176} \\ \frac{1}{6} \end{pmatrix}, \begin{pmatrix} \frac{113}{176} \\ \frac{163}{528} \end{pmatrix}, \begin{pmatrix} \frac{277}{429} \\ \frac{1}{3} \end{pmatrix}, \begin{pmatrix} \frac{277}{429} \\ \frac{2}{3} \end{pmatrix}, \begin{pmatrix} \frac{2}{3} \\ \frac{5}{6} \end{pmatrix}, \begin{pmatrix} \frac{5}{6} \\ \frac{5}{6} \end{pmatrix}, \begin{pmatrix} 1 \\ 1 \end{pmatrix}$$

and all convex combinations of any two successive points. In Figure 8.4 the solid line corresponds to the piecewise linear path of approximate zero points generated by the algorithm, while the broken line corresponds to the set of zero points of $z$.

The algorithm starts with the $s^0$-complete simplex $\tau^0 = \{\mathbf{0}\}$ in $A(s^0)$, where $s^0 = (+1,0)^\top$. Notice that although $\tau^0$ is also $(0,-1)^\top$-complete, $\tau^0$ does not lie in $A((0,-1)^\top)$. The simplex $\tau^0$ is not $s$-complete for $s = (0,+1)^\top$ since $A^{-1}_{(0,+1)^\top,\tau^0}$ is not semilexicopositive, although the system $A_{(0,+1)^\top,\tau^0}x = \mathbf{0}$, does have a solution satisfying $x_1 \geq 0,\ x_2 \geq 0$. In this way the lexicographic pivot rules determine in a unique way the direction the algorithm will follow. So the next vertex which is brought into the system is $(0,\frac{1}{6})^\top$. Remark that the direction determined by the lexicographic pivot rules is orthogonal to the set of zero points of $z$ around $\mathbf{0}$. It can easily be verified that the 0-simplex $\tau((0,\frac{1}{6})^\top)$ is not $(+1,0)^\top$-complete. However, the 1-simplex $\tau^1 = \tau((0,0)^\top,(0,\frac{1}{6})^\top)$ is $(0,0)^\top$-complete, and therefore we are

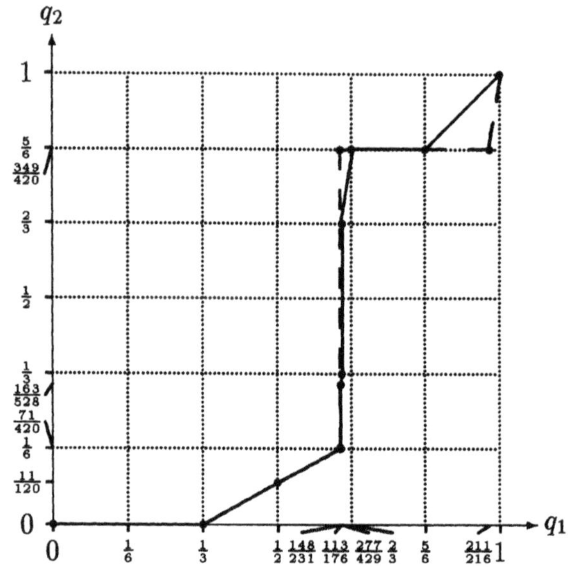


*Figure 8.4.* The P.L. path generated by the HTY algorithm.

in Case (i) of Lemma 8.3.6. The simplex $\sigma^1 = \sigma((0,0)^\top, (0, \frac{1}{6})^\top, (\frac{1}{6}, \frac{1}{6})^\top)$ is the unique 2-simplex in $A((0,0)^\top)$ having $\tau^1$ as a facet. The simplex $\tau^2 = \tau((0,0)^\top, (\frac{1}{6}, \frac{1}{6})^\top)$ is the unique $(0,0)^\top$-complete facet of the simplex $\sigma^1$ in $A((0,0)^\top)$ which is not equal to $\tau^1$. This corresponds to Case (ii) of Lemma 8.3.6, and so on.

From Figure 8.4 it can be seen that although the lexicographic pivot rules determine an initial direction to leave $\tau^0$ being orthogonal to the set of zero points of $z$ around $\mathbf{0}$, the points on the piecewise linear path of approximate zero points contained in the simplices $\tau^0, \ldots, \tau^5$ are zero points of $z$. An interesting situation occurs at the simplex $\tau^{14} = \tau((\frac{1}{2}, \frac{2}{3})^\top, (\frac{2}{3}, \frac{2}{3})^\top)$. For this simplex it holds that

$$A^{-1}_{(0,0)^\top, \tau^{14}} = \begin{pmatrix} \frac{18}{143} & \frac{48}{143} & \frac{-48}{143} \\ \frac{125}{143} & \frac{-48}{143} & \frac{48}{143} \\ \frac{5}{104} & \frac{-7}{13} & \frac{-6}{13} \end{pmatrix}.$$

The vertex brought into the system is given by $x = (\frac{2}{3}, \frac{5}{6})^\top$, with $Z(x) = (0,0)^\top$. So $y = A^{-1}_{(0,0)^\top, \tau^{14}}(1, Z(x)^\top)^\top = (\frac{18}{143}, \frac{125}{143}, \frac{5}{104})^\top$. Now there occurs a degeneracy problem since both $y_1$ and $y_2$ are positive and equal to the corresponding element of the first column of $A^{-1}_{(0,0)^\top, \tau^{14}}$. However, since the vector $\frac{(A^{-1}_{(0,0)^\top, \tau^{14}})_{1\cdot}}{y_1}$ is lexicographically larger than the vector $\frac{(A^{-1}_{(0,0)^\top, \tau^{14}})_{2\cdot}}{y_2}$, the vertex of $\tau^{14}$ corresponding to column 2 of $A_{(0,0)^\top,\tau^{14}}$, $(\frac{2}{3},\frac{2}{3})^\top$, should be replaced by the vertex $(\frac{2}{3},\frac{5}{6})^\top$, yielding the simplex $\tau^{15}$. Moreover, it has to be remarked that using the usual lexicographic pivoting rules (see Todd [1976a] and Wright [1981]) would imply that the last column of $A_{(0,0)^\top,\tau^{14}}$ should be replaced by the vertex $(\frac{2}{3},\frac{5}{6})^\top$. However, the last column of any $A_{s,\tau}$, corresponding to the unrestricted variable $\beta$, never leaves the system. Another interesting case occurs at $\tau^{20}$. The 1-simplex $\tau^{20}$ lies in $A((0,-1)^\top)$ and therefore $(0,0,-1)^\top$ is the new column brought into the system. We are in Case (iii) of Lemma 8.3.7, where $\tau^{21}$ is the unique $(0,-1)^\top$-complete facet of $\tau^{20}$. The 0-simplex $\tau^{21}$ lies in $A((-1,-1)^\top) =< e >$ and we are in Case (i) of Lemma 8.3.7. The algorithm now terminates with the $(0,-1)^\top$-complete simplex $< e >$ . It is clear from Figure 8.4 that the approximate zero points lying in the $s$-complete $(t-1)$-simplices in $A(s)$ for some sign vector $s$ with $t = |I^0(s)|$ are everywhere close to the zero points of $z$ themselves. It should be remarked that this is caused by the fact that the function $z$ given in the example has such an easy structure. For more complicated correspondences a finer mesh size might be needed to obtain a good accuracy.

The approximate supply-constrained equilibrium, being the constrained equilibrium without binding demand rationing and with at least one market without rationing, corresponds to $q^S = (\frac{1}{2},\frac{11}{120})^\top$, and the total excess demand equals zero in this point, $z(q^S) = (0,0)^\top$. This constrained equilibrium is also the Drèze equilibrium if commodity 1 is assumed to be the numeraire commodity as in Drèze's original model. If commodity 2 is assumed to be the numeraire commodity, then the approximate Drèze equilibrium, being also the constrained equilibrium without binding supply rationing and with at least one market without rationing, corresponds to $q^D = (\frac{277}{429},\frac{1}{2})$. The total excess demand in this point is close to zero, $z(q^D) = (-0.0255, 0.0480)^\top$.

## 8.6. A constructive proof for Browder's fixed point theorem

As we see in Theorem 8.4.3, any correspondence satisfying Assumption 8.2.1 has a connected set of zero points linking $\mathbf{0}$ and $e$ of $U^n$. First we prove Kakutani's fixed point theorem by using Theorem 8.4.3.

**Theorem 8.6.1** *For given $n \geq 2$, let $\phi : S^n \mapsto S^n$ be a nonempty-valued, convex-valued and compact-valued u.s.c. point-to-set mapping. Then there exists at least one point $x^* \in S^n$ such that $x^* \in \phi(x^*)$.*

Proof: Let $Q(t) = \{x \in U^n \mid \sum_{i=1}^n x_i = nt\}$ for $t \in [0,1]$. It is clear that $S^n = Q(\frac{1}{n})$. Define the function $r : \mathbf{R}^n \mapsto S^n$ by

$$r(x) = \arg\min\{||x-y||^2 \mid y \in S^n\}.$$

Define the function $h : U^n \times [0,1] \mapsto U^n$ by

$$h(x,t) = \arg\min\{||x-y||^2 \mid y \in Q(t)\}.$$

It follows from Corollary 1.6.13 that $h$ is a continuous function since $Q$ is a continuous point-to-set mapping with nonempty, convex and compact values. Now we define the point-to-set mapping $\Gamma : U^n \mapsto U^n$ by

$$\Gamma(x) = \{h(y, \frac{\sum_{i \in I_n} x_i}{n}) \mid y \in \phi(r(x))\}.$$

It is clear that $\Gamma(x) = \phi(x)$ for all $x \in S^n$. Moreover, $\Gamma$ is a nonempty-valued, convex-valued and compact-valued u.s.c. mapping, since $\phi$ is a nonempty-valued, convex-valued and compact-valued u.s.c. mapping, and $h$ is continuous. Note that for each $t \in [0,1]$ and each $x \in Q(t)$, we have $\Gamma(x) \subseteq Q(t)$. Define a correspondence $\psi : U^n \mapsto \mathbb{R}^n$ by $\psi(x) = \Gamma(x) + \{-x\}$. It can be easily checked that all conditions of Theorem 8.4.3 are satisfied. More precisely, Assumption 8.2.1 (a) is trivially satisfied. For each $x \in U^n$, $x_i = 0$ implies $z_i \geq 0$ for all $z \in \psi(x)$ and $x_i = 1$ implies $z_i \leq 0$ for all $z \in \psi(x)$. For each $x \in U^n$, we have $z^\top e = 0$ for all $z \in \psi(x)$. Therefore, Assumptions 8.2.1 (b) and (c) are also satisfied. So there exists a connected set $C$ in $U^n$ such that $\mathbf{0} \in C$, $e \in C$, and $0^n \in \psi(x)$ for all $x \in C$. Hence for all $x \in C$, $x \in \Gamma(x)$ by the definition of $\psi$. Since $C \cap S^n \neq \emptyset$, there is a point $x^* \in C \cap S^n$ such that $x^* \in \phi(x^*)$. □

Next we give a proof for Browder's fixed point theorem.

**Theorem 8.6.2** *For given $n \geq 2$, let $\phi : S^n \times [0,1] \mapsto S^n$ be a nonempty-valued, convex-valued and compact-valued u.s.c. point-to-set mapping. Then there exists a connected subset $T$ of $S^n \times [0,1]$ such that for every $(x,t) \in T$, $x \in \phi(x,t)$, $T \cap S^n \times \{0\} \neq \emptyset$ and $T \cap S^n \times \{1\} \neq \emptyset$.*

Proof: Without loss of generality we will prove the theorem in the set $S^n \times [\frac{1}{4}, 1]$. Define the point-to-set mapping $\psi : U^n \mapsto \mathbf{R}^n$ by

$$\psi(x) = \phi(\frac{x}{\sum_{i=1}^n x_i}, \sum_{i=1}^n x_i) - \{\frac{x}{\sum_{i=1}^n x_i}\}$$

for $x \in U^n$ and $\frac{1}{4} \leq \sum_{i=1}^n x_i \leq 1$ (the above construction of $\psi$ is due to Herings). For $0 < \sum_{i=1}^n x_i \leq \frac{1}{4}$ and $x \in U^n$, define

$$\psi(x) = 4(\sum_{i=1}^n x_i)\psi(\frac{x}{4\sum_{i=1}^n x_i}).$$

Moreover, define $\psi(0^n) = \{0^n\}$. We still have to extend $\psi$ to the remaining part of $U^n$.

Let $Q(t) = \{x \in U^n \mid \sum_{i=1}^n x_i = t\}$ for $t \in [1, n]$. It is clear that $S^n = Q(1)$. Define the function $r : \mathbb{R}^n \mapsto S^n$ by

$$r(x) = \arg\min\{||x - y||^2 \mid y \in S^n\}.$$

Define the function $h : U^n \times [1, n] \mapsto U^n$ by

$$h(x, t) = \arg\min\{||x - y||^2 \mid y \in Q(t)\}.$$

It follows from Corollary 1.6.13 that $h$ is a continuous function since $Q$ is a continuous point-to-set mapping with nonempty, convex and compact values. Now we define the point-to-set mapping $\Gamma : U^n \mapsto U^n$ by

$$\Gamma(x) = \{h(y, \sum_{i=1}^n x_i) \mid y \in \phi(r(x), 1)\}.$$

It is clear that $\Gamma(x) = \phi(x, 1)$ for all $x \in U^n$ with $\sum_{i=1}^n x_i = 1$. Moreover, $\Gamma$ is a nonempty-valued, convex-valued and compact-valued u.s.c. mapping, since $\phi$ is a nonempty-valued, convex-valued and compact-valued u.s.c. mapping, and $h$ is continuous. Note that for each $t \in [1, n]$ and each $x \in Q(t)$, we have $\Gamma(x) \subseteq Q(t)$. For $x \in U^n$ and $1 \leq \sum_{i=1}^n x_i \leq n$, define

$$\psi(x) = \Gamma(x) - \{x\}.$$

It can be easily checked that all conditions of Theorem 8.4.3 are satisfied by $\psi$. More precisely, Assumption 8.2.1 (a) is trivially satisfied. For each $x \in U^n$, $x_i = 0$ implies $z_i \geq 0$ for all $z \in \psi(x)$ and $x_i = 1$ implies $z_i \leq 0$ for all $z \in \psi(x)$. For each $x \in U^n$, we have $z^\top e = 0$ for all $z \in \psi(x)$. Therefore, Assumptions 8.2.1 (b) and (c) are also satisfied. So there exists a connected set $C$ in $U^n$ such that $0^n \in C$, $1^n \in C$, and $0^n \in \psi(x)$ for all $x \in C$. Hence for all $x \in C$ with $\frac{1}{4} \leq \sum_{i=1}^n x_i \leq 1$, we have

$$\frac{x}{\sum_{i=1}^n x_i} \in \phi(\frac{x}{\sum_{i=1}^n x_i}, \sum_{i=1}^n x_i)$$

by the definition of $\psi$. This completes the proof. □

Thus, Browder's fixed point theorem can be proved constructively. In Freidenfelds [1974] a constructive proof for a simpler version of this theorem is given in which a function is considered. We also mention that Herings and Talman [1998] and Guo and Yamamoto [1996] have given two alternative non-constructive proofs for Theorem 8.4.3 by using Browder theorem.

Their results together with the above one imply that Theorem 8.4.3 is equivalent to Browder theorem. The following exercise is left to the interested reader. In order to prove it, we need to extend the mapping $\psi$ from $S^n$ to $U^n$ such that Assumption 8.2.1 is satisfied at every point in $U^n$.

**Example 8.6.3** *Let $\psi : S^n \mapsto \mathbb{R}^n$ be a correspondence satisfying Assumption 8.2.1 in which $U^n$ is replaced by $S^n$. Then there exists at least one point $x^* \in S^n$ such that $\mathbf{0} \in \psi(x^*)$.*

Finally, we turn to some unresolved problems. In this chapter all results have been built upon the unit cube $U^n$. It is still an open question how to extend these results, in particular Theorem 8.4.3, to general polytopes or even to general convex compact sets. It would be of great interest to investigate this.

CHAPTER 9

# The Computation of Stationary Points on General Polytopes

## 9.1. Introduction

In this chapter we introduce an algorithm to solve the nonlinear stationary point problem on general polytopes in $\mathbb{R}^n$. Recall that given an arbitrary polytope P and a function $f : \mathrm{P} \mapsto \mathbb{R}^n$, the problem of stationary point (or variational inequality) for $f$ on P is to find a point $x^* \in \mathrm{P}$ such that

$$(x^* - x)^\top f(x^*) \geq 0, \quad \text{for all} \quad x \in \mathrm{P}.$$

The algorithm is applied to any upper semi-continuous point-to-set mapping with nonempty, convex and compact values. Therefore, we are able to give a constructive proof for Stationary Point Theorems I and II discussed in Chapter 1. Briefly speaking, starting with an arbitrary point of P, the algorithm generates a sequence of adjacent simplices of varying dimension in a specific triangulation of P. This sequence contains a piecewise linear path of points leading to an approximate stationary point of $f$.

The stationary point problem arises, for example, from economic equilibrium problems, non-cooperative games, traffic assignment problems and nonlinear optimization problems. Let us now give a short review of simplicial methods for solving stationary point problems. In the case where $f$ is affine, Eaves [1978] proposed a pivoting algorithm for tracing a piecewise linear path of stationary points with respect to a specific expanding set. In this case, no simplicial subdivision is needed. Van der Laan and Talman [1987b] introduced a simplicial algorithm for the nonlinear stationary point problem with lower and upper bounds. Yamamoto [1987b] proposed a pivoting algorithm for the linear stationary point problem on a polytope. Doup and Talman [1987b] developed a simplicial algorithm for the nonlinear stationary point problem on the cross product of several unit simplices. Talman and Yamamoto [1989] proposed the most general simplicial algorithm for the nonlinear stationary point problem on general polytopes. Dai et al [1991] extended the algorithm of Talman and Yamamoto to an unbounded polyhedron. We also refer to Doup [1988] and Dai [1991]. In this chapter the algorithm we will present was introduced by Talman and Yamamoto, whereas the lexicographic operation system was proposed by Fujishige and Yang [1998]. By using lexicographic systems the degeneracy problems facing the method of Talman and Yamamoto as well as many other simplicial methods will be resolved here.

This chapter is organized as follows. In Section 2 we introduce several basic theorems for resolving degeneracy problems. In Section 3 we consider the problem of finding a zero point of a continuous function $f : \mathrm{P} \mapsto \mathbb{R}^n$ and give a sufficient condition for the existence of zero points. As a result, the well-known multivariate mean value theorem will be generalized. In Section 4 we introduce the algorithm of Talman and Yamamoto for computing a stationary point of a continuous function $f : \mathrm{P} \mapsto \mathbb{R}^n$. Finally in Section 5 a specific triangulation which underlies the above algorithms is introduced. This triangulation is called $V$-triangulation and was developed by Talman and Yamamoto [1989].

## 9.2. Basic theorems for resolving degeneracy problems

In this section we present several basic results for resolving the degeneracy problem in simplicial algorithms. The reader is also referred to related results in Section 4.2. Consider an arbitrary full-dimensional polytope

$$\mathrm{P} = \{x \in \mathbf{R}^n \mid a^{i\top} x \le b_i, \quad \text{for all } i \in I_m\}.$$

We assume that P is simple and no constraints are redundant. For each subset $I$ of $I_m$, define

$$F(I) = \{x \in \mathrm{P} \mid a^{i\top} x = b_i, \quad \text{for all } i \in I\}.$$

Then $F(I)$ is a face of P unless it is empty. Note that $F(\emptyset) = \mathrm{P}$. Let

$$\mathcal{I} = \{I \subseteq I_m \mid F(I) \text{ is a nonempty face of } \mathrm{P}\}.$$

Under the above assumption that $dim(\mathrm{P}) = n$, P is a simple polytope, and the linear inequalities defining P are non-redundant, we have the following observations:

(i) for each face $F$ of $P$ the set $I \in \mathcal{I}$ with $F = F(I)$ is unique and identical with the set $\{i \in I_m \mid a^{i\top} x = b_i, \text{ for all } x \in F\}$;

(ii) $dim(F(I)) = n - |I|$ for any $I \in \mathcal{I}$ and $F(I)$ is a vertex of $P$ if $I \in \mathcal{I}$ with $|I| = n$;

(iii) for some $I \in \mathcal{I}$, $G$ is a facet of $F(I)$ if and only if $G = F(I \cup \{h\})$ for some $h \notin I$ with $I \cup \{h\} \in \mathcal{I}$;

(iv) for any $I \in \mathcal{I}$ all the vectors $a^j$ with $j \in I$ are linearly independent. Moreover, if $I \in \mathcal{I}$, then $I \setminus \{k\} \in \mathcal{I}$ for any $k \in I$.

Now we are ready to present one of our main theorems.

**Theorem 9.2.1** *Given any $n$-dimensional simple polytope* $\mathrm{P} = \{x \in \mathbf{R}^n \mid a^{i\top} x \leq b_i, \text{ for all } i \in I_m\}$ *with non-redundant constraints and any vector* $c \in \mathbf{R}^n$, *there exists a unique index set* $J \in \mathcal{I}$ *with* $J = \{j_1, \cdots, j_n\}$ *and* $|J| = n$ *such that the matrix*

$$D^{-1} = \begin{bmatrix} 1 & 1 & 1 & \cdots & 1 \\ c & a^{j_1} & a^{j_2} & \cdots & a^{j_n} \end{bmatrix}^{-1}$$

*exists and is lexico positive.*

Proof: Consider the linear programming (LP) problem

$$\begin{array}{ll} \max & -c^\top x, \\ s.t. & x \in \mathrm{P}. \end{array} \tag{9.2.1}$$

Then there exists an optimal solution $x^* \in \mathrm{P}$ which is a vertex of P since P is bounded. Its dual problem (DLP) is

$$\begin{array}{ll} \min & b^\top \lambda, \\ s.t. & \sum_{j \in I_m} \lambda_j a^j = -c, \\ & \lambda_j \geq 0, \text{ for all } j \in I_m. \end{array} \tag{9.2.2}$$

Let

$$\begin{array}{rcl} I(x^*) & = & \{j \in I_m \mid b_j - a^{j\top} x^* = 0\} \\ & = & \{j_1, j_2, \cdots, j_n\}. \end{array}$$

Clearly, $I(x^*) \in \mathcal{I}$ and $|I(x^*)| = n$ since $x^*$ is a vertex of P. Two cases have to be considered.

Case (1). The dual problem (9.2.2) is non-degenerate. That is to say, every basic feasible solution $\lambda^*$ of the problem (9.2.2) has exactly $n$ components being positive. So we have $\lambda_j^* > 0$ for all $j \in I(x^*)$ and $\lambda_j^* = 0$ for all $j \in I_m \setminus I(x^*)$. Note that the $n$ vectors $a^{j_1}$, $\cdots$, $a^{j_n}$ are linearly independent. Now we should prove that $a^{j_2} - a^{j_1}$, $a^{j_3} - a^{j_2}$, $\cdots$, $a^{j_n} - a^{j_{n-1}}$, $c - a^{j_n}$ are linearly independent. Suppose to the contrary that there exists a nonzero vector $(\beta_1, \beta_2, \cdots, \beta_n)$ such that

$$\beta_1(a^{j_2} - a^{j_1}) + \beta_2(a^{j_3} - a^{j_2}) + \cdots + \beta_{n-1}(a^{j_n} - a^{j_{n-1}}) + \beta_n(c - a^{j_n}) = \mathbf{0}. \tag{9.2.3}$$

Since $c = -\sum_{j \in I(x^*)} \lambda_j^* a^j$, it follows from equation (9.2.3) that

$$(-\beta_1 - \beta_n \lambda_{j_1}^*) a^{j_1} + (\beta_1 - \beta_2 - \beta_n \lambda_{j_2}^*) a^{j_2} + \cdots + (\beta_{n-1} - \beta_n - \beta_n \lambda_{j_n}^*) a^{j_n} = \mathbf{0}.$$

Because $a^{j_1}$, $\cdots$, $a^{j_n}$ are linearly independent, we have

$$\begin{array}{rcl} -\beta_1 - \beta_n \lambda_{j_1}^* & = & 0, \\ \beta_1 - \beta_2 - \beta_n \lambda_{j_2}^* & = & 0, \\ & \vdots & \\ \beta_{n-1} - \beta_n - \beta_n \lambda_{j_n}^* & = & 0. \end{array}$$

Thus, $\beta_1 = -\beta_n\lambda^*_{j_1}$, $\beta_2 = -\beta_n(\lambda^*_{j_1} + \lambda^*_{j_2})$, $\cdots$, $\beta_n = -\beta_n(\lambda^*_{j_1} + \cdots + \lambda^*_{j_n})$. Since $\lambda^*_j > 0$ for all $j \in I(x^*)$, we have $\beta_1 = \cdots = \beta_n = 0$, yielding a contradiction. It follows that $a^{j_1}$, $a^{j_2}$, $\cdots$, $a^{j_n}$, $c$ are affinely independent. Thus the matrix

$$\begin{bmatrix} 1 & 1 & 1 & \cdots & 1 \\ c & a^{j_1} & a^{j_2} & \cdots & a^{j_n} \end{bmatrix}$$

is nonsingular. By taking $\lambda_0 = 1/(1+\sum_{i=1}^n \lambda^*_{j_i})$ and $\lambda_i = \lambda^*_{j_i}/(1+\sum_{i=1}^n \lambda^*_{j_i})$ for $i \in I_n$, we have

$$\begin{aligned} \textstyle\sum_{i=0}^n \lambda_i &= 1,\ \lambda_i > 0 \text{ for all } i, \\ \textstyle\sum_{i=1}^n \lambda_i a^{j_i} + \lambda_0 c &= \mathbf{0}. \end{aligned}$$

So we conclude that

$$\begin{bmatrix} 1 & 1 & 1 & \cdots & 1 \\ c & a^{j_1} & a^{j_2} & \cdots & a^{j_n} \end{bmatrix}^{-1}$$

exists and is lexico positive.

Case (2). The dual problem (9.2.2) is degenerate. That is to say, some basic variables of a basic feasible solution to (9.2.2) are zero. So there may exist some element $i \in I(x^*)$ with $\lambda^*_i = 0$. In this case we consider the perturbed linear programming (PLP) problem

$$\begin{array}{ll} \max & -(c^\top - (\epsilon, \epsilon^2, \cdots, \epsilon^n))x, \\ s.t. & x \in \mathrm{P}. \end{array} \tag{9.2.4}$$

Its dual problem (DPLP) is

$$\begin{array}{ll} \min & b^\top\lambda, \\ s.t. & \sum_{j\in I_m} \lambda_j a^j = -(c - (\epsilon, \epsilon^2, \cdots, \epsilon^n)^\top), \\ & \lambda_j \geq 0, \text{ for all } j \in I_m. \end{array} \tag{9.2.5}$$

It is well-known in linear programming (see Dantzig [1963]) that there exists a positive number $\epsilon_0$ such that for any $\epsilon \in (0, \epsilon_0)$, the DPLP is nondegenerate. Let $x^*$ be an optimal solution of the PLP which is a vertex of P, and $\lambda^*$ be an optimal solution of the DPLP. Let

$$\begin{aligned} I(x^*) &= \{j \in I_m \mid b_j - a^{j\top}x^* = 0\} \\ &= \{j_1, j_2, \cdots, j_n\}. \end{aligned}$$

Clearly, $I(x^*) \in \mathcal{I}$ and $|I(x^*)| = n$ since $x^*$ is a vertex of P. The nondegeneracy implies that $\lambda^*_{j_i} > 0$ for all $i \in I_n$ such that

$$\sum_{i=1}^n \lambda^*_{j_i} a^{j_i} = -c + (\epsilon, \epsilon^2, \cdots, \epsilon^n)^\top. \tag{9.2.6}$$

It follows from the argument in Case (1) that the matrix

$$\bar{B} = \begin{bmatrix} 1 & 1 & 1 & \cdots & 1 \\ c - (\epsilon, \epsilon^2, \cdots, \epsilon^n)^\top & a^{j_1} & a^{j_2} & \cdots & a^{j_n} \end{bmatrix}^{-1}$$

exists and is lexico positive. Let $l = (l_1, l_2, \cdots, l_{n+1})^\top = (1, 0, \cdots, 0)^\top + \bar{B}(0, \epsilon, \epsilon^2, \cdots, \epsilon^n)^\top$ and let

$$B = \begin{bmatrix} 1 & 1 & 1 & \cdots & 1 \\ c & a^{j_1} & a^{j_2} & \cdots & a^{j_n} \end{bmatrix}.$$

It is easy to see that

$$\bar{B}B = \begin{bmatrix} l_1 & 0 & \cdots & 0 & 0 \\ l_2 & 1 & \cdots & 0 & 0 \\ & \vdots & & \vdots & \\ l_n & 0 & \cdots & 1 & 0 \\ l_{n+1} & 0 & \cdots & 0 & 1 \end{bmatrix}.$$

Since the matrix on the right hand is nonsingular, the matrix $B$ must also be nonsingular.

Let $W = B^{-1}$. We will show that $W$ is lexico positive. Suppose to the contrary that the first nonzero entry of the $i$th row of $W$ is less than zero. In this case there exists a positive number $\epsilon < \epsilon_0$ such that the $i$th element $\mu_i$ of the vector

$$(\mu_1, \mu_2, \cdots, \mu_{n+1})^\top = W(1, \epsilon, \cdots, \epsilon^n)^\top \tag{9.2.7}$$

is less than zero. On the other hand it follows from equation (9.2.6) that

$$\begin{bmatrix} 1 & 1 & 1 & \cdots & 1 \\ c & a^{j_1} & a^{j_2} & \cdots & a^{j_n} \end{bmatrix} \begin{pmatrix} 1 \\ \lambda^* \end{pmatrix} = \begin{bmatrix} d \\ \epsilon \\ \vdots \\ \epsilon^n \end{bmatrix}, \tag{9.2.8}$$

where $d = 1 + \sum_{i=1}^n \lambda^*_{j_i} > 1$. Premultiply both sides of equation (9.2.8) by $W$. Then we obtain

$$\begin{pmatrix} 1 \\ \lambda^* \end{pmatrix} = W(d, \epsilon, \cdots, \epsilon^n)^\top. \tag{9.2.9}$$

It is clear from (9.2.7) and (9.2.9) that $\lambda^*_{j_i} \leq \mu_i < 0$, yielding a contradiction.

The uniqueness can easily be shown as follows. Consider the above two cases respectively. In the above Case (1), we first note that $x^*$ is the unique optimal solution of the LP (9.2.1). In other words, if the DLP (9.2.2) is non-degenerate, the LP (9.2.1) has a unique optimal solution.

Suppose now that there exists a different set $I \in \mathcal{I}$ with $I = \{i_1, i_2, \cdots, i_n\}$ and $|I| = n$ such that the matrix

$$\begin{bmatrix} 1 & 1 & 1 & \cdots & 1 \\ c & a^{i_1} & a^{i_2} & \cdots & a^{i_n} \end{bmatrix}^{-1} \tag{9.2.10}$$

is lexico positive. Then the linear system

$$a^{i_t \top} x = b_{i_t}, \quad \text{for all } t \in I_n$$

has a unique solution, say $z^*$ being a vertex of P. It follows from (9.2.10) that $-c = \sum_{t \in I_n} \lambda_{i_t} a^{i_t}$ with $\lambda_{i_t} \geq 0$ for all $t \in I_n$. Let $\lambda_t = 0$ for all $t \in I_m \setminus I$. It is clear that the vector $\lambda$ is a basic feasible solution of the DLP (9.2.2). The complementary slackness condition is satisfied. Hence $z^*$ is another optimal solution of the LP (9.2.1), yielding a contradiction.

The Case (2) above can be shown in the same way. □

<u>Note 1</u>: There is a bijective mapping from the set of the vertices of the polytope P to the set $\{I \in \mathcal{I} \mid |I| = n\}$. So in the above theorem the uniqueness of the index set $J$ is equivalent to the uniqueness of a vertex of P.

<u>Note 2</u>: If we change the ordering of the columns in the matrix

$$D = \begin{bmatrix} 1 & 1 & 1 & \cdots & 1 \\ c & a^{j_1} & a^{j_2} & \cdots & a^{j_n} \end{bmatrix},$$

the theorem still holds.

Following the same line of the proof, we can derive the following result.

**Theorem 9.2.2** *Given any $n$-dimensional simple polytope* $\mathrm{P} = \{x \in \mathbf{R}^n \mid a^{i\top} x \leq b_i$, *for all* $i \in I_m\}$ *with only non-redundant constraints and any vector* $c \in \mathbf{R}^n$, *there exists a unique vertex* $x^*$ *of* P *such that the matrix*

$$D^{-1} = \begin{bmatrix} 1 & 0 & \cdots & 0 & 0 \\ c & a^{j_1} & a^{j_2} & \cdots & a^{j_n} \end{bmatrix}^{-1}$$

*exists and is lexico positive, where* $x^*$ *is the solution of the* $n$ *linear equations* $a^{j_t \top} x = b_{j_t}$, $t \in I_n$, *and* $\{j_1, \cdots, j_n\}$ *is a subset of* $I_m$ *with cardinality* $n$.

<u>Note 3</u>: If we change the ordering of the columns in the above matrix $D$, the theorem still holds.

In the previous results we assumed that the polytope P is $n$-dimensional and simple and that none of the constraints $a^{i\top}x \le b_i$ is redundant. Similarly, we can extend Theorem 9.2.1 to a lower-dimensional polytope in $\mathbf{R}^n$.

**Theorem 9.2.3** *Let a polytope* P *be defined by* $\mathrm{P} = \{x \in \mathbf{R}^n \mid a^{i\top}x \le b_i,\ i \in I_m \text{ and } c^{j\top}x = d_j,\ j \in I_{m^1}\}$ *where* $0 \le m^1 \le n$. *Assume that* P *is simple and* $(n - m^1)$*-dimensional and that no constraints are redundant. For any* $c \in \mathbf{R}^n$, *there exists a unique vertex* $x^*$ *of* P *such that the following matrix*

$$D^{-1} = \begin{bmatrix} 1 & 0 & 0 & \cdots & 0 & 0 & \cdots & 0 \\ c & a^{j_1} & a^{j_2} & \cdots & a^{j_{n-m^1}} & c^1 & \cdots & c^{m^1} \end{bmatrix}^{-1}$$

*exists and the first* $n-m^1+1$ *rows of this matrix are all lexico positive, where* $x^*$ *is the solution of the* $n$ *linear equations* $c^{j\top}x = d_j$, $j \in I_{m^1}$, $a^{j_t\top}x = b_{j_t}$, $t \in I_{n-m^1}$, *and* $\{j_1, \cdots, j_{n-m^1}\}$ *is a subset of* $I_m$ *with cardinality* $n - m^1$.

<u>Note 4</u>: If we change the ordering of the columns in the above matrix $D$, the theorem still holds but we should say that all the rows of $D^{-1}$ corresponding to the columns $(1, c^\top)^\top$ and $(0, a^{j_i\top})^\top$ for all $i \in I_{n-m^1}$ are lexico positive.

We discuss two applications of Theorem 9.2.3. In these applications we can easily compute the matrix $D^{-1}$. In fact we will give the formula of $D^{-1}$. First, let us define the function bi : $\mathbf{R} \mapsto \{1, -1\}$ by

$$\mathrm{bi}(x) = \begin{cases} 1, & \text{if } x \ge 0, \\ -1, & \text{if } x < 0. \end{cases}$$

Wright [1981] proved the following algebraic result on the cube which is used to resolve the degeneracy problem for his octahedral or $2^n$-ray simplicial algorithm.

**Corollary 9.2.4** *Given any vector* $c \in \mathbf{R}^n$, *there exists a sign vector* $s \in \mathbf{R}^n$, *defined by*

$$s_i = \begin{cases} 1, & \text{if } c_i \le 0, \\ -1, & \text{if } c_i > 0, \end{cases}$$

*such that the matrix*

$$\begin{bmatrix} 1 & 1 & 1 & \cdots & 1 \\ c & s_1e(1) & s_2e(2) & \cdots & s_ne(n) \end{bmatrix}^{-1}$$

*exists and is lexico positive.*

Let $W^n$ denote the $n$-dimensional cube $\{x \in \mathbf{R}^n \mid x_i \leq q_i$ and $-x_i \leq -p_i, i \in I_n\}$ with $-\infty < p_i < q_i < +\infty$ for all $i \in I_n$. We can rewrite $W^n$ as

$$W^n = \{x \in \mathbf{R}^n \mid a^{i\top} x \leq b_i,\ i \in I_{2n}\},$$

where $a^i = e(i)$, $a^{n+i} = -e(i)$, $b_i = q_i$ and $b_{n+i} = -p_i$ for $i \in I_n$. Applying Theorem 9.2.1 to this polytope $W^n$, we immediately get the above corollary.

When Theorem 9.2.3 is applied to the adaptive simplicial algorithm in Chapter 7, the polytope $\mathrm{P}(t)$ is defined as follows. Let $v$ be a positive vector in $\mathbf{R}^n$ the sum of whose components is equal to one. We rearrange the components of $v$ in the decreasing order to obtain the vector $u$. For $t \in (0, 1/2]$, let

$$p_i(t) = u_i t^{i-1} / \sum_{j \in I_n} u_j t^{j-1},\ i \in I_n.$$

It is clear that $p_1(t) \geq p_2(t) \geq \cdots \geq p_n(t)$ for $t \in (0, 1/2]$. For each $t \in (0, 1/2]$, the polytope $\mathrm{P}(t)$ is defined by

$$\begin{aligned} \mathrm{P}(t) \quad = \quad & \{x \in \mathbf{R}^n \mid \textstyle\sum_{i \in I_n} x_i = 1, \\ & \textstyle\sum_{j \in J} x_j \leq \sum_{h=1}^{|J|} p_h(t) \ \text{ for any proper subset } J \text{ of } I_n \ \}. \end{aligned}$$

In this case, we have the following nice characterization of Theorem 9.2.3. This result can be viewed as the dual of Lemma 7.5.5.

**Corollary 9.2.5** *For the polytope* $\mathrm{P}(t)$ *defined above, and any* $c \in \mathbf{R}^n$, *the vertex* $x^*$ *of Theorem 9.2.3 is the unique solution of the* $n$ *linear equations* $\sum_{h \in I_n} x_i = 1$, $\sum_{h \in K_j} x_h = \sum_{h=1}^{j} p_h(t)$, $j \in I_{n-1}$, *where* $K_j = \{i_1, \cdots, i_j\}$ *for each* $j \in I_n$, *and* $(i_1, i_2, \cdots, i_n)$ *is the permutation of* $(1, 2, \cdots, n)$ *such that* $c_{i_1} \leq c_{i_2} \leq \cdots \leq c_{i_n}$, *and* $c_{i_s} = c_{i_t}$ *with* $s < t$ *implies* $i_s < i_t$.

This is clear because

$$D^{-1} = [l_{j,i}] = \begin{bmatrix} 1 & 0 & \cdots & 0 & 0 \\ c & \sum_{h \in K_1} e(h) & \cdots & \sum_{h \in K_{n-1}} e(h) & \sum_{h \in K_n} e(h) \end{bmatrix}^{-1}$$

exists and the first $n$ rows of this matrix are all lexico positive, where $l_{j,i}$ is given by

$$\begin{aligned} & l_{1,1} = 1,\ l_{1,j} = 0,\ j = 2, \cdots, n+1, \\ & l_{j,1} = c_{i_j} - c_{i_{j-1}},\ j = 2, \cdots, n, \\ & l_{j,i_j+1} = -1,\ l_{j,i_{j-1}+1} = 1,\ l_{j,t} = 0,\ j = 2, \cdots, n, t \neq i_j + 1,\ t \neq i_{j-1} + 1, \\ & l_{n+1,1} = -c_{i_n},\ l_{n+1,i_n+1} = 1,\ l_{n+1,t} = 0,\ t \neq 1,\ t \neq i_n + 1. \end{aligned}$$

We remark that the above result also holds for a class of polytopes, called base polytopes of sub-modular systems (see e.g., Fujishige [1991]).

In fact the above polytope P(t) is a special kind of base polytope. Denote by $2^{I_n}$ the collection of all subsets of $I_n$. A function $f : 2^{I_n} \mapsto \mathbf{R}$ is called strictly sub-modular if for any $X, Y \in 2^{I_n}$ with $X \not\subseteq Y$ and $Y \not\subseteq X$ we have

$$f(X) + f(Y) > f(X \cap Y) + f(X \cup Y).$$

We assume $f(\emptyset) = 0$. Define

$$\mathrm{B}(f) = \{x \in \mathbf{R}^n \mid \sum_{h \in X} x_h \le f(X),\ X \in 2^{I_n},\ X \ne I_n,\ \sum_{h \in I_n} x_h = f(I_n)\}.$$

$\mathrm{B}(f)$ is called the base polyhedron associated with $(2^{I_n}, f)$. It can be shown that $\mathrm{B}(f)$ is a simple $(n-1)$-dimensional polytope with non-redundant constraints.

Let us now consider another class of interesting polytopes. Denote by $3^{I_n}$ the set of all the ordered pairs of disjoint subsets of $I_n$, i.e., $3^{I_n} = \{(X,Y) \mid X, Y \subseteq I_n, X \cap Y = \emptyset\}$. For any $(X_1, Y_1)$, $(X_2, Y_2) \in 3^{I_n}$ we write $(X_1, Y_1) \sqsubseteq (X_2, Y_2)$ if and only if $X_1 \subseteq X_2$ and $Y_1 \subseteq Y_2$. A function $f : 3^{I_n} \mapsto \mathbf{R}$ is called strictly bisubmodular if it satisfies

$$f(X_1,Y_1) + f(X_2,Y_2) > f(X_1 \cap X_2, Y_1 \cap Y_2) + f((X_1 \cup X_2) \setminus (Y_1 \cup Y_2), (Y_1 \cup Y_2) \setminus (X_1 \cup X_2)),$$

for any $(X_1, Y_1)$, $(X_2, Y_2) \in 3^{I_n}$ with $(X_1, Y_1) \not\sqsubseteq (X_2, Y_2)$ and $(X_2, Y_2) \not\sqsubseteq (X_1, Y_1)$. It is assumed that $f(\emptyset, \emptyset) = 0$. Define

$$\mathrm{P}_*(f) = \{x \in \mathbf{R}^n \mid \sum_{h \in X} x_h - \sum_{h \in Y} x_h \le f(X,Y),\ (X,Y) \in 3^{I_n}\}.$$

The set $\mathrm{P}_*(f)$ is called the bisubmodular polyhedron associated with $(3^{I_n}, f)$. It can be shown that $\mathrm{P}_*(f)$ is a simple full-dimensional polytope with non-redundant constraints (see Ando and Fujishige [1996]). Now we have the following nice characterization of Theorem 9.2.3.

**Corollary 9.2.6** *For the polytope $\mathrm{P}_*(f)$ defined above and any $c \in \mathbf{R}^n$, let $(i_1, i_2, \cdots, i_n)$ be the permutation of $(1, 2, \cdots, n)$ such that $|c_{i_1}| \ge |c_{i_2}| \ge \cdots \ge |c_{i_n}|$ satisfying the following conditions:*
*(a) $c_{i_s} = c_{i_t} \le 0$ and $s < t$ imply $i_s < i_t$;*
*(b) $c_{i_s} = c_{i_t} > 0$ and $s < t$ imply $i_s > i_t$;*
*(c) $c_{i_s} = -c_{i_t} > 0$ implies $s > t$.*
*Define $K_j = X_j \cap Y_j = \{i_1, \cdots, i_j\}$ for each $j \in I_n$, where*

$$X_j = \{i_l \mid \mathrm{bi}(-c_{i_l}) = 1 \ \ \textit{for some } 1 \le l \le j\},$$

$$Y_j = \{i_l \mid \mathrm{bi}(-c_{i_l}) = -1 \ \ \textit{for some } 1 \le l \le j\}.$$

*Define $A(K_j) = \sum_{h \in X_j} e(h) - \sum_{h \in Y_j} e(h)$. Then the vertex $x^*$ of Theorem 9.2.3 is the unique solution of the $n$ linear equations $\sum_{h \in X_j} x_h - \sum_{h \in Y_j} x_h = f(X_j, Y_j)$ for all $j \in I_n$.*

This can be seen since the following matrix

$$D^{-1} = [l_{j,i}] = \begin{bmatrix} 1 & 0 & \cdots & 0 & 0 \\ c & \sum_{h\in K_1} e(h) & \cdots & \sum_{h\in K_{n-1}} e(h) & \sum_{h\in K_n} e(h) \end{bmatrix}^{-1}$$

exists and is lexico positive, where $l_{j,i}$ is given by

$$\begin{aligned}
&l_{1,1} = 1,\ l_{1,j} = 0,\ j = 2, \cdots, n+1, \\
&l_{j,1} = \mathrm{bi}(-c_{i_j})c_{i_j} - \mathrm{bi}(-c_{i_{j-1}})c_{i_{j-1}},\ j = 2, \cdots, n, \\
&l_{j,i_j+1} = -\mathrm{bi}(-c_{i_j}),\ l_{j,i_{j-1}+1} = \mathrm{bi}(-c_{i_{j-1}}),\ l_{j,t} = 0, \\
&\qquad j = 2, \cdots, n, t \neq i_j + 1,\ t \neq i_{j-1} + 1, \\
&l_{n+1,1} = -\mathrm{bi}(-c_{i_n})c_{i_n},\ l_{n+1,i_n+1} = \mathrm{bi}(-c_{i_n}), \\
&l_{n+1,t} = 0,\ t \neq 1,\ t \neq i_n + 1.
\end{aligned}$$

## 9.3. An algorithm for finding zero points

Let $f : \mathrm{P} \mapsto \mathbf{R}^n$ be a continuous function, where P is a polytope as described in Theorem 9.2.1. Our objective is to find a point $x^* \in \mathrm{P}$ so that $f(x^*) = \mathbf{0}$. Such a point $x^*$ is called a zero point. We will also provide a sufficient condition for the existence of zero points.

Let $v$ be any point of P which is usually an initial guess of a zero point. The point $v$ will be the starting point of the algorithm. For any face $F(I)$ of P with $I \in \mathcal{I}$ which does not contain $v$ we define $vF(I)$ to be the convex hull of $v \cup F(I)$, i.e.,

$$vF(I) = \{x \mid x = \alpha v + (1-\alpha)z \text{ for some } z \in F(I) \text{ and } \alpha \in [0,1]\}.$$

Let $t = n - |I| + 1$. Then we have $dim(vF(I)) = dim(F(I)) + 1 = t$. Figure 9.1 shows two examples of sets $F(I)$ and $vF(I)$ for two different starting points, where P is defined by four constraints. Let $\mathcal{T}$ be a triangulation of P such that each set $vF(I)$ is subdivided into $t$-simplices. For example we can take the $V$-triangulation from Talman and Yamamoto [1989] which will be introduced in Section 5. Since $\mathcal{T}$ is finite and P is compact, it can be shown that every facet $\tau$ of an $(n-|I|+1)$-simplex $\sigma$ in $vF(I)$ either lies in the boundary of $vF(I)$ and is only a facet of $\sigma$ or is a facet of exactly one other $(n-|I|+1)$-simplex in $vF(I)$. Let $\bar{f}$ denote the P.L. approximation of $f$ with respect to $\mathcal{T}$.

Let $\tau(x^1, \cdots, x^t)$ be a $(t-1)$-simplex in $vF(I)$ where $I \in \mathcal{I}$ with $I = \{i_{t+1}, \cdots, i_{n+1}\}$, $t = n - |I| + 1$. The $(n+1) \times (n+1)$ matrix

$$A_{\tau,I} = \begin{bmatrix} 1 & \cdots & 1 & 1 & \cdots & 1 \\ \bar{f}(x^1) & \cdots & \bar{f}(x^t) & a^{i_{t+1}} & \cdots & a^{i_{n+1}} \end{bmatrix}$$

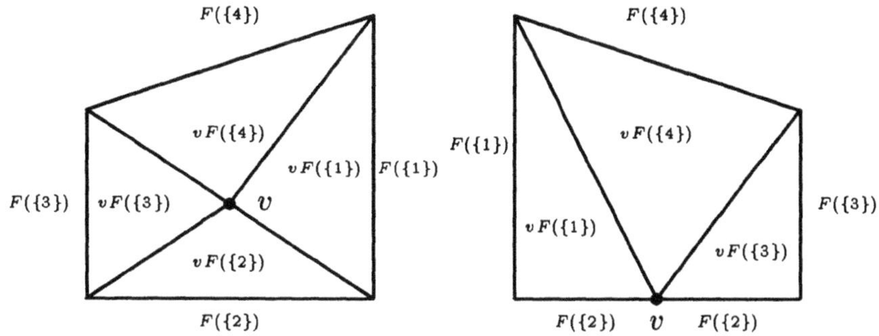


*Figure 9.1.* Sets $F(I)$ and $vF(I)$ of the polytope

is called the *label matrix* of $\tau$ with respect to $I$. The simplex $\tau$ is said to be $I$*-complete* if $A^{-1}_{\tau,I}$ exists and is lexico positive. In particular, an $n$-simplex $\sigma(x^1, \cdots, x^{n+1})$ is called $\emptyset$*-complete* if

$$A^{-1}_{\tau,\emptyset} = \begin{bmatrix} 1 & 1 & \cdots & 1 \\ f(x^1) & f(x^2) & \cdots & f(x^{n+1}) \end{bmatrix}^{-1}$$

exists and is lexico positive.

**Lemma 9.3.1** *Let $x^1 = v$ and $\tau = \{x^1\}$. Then there exists a unique set $I \in \mathcal{I}$ with $|I| = n$ such that $\tau$ is a 0-simplex in $vF(I)$ and $\tau$ is $I$-complete.*

Proof: It is a straightforward result of Theorem 9.2.1. □

**Lemma 9.3.2** *Let $\sigma$ be a $t$-simplex in $vF(I)$ where $I \in \mathcal{I}$ and $t = n - |I| + 1$. If $\sigma$ has an $I$-complete facet $\tau$, then exactly one of the following two cases occurs:*
*(i) The simplex $\sigma$ is an $\bar{I}$-complete simplex in $vF(\bar{I})$ where $\bar{I} = I \setminus \{l\}$ for precisely one index $l \in I$, and no facet of $\sigma$ other than $\tau$ is $I$-complete;*
*(ii) The simplex $\sigma$ has exactly one other $I$-complete facet $\bar{\tau}$.*

Proof: Let $x^+$ be the vertex of $\sigma$ opposite to $\tau$, and let $y = A^{-1}_{\tau,I}(1, f(x^+)^\top)^\top$. Let $K = \{i \in I_{n+1} \mid y_i > 0\}$. We first prove $|K| > 0$. Since $A_{\tau,I}A^{-1}_{\tau,I} = \mathbf{I}$, it holds that $\sum_{i=1}^{n+1} y_i = 1$. Clearly, $K$ is not empty. Consider the ratio vectors $(1/y_j)(A^{-1}_{\tau,I})_{j\cdot}$ for all $j \in K$. Choose $k \in K$ such that the $k$-th ratio vector is the minimum in the lexico order over all such ratio vectors. Since $A^{-1}_{\tau,I}$ is regular, $k$ is uniquely determined. Now, we consider the following two cases (1) and (2).

(1) If $k \in I_{n+1} \setminus I_t$, then let $l = i_k$ and $\bar{I} = I \setminus \{l\}$. Clearly, $\bar{I} \in \mathcal{I}$ and $\sigma$ is in $vF(\bar{I})$. Let $B$ be the matrix obtained from $A_{\tau,I}$ by replacing its $k$-th column by $(1, f(x^+)^\top)^\top$. It follows from Lemma 8.3.5 that $B^{-1}$ exists and

is lexico positive. By reordering the columns of $B$ we get $A_{\sigma,\bar{I}}$ whose inverse exists and is lexico positive by Note 2. So $\sigma$ is $\bar{I}$-complete.

(2) If $k \in I_t$, then let $\bar{\tau}$ be the facet of $\sigma$ opposite to the vertex $x^k$. Using Lemma 8.3.5, it follows from the choice of $k$ that $A_{\bar{\tau},I}^{-1}$ exists and is lexico positive. Hence $\bar{\tau}$ is an $I$-complete $(t-1)$-simplex in $vF(I)$.

It follows immediately from Lemma 8.3.5 that if any column other than the $k$-th column is replaced, then the inverse of the new matrix is not lexico positive. □

**Lemma 9.3.3** *Let $\sigma$ be an $I$-complete $(t-1)$-simplex in $vF(I)$ where $I \in \mathcal{I}$ and $t = n - |I| + 1$. If $\sigma$ is in $vF(\bar{I})$ where $\bar{I} = I \cup \{l\} \in \mathcal{I}$ for some $l \in I_m \setminus I$, then exactly one of the following two cases occurs:*
*(i) There exists a unique set $J \in \mathcal{I}$ with $|I| = |J|$ so that $\sigma$ is in $vF(J)$ and is $J$-complete;*
*(ii) There exists exactly one facet $\tau$ of $\sigma$ which is in $vF(\bar{I})$ and is $\bar{I}$-complete.*

Proof: Let $x = (1, a^{l\top})^\top$ and $y = A_{\sigma,I}^{-1}x$. Let $K = \{i \in I_{n+1} \mid y_i > 0\}$. Choose $k$ as in the proof of Lemma 9.3.2.

(1) If $k \in I_{n+1} \setminus I_t$, then let $p = i_k$ and $J = I \cup \{l\} \setminus \{p\}$. Clearly, $J \in \mathcal{I}$, $|J| = |I|$ and $\sigma$ is in $vF(J)$. Let $B$ be the matrix obtained from $A_{\sigma,I}$ by replacing its $k$-th column by $x$. It follows from Lemma 8.3.5 that $B^{-1}$ exists and is lexico positive. It is clear that $A_{\sigma,J} = B$. So $\sigma$ is $J$-complete.

(2) If $k \in I_t$, then let $\tau$ be the facet of $\sigma$ opposite to the vertex $x^k$. Clearly, $\tau$ is in $vF(\bar{I})$. Let $B$ be the matrix obtained from $A_{\sigma,I}$ by replacing its $k$-th column by $x$. It follows from Lemma 8.3.5 that $B^{-1}$ exists and is lexico positive. By reordering the columns of $B$ we get $A_{\tau,\bar{I}}$ whose inverse exists and is lexico positive by Note 2. So $\tau$ is $\bar{I}$-complete.

Again it follows from Lemma 8.3.5 that if any other column is replaced, then the inverse of the new matrix is not lexico positive. □

In order to guarantee the existence of a zero point, we impose the following condition on $f$:

**Condition 1** The function $f : \mathrm{P} \mapsto \mathbf{R}^n$ is a continuous function such that $a^{i\top} f(x) \geq 0$ for each $i \in I_m$ and each $x \in F(\{i\})$.

**Lemma 9.3.4** *Suppose that $f$ satisfies Condition 1. Let $\tau$ be an $I$-complete $(t-1)$-simplex where $I \in \mathcal{I}$ and $t = n - |I| + 1$. If $I = \emptyset$ or $\tau$ lies in $F(I)$, then $\tau$ contains a zero point of $\bar{f}$.*

Proof: Let $\beta$ be the first column of $A_{\tau,I}^{-1}$. It is clear that $\sum_{i=1}^{n+1} \beta_i = 1$ and $\beta_i \geq 0$ for all $i \in I_{n+1}$. If $I = \emptyset$, let $\bar{x} = \sum_{i=1}^{n+1} \beta_i x^i$. It follows from $A_{\tau,\emptyset}\beta = (1, 0^{n\top})^\top$ that $\bar{f}(\bar{x}) = \sum_{i=1}^{n+1} \beta_i f(x^i) = \mathbf{0}$. If $\tau$ lies in $F(I)$, we first prove that $\sum_{i \in I_t} \beta_i > 0$. Suppose on the contrary that $\beta_i = 0$ for all $i \in I_t$. This implies that $\sum_{j=t+1}^{n+1} \beta_j = 1$. Furthermore we have that $\sum_{j=t+1}^{n+1} \beta_i a^{i_j} = \mathbf{0}$ from $\sum_{j=1}^{t} \beta_j f(x^j) + \sum_{j=t+1}^{n+1} \beta_j a^{i_j} = \mathbf{0}$. But this contradicts the fact that the vectors $a^{i_{t+1}}, \cdots, a^{i_{n+1}}$ are linearly independent. Let $\lambda_i = \beta_i / \sum_{j \in I_t} \beta_j$ for all $i \in I_{n+1}$. Clearly, $\lambda_i \geq 0$ for all $i \in I_{n+1}$ and $\sum_{i \in I_t} \lambda_i = 1$. Let $\bar{x} = \sum_{i \in I_t} \lambda_i x^i$. Then $\bar{f}(\bar{x}) = \sum_{i \in I_t} \lambda_i f(x^i)$. It follows from $A_{\tau,I}\beta = (1, 0^{n\top})^\top$ that

$$\bar{f}(\bar{x}) = -\sum_{j=t+1}^{n+1} \lambda_j a^{i_j}. \tag{9.3.11}$$

Since $\tau$ lies in $F(I)$, it holds that $a^{i\top} f(x^j) \geq 0$ for all $i \in I$ and all $j \in I_t$ by Condition 1. This also implies that

$$a^{j\top} \bar{f}(\bar{x}) \geq 0, \; \forall j \in I. \tag{9.3.12}$$

Note that $I = \{i_{t+1}, \cdots, i_{n+1}\}$. It follows from (9.3.11) and (9.3.12) that $\bar{f}(\bar{x})$ lies in both a cone and its polar cone. This implies that $\bar{f}(\bar{x}) = \mathbf{0}$. □

The following result implies that a zero point of $\bar{f}$ is an approximate zero point of $f$.

**Lemma 9.3.5** *Let $\gamma = \sup\{diam f(\sigma) \mid \sigma \in \mathcal{T}\}$. Let $\bar{x}$ be a zero point of $\bar{f}$. Then $\bar{x}$ is an approximate zero point of $f$, i.e., $||f(\bar{x})|| \leq \gamma$.*

Proof: Let $x^1, \cdots, x^t$ be the vertices of a $(t-1)$-simplex of $\mathcal{T}$ containing $\bar{x}$. Then $\bar{f}(\bar{x}) = \sum_{i=1}^{t} \lambda_i f(x^i)$ and $\bar{x} = \sum_{i=1}^{t} \lambda_i x^i$ where $\sum_{i=1}^{t} \lambda_i = 1$ and $\lambda_i \geq 0$ for all $i \in I_t$. Moreover it holds that

$$\begin{aligned} ||f(\bar{x})|| &= ||\textstyle\sum_{i=1}^{t} \lambda_i (f(\bar{x}) - f(x^i))|| \\ &\leq \textstyle\sum_{i=1}^{t} \lambda_i ||f(\bar{x}) - f(x^i)|| \\ &\leq \gamma. \end{aligned}$$

□

We will construct a graph $G = (V, A)$ where $V$ denotes the set of nodes and $A$ denotes the set of edges. Each $I$-complete $(n - |I|)$-simplex is a *node* in $V$. An $I$-complete $(n-|I|)$-simplex $\tau^1$ in $vF(I)$ and an $J$-complete $(n - |J|)$-simplex $\tau^2$ in $vF(J)$ are said to be *adjacent* complete simplices if $I = J = L$ and $\tau^1$ and $\tau^2$ are both facets of an $(n - |L| + 1)$-simplex $\sigma$ in $vF(L)$ or $\tau^1$ is a facet of $\tau^2$ and $\tau^2$ is an $(n - |I| + 1)$-simplex in $vF(I)$, or $\tau^2$ is a facet of $\tau^1$ and $\tau^1$ is an $(n-|J|+1)$-simplex in $vF(J)$. Two adjacent complete simplices $\tau^1$ and $\tau^2$ are connected by an edge $e = \{\tau^1, \tau^2\} \in A$. The degree of a node $\tau$ in $G$ is defined to be the number of nodes connected with it, denoted by $deg(\tau)$. A path in $G$ from node $\tau^0$ to node $\tau^l$ is defined as a sequence of the form $(\tau^0, e_1, \tau^1, \cdots, e_l, \tau^l)$ where $\tau^0, \tau^1, \cdots, \tau^l$ are nodes and $e_1, \cdots, e_l$ are edges, such that $e_i = \{\tau^{i-1}, \tau^i\}$ for $i \in I_l$. A path is simple if all its nodes and edges are different. Summarizing the above results, we have the following theorem.

**Theorem 9.3.6** *Suppose that $f : \mathrm{P} \mapsto \mathbf{R}^n$ satisfies Condition 1. Let $\mathcal{T}$ be a triangulation of* P *such that every set $vF(I)$ with $I \in \mathcal{I}$ is subdivided into $(n-|I|+1)$-simplices. Starting with $\tau^0 = \{v\}$, there is a finite sequence of adjacent $J$-complete simplices for varying $J \in \mathcal{I}$ which leads to an $I$-complete $(n-|I|)$-simplex $\tau$ for which $I = \emptyset$ or $\tau \subset F(I)$. In either case, there exists $y \in \tau$ such that $\bar{f}(y) = \mathbf{0}$.*

Proof: By Lemma 9.3.1, $\{v\}$ is an $I$-complete 0-simplex in $vF(I)$ for some unique set $I \in \mathcal{I}$ with $|I| = n$. Since $\{v\}$ lies in the boundary of $vF(I)$, there exists a unique 1-simplex $\sigma$ in $vF(I)$ having $\tau$ as its facet. By Lemma 9.3.2 either $\sigma$ is an $\bar{I}$-complete simplex in $vF(\bar{I})$ where $\bar{I} = I\setminus\{l\}$ for some unique $l \in I$ and no facet of $\sigma$ other than $\{v\}$ is $I$-complete, or $\sigma$ has exactly one other $I$-complete facet $\bar{\tau}$. Hence there exists a unique adjacent complete simplex to $\{v\}$. That is, $deg(\{v\}) = 1$.

Suppose that $\tau$ is an $I$-complete $(n-|I|)$-simplex lying on $F(I)$. This implies that $\tau$ lies in the boundary of $vF(I)$. Then there exists a unique $(n-|I|+1)$-simplex $\sigma$ in $vF(I)$ having $\tau$ as its facet. By Lemma 9.3.2 either $\sigma$ is an $\bar{I}$-complete simplex in $vF(\bar{I})$ where $\bar{I} = I\setminus\{l\}$ for some unique $l \in I$ and no facet of $\sigma$ other than $\tau$ is $I$-complete, or $\sigma$ has exactly one other $I$-complete facet $\bar{\tau}$. Hence there exists a unique adjacent complete simplex to $\tau$. Again $deg(\tau) = 1$.

Suppose that $\tau$ is an $\emptyset$-complete $n$-simplex. Then $\tau$ lies in $vF(\{l\})$ for a unique element $l \in I_m$. Moreover, $\tau$ has exactly one facet $\sigma$ which is an $\{l\}$-complete simplex in $vF(\{l\})$. Clearly, $\sigma$ is a unique adjacent complete simplex to $\tau$. Again $deg(\tau) = 1$.

In all other cases, we will prove that if $\tau$ is an $I$-complete $(n-|I|)$-simplex for some $I \in \mathcal{I}$, $\tau$ has exactly two adjacent complete simplices. There are two possibilities: either $\tau$ lies in the interior of $vF(I)$ or $\tau$ lies in the boundary of $vF(I)$. If $\tau$ lies in the interior of $vF(I)$, then $\tau$ is a facet of exactly two $(n-|I|+1)$-simplices in $vF(I)$. It follows from Lemma 9.3.2 that $\tau$ is adjacent to exactly two complete simplices. If $\tau$ lies in the boundary of $vF(I)$, then there exists exactly one $(n-|I|+1)$-simplex $\sigma$ in $vF(I)$ having $\tau$ as its facet. By Lemma 9.3.2 either $\sigma$ is an $\bar{I}$-complete simplex in $vF(\bar{I})$ for some unique $\bar{I} \in \mathcal{I}$ with $|\bar{I}| = |I|-1$ and has no other $I$-complete facets, or $\sigma$ has exactly one other $I$-complete facet. This yields one adjacent complete simplex to $\tau$. On the other hand, since $\tau$ lies in the boundary of $vF(I)$, $\tau$ lies in $vF(\tilde{I})$ for some unique set $\tilde{I} \in \mathcal{I}$ with $|\tilde{I}| = |I|+1$. By Lemma 9.3.3 either $\tau$ is $J$-complete for some unique set $J \in \mathcal{I}$ with $|J| = |I|$, or $\tau$ has exactly one $\tilde{I}$-complete facet. In the former case, $\tau$ lies in $vF(\tilde{I})$ and hence there exists exactly one simplex $\bar{\sigma}$ in $vF(\tilde{I})$ having $\tau$ as its facet. It follows again from Lemma 9.3.2 that there exists exactly one other complete simplex adjacent to $\tau$. This concludes that $\tau$ has exactly two adjacent complete simplices. In other words, we have $deg(\tau) = 2$.

As shown above, the degree of each node in the graph $G = (V, A)$ is at most two. Note that since the number of simplices in $P$ is finite, the number of nodes in $G$ must also be finite. Since $deg(\{v\}) = 1$, it is easy to see from Theorem 1.4.14 that there exists a finite simple path starting from $\{v\}$. The end node of this path must be a node $\tau$ of degree one. As shown above, $\tau$ must be an $I$-complete $(n - |I|)$-simplex for which $|I| = 0$ or $\tau \subset F(I)$. By Lemma 9.3.4 in either case there exists a point $y \in \tau$ such that $\bar{f}(y) = \mathbf{0}$. □

In this theorem it is shown that starting with $v$, the algorithm leads to an approximate zero point $y$ of $f$ (note that $y$ is a zero point of $\bar{f}$) within a finite number of steps. If the accuracy of this approximation is not good enough, the algorithm can restart from $y$ with a finer triangulation of P in the hope that a better approximation will be found within a smaller number of steps. Define the mesh size $\delta = \sup\{diam(\sigma) \mid \sigma \in \mathcal{T}\}$ for the triangulation $\mathcal{T}$ of P. Since P is compact and $f$ is continuous on P, the error $\gamma$ in Lemma 9.3.5 goes to zero as the mesh size $\delta$ of the triangulation $\mathcal{T}$ converges to zero. So a zero point of $f$ on P will be found in this way. This leads to the next result.

**Corollary 9.3.7** *Let $f : \mathrm{P} \mapsto \mathbf{R}^n$ satisfy Condition 1. Let $y^k$ be a zero point of $\bar{f}$ generated by the algorithm with respect to the mesh size $\delta^k$ of the triangulation $\mathcal{T}$ for $k = 1, 2, \cdots$. Suppose that $\delta^k$ converges to zero as $k$ goes to infinity. Then the sequence $\{y^k \mid k = 1, 2, \cdots\}$ must have a convergent subsequence and any cluster point of this subsequence is a zero point of $f$ on* P.

So we have established the following basic existence theorem.

**Theorem 9.3.8** *Given any $n$-dimensional simple polytope $\mathrm{P} = \{x \in \mathbf{R}^n \mid a^{i\top} x \leq b_i$, for all $i \in I_m\}$ with non-redundant constraints. Let $f : \mathrm{P} \mapsto \mathbf{R}^n$ satisfy Condition 1. Then there exists at least one $x^* \in \mathrm{P}$ such that $f(x^*) = \mathbf{0}$.*

This theorem implies the following well-known multivariate mean value theorem. See, for example, Istratescu [1981].

**Corollary 9.3.9** (**Multivariate Mean Value Theorem**) *Let $f$ be a continuous function from the $n$-dimensional cube $W^n$ as defined before into $\mathbf{R}^n$. Then there exists a point $x^* \in W^n$ such that $f(x^*) = \mathbf{0}$ if the following condition holds for any $x \in W^n$ :*
*(a) $f_i(x_1, \cdots, x_{i-1}, p_i, x_{i+1}, \cdots, x_n) \leq 0$;*
*(b) $f_i(x_1, \cdots, x_{i-1}, q_i, x_{i+1}, \cdots, x_n) \geq 0$, for all $i \in I_n$.*

It should be noted that in the above discussions, because of the use of Theorem 9.2.1, we need not make any non-degeneracy assumptions. In a similar way, we can also discuss the applications of these results to other simplicial algorithms. We leave it to the interested reader.

## 9.4. An algorithm for finding stationary points

In this section we present an algorithm for computing a stationary point of a continuous function $f : \mathrm{P} \mapsto \mathbf{R}^n$. Without loss of any generality, we consider an arbitrary $(n-1)$-dimensional polytope

$$\mathrm{P} = \{x \in \mathbf{R}^n \mid a^{i\top} x \le b_i, \text{ for all } i \in I_m \text{ and } c^\top x = d\}.$$

We assume that P is simple and no constraints are redundant. For each subset $I$ of $I_m$, define

$$F(I) = \{x \in \mathrm{P} \mid a^{i\top} x = b_i, \text{for all}\, i \in I\}.$$

Then $F(I)$ is a face of P unless it is empty. Note that $F(\emptyset) = \mathrm{P}$. Let

$$\mathcal{I} = \{I \subset I_m \mid F(I) \text{ is a nonempty face of } \mathrm{P}\}.$$

Under the above assumption that $dim(\mathrm{P}) = n-1$, P is a simple polytope, and the linear inequalities defining P are non-redundant, we have the following observations:

(i) for each face $F$ of P the set $I \in \mathcal{I}$ with $F = F(I)$ is unique and identical with the set $\{i \in I_m \mid a^{i\top} x = b_i, \text{ for all}\, x \in F\}$;

(ii) $dim(F(I)) = n - 1 - |I|$ for any $I \in \mathcal{I}$ and $F(I)$ is a vertex of P if $I \in \mathcal{I}$ with $|I| = n-1$;

(iii) for some $I \in \mathcal{I}$, $G$ is a facet of $F(I)$ if and only if $G = F(I \cup \{h\})$ for some $h \notin I$ with $I \cup \{h\} \in \mathcal{I}$;

(iv) for any $I \in \mathcal{I}$ all the vectors $a^j$ with $j \in I$ and $c$ are linearly independent. Moreover, if $I \in \mathcal{I}$, then $I \setminus \{k\} \in \mathcal{I}$ for any $k \in I$.

For each index set $I \in \mathcal{I}$, define the cone $F^*(I)$ by

$$F^*(I) = \{y \in \mathbf{R}^n \mid y = \sum_{h \in I} \mu_h a^h + \beta c,\ \mu_h \ge 0 \text{ for all } h \in I \text{ and } \beta \in \mathbf{R}\}.$$

Now we have the following simple but important observation.

**Lemma 9.4.1** *Let $f : \mathrm{P} \mapsto \mathbf{R}^n$ be a continuous function. Then $x^* \in \mathrm{P}$ is a stationary point of $f$ on P if and only if $x^* \in F(I)$ and $f(x^*) \in F^*(I)$ for some $I \in \mathcal{I}$.*

Let $v$ be any point of P which is usually an initial guess of a stationary point. The point $v$ will be the starting point of the algorithm. For any face $F(I)$ of P with $I \in \mathcal{I}$ which does not contain $v$ we define $vF(I)$ to be the convex hull of $v \cup F(I)$, i.e.,

$$vF(I) = \{x \mid x = \alpha v + (1-\alpha)z \text{ for some } z \in F(I) \text{ and } \alpha \in [0,1]\}.$$

Let $t = n - |I|$. Then $dim(vF(I)) = dim(F(I)) + 1 = t$. Let $\mathcal{T}$ be a triangulation of P such that each set $vF(I)$ is subdivided into $t$-simplices. The $V$-triangulation to be introduced in the next section will do this job. Again, let $\bar{f}$ be the P.L. approximation of $f$ with respect to $\mathcal{T}$.

Let $\tau(x^1, \cdots, x^t)$ be a $(t-1)$-simplex in $vF(I)$ where $I \in \mathcal{I}$ with $I = \{i_{t+1}, \cdots, i_{n+1}\}$, $t = n - |I|$. The $(n+1) \times (n+1)$ matrix

$$A_{\tau,I} = \begin{bmatrix} 1 & \cdots & 1 & 0 & \cdots & 0 & 0 \\ -f(x^1) & \cdots & -f(x^t) & a^{i_{t+1}} & \cdots & a^{i_{n+1}} & c \end{bmatrix}$$

is called the *label matrix* of $\tau$ with respect to $I$. The simplex $\tau$ is said to be *$I$-complete* if $A_{\tau,I}^{-1}$ exists and the first $n$ rows of this matrix are all lexico positive. In particular, an $(n-1)$-simplex $\sigma(x^1, \cdots, x^n)$ is called *$\emptyset$-complete* if

$$A_{\tau,\emptyset}^{-1} = \begin{bmatrix} 1 & 1 & \cdots & 1 & 0 \\ -f(x^1) & -f(x^2) & \cdots & -f(x^n) & c \end{bmatrix}^{-1}$$

exists and the first $n$ rows of this matrix are all lexico positive. The reader may have realized that the lexicographic operation system here is more general than that introduced in the previous section.

**Lemma 9.4.2** *Let $x^1 = v$ and $\tau = \{x^1\}$. Then there exists a unique set $I \in \mathcal{I}$ with $|I| = n - 1$ such that $\tau$ is a 0-simplex in $vF(I)$ and $\tau$ is $I$-complete.*

Proof: It is a straightforward result of Theorem 9.2.3. □

We point out that the following proofs are similar to those in the previous section but there are some subtle differences.

**Lemma 9.4.3** *Let $\sigma$ be a $t$-simplex in $vF(I)$ where $I \in \mathcal{I}$ and $t = n - |I|$. If $\sigma$ has an $I$-complete facet $\tau$, then exactly one of the following two cases occurs:*

*(i) The simplex $\sigma$ is an $\bar{I}$-complete simplex in $vF(\bar{I})$ where $\bar{I} = I \setminus \{l\}$ for precisely one index $l \in I$, and no facet of $\sigma$ other than $\tau$ is $I$-complete;*

*(ii) The simplex $\sigma$ has exactly one other $I$-complete facet $\bar{\tau}$.*

Proof: Let $x^+$ be the vertex of $\sigma$ opposite to $\tau$, and let $y = A_{\tau,I}^{-1}(1, -f(x^+)^\top)^\top$ Let $K = \{i \in I_n \mid y_i > 0\}$. We first prove $|K| > 0$. Since $A_{\tau,I}A_{\tau,I}^{-1} = \mathbf{I}$, it holds that $\sum_{i=1}^t y_i = 1$. Hence $K$ is not empty.

Consider the ratio vectors $(1/y_j)(A_{\tau,I}^{-1})_{j\cdot}$ for all $j \in K$. Choose $k \in K$ such that the $k$-th ratio vector is the minimum in the lexicographic order over all such ratio vectors. Since $A_{\tau,I}^{-1}$ is regular, $k$ is uniquely determined. Now, we consider the following two cases (1) and (2).

(1) If $k \in I_n \setminus I_t$, then let $l = i_k$ and $\bar{I} = I \setminus \{l\}$. Clearly, $\bar{I} \in \mathcal{I}$ and $\sigma$ is in $vF(\bar{I})$. Let $B$ be the matrix obtained from $A_{\tau,I}$ by replacing its $k$-th column by $(1, -f(x^+)^\top)^\top$. It follows from Lemma 8.3.5 that $B^{-1}$ exists and is lexico positive. By reordering the columns of $B$ we get $A_{\sigma,\bar{I}}$ whose inverse exists and is lexico positive by Note 2. So $\sigma$ is $\bar{I}$-complete.

(2) If $k \in I_t$, then let $\bar{\tau}$ be the facet of $\sigma$ opposite to the vertex $x^k$. Using Lemma 8.3.5, it follows from the choice of $k$ that $A_{\bar{\tau},I}^{-1}$ exists and is lexico positive. Hence $\bar{\tau}$ is an $I$-complete $(t-1)$-simplex in $vF(I)$.

It follows immediately from Lemma 8.3.5 that if any column other than the $k$-th column is replaced, then the inverse of the new matrix is not lexico positive. □

**Lemma 9.4.4** *Let $\sigma$ be an $I$-complete $(t-1)$-simplex in $vF(I)$ where $I \in \mathcal{I}$ and $t = n - |I|$. If $\sigma$ is in $vF(\bar{I})$ where $\bar{I} = I \cup \{l\} \in \mathcal{I}$ for some $l \in I_m \setminus I$, then exactly one of the following two cases occurs:*
*(i) There exists a unique set $J \in \mathcal{I}$ with $|I| = |J|$ so that $\sigma$ is in $vF(J)$ and is $J$-complete;*
*(ii) There exists exactly one facet $\tau$ of $\sigma$ which is in $vF(\bar{I})$ and is $\bar{I}$-complete.*

Proof: Let $x = (0, a^{l\top})^\top$ and $y = A_{\sigma,I}^{-1}x$. Let $K = \{i \in I_n \mid y_i > 0\}$. Notice that $y \neq 0^{n+1}$ and $A_{\sigma,I}y = x$. If there is an index $h \in I_t$ such that $y_h < 0$, there must exist another index $i \neq h \in I_t$ such that $y_i > 0$ since $\sum_{j \in I_t} y_j = 0$. In this case $K \neq \emptyset$. Suppose that $y_i = 0$ for all $i \in I_t$. This implies that $a^l = \sum_{h=t+1}^{n} y_h a^{i_h} + y_{n+1}c$. So, the vectors $a^l$, $c$, and $a^{i_h}$ for all $h = t+1, \cdots, n+1$, are linearly dependent. This is a contradiction. Therefore $K$ must be nonempty. Choose $k$ as in the proof of Lemma 9.3.2.

(1) If $k \in I_n \setminus I_t$, then let $p = i_k$ and $J = I \cup \{l\} \setminus \{p\}$. Clearly, $J \in \mathcal{I}$, $|J| = |I|$ and $\sigma$ is in $vF(J)$. Let $B$ be the matrix obtained from $A_{\sigma,I}$ by replacing its $k$-th column by $x$. It follows from Lemma 8.3.5 that $B^{-1}$ exists and is lexico positive. It is clear that $A_{\sigma,J} = B$. So $\sigma$ is $J$-complete.

(2) If $k \in I_t$, then let $\tau$ be the facet of $\sigma$ opposite to the vertex $x^k$. Clearly, $\tau$ is in $vF(\bar{I})$. Let $B$ be the matrix obtained from $A_{\sigma,I}$ by replacing its $k$-th column by $x$. It follows from Lemma 8.3.5 that $B^{-1}$ exists and is lexico positive. By reordering the columns of $B$ we get $A_{\tau,\bar{I}}$ whose inverse exists and is lexico positive by Note 2. So $\tau$ is $\bar{I}$-complete.

Again it follows from Lemma 8.3.5 that if any other column is replaced, then the inverse of the new matrix is not lexico positive. □

**Lemma 9.4.5** *Let $\tau$ be an $I$-complete $(t-1)$-simplex where $I \in \mathcal{I}$ and $t = n - |I|$. If $I = \emptyset$ or $\tau$ lies in $F(I)$, then $\tau$ contains a stationary point of $\bar{f}$.*

Proof: Let $\beta$ be the first column of $A_{\tau,I}^{-1}$. It is clear that $\sum_{i=1}^{t} \beta_i = 1$ and $\beta_i \geq 0$ for all $i \in I_n$. If $I = \emptyset$, let $\bar{x} = \sum_{i=1}^{n} \beta_i x^i$. It follows from $A_{\tau,\emptyset}\beta = (1, 0^{n\top})^\top$ that $\bar{f}(\bar{x}) = \sum_{i=1}^{n} \beta_i f(x^i) = c$. If $\tau$ lies in $F(I)$, we have that

$$\bar{f}(\bar{x}) = \sum_{i=1}^{t} f(x^i) = \sum_{h=t+1}^{n} \beta_h a^{i_h} + \beta_{n+1} c.$$

In both cases $\bar{f}(\bar{x}) \in F^*(I)$. It follows from Lemma 9.4.1 that $\bar{x}$ is a stationary point of $\bar{f}$. □

The following result implies that a stationary point of $\bar{f}$ is an approximate stationary point of $f$.

**Lemma 9.4.6** *Let $\gamma = \sup\{diam f(\sigma) \mid \sigma \in \mathcal{T}\}$. Let $x^*$ be a stationary point of $\bar{f}$, i.e., $x^* \in F(I)$ and $\bar{f}(x^*) \in F^*(I)$ for some $I \in \mathcal{I}$. Then $x^*$ is an approximate stationary point of $f$, i.e., there is $y \in F^*(I)$ with $||f(x^*) - y|| \leq \gamma$.*

Proof: Let $x^1, \cdots, x^t$ be the vertices of a $(t-1)$-simplex of $\mathcal{T}$ containing $x^*$. Then $\bar{f}(x^*) = \sum_{i=1}^{t} \lambda_i f(x^i)$ and $x^* = \sum_{i=1}^{t} \lambda_i x^i$ where $\sum_{i=1}^{t} \lambda_i = 1$ and $\lambda_i \geq 0$ for all $i \in I_t$. Let $y = \bar{f}(x^*)$. Clearly, $y \in F^*(I)$. Moreover it follows that

$$\begin{aligned} ||\bar{f}(x^*) - f(x^*)|| &= ||\textstyle\sum_{i=1}^{t} \lambda_i (f(x^i) - f(x^*))|| \\ &\leq \textstyle\sum_{i=1}^{t} \lambda_i ||f(x^i) - f(x^*)|| \\ &\leq \gamma. \end{aligned}$$

□

Construct a graph $G = (V, A)$ in the same way as in the previous section. We have the following theorem.

**Theorem 9.4.7** *Let $f : \mathrm{P} \mapsto \mathbf{R}^n$ be a continuous function. Let $\mathcal{T}$ be a triangulation of $\mathrm{P}$ such that every set $vF(I)$ with $I \in \mathcal{I}$ is subdivided into $(n - |I|)$-simplices. Starting with $\tau^0 = \{v\}$, there exists a finite sequence of adjacent $J$-complete simplices for varying $J \in \mathcal{I}$ which leads to an $I$-complete $(n - |I|)$-simplex $\tau$ for which $I = \emptyset$ or $\tau \subset F(I)$. In either case, there exists $y \in \tau$ such that $y$ is a stationary point of $\bar{f}$.*

Proof: The proof is the same as that in the previous section. □

In this theorem it is shown that starting with $v$, the algorithm leads to an approximate stationary point $y$ of $f$ (note that $y$ is a stationary point of

$\bar{f}$) within a finite number of steps. If the accuracy of this approximation is not good enough, the algorithm can restart from $y$ with a finer triangulation of P in the hope that a better approximation will be found within a smaller number of steps. Define the mesh size $\delta = \sup\{diam(\sigma) \mid \sigma \in \mathcal{T}\}$ for the triangulation $\mathcal{T}$ of P. Since P is compact and $f$ is continuous on P, the error $\gamma$ in Lemma 9.4.6 goes to zero as the mesh size $\delta$ of the triangulation $\mathcal{T}$ converges to zero. So a stationary point of $f$ on P will be found in this way. This leads to the next result.

**Corollary 9.4.8** *Let $f : \mathrm{P} \mapsto \mathbf{R}^n$ be a continuous function. Let $y^k$ be a stationary point of $\bar{f}$ generated by the algorithm with respect to the mesh size $\delta^k$ of the triangulation $\mathcal{T}$ for $k = 1, 2, \cdots$ . Suppose that $\delta^k$ converges to zero as $k$ goes to infinity. Then the sequence $\{y^k \mid k = 1, 2, \cdots\}$ must have a convergent subsequence and any cluster point of this subsequence is a stationary point of $f$ on* P.

Notice that the above algorithm is equally applicable to any upper semi-continuous point-to-set mapping with nonempty, convex and compact values.

## 9.5. The $V$-triangulation of polytopes

Now we introduce the $V$-triangulation of general polytopes which underlies the algorithms presented in the last two sections. This triangulation is a generalization of the $V$-triangulation of the cross product of unit simplices proposed in Doup and Talman [1987b]. Without loss of generality, we consider a full-dimensional polytope $\mathrm{P} = \{x \in \mathbf{R}^n \mid a^{i\top}x \leq b_i, i = 1, \cdots, m\}$. Let P satisfy all the conditions imposed in Sections 2 and 3. The triangulation of each set $vF(I)$, $v \notin F(I)$, is completely determined by projections of $v$ on the faces of $F(I)$ and on $F(I)$ itself. The projection of the starting point $v$ on a face $F(I)$ of P can be any point in $int(F(I))$ and is denoted by $v(I)$. Clearly, if $I \in \mathcal{I}$ consists of $n$ elements, then $F(I) = \{v(I)\}$ is a vertex of P. These projection points points are automatically generated in the course of the algorithm, but once a projection $v(I)$ has been chosen it is fixed during the rest course of the algorithm. For each $I \in \mathcal{I}$, for which $v \notin F(I)$, let $v(I(n))$ be a vertex of $F(I)$, i.e., $|I(n)| = n$ and $I \subseteq I(n) \in \mathcal{I}$. Let $(\gamma_1, \cdots, \gamma_{t-1}) = \gamma(I(n) \setminus I)$ be a permutation of the $t - 1 = n - |I|$ elements of the set $I(n) \setminus I$. Then the subset $vF(I, \gamma(I(n) \setminus I))$ of $vF(I)$ is defined to be the convex hull of $v$, $v(I(n))$, $v(I(n) \setminus \{\gamma_1\}$, $\cdots$, $v(I(n) \setminus \{\gamma_1, \cdots, \gamma_{t-2}\}$, and $v(I)$. In other words,

$$vF(I, \gamma(I(n) \setminus I)) = \{\, x \in \mathrm{P} \mid \begin{array}{l} x = v + \sum_{h=0}^{t-1} \beta_h q(h), \\ 0 \leq \beta_{t-1} \leq \cdots \beta_1 \leq \beta_0 \leq 1 \,\} \end{array}$$

where $q(0) = v(I(n)) - v$, and for $h \in I_{t-1}$,

$$q(h) = v(I \cup \{\gamma_{h+1}, \cdots, \gamma_{t-1}\}) - v(I \cup \{\gamma_h, \cdots, \gamma_{t-1}\}).$$

The above definition leads to the following lemma.

**Lemma 9.5.1** *Let $\mathcal{G}(I)$ be the collection of the sets $vF(I, \gamma(I(n) \setminus I))$ over all permutations $\gamma(I(n) \setminus I)$ and over all vertices $v(I(n))$ of $F(I)$. Then $\mathcal{G}(I)$ is a simplicial subdivision of the set $vF(I)$. Furthermore, let $\mathcal{G}$ be the collection of the sets $vF(\{i\}, \gamma(I(n) \setminus \{i\}))$ over all permutations $\gamma(I(n) \setminus \{i\})$ and over all vertices $v(I(n))$ of $F(\{i\})$ and over all $i \in I_m$. Then $\mathcal{G}$ is a simplicial subdivision of the polytope* P.

The above simplicial subdivision of the polytope P is called $V$-triangulation. In order to get a triangulation of P with arbitrary small mesh size, we need to introduce the following definition.

**Definition 9.5.2** *Let $d \in \mathbb{N}$ be given. The set $G^d(I, \gamma(I(n) \setminus I))$ is the collection of $t$-simplices $\sigma(y^1, \pi)$ in $vF(I, \gamma(I(n) \setminus I))$ with vertices $y^1, \cdots, y^{t+1}$ such that*
*(a) $y^1 = v + \sum_{h=0}^{t-1} a(h)q(h)/d$ where $a(0), \cdots, a(t-1)$ are integers such that $0 \leq a(t-1) \leq \cdots \leq a(0) \leq d-1$;*
*(b) $\pi = (\pi(1), \cdots, \pi(t))$ is a permutation of $(0, \cdots, t-1)$ such that $k < l$ if for some $h \in \{0, \cdots, t-2\}$, $\pi(k) = h$, $\pi(l) = h+1$, and $a(h) = a(h+1)$;*
*(c) $y^{i+1} = y^i + q(\pi(i))/d$, $i \in I_t$.*

This leads to the following theorem.

**Theorem 9.5.3** *The set $G^d(I, \gamma(I(n) \setminus I))$ is a simplicial subdivision of the set $vF(I, \gamma(I(n) \setminus I))$ with refinement factor $d$. Let $V(d)$ be the collection of the sets $G^d(\{i\}, \gamma(I(n) \setminus \{i\}))$ over all permutations $\gamma(I(n) \setminus \{i\})$ and over all vertices $v(I(n))$ of $F(\{i\})$ and over all $i \in I_m$. Then $V(d)$ is a simplicial subdivision of the polytope* P *having mesh size less than or equal to* diam(P)/$d$.

For $n = 2$ and $d = 1$ or 2, the $V(d)$-triangulation of a polytope P with four constraints is illustrated in Figures 9.2 and 9.3.

The following result gives a sufficient and necessary condition for checking whether a simplex $\tau$ lies in the boundary of the set $vF(I, \gamma(I(n) \setminus I))$ or not.

**Lemma 9.5.4** *The facet $\tau$ opposite to the vertex $y^k$ of a $t$-simplex $\sigma(y^1, \pi)$ in the $t$-dimensional set $vF(I, \gamma(I(n) \setminus I))$ lies in the boundary of this set if and only if one of the following cases occurs:*
*(i) $k = 1$, $\pi(1) = 0$, and $a(0) = d-1$;*
*(ii) $1 < k < t+1$, $\pi(k) = h+1$, $\pi(k-1) = h$, and $a(h) = a(h+1)$ for some $h \in \{0, 1, \cdots, t-2\}$;*
*(iii) $k = t+1$, $\pi(t) = t-1$, and $\pi(t-1) = 0$.*

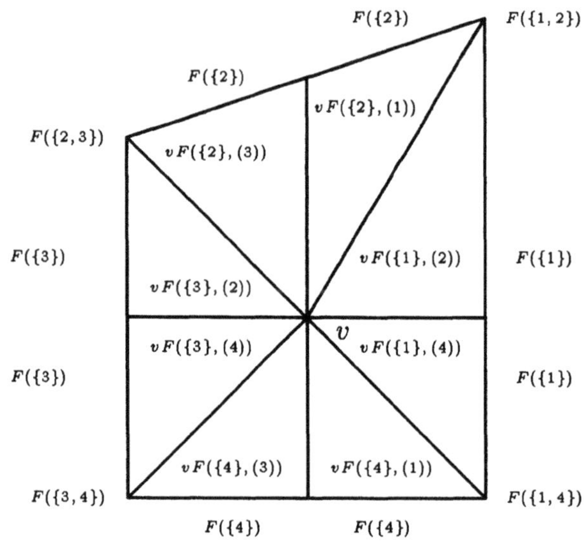


*Figure 9.2.* The $V(d)$ triangulation of P for $n = 2$ and $d = 1$.

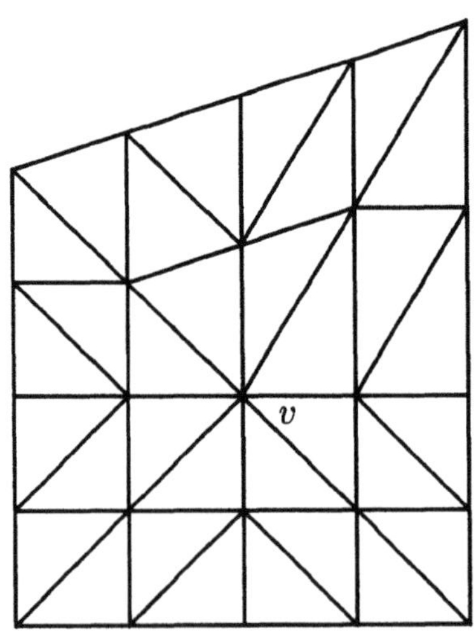


*Figure 9.3.* The $V(d)$-triangulation of P for $n = 2$ and $d = 2$.

CHAPTER 10

# The Computation of Antipodal Fixed Points

## 10.1. Introduction

In this chapter we consider the problem of finding zero points of a continuous function $f$ from the $n$-dimensional symmetric cube $C^n$ into $\mathbf{R}^n$. It will be shown in a constructive way that there exists a zero point of $f$ in $C^n$ if for every $x \in bd(C^n)$, $f(-x) \neq \alpha f(-x)$ for all $\alpha > 0$. It will be argued that this result is so strong that it implies several powerful fixed point theorems, including Borsuk-Ulam's theorem and Brouwer's fixed point theorem. We will introduce two algorithms to compute zero points of the continuous function. One is an integer labeling algorithm and the other a vector labeling algorithm. The integer labeling algorithm will lead to a constructive proof for Tucker's theorem. The algorithm we introduce here is a modification of the algorithms developed by Todd and Wright [1980], Freund and Todd [1981], and van der Laan [1984] in the sense that the modified algorithm can start with an arbitrarily chosen point in $C^n$, whereas the existing methods can only start with the point $0^n$. The modified algorithm will be built upon a new triangulation of $C^n$, called the $AS$-triangulation.

This chapter is organized as follows. In Section 2 we recall Borsuk-Ulam's theorem and Tucker's theorem. We also discuss the implications of Borsuk-Ulam's theorem. In Section 3 we present a new triangulation, called the $AS$-triangulation which underlies the algorithms. In Section 4 we give the integer labeling algorithm and give a constructive proof for Tucker's theorem. In Section 5 we introduce the vector labeling algorithm and discuss how to resolve the degeneracy problem. Furthermore an extension of Borsuk-Ulam's theorem to a point-to-set mapping will be given.

## 10.2. Borsuk-Ulam's theorem and Tucker's theorem

We first introduce some notation. For a positive integer $k$, let $C^n(k) = \{x \in \mathbf{R}^n \mid ||x||_\infty \leq k\}$. When $k = 1$, we simply write $C^n(1)$ as $C^n$. For a positive integer $n$, let $K_n = I_n \cup \{-i \mid i \in I_n\}$. A triangulation $\mathcal{T}$ of $C^n$ is said to be a *centrally symmetric triangulation* of $C^n$ if $-\sigma \in \mathcal{T}^+$ for every $\sigma \in \mathcal{T}^+$. Recall that $\mathcal{T}^+$ denotes the collection of all faces of simplices of $\mathcal{T}$. It is easy to see that both $K'$-triangulation and $J_1$-triangulation are centrally symmetric triangulations.

A continuous function $f : C^n \mapsto \mathbb{R}^n$ is said to have *the antipodal property* if $f(x) = -f(-x)$ for all $x \in bd(C^n)$. The following theorem is a variation of Borsuk-Ulam's theorem and can be found in Todd and Wright [1980], and van der Laan [1984].

**Theorem 10.2.1** *Let $f : C^n \mapsto \mathbb{R}^n$ be a continuous function satisfying the antipodal property. Then there exists a point $x^* \in C^n$ such that $f(x^*) = \mathbf{0}$.*

The goal of this chapter is to compute a zero point stated in the above theorem and therefore to provide a constructive proof for the theorem. We will also call zero points in the above theorem *antipodal fixed points.* In the following sections we will introduce two algorithms to compute zero points. One is using integer labeling and the other using vector labeling. The latter one is more efficient, while the first one is easily understandable. In this section and Section 4 we will restrict ourselves to integer labeling.

**Definition 10.2.2** *Let $f : C^n \mapsto \mathbb{R}^n$ be a continuous function. For each $x \in C^n$ let*

$$J(x) = \{\, i \in I_n \mid |f_i(x)| = ||f(x)||_\infty \}.$$

*Then the labeling function $l : C^n \mapsto K_n$ is given by for $k$ being the smallest index in $J(x)$*

$$l(x) = \begin{cases} k, & \text{if} \quad f_k(x) \geq 0, \\ -k, & \text{if} \quad f_k(x) < 0. \end{cases}$$

As the following lemma shows, a 1-simplex $\tau(w^1, w^2)$ with $l(w^1)+l(w^2) = 0$ provides a good approximation of a zero point of $f$. Such a 1-simplex will be called a *complementary* 1*-simplex.*

**Lemma 10.2.3** *Let $\mathcal{T}$ be a triangulation of $C^n$. Suppose that the mesh size is so small that for some $\epsilon > 0$, $||f(x) - f(y)||_\infty < \epsilon$ for any two points $x$ and $y$ in any simplex of $\mathcal{T}$. Then it holds $||f(x^*)||_\infty < \epsilon$ for any $x^*$ in a complementary* 1*-simplex.*

The proof is quite easy and left as an exercise. One may wonder what condition can ensure the existence of a complementary 1-simplex. The following combinatorial theorem of Tucker [1945] and Lefschetz [1949] provides an answer for it. Precisely, the following result was proved for $n = 2$ by Tucker [1945]; for the general case, see Lefschetz [1949, pp.134-141].

**Theorem 10.2.4 (Tucker Theorem)** *Let $\mathcal{T}$ be a centrally symmetric triangulation of $C^n$. If a labeling function $l : C^n \mapsto K_n$ is complementary on $bd(C^n)$, i.e.,*

$$l(x) = -l(-x)$$

*for all $x \in bd(C^n)$. Then there exists a complementary* 1*-simplex $\tau(w^1, w^2)$.*

Cohen [1979] gave an inductive and semi-constructive proof for Tucker's theorem. In Section 4 we will describe an algorithm to find a complementary 1-simplex. Now let us see how Theorem 10.2.1 can be proved via Tucker's theorem. Let $f$ satisfy the condition of Theorem 10.2.1. Then $f$ induces a labeling function $l$ which satisfies the complementary property of Theorem 10.2.4. Hence a centrally symmetric triangulation for $C^n$ will guarantee the existence of a complementary 1-simplex. By Lemma 10.2.3 this simplex gives a good approximation of a zero point. By taking a sequence of centrally symmetric triangulations for $C^n$ with mesh sizes tending to zero, the sequence of approximate zero points contains a subsequence that converges to a point $x^*$ with $f(x^*) = \mathbf{0}$.

Let $Y^n = \{x \in \mathbb{R}^{n+1} \mid ||x||_2 = (\sum_{i=0}^n x_i^2)^{\frac{1}{2}} = 1\}$, $Y_+^n = \{x \in Y^n \mid x_0 \geq 0\}$ and $Y_-^n = \{x \in Y^n \mid x_0 \leq 0\}$. The following famous antipodal fixed point theorem was conjectured by Ulam, proved by Borsuk [1933] and is called Borsuk-Ulam's theorem. The reader is referred to Tucker [1945] and Lefschetz [1949] for other alternative formulations of this theorem.

**Theorem 10.2.5 (Borsuk-Ulam Theorem)** *Let $h : Y^n \mapsto \mathbb{R}^n$ be a continuous function. Then there exists a point $z \in Y^n$ such that $h(z) = h(-z)$.*

Proof: We follow a proof of Todd and Wright [1980]. Let $g(x) = h(x) - h(-x)$ for all $x \in Y^n$. Then we have $g(x) = h(x) - h(-x) = -(h(-x) - h(x)) = -g(-x)$. Define the function $p : Y_+^n \mapsto C^n$ by

$$p(x) = \beta(x_1, \cdots, x_n)^\top$$

for $x \in Y_+^n$ with $x \neq (1, 0, \cdots, 0)^\top$, where

$$\beta = ||(x_1, \cdots, x_n)||_2/[(1 + x_0)||(x_1, \cdots, x_n)||_\infty],$$

with $p(1, 0, \cdots, 0) = (0, \cdots, 0)^\top$. It can be easily checked that $p$ is a bijection. In fact, $p$ has an inverse defined by

$$p^{-1}(z) = ((1 - ||z||_\infty^2)||z||_2/||z||_\infty, 2z_1, \cdots, 2z_n)^\top \frac{||z||_\infty}{(1 + ||z||_\infty^2)||z||_2}$$

for each non-zero vector $z = (z_1, \cdots, z_n)^\top \in C^n$. In addition, define

$$p^{-1}(0, \cdots, 0) = (1, 0, \cdots, 0)^\top.$$

It is readily seen that $p$ is a homeomorphism with the property that if $z \in bd(C^n)$, $p^{-1}(-z) = -p^{-1}(z)$. By applying Theorem 10.2.1 to $gp^{-1} : C^n \mapsto \mathbb{R}^n$, we have a point $y \in C^n$ such that $gp^{-1}(y) = 0^n$. Let $z = p^{-1}(y)$. We obtained the desired result. □

Meyerson and Wright [1979] and Bárány [1980] proposed algorithms for Borsuk-Ulam's theorem by using vector labeling. However, their algorithms provide no tractable way to cope with degeneracy problems.

As a matter of fact the algorithms to be presented will be able to compute zero points for a large class of continuous functions. To see this, we first introduce the following non-parallel condition.

**Definition 10.2.6** *A continuous function $f : C^n \mapsto \mathbb{R}^n$ is non-parallel if for any two antipodal points $x$ and $-x$ on the boundary of $C^n$,*

$$f(x) \neq \alpha f(-x) \text{ for all } \quad \alpha \geq 0.$$

*If $f(x) \neq \alpha f(-x)$ for all $\alpha > 0$, $f$ is called a weakly non-parallel function.*

Now we have the following powerful existence theorem. This theorem can be found in Krasnosell'skii [1964]. Nevertheless, we follow the proof of van der Laan [1984] which is very suitable for computational purpose.

**Theorem 10.2.7** *Let $f : C^n \mapsto \mathbb{R}^n$ be a continuous function with the weak non-parallel property for any two antipodal points on the boundary of $C^n$. Then there exists an $x^* \in C^n$ with $f(x^*) = \mathbf{0}$.*

Proof: We extend the function $f$ on $C^n$ to a function $g$ on $C^n(k)$ for some $k > 1$, such that $g(x) = -g(-x)$ for any $x \in bd(C^n(k))$. To do so, define $g : C^n(k) \mapsto \mathbb{R}^n$ by

$$g(x) = \begin{cases} f(x), & \text{if} \quad x \in C^n, \\ f(y) - \beta(x) f(-y), & \text{if} \quad x \in C^k(k) \setminus C^n, \end{cases}$$

where $y = \frac{x}{||x||_\infty}$ and $\beta(x) = \frac{||x||_\infty - 1}{k-1}$. Obviously, $||y||_\infty = \frac{||x||_\infty}{||x||_\infty} = 1$ and so $y \in bd(C^n)$ and $g$ is well defined. Moreover $\beta(x)$ goes to zero and $g(x)$ goes to $f(x)$ as $||x||_\infty$ tends to 1. This means that $g$ is continuous. Finally, if $x \in bd(C^n(k))$, we have $||x||_\infty = k$ and $\beta(x) = 1$. Thus,

$$g(x) = f(y) - f(-y) = -(f(-y) - f(y)) = -g(-x)$$

for $x \in bd(C^n(k))$. Therefore $g$ satisfies the condition of Theorem 10.2.1 and has a zero point $x^* \in C^n(k)$. We have to show that $x^* \in C^n$ and therefore $g(x^*) = f(x^*) = \mathbf{0}$. Suppose to the contrary that $x^* \notin C^n$. Then

$$g(x^*) = f(y^*) - \beta(x^*) f(-y^*) = \mathbf{0},$$

where $y^* = \frac{x^*}{||x^*||_\infty} \in bd(C^n)$ and $\beta(x^*) = \frac{||x^*||_\infty - 1}{k-1} > 0$. This contradicts the non-parallel property. So $x^* \in C^n$ and $f(x^*) = \mathbf{0}$. □

From this, we obtain the following corollary.

**Corollary 10.2.8** *Let $f : U \mapsto \mathbf{R}^n$ be a continuous function where $U$ is an $n$-dimensional, convex and compact subset of $\mathbf{R}^n$. Suppose that there exists some point $v \in int(U)$ such that for every $x \in bd(U)$, $f(x) \neq \alpha f(y)$ for all $\alpha > 0$, where $y$ is the unique point on $bd(U)$ and is on the line through $v$ and $x$. Then there exists at least one point $x^* \in U$ with $f(x^*) = \mathbf{0}$.*

Proof: Since $U$ is convex and compact and $v \in int(U)$, there exists a homeomorphism $h : U \mapsto C^n$. □

The next examples show that Theorem 10.2.7 is indeed a powerful theorem.

**Example 1**: Let $f : C^n \mapsto int(C^n)$ be a continuous function. Then by Brouwer theorem there exists a point in $x^* \in C^n$ such that $f(x^*) = x^*$. This also follows immediately from Theorem 10.2.7 by considering $g : C^n \mapsto \mathbf{R}^n$ as $g(x) = x - f(x)$.

**Example 2**: Let $f : C^n \mapsto \mathbb{R}^n$ be a continuous function satisfying the antipodal property. Then by Theorem 10.2.1 there exists a zero point in $C^n$. Excluding a trivial case that there exists a zero point on the boundary of $C^n$, Theorem 10.2.7 includes Theorem 10.2.1. Clearly, since $f(x) = -f(-x) \neq \mathbf{0}$ for all $x \in bd(C^n)$, we have that $f(x) \neq \alpha f(-x)$ for all $\alpha > 0$.

The next example given by van der Laan [1984] shows that the weak non-parallel condition is a necessary and sufficient condition in the case that $f$ is linear.

**Example 3**: Let $f : \mathbb{R}^n \mapsto \mathbb{R}^n$ be given by $f(x) = Ax + b$ where $A$ is a regular $n \times n$ matrix and $b \in \mathbb{R}^n$. Then $f(x^*) = \mathbf{0}$ for some $x^* \in C^n$ if and only if $f(x) \neq \alpha f(-x)$ for every $x \in bd(C^n)$ and for all $\alpha > 0$. From Theorem 10.2.7 we know that $f(x^*) = \mathbf{0}$ for some $x^* \in C^n$ if the weak non-parallel property holds. Conversely, we now assume that $f(x^*) = \mathbf{0}$ for some $x^* \in C^n$. Suppose to the contrary that there exists $y \in bd(C^n)$ and some $\alpha > 0$ such that $f(y) = \alpha f(-y)$. Since $f(x^*) = Ax^* + b = \mathbf{0}$, we have $x^* = -A^{-1}b$. Moreover,

$$Ay + b = -\alpha Ay + \alpha b$$

implies

$$(1 + \alpha)A^{-1}Ay = -(1 - \alpha)A^{-1}b = (1 - \alpha)x^*.$$

So we have $y = \frac{1-\alpha}{1+\alpha}x^*$. Since $x^* \in C^n$ and $\alpha > 0$, this contradicts that $y \in bd(C^n)$.

## 10.3. The *AS*-triangulation of $C^n$

In this and next sections we present an algorithm for finding a zero point of a continuous function on $C^n$ satisfying the antipodal property. To do so, we need to search for a complementary 1-simplex in $C^n$. Lemma 10.2.3 indicates that when using the $2n$-labeling rule given in Definition 10.2.2, any complementary 1-simplex in $C^n$ yields an approximate zero point of $f$. The algorithm we present here has been given in Todd and Wright [1980], Freund and Todd [1981] and van der Laan [1984]. It is a modification of the $2n$-ray algorithm of van der Laan and Talman [1981] and Reiser [1981] by using a crucial step, called the reflection step. Because of this it is called the *reflection $2n$-ray algorithm*. It should be pointed out that all these authors have used the $K'$-triangulation or the $J_1$-triangulation with center $0^n$ of $C^n$ to underlie the algorithm. Doing so will force the algorithm to start at the origin $0^n$ of $C^n$. Thus, the obtained approximate zero point cannot be utilized when refining the triangulation. To break this impasse we will modify the reflection $2n$-ray algorithm and present a new triangulation of $C^n$ such that the algorithm can start anywhere in $C^n$. Thus, when an approximate zero point $v$ in a coarse triangulation of $C^n$ is found, we can restart the algorithm from $v$ in a finer triangulation in the hope that a better approximate zero point will be found. To do so, we have to relax the definition of the centrally symmetric triangulation. A triangulation $\mathcal{T}$ of $C^n$ is said to be an *antipodal symmetric triangulation* of $C^n$ if for any $\sigma \subseteq bd(C^n)$, $\sigma \in \mathcal{T}^+$ implies $-\sigma \in \mathcal{T}^+$. We will be able to show the following result in the next section.

**Theorem 10.3.1** *Let $\mathcal{T}$ be an antipodal symmetric triangulation for $C^n$ and let $l : C^n \mapsto K_n$ be a labeling function which is complementary on $bd(C^n)$, i.e.,*

$$l(x) = -l(-x)$$

*for any point $x \in bd(C^n)$. Then there exists a complementary 1-simplex $\tau(w^1, w^2)$.*

Now we are ready to introduce an antipodal symmetric triangulation of $C^n$. Let $v \in C^n$ be a starting point of the algorithm. Take an initial guess of an approximate zero point as $v$. For each sign vector $s \in \{-1, 0, +1\}^n$, define the face $F(s)$ of $C^n$ by

$$F(s) = \{x \in C^n \mid x_i = 1 \text{ for } \ s_i = 1; x_i = -1 \text{ for } \ s_i = -1\}.$$

It is clear that $dim(F(s)) = |I^0(s)|$ and $F(0^n) = C^n$. For a face $F(s)$ of $C^n$ which does not contain $v$, let $vF(s)$ denote the convex hull of $\{v\} \cup F(s)$. Define $S(v) = \{s \in \{-1, 0, 1\}^n \mid v \in F(s)\}$.

Define $v(s)$ to be the projection of $v$ on a face $F(s)$ not containing $v$ with $-s \notin S(v)$ by $v(s) = s$. For $s \in S(v)$, define $v(s) = v$ and $v(-s) = -v$.

We say that a vector $s^1 \in \{-1, 0, 1\}^n$ is conformed to a vector $s^2 \in \{-1, 0, 1\}^n$, denoted by $s^1 \sqsubset s^2$, if $s^1_i \neq 0$ implies $s^2_i = s^1_i$ and $|I^0(s^1)| = |I^0(s^2)| + 1$. For a general vector $s \in \{-1, 0, 1\}^n$ for which $v \notin F(s)$, let

$s(n) \in \{-1,0,1\}^n$ be a vector such that no component of $s(n)$ is zero and $s_i \neq 0$ implies $s_i(n) = s_i$. For $s \in \{-1,0,+1\}^n$ and a vector $s(n)$, a conformed chain of $(s, s(n))$ denoted by $\gamma(s, s(n))$ is a sequence of $t = |I^0(s)| + 1$ vectors $s^1, \cdots, s^t$ such that

$$s = s^1 \sqsubset s^2 \sqsubset \cdots \sqsubset s^t = s(n).$$

For a conformed chain $\gamma(s, s(n))$, the subset $vF(s|\gamma(s, s(n)))$ is the convex hull of $v$ and the projections $v(s(n))$, $v(s^{t-1})$, $\cdots$, $v(s^2)$ and $v(s)$, i.e.,

$$vF(s|\gamma(s,s(n))) = \{x \in C^n \mid x = v + \textstyle\sum_{j=0}^{t-1} \beta_j q(j), \; 0 \leq \beta_{t-1} \leq \cdots \leq \beta_1 \leq \beta_0 \leq 1\},$$

where $q(0) = v(s(n)) - v$, and for $j = 1, \cdots, t-1$

$$q(j) = v(s^{t+1-j}) - v(s^{t-j}).$$

It should be noticed that $dim(vF(s|\gamma(s, s(n)))$ may be less than $|I^0(s)| + 1$ for some $\gamma(s, s(n))$. In this case the set $vF(s|\gamma(s, s(n))$ is said to be degenerate. Let $SF(1) = \{s \in \{-1,0,+1\}^n \mid v \notin F(s) \text{ and } |I^0(s)| = n-1\}$.

**Lemma 10.3.2** *The collection of the sets $vF(s|\gamma(s, s(n)))$ over all conformed chains $\gamma(s, s(n))$ and over all vertices $F(s(n))$ of $F(s)$ is a triangulation of the set $vF(s)$. Moreover, the collection of the sets $vF(s|\gamma(s, s(n)))$ over all conformed chains $\gamma(s, s(n))$ and over all vertices $F(s(n))$ of $F(s)$ and over all $s \in SF(1)$, is a triangulation of $C^n$.*

Clearly, this triangulation is closely related to the $V$-triangulation in the previous chapter. However, because of its particular nature, we will give it a new name and call it the *AS-triangulation* of $C^n$. For $n = 2$ and $v = (1, 1/2)^\top$, $(1, -1)^\top$, and $(1/2, -1/2)^\top$, the triangulations are illustrated in Figures 10.1, 10.2, and 10.3, respectively.

A triangulation with arbitrary small mesh size is obtained by triangulating each simplex $vF(s|\gamma(s, s(n)))$ into $dim(vF(s|\gamma(s, s(n))))$-dimensional simplices. This is expressed in the following definition.

**Definition 10.3.3** *Let $d$ be a positive integer. The set $AS^d(s|\gamma(s, s(n)))$ is the collection of $t$-simplices $\sigma(y^1, \pi)$ in $vF(s|\gamma(s, s(n)))$ with vertices $y^1$, $\cdots$, $y^{t+1}$ such that*
*(a) $y^1 = v + \sum_{j=0}^{t-1} a(j) d^{-1} q(j)$, where $a(0), a(1), \cdots, a(t-1)$ are integers satisfying $0 \leq a(t-1) \leq \cdots \leq a(0) \leq d-1$;*
*(b) $\pi = (\pi(1), \cdots, \pi(t))$ is a permutation of $(0, 1, \cdots, t-1)$ such that $k < l$ if for some $h \in \{0, 1, \cdots, t-2\}$, $\pi(k) = h$, $\pi(l) = h+1$, and $a(h) = a(h+1)$;*
*(c) $y^{i+1} = y^i + d^{-1} q(\pi(i))$ for all $i \in I_t$.*

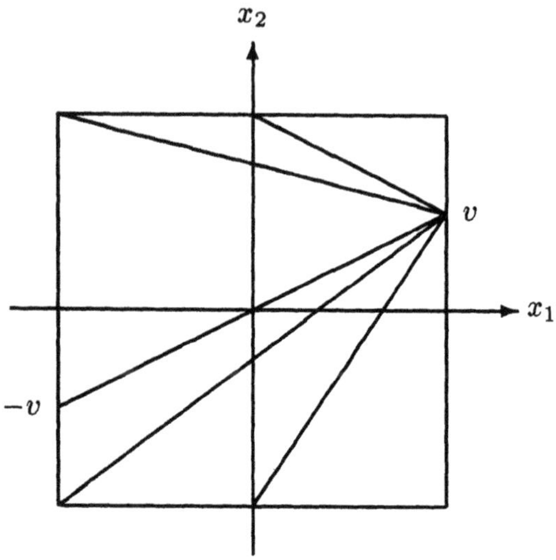


*Figure 10.1.* The $AS$-triangulation of $C^2$ with $v = (1, 1/2)^\top$.

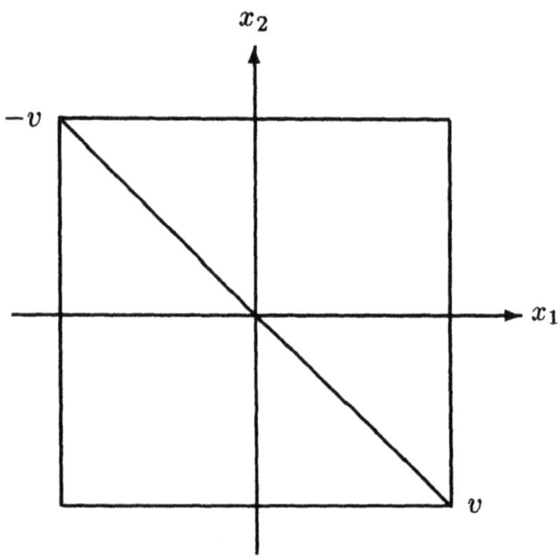


*Figure 10.2.* The $AS$-triangulation of $C^2$ with $v = (1, -1)^\top$.

The set $AS^d(s|\gamma(s, s(n)))$ is a simplicial subdivision of $vF(s|\gamma(s, s(n)))$ with refinement factor $d$. The collection of the sets $AS^d(s|\gamma(s, s(n)))$ over all conformed chains $\gamma(s, s(n))$ and over all vertices $F(s(n))$ of $F(s)$ and over all $s \in SF(1)$, is a triangulation of $C^n$ with refinement factor $d$, denoted by $AS(d)$. It is not difficult to show that the mesh size of the $AS(d)$-triangulation of $C^n$ tends to zero as $d$ goes to infinity. Moreover, we have the following theorem.

**Theorem 10.3.4** *For any given $d \in \mathbf{N}$, the $AS(d)$-triangulation of $C^n$ is an antipodal symmetric triangulation of $C^n$.*

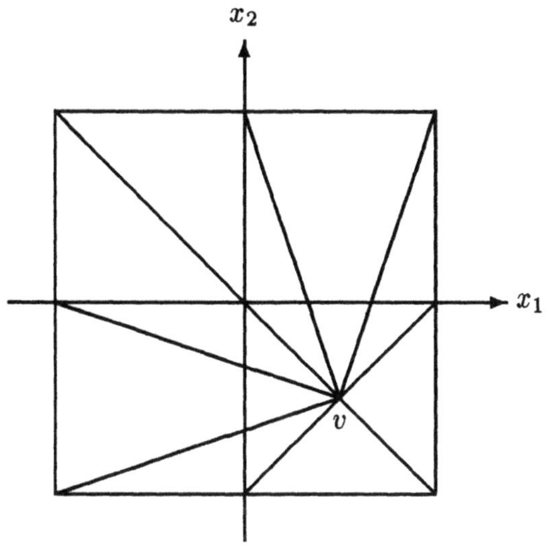


*Figure 10.3.* The $AS$-triangulation of $C^2$ with $v = (1/2, -1/2)^\top$.

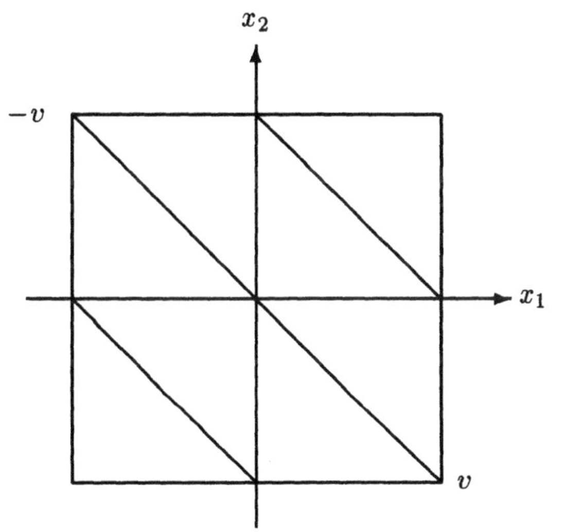


*Figure 10.4.* The $AS(d)$-triangulation of $C^2$ with $v = (1, -1)^\top$ and $d = 2$.

The $AS(d)$-triangulation of $C^n$ will underlie the algorithms to be presented in the next two sections. For $n = 2$, $v = (1, -1)^\top$ and $d = 2$, the $AS(d)$-triangulation of $C^2$ is illustrated in Figure 10.4.

## 10.4. The reflection $2n$-ray algorithm with integer labeling

In this section we present an algorithm to find a complementary 1-simplex in the $AS(d)$-triangulation of $C^n$ for any given $d \in \mathbb{N}$ and any given labeling function $l : C^n \mapsto K_n$ which is complementary on $bd(C^n)$. To describe the

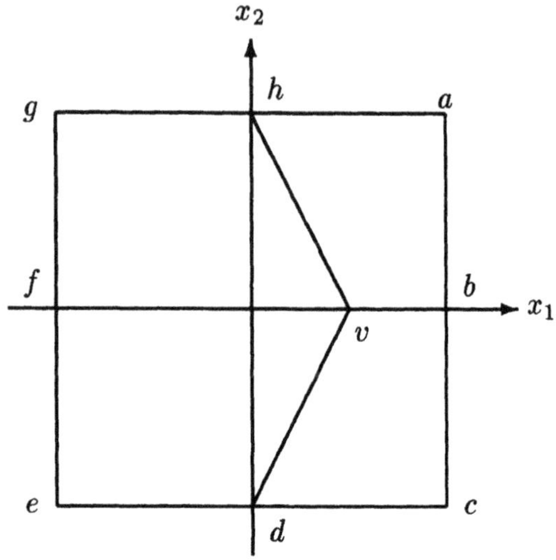


*Figure 10.5.* The sets $A(s)$ of $C^2$ with $v = (1/2, 0)^\top$.

algorithm we first introduce the following definition.

**Definition 10.4.1** *For some non-zero $s \in \{-1, 0, +1\}^n$, with $t = n - |I^0(s)|$, a $(t-1)$-simplex $\tau(y^1, \cdots, y^t)$ is called s-complete if*

$$\{l(y^i) \mid i = 1, \cdots, t\} = \{is_i \mid s_i \neq 0\}.$$

Furthermore, we introduce some additional notation. For each $s \in \{-1, 0, +1\}^n$, define the subset $A(s)$ of $C^n$ by

$$A(s) = co(\{v(s') \mid s' \in \{-1, 0, +1\}^n,\ 0^n \leq s' \leq s\}).$$

For $s \notin S(v) \cup -S(v)$ we have $dim(A(s)) = n - |I^0(s)|$. It is easy to check that the $AS(d)$-triangulation induces a triangulation for every set $A(s)$. We demonstrate sets $A(s)$ of $C^2$ with $v = (1/2, 0)^\top$ in Figure 10.5, where

$$\begin{array}{rr} A((1,1)^\top) = co(\{v,h,a,b\}), & A((1,-1)^\top) = co(\{v,b,c,d\}), \\ A((-1,-1)^\top) = co(\{v,d,e,f\}), & A((-1,1)^\top) = co(\{v,f,g,h\}), \\ A((1,0)^\top) = co(\{v,b\}), & A((0,-1)^\top) = co(\{v,d\}), \\ A((-1,0)^\top) = co(\{v,f\}), & A((0,1)^\top) = co(\{v,h\}). \end{array}$$

From now on, let us assume that the labeling function $l : C^n \mapsto K_n$ is complementary on $bd(C^n)$. Now we construct a graph $G = (V, A)$ as follows, where $V$ is the set of nodes and $A$ is the set of arcs. We say that a simplex $\sigma$ is a *node* if $\sigma = \{v\}$ or $\sigma$ has an $s$-complete facet $\tau$ and $\sigma \subseteq A(s)$.

**Definition 10.4.2** *Two nodes $\sigma^1$ and $\sigma^2$ are adjacent if one of the following cases occurs:*

*(a) they share a common s-complete facet $\tau$ and they are both in $A(s)$;*
*(b) one, say, $\sigma^1$, has an s-complete facet $\tau \subseteq bd(C^n)$ and $-\tau \subseteq bd(C^n)$ is a $-s$-complete facet of $\sigma^2$;*
*(c) one, say, $\sigma^1$, is an s-complete facet of the other, say, $\sigma^2$;*
*(d) one, say, $\sigma^1 = \{v\}$, is s-complete with $A(s) = \{v\}$ and $\sigma^2$ has a $-s$-complete facet $\{-v\}$.*

The *degree* of a node $\sigma$ is the number of nodes adjacent to it in $C^n$, denoted by $deg(\sigma)$.

**Lemma 10.4.3** *Let $G = (V, A)$ be the graph defined above and let $\sigma$ be a node of $G$. Then*
*(i) $deg(\sigma) = 1$ when $\sigma = \{v\}$;*
*(ii) $deg(\sigma) = 1$ when $\sigma$ contains a complementary 1-simplex;*
*(iii) $deg(\sigma) = 2$ in all other cases.*

Proof: (i) If $\sigma = \{v\}$, let $s = s(\{l(v)\})$. When $s \in S(v)$, then $\sigma$ lies in the 0-dimensional set $A(s)$ in which $\sigma$ is $s$-complete. Let $\sigma^1$ be the 1-simplex in $A(-s)$ having $-\sigma$ as its facet. Clearly, $-\sigma$ is $-s$-complete since $l(-v) = -l(v)$. This means that $\sigma^1$ is a node since its facet $-\sigma$ is $-s$-complete. Moreover, $\sigma^1$ is the only node adjacent to $\sigma$. Thus, $deg(\sigma) = 1$. When $s \notin S(v)$, then $\sigma$ lies in the 1-dimensional set $A(s)$ in which $\sigma$ is $s$-complete. Let $\sigma^1$ be the 1-simplex in $A(s)$ having $\sigma$ as its facet. Clearly, $\sigma^1$ is a node since its facet $\sigma$ is $s$-complete. Moreover, $\sigma^1$ is the only node adjacent to $\sigma$. Thus, $deg(\sigma) = 1$.

(ii) If a $t$-simplex $\sigma$ is a node and has a face being a complementary 1-simplex, then $\sigma$ has an $s$-complete $(t-1)$-simplex $\tau_1$ as its facet, where $\sigma \subseteq A(s)$ and $t = n - |I^0(s)|$. This implies that the $t+1$ vertices of $\sigma$ contains $t+1$ distinct labels and there exists exactly one non-zero component $s_j$ of $s$ such that $\{l(y^i) | i \in I_{t+1}\} = \{is_i \mid s_i \neq 0\} \cup \{-js_j\}$. First note that $\tau_1$ is the unique facet of $\sigma$ being $s$-complete. If $\tau_1 \not\subseteq bd(A(s))$, then $\tau_1$ is a common facet of exactly two $t$-simplices $\sigma$ and $\sigma^1$. Clearly $\sigma^1$ is a node and adjacent to $\sigma$. So, $deg(\sigma) = 1$. If $\tau_1$ is on the boundary of $C^n$, then by the antipodal property we have a $t$-simplex $\sigma^1$ in $A(-s)$ having $-\tau_1$ as its $-s$-complete facet. So $\sigma^1$ is a node and adjacent to $\sigma$. Again $deg(\sigma) = 1$. If $\tau_1$ lies on the boundary of $A(s)$ and $\tau_1 \subseteq A(s')$ for some $s' \sqsubset s$, then $\tau_1$ has a facet being $s'$-complete and so $\tau_1$ is a node. Consequently $deg(\sigma) = 1$.

(iii) In all other cases, let a $t$-simplex $\sigma$ be a node. We distinguish two cases.

Case (1): If the label set of $t$-simplex $\sigma$ is equal to $\{s_i \mid s_i \neq 0\}$, then $\sigma$ has exactly two $s$-complete $s$-facets $\tau_1$ and $\tau_2$. If $\tau_1 \subseteq bd(C^n)$, then by the antipodal property, we have a $t$-simplex $\sigma^1$ in $A(-s)$ having $-\tau_1$ as its $-s$-complete facet. So $\sigma^1$ is a node and adjacent to $\sigma$. If $\tau_1$ lies on the boundary of $A(s)$ and $\tau_1 \subseteq A(s')$ for some $s' \sqsubset s$, then $\tau_1$ has a facet being $s'$-complete and so is a node. If $\tau_1 \not\subseteq bd(A(s))$, then $\tau_1$ is a common $s$-complete facet of

precisely two $t$-simplices $\sigma$ and $\sigma^1$. So $\sigma^1$ is a node and adjacent to $\sigma$. The same argument applies to $\tau_2$. Hence we have $deg(\sigma) = 2$.
Case (2): If the label set of $\sigma$ contains $t+1$ distinct labels, then $\{l(y^i) \mid i \in I_{t+1}\} = \{is'_i \mid s'_i \neq 0\}$ for some $s'$ with $s \sqsubset s'$. Clearly $\sigma \subseteq A(s')$ and $\sigma$ is $s'$-complete. When $dim(A(s')) = t+1$, then let $\sigma'$ be the $(t+1)$-simplex in $A(s')$ having $\sigma$ as its facet. Then $\sigma'$ is a node and adjacent to $\sigma$. When $dim(A(s')) = t$, this may happen only if $s, s' \in -S(v)$ and $t = 1$. Let $y^1$ and $y^2$ be the vertices of $\sigma$ with $l(y^1) = l_1$ and $l(y^2) = l_2$. Clearly, $l_1 \neq l_2$. Let $s = s(\{l_1\})$. Then $\{y^2\}$ is $s(\{l_2\})$-complete. Let $\sigma^1$ be the 1-simplex in $A(s)$ having $\{y^2\}$ as its facet. Clearly, $\sigma^1$ is a node since $\sigma^1 \subseteq A(s(\{l_2\}))$ and its facet $\{y^2\}$ is $s(\{l_2\})$-complete. Now let $\tau$ be the unique $s$-complete facet of $\sigma$. Following the argument in (ii), we have $deg(\sigma) = 2$. □

By applying Theorem 1.4.14, we have the following convergence theorem which leads to a constructive proof of Theorem 10.3.1 and Tucker theorem.

**Theorem 10.4.4** *Let $d \in \mathbb{N}$ be given. Then there exists a finite sequence of adjacent simplices $\sigma^1, \cdots, \sigma^M$ in the $AS(d)$-triangulation of $C^n$ with refinement factor $d$, where $\sigma^1 = \{v\}$ and $\sigma^M$ contains a complementary 1-simplex.*

Let $\mathcal{K}$ be the collection of subsets $L$ of $K_n$ such that

$$j \in L \text{ implies } \quad -j \notin L.$$

Define the function $s : \mathcal{K} \mapsto \{-1, 0, +1\}^n$ by

$$s_i(L) = \begin{cases} -1 & \text{if } \ -i \in L, \\ 0 & \text{if } \ i \notin L \text{ and } \ -i \notin L, \\ 1 & \text{if } \ i \in L. \end{cases}$$

Now we can give an explicit statement of the reflection algorithm with an arbitrary starting point that generates the sequence stated in Theorem 10.4.4.

The Reflection $2n$-Ray (R2R) Algorithm

**Step (0)** Choose any point $v \in C^n$ and any element $d \in \mathbb{N}$ to fix the triangulation $AS(d)$ of $C^n$. Set $c = v$, $\sigma = \{c\}$, $L = \emptyset$, and $s = s(L)$.

**Step (1)** Compute $l(c)$. If $l(c)$ is the complement of the label of some vertex of $\sigma$, then stop with the complementary 1-simplex of $\sigma$. Otherwise, if $l(c) \notin L$, then go to Step (2); if not, search for the vertex $w \neq c$ of $\sigma$ such that $l(w) = l(c)$. Go to Step (3).

**Step (2)** Set $J = L \cup \{l(c)\}$. If $|L| \geq 1$ and $s(J) \notin -S(v)$, or if $|L| = 0$ and $s(J) \notin S(v)$, then let $\tau$ be the unique $(t+1)$-simplex of $A(s(J))$

having $\sigma$ as its facet and let $w$ be the vertex of $\tau$ not in $\sigma$. Let $\sigma = \tau$, $c = w$, $L = L \cup \{l(c)\}$, $s = s(L)$ and go to Step (1). If $s(J) \in -S(v)$ and $|L| = 1$, then let $\tau \neq \sigma$ be the 1-simplex in $A(s)$ having $\{c\}$ as its facet and let $w$ be the vertex of $\tau$ not in $\sigma$. Set $\sigma = \tau$, $L = \{l(c)\}$, $s = s(L)$, $c = w$ and go to Step (1). If $|L| = 0$ and $s(J) \in S(v)$, then set $\tau = \sigma$, $L = L \cup \{l(c)\}$ and go to Step (4).

**Step (3)** Let $\tau$ be the facet of $\sigma$ opposite $w$. If $\tau \subseteq A(s) \cap bd(C^n)$ go to Step (4). If for some $j \in L$, $\tau \subseteq A(s(L \setminus \{j\}))$, go to Step (5). Otherwise, let $\sigma' \neq \sigma$ be the simplex in $A(s)$ having $\tau$ as its facet and let $x$ be its vertex not in $\tau$. Set $\sigma = \sigma'$, $c = x$ and go to Step (1).

**Step (4)** (Reflection steps) Since $l$ is complementary on $bd(C^n)$, we have that $l(-x) = -l(x)$ for any vertex $x$ of $\tau$. Let $\sigma'$ be the simplex in $A(-s)$ having $-\tau$ as its facet. Let $w$ be the vertex of $\sigma'$ not in $-\tau$. Set $\sigma = \sigma'$, $s = -s$, $L = -L$ and $c = w$. Go to Step (1).

**Step (5)** Let $x$ be the vertex of $\tau$ with $l(x) = j$. Set $\sigma = \tau$, $L = L \setminus \{j\}$, $s = s(L)$, $c = x$ and go to Step (3).

We illustrate the algorithm by three examples. In Figure 10.6, the algorithm starts with $v = (-1,-1)^\top$ and $l(v) = 1$, and stops with a complementary 1-simplex $\tau(w^1, w^2)$. Note that there is one reflection step. In Figure 10.7, the algorithm starts with $v = (-1,-1)^\top$ and $l(v) = -2$, and stops with a complementary 1-simplex $\tau(w^1, w^2)$. There are two reflections in the searching process. In Figure 10.8, the algorithm starts with $v = (-1/2, 0)^\top$ and $l(v) = 1$, and stops with a complementary 1-simplex $\tau(w^1, w^2)$. There are two reflections in the algorithm.

## 10.5. The reflection $2n$-ray algorithm with vector labeling

In this section we introduce the reflection $2n$-ray algorithm with vector labeling. All the results of this section are drawn from Todd and Wright [1980]. The only modification here is that due to the introduction of the $AS$-triangulation we are able to allow the algorithm to start from any point of $C^n$ whereas in Todd and Wright [1980] the algorithm can only start with the origin $0^n$. To simplify the analysis, we take the starting point $v$ in the interior of $C^n$. Let $\tau(x^1, \cdots, x^t)$ be a $(t-1)$-simplex in $A(s)$ with $t = n - |I^0(s)|$. Define

$$L_\tau = [\ f(x^1) \quad f(x^2) \quad \cdots f(x^t)\ ].$$

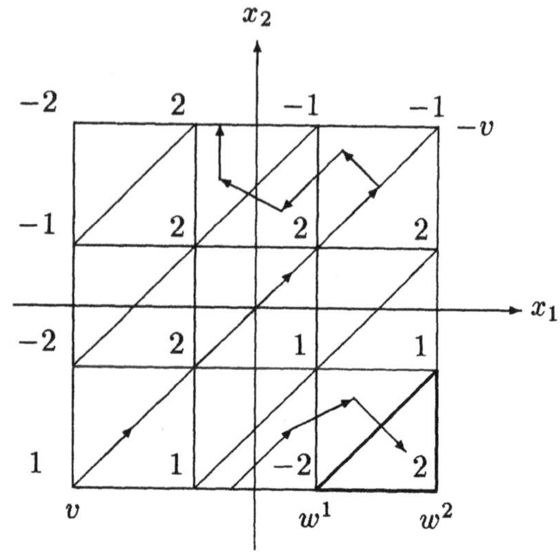


*Figure 10.6.* Illustration of the R2R algorithm with $v = (-1, -1)^\top$ and $d = 3$.

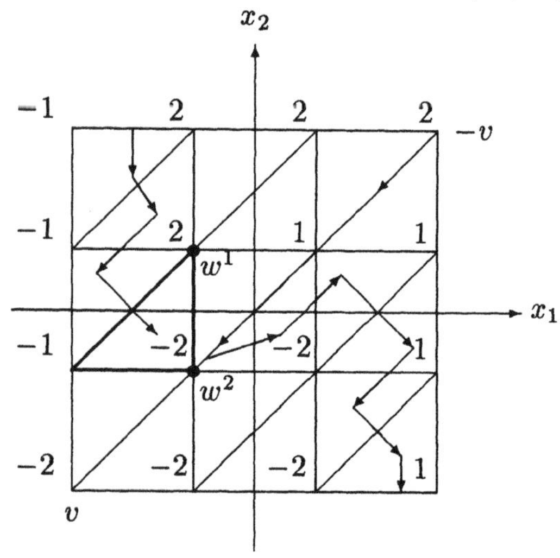


*Figure 10.7.* Illustration of the R2R algorithm for $v = (-1, -1)^\top$ and $d = 3$.

The $(2n+1) \times (2n+1)$ matrix

$$M_{\tau,s} = \begin{bmatrix} 1 & 1^{t\top} & 0^{2n-t\top} \\ -1^n & L_\tau & \\ & & \mathbf{I}_s \\ -1^n & -L_\tau & \end{bmatrix}$$

is called the *label matrix* of $\tau$ with respect to $s$, where $\mathbf{I}_s$ is the $2n \times 2n$ identity matrix with column $i$ deleted for $s_i = 1$, and column $n+i$ deleted for $s_i = -1$.

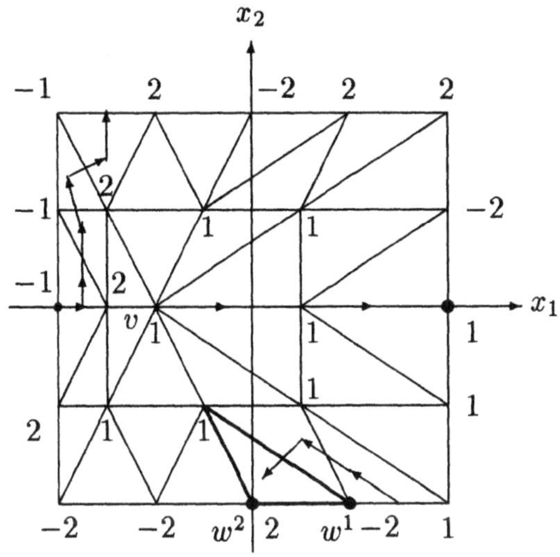


*Figure 10.8.* Illustration of the R2R algorithm for $v = (-1/2, 0)^\top$ and $d = 2$.

Next, we define the $(2n+1) \times (n+1)$ matrix $B$ by

$$B = \begin{bmatrix} 1 & 0^{n\top} \\ 0^n & \mathbf{I} \\ 0^n & \mathbf{I} \end{bmatrix}$$

where $\mathbf{I}$ is the identity matrix of rank $n$.

The simplex $\tau$ is said to be *s-complete* or *very complete* if $M_{\tau,s}^{-1}$ exists and there is an $(2n+1) \times (n+1)$ matrix $W$ such that

$$M_{\tau,s} W = B, \ W \succ \mathbf{0}.$$

From now on we will also use an alternative representation for the matrix $M_{\tau,s}$. Let $T = \{1, 2, \cdots, t\}$, and define $\pi : T \mapsto \{1, 2, \cdots, 2n\}$ by setting $\pi(i) = j$ if the $i$-th non-zero component of $s$ is $s_j = 1$, and $\pi(i) = n + j$ if the $i$-th non-zero component of $s$ is $s_j = -1$. Let $A$ be the $(2n+1) \times (2n+1)$ matrix

$$\begin{bmatrix} 1 & 0 \\ -1^{2n} & \mathbf{I} \end{bmatrix}.$$

Construct a matrix by replacing column $\pi(i)$ of $A$ by $(1, f^\top(x^i), -f^\top(x^i))^\top$ for each $i \in T$. This matrix is also denoted by $M_{\tau,s}$. Note that this operation does not change the system

$$M_{\tau,s} W = B, \ W \succ \mathbf{0}$$

except that some rows of $W$ are permuted. In the following we denote by $u^i$ the $i$-th unit vector in $\mathbb{R}^{2n+1}$, $i \in \{0, 1, \cdots, 2n\}$.

**Lemma 10.5.1** *If $\tau$ is an $s$-complete $(t-1)$-simplex lying on the boundary of $C^n$, then $-\tau$ is $-s$-complete.*

Proof: Since $f(-x) = -f(x)$ for $x \in bd(C^n)$, we know that $-\tau$ lies in $A(-s) \cap bd(C^n)$ and $M_{-\tau,-s} = QM_{\tau,s}Q$, where

$$Q = \begin{bmatrix} 1 & 0 & 0 \\ 0 & 0 & \mathbf{I} \\ 0 & \mathbf{I} & 0 \end{bmatrix}$$

with $\mathbf{I}$ the $n \times n$ identity matrix. Let $W$ be a lexico positive matrix so that $M_{\tau,s}W = B$, and let $Y = QW = Q^{-1}W$. Clearly $Y \succ \mathbf{0}$. Note that

$$M_{-\tau,-s}Y = (QM_{\tau,s}Q)(Q^{-1}W) = QM_{\tau,s}W = QB = B.$$

Thus, $-\tau$ is $-s$-complete. □

A $t$-simplex $\sigma =< x^+, x^1, \cdots, x^t >$ is called a *terminal simplex* if there is a vector $\mu \in \mathbb{R}^{t+1}_+$ such that $\sum_{i=0}^t \mu_i = 1$ and $\mu_0 f(x^+) + \sum_{j=1}^t \mu_j f(x^j) = 0^n$. That is, $\bar{x} = \mu_0 x^+ + \sum_{i=1}^t \mu_i x^i$ is an approximate zero point of $f$. Let $\bar{f}$ denote the P.L. approximation of $f$ with respect to the $AS$-triangulation of $C^n$.

**Lemma 10.5.2** *Let $\tau$ be an $s$-complete facet of a $t$-simplex $\sigma$ in $A(s)$. If $\sigma$ is not a terminal simplex, then exactly one of the following cases occurs:*
*(i) $\sigma$ has exactly two $s$-complete facets, $\tau$ and $\bar{\tau}$;*
*(ii) $\sigma$ is $\bar{s}$-complete for some unique $\bar{s}$ with $s \sqsubset \bar{s}$.*

Proof: Since $\tau$ is $s$-complete, there exists a lexico positive matrix $W$ such that $M_{\tau,s}W = B$. Let $x^+$ be the vertex of $\sigma$ not contained in $\tau$ and let

$$y = M_{\tau,s}^{-1}(1, f^\top(x^+), -f^\top(x))^\top.$$

Define the subset $J$ of $\{0, 1, \cdots, 2n\}$ by

$$J = \{j \mid \frac{w_{j,0}}{y_j} = \min\{\frac{w_{l,0}}{y_l} \mid y_l > 0\}\}.$$

Equation 0 of the system $M_{\tau,s}y = (1, f^\top(x^+), -f^\top(x))^\top$ shows that $y_0 + \sum_{i=1}^t y_{\pi(i)} = 1$. So, $J \neq \emptyset$.

Suppose that $0 \in J$. We will show that $\sigma$ is a terminal simplex. Construct $M$ by replacing column 0 of $M_{\tau,s}$ by $(1, f^\top(x^+), -f^\top(x))^\top$. An LP pivot operation applied to column 0 of $W$ using $y_0$ as a pivot element and $y$ as a pivot column will give a vector $z \in \mathbb{R}^{2n+1}_+$ such that $Mz = u^0$. Equation 0 of this system is

$$z_0 + \sum_{i=1}^t z_{\pi(i)} = 1.$$

If $l \in \pi(T)$, then the $l$-th equation of $Mz = u^0$ is either

$$z_0 f_l(x^+) + \sum_{i=1}^{t} z_{\pi(i)} f_l(x^i) = 0$$

or

$$-z_0 f_{l-n}(x^+) - \sum_{i=1}^{t} z_{\pi(i)} f_{l-n}(x^i) = 0.$$

If $l \notin \pi(T)$ and $n + l \notin \pi(T)$, then the $l$-th and $(n + l)$-th equations are

$$\begin{aligned} z_0 f_l(x^+) + \textstyle\sum_{i=1}^{t} z_{\pi(i)} f_l(x^i) + z_l = 0 \\ -z_0 f_l(x^+) - \textstyle\sum_{i=1}^{t} z_{\pi(i)} f_l(x^i) + z_{n+l} = 0. \end{aligned}$$

Since $z_l \geq 0$, and $z_{n+l} \geq 0$, we then also obtain

$$z_0 f_l(x^+) + \sum_{i=1}^{t} z_{\pi(i)} f_l(x^i) = 0.$$

This shows that $\sigma$ is indeed a terminal simplex.

Suppose that $j \in J$, $1 \leq j \leq n$, and $n + j \in \pi(T)$. Construct $M$ by replacing column $j$ of $M_{T,s}$ by $(1, f^\top(x^+), -f^\top(x^+))^\top$. Let $z$ be a non-negative vector such that $Mz = u^0$. The $j$-th and $(n + j)$-th equations of this system show that $z_0 = 0$. An argument similar to that given above shows that $\sigma$ is a terminal simplex. The same type of argument shows that if $n + j \in J$, $1 \leq j \leq n$, and $j \in \pi(T)$, then $\sigma$ is a terminal simplex.

Suppose that $\{j, n + j\} \subseteq J$, $1 \leq j \leq n$. We will prove that $\sigma$ is again a terminal simplex. By the previous case, we can assume that $\{j, n + j\} \cap \pi(T) = \emptyset$. Construct $M$ by replacing column $j$ of $M_{T,s}$ by

$$(1, f^\top(x^+), -f^\top(x^+))^\top,$$

and let $z$ be the non-negative vector such that $Mz = u^0$. Then $z_{n+j} = 0$, so the $(n + j)$-th equation is

$$-z_0 - \sum_{i=1}^{t} z_{\pi(i)} f_j(x^i) - z_j f_j(x^+) = 0.$$

Construct $M'$ by replacing column $n+j$ of $M_{T,s}$ by $(1, f^\top(x^+), -f^\top(x^+))^\top$, and let $z'$ be the non-negative vector such that $M'z' = u^0$. It is easy to check that $z_i = z'_i$ for $i \neq j$, $i \neq n + j$. Furthermore, $z'_j = 0$, and $z'_{n+j} = z_j$. The $j$-th equation of $M'z' = u^0$ is

$$-z'_0 + \sum_{i=1}^{t} z'_{\pi(i)} f_j(x^i) + z'_{n+j} f_j(x^+) = 0.$$

These equations together imply that $z_0 = z_0' = 0$. Then the arguments of the previous paragraphs show that $\sigma$ is a terminal simplex.

Next we construct an $(n+1)$-element set $H \subseteq \{0, 1, \cdots, 2n\}$ with the following properties:

(a) $0 \in H$;

(b) $J \subseteq H$;

(c) For each $j \in I_n$, $H$ contains exactly one of $\{j, n+j\}$.

Let the $(n+1) \times (n+1)$ matrix $M_H$ be constructed by deleting from $M_{T,s}$ all rows and columns whose indices are not in $H$. Let $W_H$ and $B_H$ consist of those rows of $W$ and $B$ respectively whose indices are in $H$. It can be verified that $M_H W_H = B_H$ and the rows of $B_H$ can be permuted to give the identity matrix. This means that $W_H$ is regular. So it is possible to choose a unique $j \in J$ such that $y_j > 0$ and the ration vector

$$[\frac{w_{j,0}}{y_j}, \cdots, \frac{w_{j,n}}{y_j}]$$

is lexico minimum. Construct $M'$ by replacing column $j$ of $M_{T,s}$ by

$$(1, f^\top(x^+), -f^\top(x^+))^\top,$$

and use an LP pivot operation to construct a lexico positive matrix $W'$ such that $M'W' = B$. Note that if we had selected any other $j \in J$, the corresponding $W'$ would not be lexico positive. If $j \in \pi(T)$, then this system shows that the facet of $\sigma$ not containing $x^h$ with $\pi(h) = j$ is $s$-complete; thus, conclusion $(i)$ holds. If $j \notin \pi(T)$, define $\bar{s} \in \mathbf{R}^n$ by letting $\bar{s}_j = 1$ if $1 \le j \le n$, $\bar{s}_{j-n} = -1$ if $n+1 \le j \le 2n$, and $\bar{s}_i = s_i$ otherwise. Clearly, $s \sqsubset \bar{s}$. Then $\sigma$ is a facet of a unique simplex $\sigma'$ of $A(\bar{s})$, and the system $M'W' = B$ demonstrates that $\sigma$ is $\bar{s}$-complete. Thus, conclusion $(ii)$ holds. □

By applying the proof of the above lemma to

$$M_{\emptyset,0^n} = \begin{bmatrix} 1 & 0 & 0 \\ -1^n & \mathbf{I} & 0 \\ -1^n & 0 & \mathbf{I} \end{bmatrix}$$

and

$$W = \begin{bmatrix} 1 & 0 & 0 \\ 1^n & \mathbf{I} & 0 \\ 1^n & 0 & \mathbf{I} \end{bmatrix}$$

we have $M_{\emptyset,0^n} W = B$ and $W \succ \mathbf{0}$. This shows that if $f(v) \neq 0^n$, then the 0-simplex $< v >$ is $s$-complete for some unique $s \in \{-1, 0, +1\}^n$ with $|I^0(s)| = n - 1$. If $f(v) = 0^n$, then $< v >$ is a terminal simplex.

**Lemma 10.5.3** *Let $\tau$ be an s-complete facet of a t-simplex $\sigma$ in $A(s)$. Let $\tau \subseteq A(q)$ for some sign vector $q$ with $q \sqsubset s$. If $\sigma$ is not a terminal simplex, then exactly one of the following cases occurs:*
*(i) $\tau$ has a unique facet which is q-complete;*
*(ii) there is a unique t-simplex $\sigma' \neq \sigma$ so that $\tau$ is p-complete for some unique sign vector with $p \neq s$ and $|I^0(p)| = |I^0(s)|$.*

Proof: Since $\tau$ is $s$-complete, there exists a lexico positive matrix $W$ such that $M_{\tau,s}W = B$. There is an index $r$ such that $s_r \neq 0$ and $q_r = 0$. If $s_r = 1$, let $y = M_{\tau,s}^{-1}u^r$, and if $s_r = -1$, let $y = M_{\tau,s}^{-1}u^{n+r}$. For simplicity, we consider only the first case since the proof for the second case is similar.

Define the subset $J$ of $\{0, 1, \cdots, 2n\}$ by

$$J = \{j \mid \frac{w_{j,0}}{y_j} = \min\{\frac{w_{l,0}}{y_l} \mid y_l > 0\}\}.$$

As shown before, $J \neq \emptyset$.

Suppose that $0 \in J$. Construct $M$ by replacing column 0 of $M_{\tau,s}$ by $u^r$ and let $z$ be the non-negative vector so that $Mz = u^0$. Equation 0 of this system is

$$\sum_{i=1}^{t} z_{\pi(i)} = 1.$$

If $l \in \pi(T)$, $l \neq r$, the $l$-th equation of $Mz = u^0$ is either

$$\sum_{i=1}^{t} z_{\pi(i)} f_l(x^i) = 0$$

or

$$-\sum_{i=1}^{t} z_{\pi(i)} f_{l-n}(x^i) = 0.$$

If $l \notin \pi(T)$ and $n + l \notin \pi(T)$, $1 \leq l \leq n$, then the $l$-th and $(n + l)$-th equations are

$$\begin{aligned} \textstyle\sum_{i=1}^{t} z_{\pi(i)} f_l(x^i) + z_l &= 0 \\ -\textstyle\sum_{i=1}^{t} z_{\pi(i)} f_l(x^i) + z_{n+l} &= 0. \end{aligned}$$

Since $z_l \geq 0$, and $z_{n+l} \geq 0$, we have

$$\sum_{i=1}^{t} z_{\pi(i)} f_l(x^i) = 0.$$

Equations $r$ and $n+r$ are similar, with $z_0$ replacing $z_l$ in equation $r$. The two equations mean that $\sum_{i=1}^{t} z_{\pi(i)} f_r(x^i) = 0$. This shows that $\sigma$ is a terminal simplex.

Suppose that $j \in \pi(T)$, $j \neq r$, $1 \leq j \leq n$, and $n+j \in J$. An argument similar to that used in Lemma 10.5.2 shows that $\tau$ is a terminal simplex. Similarly, if $j \in J$, $1 \leq j \leq n$ and $n+j \in \pi(T)$, $\tau$ can be also shown to be a terminal simplex.

Suppose that $\{j, n+j\} \subseteq J$, $1 \leq j \leq n$ and $\{j, n+j\} \cap \pi(T) = \emptyset$. Again, an argument similar to that used in Lemma 10.5.2 shows that $\tau$ is a terminal simplex.

Finally suppose that $\{r, n+r\} \subseteq J$. Note that by assumption $r \in \pi(T)$ and $1 \leq r \leq n$. We claim that $\pi(T) \not\subseteq J$. For suppose that $\pi(T) \subseteq J$. Construct $M$ by replacing column $n+r$ of $M_{\tau,s}$ by $u^r$. Let $z$ be a non-negative vector such that $Mz = u^0$. Then $z_{\pi(i)} = 0$ for all $i = 1, \cdots, t$. Equation $n+r$ of $Mz = u^0$ is $-z_0 = 0$. This contradicts equation 0 of the system, namely, $z_0 + \sum_{i=1}^{t} z_{\pi(i)} = 1$. So, there is an $l \in \pi(T) \setminus J$. Modify $M_{\tau,s}$ by interchanging columns $r$ and $l$, and modify $W$ by interchanging rows $r$ and $l$. The new system $M_{\tau,s}W = B$ still defines $\tau$ as a very complete facet of $\sigma$, and $r \notin J$.

Since $\tau$ is not a terminal simplex, an $(n+1)$-element set $H \subseteq \{0, 1, \cdots, 2n\}$ with the following properties can be constructed:

(a) $0 \in H$;

(b) $J \subseteq H$;

(c) For each $j = 1, \cdots, n$, $H$ contains exactly one of $\{j, n+j\}$.

Construct $M_H$, $W_H$ and $B_H$ as in the proof of Lemma 10.5.2. Since $A_H W_H = B_H$, $W_H$ is regular. So it is possible to choose a unique $j \in J$ such that $y_j > 0$ and the ration vector

$$[\frac{w_{j,0}}{y_j}, \cdots, \frac{w_{j,n}}{y_j}]$$

is lexico minimum.

Construct $M'$ from $M_{\tau,s}$ by interchanging columns $j$ and $r$ of $M_{\tau,s}$, and replacing the new column $r$ by $u^r$. There is a lexico positive $(2n+1) \times (n+1)$ matrix $W'$ such that $M'W' = B$. Note that if we had selected any other $j \in J$, the corresponding $W'$ would not be lexico positive.

Suppose that $j \notin \pi(T)$. Define $\pi' : T \mapsto \{1, \cdots, 2n\}$ by letting $\pi'(i) = \pi(i)$ if $\pi(i) \neq r$, and $\pi'(i) = j$ if $\pi(i) = r$. Define $p \in \{-1, 0, +1\}^n$ by letting $p_j = 1$ if $1 \leq j \leq n$, $p_{n-j} = -1$ if $n+1 \leq j \leq 2n$, and $p_i = q_i$ otherwise. Then there exists a unique simplex $\sigma' \neq \sigma$ in $A(p)$ so that $\tau$ is a facet of $\sigma'$. Note that $M' = M_{\tau,q}$, $M_{\tau,q}W' = B$ and $W' \succ \mathbf{0}$. Thus, conclusion (ii) holds.

Now suppose that $j = \pi(h)$ for some $h \in T$. Let $\pi'$ be defined as in the previous paragraph. Define $\rho : \{1, \cdots, t-1\} \mapsto \{1, \cdots, 2n\}$ by letting

$\rho(i) = \pi'(i)$ if $1 \leq i \leq h-1$, and $\rho(i) = \pi'(i+1)$ if $h \leq i \leq t-1$. Let $\tau'$ be the facet of $\tau$ not containing $x^h$. Clearly, $M_{\tau',q} = M'$. Then the system $M'W' = B$, the sign vector $q$ and the function $\rho$ show that the facet $\tau'$ of $\tau$ not containing $x^h$ is $q$-complete. Thus, conclusion $(i)$ holds. □

An $s$-complete simplex is called a *node*. Two nodes $\sigma$ and $\sigma'$ are *adjacent* if one of the following cases occurs:

(a) $\sigma \subseteq bd(C^n)$ and $\sigma = -\sigma'$;

(b) $\sigma$ and $\sigma'$ are very complete facets of a common simplex;

(c) one node is a very complete facet of the other.

Summarizing the above results and applying Theorem 1.4.14, we have the following convergence theorem.

**Theorem 10.5.4** *For any given positive integer $d$, there exists a finite sequence of distinct simplices $\tau^0, \cdots, \tau^m$ in the triangulation $AS(d)$ of $C^n$ such that*
*(i) $\tau^0$ is the zero simplex $< v >$;*
*(ii) $\tau^m$ is a terminal simplex;*
*(iii) $\tau^i$ is a very complete facet for $i = 0, \cdots, m-1$;*
*(v) $\tau^i$ is adjacent to $\tau^{i-1}$ and $\tau^{i+1}$ for $i = 1, \cdots, m-1$.*
*The terminal simplex contains a point $y$ such that $\bar{f}(y) = 0^n$. That is, $y$ is an approximate zero point of $f$.*

Now we wind up this chapter with an extension of Theorem 10.2.1 to a point-to-set mapping. Since the proof is quite similar to the proof of Kakutani's fixed point theorem, we leave it as an exercise.

**Theorem 10.5.5** *Let $\psi : C^n \mapsto \mathbb{R}^n$ be a nonempty-valued, compact-valued, and convex-valued u.s.c. point-to-set mapping. If for every $x \in bd(C^n)$, there exist $y \in \psi(x)$ and $z \in \psi(-x)$ such that $y = -z$, then there is at least one point $x^* \in C^n$ such that $0^n \in \psi(x^*)$.*

# CHAPTER 11

# Kuhn's Method for Finding All Roots of Univariate Polynomials

## 11.1. Introduction

In this chapter we consider the problem of finding all roots of a monic polynomial $f(z)$ of degree $n$ in the complex variable $z$ with complex numbers as coefficients. As it is well-known, the *fundamental theorem of algebra* asserts that the polynomial $f(z)$ has at least one root. See for example van der Waerden [1970]. There are many non-constructive proofs for this theorem in the literature. The problem of finding all roots of a polynomial is not only interesting in its theoretical nature but also very important in many engineering designs such as, to mention a few, control system design and telecommunication system design. Kuhn [1974, 1977] developed a remarkable algorithm which is guaranteed to find all roots of a polynomial, whereas all the existing methods before Kuhn's method are not guaranteed to find all roots. Kuhn's method sets yet another powerful example of simplicial algorithms. Briefly speaking, Kuhn's method is based on both a continuously refining triangulation of the cross product of the complex plane $\mathbb{C}$ and an interval $[-1, +\infty)$, and a special integer labeling rule. The algorithm starts with $n$ entrances at level $\mathbb{C} \times \{-1\}$, and generates $n$ paths of simplices which lead to $n$ approximate roots of any a priori chosen accuracy within a finite number of steps. By taking limit, the $n$ paths lead to $n$ roots of the polynomial $f(z)$. Kuhn's method will be used as a subroutine for finding all solutions of multivariate polynomial equations to be discussed in the next chapter.

This chapter is organized as follows. In Section 2 an integer labeling rule is introduced and one basic result linking the labeling rule and the roots of the polynomial is established. In Section 3 we describe a triangulation of $\mathbb{C} \times [-1, +\infty)$ which underlies the algorithm. In Section 4 we introduce Kuhn's algorithm in detail. In Section 5 it will be proved that the $n$ paths generated by Kuhn's algorithm converge to all roots of the polynomial. Finally, in Section 6 we analyze the complexity of Kuhn's algorithm, discuss an extension of Kuhn's algorithm for solving the complex transcendental equation, and report a number of numerical results.

## 11.2. An integer labeling rule

We first give some notation. For a complex number $z = x + iy \in \mathbb{C}$, $x$ and $y$ are the *real* and *imaginary part* of the complex number, and $i$ is the *imaginary unit* with the property $i^2 = -1$. The *modulus* or *absolute value* of the complex number $z$ is defined to be square root of $x^2 + y^2$, denoted by $|z|$. Let $f(z) = c_n z^n + c_{n-1} z^{n-1} + \cdots + c_0$ be a *monic polynomial* of degree $n$ in complex variable $z$ with complex constants $c_n, \cdots, c_0$. By definition $c_n = 1$. The goal is to find the $n$ roots of $f(z)$, counting multiplicities. It is well-known that $f(z)$ can be expressed as $f(z) = (z - \xi_1) \cdots (z - \xi_n)$, where $\xi_1, \cdots, \xi_n$ are the $n$ roots of $f(z)$. Now we describe Kuhn's integer labeling rule.

**Definition 11.2.1** *Let $z = x + iy$, $w = u + iv$ and $f_{-1}(z) = z^n$ and $f_k(z) = f(z)$ for all $k \in \mathbb{Z}_+$. If $f_k(z) = u + iv \neq 0$, then choose* $\arg f(z)$ *to be the unique angle $\alpha$ such that $-\pi < \alpha \leq \pi$, $\cos(\alpha) = \frac{u}{(u^2+v^2)^{1/2}}$ and $\sin(\alpha) = \frac{v}{(u^2+v^2)^{1/2}}$. For each $(z, k) \in \mathbb{C} \times \{k\}$, the label of $(z, k)$ induced by $f_k(z)$ is given by*

$$l(z,k) = \begin{cases} 1, & if \quad -\frac{\pi}{3} \leq \arg f_k(z) \leq \frac{\pi}{3} \ \ or \ \ f_k(z) = 0, \\ 2, & if \quad \frac{\pi}{3} < \arg f_k(z) \leq \pi, \\ 3, & if \quad -\pi < \arg f_k(z) < -\frac{\pi}{3}. \end{cases}$$

It follows from the definition that if $k \geq 0$, then the label of $(z, k)$ is determined by $f(z)$, while if $k = -1$, then the label of $(z, k)$ is determined by $z^n$. This definition is illustrated geometrically in Figure 11.1, where open boundaries are indicated by thin lines and closed boundaries by thick lines. Clearly the label of a point $(z, k)$ is determined by the area in which $f_k(z)$ falls. Note that the image of a straight line of $z$-plane under $f_k$ is a smooth curve on $w$-plane and is not necessarily a straight line. Obviously a triangle $< a, b, c >$ in the $z$-plane which carries all three labels 1, 2, 3, contains an approximate root of $f(z)$, since the origin of $w$-plane is contained by the image of the triangle under $f$. It is readily seen that as the mesh size of the triangle becomes smaller and smaller, the accuracy of the root of $f(z)$ approximated by the triangle is becoming higher and higher. This is the basic idea of Kuhn's algorithm.

Let $\mathcal{T}$ be a continuously refining (CR) triangulation of $\mathbb{C} \times [-1, +\infty)$ such that each level $\mathbb{C} \times \{k\}$ is triangulated by $\mathcal{T}$ for $k = -1, 0, \cdots$. A 2-simplex of $\mathcal{T}$ is called to be *completely labelled* (c.l.) if it carries all three labels 1, 2 and 3. A 3-simplex is called to be *completely labelled* (c.l.) if one of its facets is c.l. Furthermore, a 1-simplex of $\mathcal{T}$ on $\mathbb{C} \times \{-1\}$ is called to be *almost completely labelled* (a.c.l.) if its label set is equal to $\{1, 2\}$. A 2-simplex of $\mathcal{T}$ on $\mathbb{C} \times \{-1\}$ is called to be *almost completely labelled* (a.c.l.) if one of its facets is a.c.l. Now we can establish the following lemma.

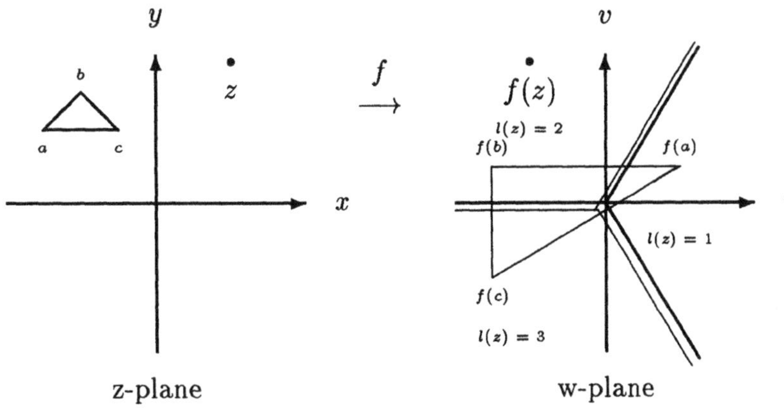


*Figure 11.1.* The image of the points in $z$-plane under $w = f(z)$.

**Lemma 11.2.2** *If a 2-simplex $\sigma =< (z_1, k), (z_2, k), (z_3, k) >$ is a completely labelled simplex induced by $f(z)$ and $|f(z_i) - f(z_j)| \leq \epsilon$ for all $i, j \in I_3$, then it holds that*

$$|f(z_i)| \leq 2\epsilon/\sqrt{3}$$

*for all $i \in I_3$.*

Proof: In Figure 11.2 the values of $f(z)$ are illustrated in the $w$-plane. Areas which induce 1, 2 and 3 are labelled accordingly. By the assumption of the lemma, the shaded region exhibits all $w = f(z)$ with $l(z, k) = 1$ which are within $\epsilon$ of areas 2 and 3. Therefore $|f(z_1)| \leq 2\epsilon/\sqrt{3}$. Similarly we can prove for $i = 2$ and 3. □

## 11.3. A triangulation of $\mathbb{C} \times [-1, \infty)$

Now we introduce a *CR* triangulation of $\mathbb{C} \times [-1, +\infty)$ which underlies Kuhn's algorithm. The triangulation of $\mathbb{C} \times [-1, +\infty)$ will be denoted by $\mathcal{T}$. The half of the 3-dimensional space expressed as

$$\mathbb{C} \times [-1, \infty) = \{(z, d) \mid z = x + iy \in \mathbb{C},\ d \geq -1\}$$

contains a sequence of replicas $C_d = \mathbb{C} \times \{d\}$ of the complex plane $\mathbb{C}$ for $d = -1, 0, 1, \cdots$. Given a center $\bar{z} = \bar{x} + i\bar{y}$ and a grid size $h > 0$, we can define a triangulation $T_d(\bar{z}, h)$ of $C_d$ as follows. The triangulation $T_{-1}(\bar{z}, h)$ is the collection of 2-simplices $< z_1, z_2, z_3 >$ specified uniquely by

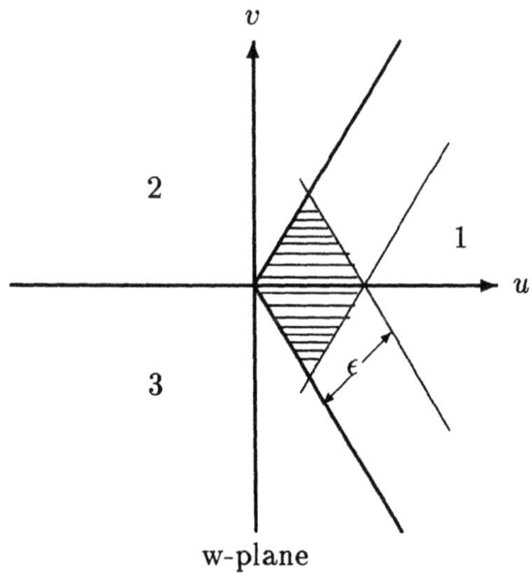


*Figure 11.2.* The three areas of $f(z)$ in $w$-plane.

the following 3 vertices

$$\begin{aligned} z_1 &= (\bar{x} + rh) + i(\bar{y} + sh), \\ z_2 &= [\bar{x} + (r+a)h] + i[\bar{y} + (s+b)h], \\ z_3 &= [\bar{x} + (r-b)h] + i[\bar{y} + (s+a)h], \end{aligned}$$

where $r$, $s$ are integers and $r+s$ is even, and $(a,b) \in \{(1,0),(0,1),(-1,0),(0,-1)\}$. The mesh size of $T_{-1}(\bar{z},h)$ is $\sqrt{2}h$. This is illustrated in Figure 11.3.

For $d \in \mathbf{Z}_+$, the triangulation $T_d(\bar{z},h)$ is the collection of 2-simplices $<(z_1,d),(z_2,d),(z_3,d)>$ specified uniquely by the following 3 vertices

$$\begin{aligned} z_1 &= (\bar{x} + rh/2^d) + i(\bar{y} + sh/2^d), \\ z_2 &= [\bar{x} + (r+a)h/2^d] + i[\bar{y} + (s+b)h/2^d], \\ z_3 &= [\bar{x} + (r-b)h/2^d] + i[\bar{y} + (s+a)h/2^d], \end{aligned}$$

where $r$, $s$ are integers and $r+s$ is odd, and $(a,b) \in \{(1,0),(0,1),(-1,0),(0,-1)\}$. The mesh size of $T_d(\bar{z},h)$ is $\sqrt{2}h/2^d$. This is illustrated in Figure 11.4.

The layer between $C_{-1}$ and $C_0$ is subdivided into 3-simplices or tetrahedra as shown in Figure 11.5 for a square. There are five 3-simplices, i.e.,

$$< A_1, D_1, A, B_1 >, < B, B_1, C, A >, < C_1, C, D_1, B_1 >,$$

$$< D, A, C, D_1 >, < B_1, A, C, D_1 > .$$

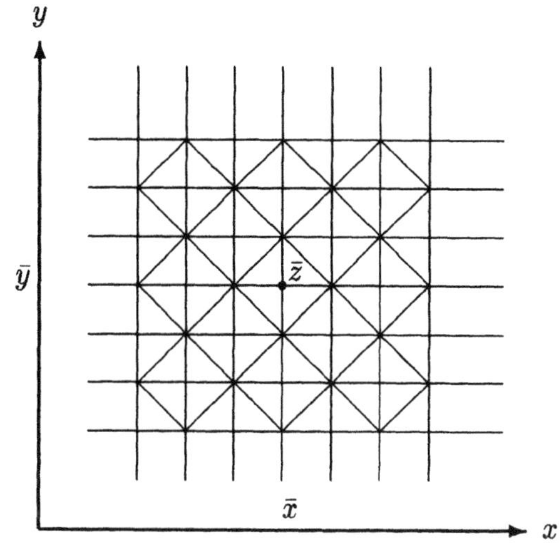


*Figure 11.3.* The triangulation of $C_{-1}$.

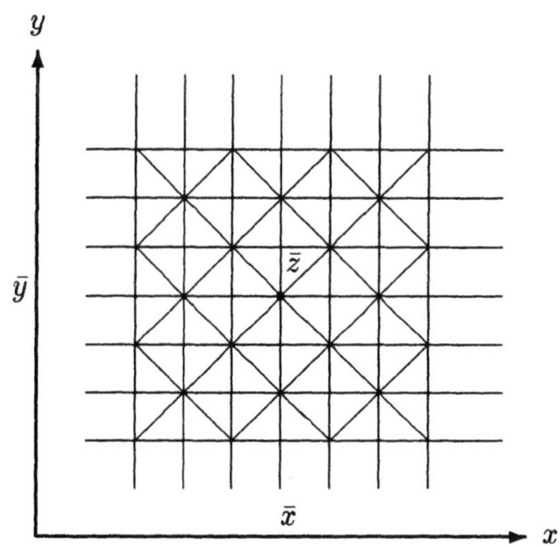


*Figure 11.4.* The triangulation of $C_d$ for $d \geq 0$.

All similar squares are treated in the same way.

The layer between $C_d$ and $C_{d+1}$, where $d \in \mathbb{Z}_+$, is subdivided into 3-simplices or tetrahedra as shown in Figure 11.6 for a square. There are

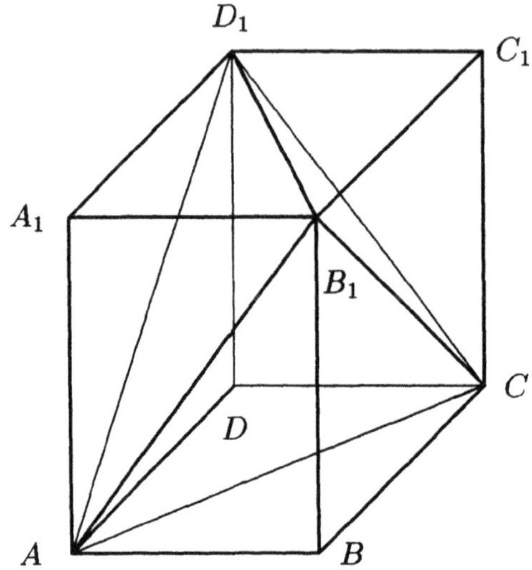


*Figure 11.5.* The triangulation of the layer between $C_{-1}$ and $C_0$.

fourteen 3-simplices, i.e.,

$$< A_1, A, F, K >, < A_1, A, E, K >, < B_1, B, F, K >, < B_1, B, G, K >,$$
$$< C_1, C, G, K >, < C_1, C, H, K >, < D_1, D, E, K >, < D_1, D, H, K >,$$
$$< A, B, F, K >, < C, D, H, K >, < B, C, G, K >, < A, D, E, K >,$$
$$< A, B, C, K >, < A, C, D, K > .$$

All similar squares are treated in like manner.

Let $\sigma =< (z_1, d_1), (z_2, d_2), (z_3, d_3) >$ be a 2-simplex in the triangulation of $\mathbb{C} \times [-1, \infty)$. Then we have $d_i = d$ or $d + 1$ for some integer $d \geq -1$, and $i \in I_3$. So we say that the three points $\{(z_1, d_1), (z_2, d_2), (z_3, d_3)\}$ lies between levels $d$ and $d + 1$. It is easy to derive the following result by the construction of the triangulation.

**Lemma 11.3.1** *If $\tau = \{(z_1, d_1), (z_2, d_2), (z_3, d_3)\}$ lies between levels $d$ and $d + 1$, then*

$$diam(\tau) = \sqrt{2}h/2^d.$$

The reader may have realized that the triangulation above is a special case of the triangulations described in Chapter 4. Here we give a complete description of the above triangulation in order for the reader to have a whole picture of Kuhn's method.

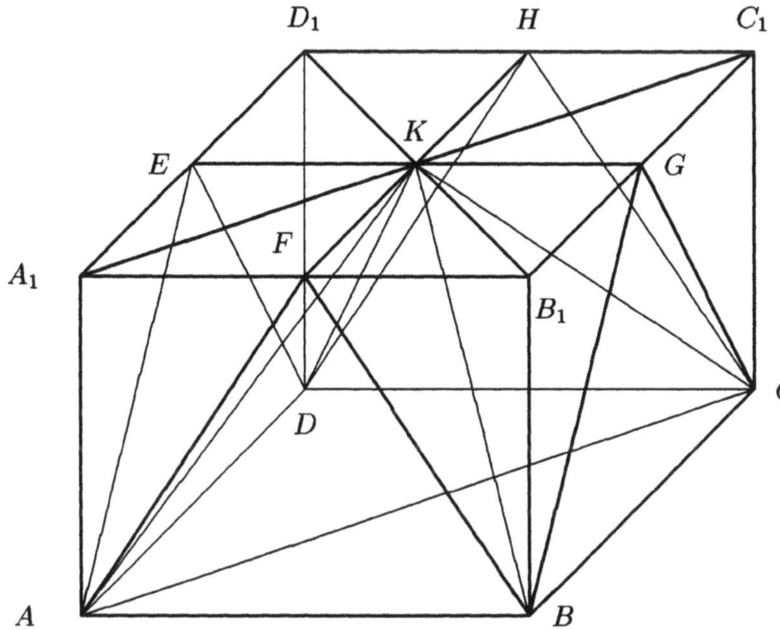


*Figure 11.6.* The triangulation of the layer between $C_d$ and $C_{d+1}$ for $d \geq 0$.

## 11.4. Kuhn's algorithm

In this section we describe Kuhn's algorithm for computing all roots of a polynomial $f(z)$. Each vertex in the triangulation described in the previous section will receive a label 1, 2, or 3. The labeling rule is given in Section 2. Here we need to point out that since we move the origin of $\mathbb{C}$ to the point $\bar{z}$, for each point in $C_{-1}$ its label is now induced by the polynomial $(z-\bar{z})^n$ rather than $z^n$.

In the plane $C_{-1}$ and for any given $m \in \mathbb{N}$, let $Q_m(\bar{z}, h)$ denote the square with corner points at $(\bar{x} \pm mh + i(\bar{y} \pm mh), -1)$. That is,

$$Q_m(\bar{z}, h) = \{(x+iy, -1) \in C_{-1} \mid -mh \leq x-\bar{x} \leq mh \quad \text{and} \quad -mh \leq y-\bar{y} \leq mh\}.$$

Then $\partial Q_m(\bar{z}, h)$ denotes the boundary of $Q_m(\bar{z}, h)$. The boundary $\partial Q_m(\bar{z}, h)$ is oriented counterclockwise and the 2-simplices inside $Q_m(\bar{z}, h)$ are oriented in the usual counterclockwise cyclic order of their vertices. The following lemma says that if $m$ is chosen properly, there exist exactly $n$ almost completely labelled 1-simplices on $\partial Q_m(\bar{z}, h)$.

**Lemma 11.4.1** *Let $m \geq \frac{3n}{2\pi}$ be an integer. Then in the counterclockwise order it holds that*
*(i) for any 1-simplex $\tau = < y^1, y^2 >$ in $\partial Q_m(z, h)$, the possibilities of the*

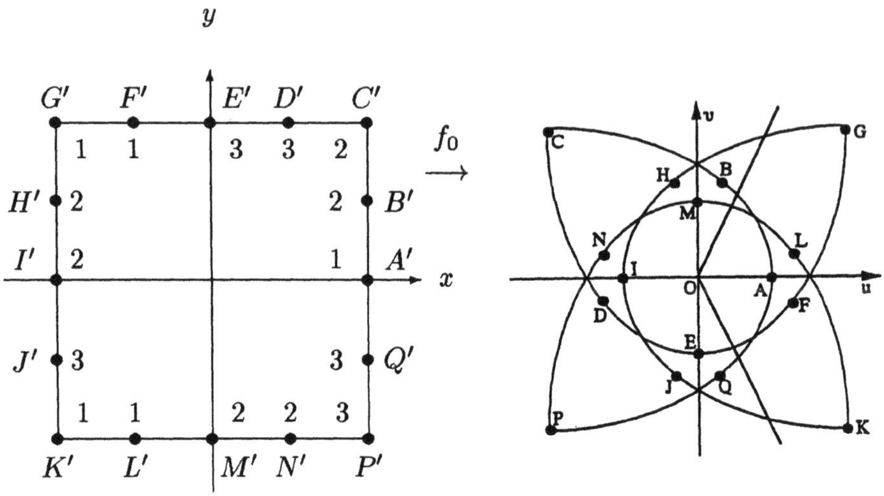


*Figure 11.7.* The image of $\partial Q_m(0,h)$ under $w = z^n$.

*labels of* $\tau$

$$(1,1),\ (2,2),\ (3,3),\ (1,2),\ (2,3),\ (3,1).$$

(*ii*) *there are exactly* $n$ *1-simplices with labels* $(1,2)$.

Proof: We consider the case $\bar{z} = 0$. For $\bar{z} \neq 0$, the proof goes similarly. Notice that the label of every point in $C_{-1}$ is induced by $w = z^n$. Since $\partial Q_m(0,h)$ is symmetric, the image of $\partial Q_m(0,h)$ under $w = z^n$ is also symmetric and circles the origin of the $w$-plane exactly $n$ times. See Figure 11.7 for $n = 3$ and $m = 2$ where $A = f_0(A')$, $B = f_0(B')$, etc.

Let $< (\tilde{z}, -1), (\bar{z}, -1) >$ be a 1-simplex in $\partial Q_m(0,h)$. Then the angle $\theta$ in Figure 11.8 can be calculated as follows.

$$\begin{aligned}
\theta &= \arg \tilde{z} - \arg \bar{z} \\
&= \arctan(\tfrac{l+1}{m}) - \arctan(\tfrac{l}{m}) \\
&= \tfrac{1}{1+\beta}\tfrac{1}{m} \\
&< \tfrac{1}{m}
\end{aligned}$$

where $\frac{1}{m} < \beta < \frac{l+1}{m}$. The third equality above follows from Lagrange's

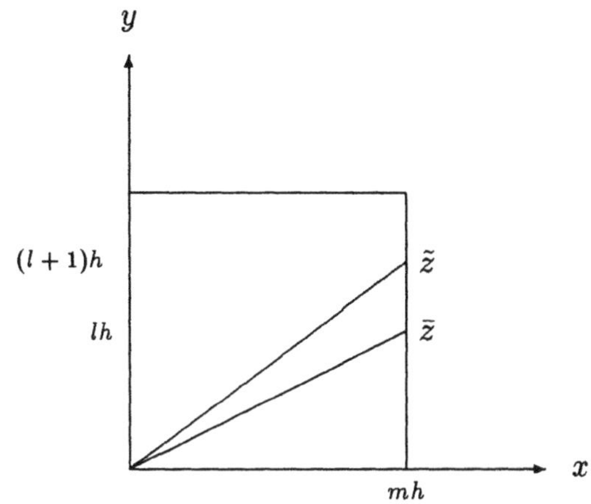


*Figure 11.8.* The angle $\theta$ in $Q_m(0,h)$.

intermediate theorem. So we have

$$\begin{aligned} 0 < \gamma = \arg(\tilde{z})^n - \arg(\bar{z})^n &= n(\arg\tilde{z} - \arg\bar{z}) \\ &< \tfrac{n}{m} \\ &\leq \tfrac{n}{\frac{3n}{2\pi}} = \tfrac{2\pi}{3}. \end{aligned}$$

This also holds for 1-simplices in $\partial Q_m(0,h)$ in horizontal case.

This means that the possibilities of the labels of $\tau =< (\bar{z}, -1), (\tilde{z}, -1) >$ are

$$(1,1),\ (2,2),\ (3,3),\ (1,2),\ (2,3),\ (3,1).$$

On the other hand, since $0 < \gamma < \frac{2\pi}{3}$ and the image of $\partial Q_m(0,h)$ under $w = z^n$ circles the origin of the $w$-plane exactly $n$ times, it is readily seen that there are $n$ 1-simplices on $\partial Q_m(0,h)$ with labels $(1,2)$ in the counterclockwise way.

Note that it is impossible to have a 1-simplex with labels $(2,1)$ in the counterclockwise way, since $(2,1)$ implies that the change of the labels of this 1-simplex jumps two label areas. □

**Lemma 11.4.2** *If $m \geq 3(1+\sqrt{2})n/4\pi$, then there exist no completely labelled 2-simplices outside $Q_m(\bar{z},h)$.*

Proof: We only consider the case $\bar{z} = 0$. Note that

$$m \geq 3(1+\sqrt{2})n/4\pi > \frac{3n}{2\pi}.$$

This means by Lemma 11.4.1 that we have

$$0 < \arg(\tilde{z})^n - \arg(\bar{z})^n < 2\pi/3$$

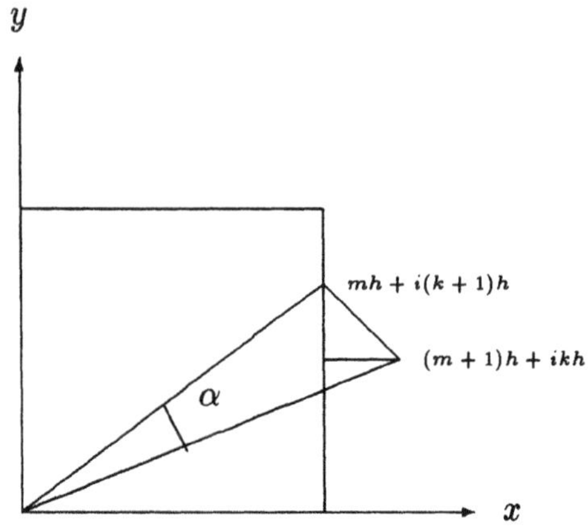


*Figure 11.9.* The angle $\theta$ for the diagonal 1-simplex case.

for any horizontal or vertical 1-simplex $\tau =< (\bar{z}, -1), (\tilde{z}, -1) >$ lying outside or on the boundary of $Q_m(0,h)$. We will show that the above relation also holds for any diagonal 1-simplex. The worst case is illustrated in Figure 11.9, where $k$ is chosen so as to maximize the angle $\alpha$.

Since $\tan\gamma = \frac{k+1}{m}$ and $\tan\beta = \frac{k}{m+1}$, we have

$$\begin{aligned}\tan\alpha &= \tan(\gamma-\beta)\\ &= \frac{\tan\gamma-\tan\beta}{1+\tan\gamma\cdot\tan\beta}\\ &= \frac{m+k+1}{m^2+k^2+m+k}.\end{aligned}$$

This can be maximized over the continuous values of $k$ when

$$k = \sqrt{2m(m+1)} - (m+1).$$

Notice that when $k = \sqrt{2m(m+1)} - (m+1)$, we have

$$0 < \tan\alpha = \frac{1}{2\sqrt{2m(m+1)} - 2m - 1} < \frac{1}{2m(\sqrt{2}-1)} = \frac{1+\sqrt{2}}{2m}.$$

So if $m \geq 3(1+\sqrt{2})n/4\pi$, it holds

$$0 < \alpha < \tan\alpha < \frac{1+\sqrt{2}}{2\lceil\frac{3(1+\sqrt{2})n}{4\pi}\rceil} = \frac{2\pi}{3n}.$$

Summarizing the above results, we have

$$0 < \arg(\tilde{z})^n - \arg(\bar{z})^n < 2\pi/3$$

for any horizontal, vertical or diagonal 1-simplex $\tau = <(\bar{z}, -1), (\tilde{z}, -1)>$ lying outside or on the boundary of $Q_m(0, h)$.

To complete the proof, suppose to the contrary that we have a completely labelled 2-simplex $\sigma = <(z, -1), (z', -1), (z'', -1)>$ outside $Q_m(0, h)$. We may assume the following

$$0 < \arg(z')^n - \arg z^n < \tfrac{2\pi}{3},$$
$$0 < \arg(z'')^n - \arg(z')^n < \tfrac{2\pi}{3},$$
$$0 < \arg(z'')^n - \arg z^n < \tfrac{2\pi}{3}.$$

If $l((z, -1)) = i$, then the first inequality implies $l((z', -1)) = i + 1$. Note that the labels of $\sigma$ follow the cyclic order 1, 2, 3. Since $\sigma$ is completely labelled, we have $l((z'', -1)) = i + 2$. This contradicts the third inequality. □

From now on we will always assume that $m \geq 3(1 + \sqrt{2})n/4\pi$. In order to prove the convergence, we construct a graph $G = (V, A)$ as follows. A simplex $\sigma$ in $\mathcal{T}$ is a *node* in $V$ if one of the following cases occurs:

(a) $\sigma$ is an almost completely labelled 1-simplex in $\partial Q_m(\bar{z}, h)$;

(b) $\sigma$ is an almost completely labelled 2-simplex in $Q_m(\bar{z}, h)$;

(c) $\sigma$ is a completely labelled 3-simplex.

Two nodes $\sigma_1$ and $\sigma_2$ are *adjacent* (i.e., $\{\sigma_1, \sigma_2\}$ is an arc in $A$) if one of the following cases occurs:

(a) $\sigma_1$ and $\sigma_2$ share a completely labelled 2-simplex as their common facet;

(b) both $\sigma_1$ and $\sigma_2$ are in $C_{-1}$ and share an almost completely labelled 1-simplex as their common facet;

(c) one of the two nodes is a facet of the other.

The degree of a node $\sigma$, denoted by $deg(\sigma)$, is the number of nodes adjacent to it.

**Lemma 11.4.3** *Let $\sigma$ be a node in $G = (V, A)$ defined above. Then if $\sigma$ is an almost completely labelled 1-simplex on $\partial Q_m(\bar{z}, h)$, we have $deg(\sigma) = 1$. Otherwise, we have $deg(\sigma) = 2$.*

Proof: Note that every 1-simplex $\sigma$ on $\partial Q_m(\bar{z}, h)$ is a facet of exactly one 2-simplex $\sigma'$ in $Q_m(\bar{z}, h)$. It is clear that if $\sigma$ is a.c.l., then $\sigma'$ is a node. Thus, $deg(\sigma) = 1$.

If $\sigma \in \mathcal{T}$ is a 3-simplex and a node, then $\sigma$ has exactly two facets being completely labelled 2-simplices $\tau_1$ and $\tau_2$. If $\tau_1$ and $\tau_2$ are not subsets of $C_{-1}$, then by Theorem 1.4.6 there are exactly one 3-simplex $\sigma_1$ sharing $\tau_1$

as its facet and exactly one 3-simplex $\sigma_2$ sharing $\tau_2$ as its facet, with $\sigma$, respectively. Clearly, $\sigma_1$ and $\sigma_2$ are two nodes adjacent to $\sigma$. If one of the two facets $\tau_1$ and $\tau_2$, say $\tau_1$, is a subset of $C_{-1}$, then by Lemma 11.4.2 $\tau_1$ lies in $Q_m(\bar{z}, h)$. So $\tau_1$ itself is a node and adjacent to $\sigma$. $\sigma_2$ as above is a 3-simplex sharing $\tau_2$ as its facet with $\sigma$ and is also a node adjacent to $\sigma$. Thus $deg(\sigma) = 2$.

If $\sigma$ is an almost completely labelled but not a completely labelled 2-simplex in $Q_m(\bar{z}, h)$, we can show that $deg(\sigma) = 2$ in a similar way as above.

Finally we consider the case in which $\sigma$ is a completely labelled 2-simplex in $Q_m(\bar{z}, h)$. In this case, $\sigma$ has exactly one almost completely labelled facet $\tau$ and there exists exactly one 3-simplex $\sigma_2$ in $\mathcal{T}$ having $\sigma$ as its facet. Since $\sigma$ is completely labelled, so is $\sigma_2$. Thus, $\sigma_2$ is a node adjacent to $\sigma$. On the other hand, if $\tau$ lies on the boundary of $Q_m(\bar{z}, h)$, then $\tau$ is a node adjacent to $\sigma$ by definition. Otherwise, if $\tau$ lies in $Q_m(\bar{z}, h)$, then by Theorem 1.4.6 there exists exactly one simplex $\sigma_1$ sharing $\tau$ as its facet with $\sigma$. Clearly, $\sigma_1$ is a node adjacent to $\sigma$. Therefore $deg(\sigma) = 2$. □

We have the following simple corollary.

**Corollary 11.4.4** *Let $G = (V, A)$ be defined as above. Then there exist simple paths or circuits in the graph $G$.*

As shown in Lemma 11.4.1, there are exactly $n$ almost completely labelled 1-simplices in $\partial Q_m(\bar{z}, h)$. Let us denote these $n$ 1-simplices by $\tau_{1,0}, \cdots, \tau_{n,0}$, respectively. And for each $i \in I_n$, let $\sigma_{i,1}$ be the unique 2-simplex in $Q_m(\bar{z}, h)$ having $\tau_{i,0}$ as its facet. In general, for each $i \in I_n$ and each $j \in \mathbb{N}$ by induction we denote a node adjacent to both $\sigma_{i,j}$ and $\sigma_{i,j+2}$ by $\sigma_{i,j+1}$. In this way we can obtain $n$ sequences of adjacent nodes in $\mathcal{T}$, $\{\sigma_{i,j} \mid j \in \mathbb{N}\}$ for all $i \in I_n$. We illustrate these sequences for $n = 3$, $m = 2$, $\bar{z} = 0$ and $h = 1$ in Figure 11.10. Kuhn's algorithm generates exactly these $n$ sequences $\{\sigma_{i,j} \mid j \in \mathbb{N}\}$ for $i = 1, \cdots, n$. As will be shown later, these $n$ sequences will converge to $n$ roots of $f(z)$, counting multiplicities. Now by using Theorem 1.4.14 we have the following theorem.

**Theorem 11.4.5** *For each $i \in I_n$, we have*
*(i) when $j \geq 1$, $dim(\sigma_{i,j}) \geq 2$;*
*(ii) there exists a number $K(i)$ such that $j \geq K(i)$ implies $dim(\sigma_{i,j}) = 3$;*
*(iii) the sequence $\{\sigma_{i,j} \mid j \in \mathbb{N}\}$ generates an infinite simple path.*

## 11.5. Convergence proofs

In this section we prove that the $n$ sequences $\{\sigma_{i,j} \mid j \in \mathbb{N}\}$ for $i = 1, \cdots, n$ described in the previous section converge to $n$ roots of $f(z)$, counting multiplicities.

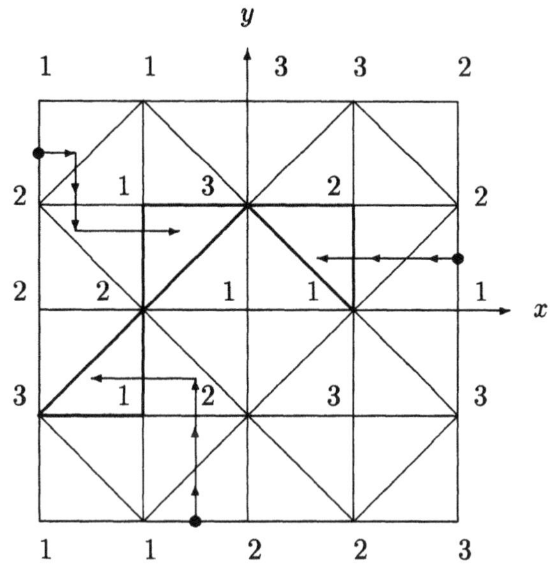


*Figure 11.10.* The $n$ simple paths in $Q_m(\bar{z}, h)$.

**Lemma 11.5.1** *For any $w \in \mathbb{C}$ with $|w| < 1$, we have*

$$|\arg(1+w)| < \frac{\pi}{2}|w|.$$

Proof: If $w$ is a real number, the conclusion is trivially true. So we assume that $w$ is a complex number.

There are four cases to occur. By symmetry, we only need to consider the following two cases. (In the remaining of the proof $|AT|$ denotes the distance between $A$ and $T$, $|AB|$, $|AH|$, $|PD|$ likewise.)
Case 1: In this case $0 \le \arg w \le \frac{\pi}{2}$, see Figure 11.11. Since

$$\sin\theta = h \le |AT| \le |AB| = |w|,$$

this implies

$$\theta \le \arcsin|w|.$$

That is to say,

$$|\arg(1+w)| \le \arcsin|w|.$$

Case 2: In this case $\frac{\pi}{2} \le \arg w \le \pi$, see Figure 11.12. Since

$$\sin\theta = |PD| = |AH| \le |AB| = |w|,$$

this means

$$\theta \le \arcsin|w|.$$

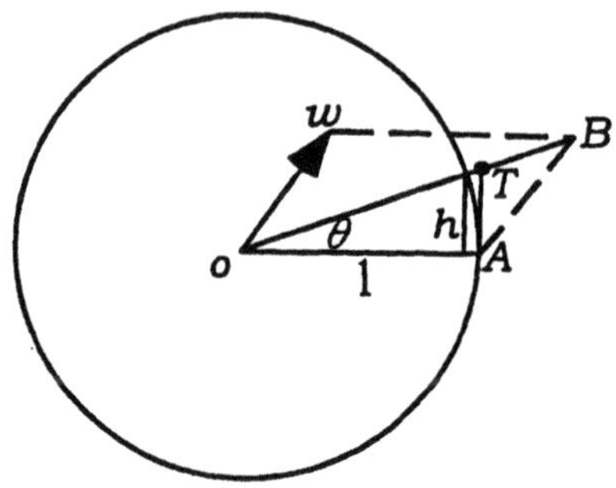


*Figure 11.11.* The first case for $w$ in $z$-plane.

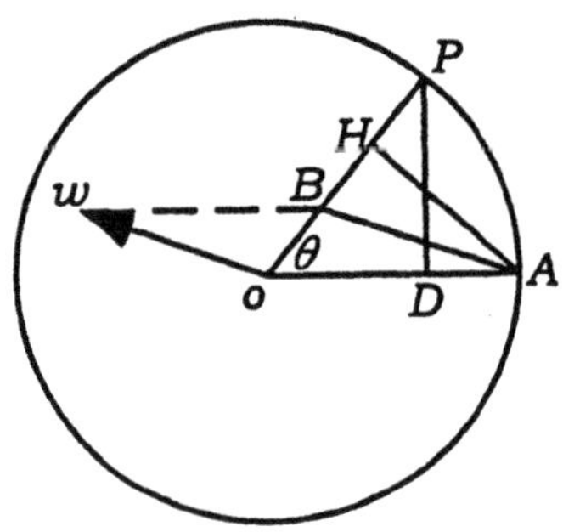


*Figure 11.12.* The second case for $w$ in $z$-plane.

That is to say,

$$|\arg(1+w)| \leq \arcsin|w|.$$

By using the following well-known inequality from mathematical analysis

$$\theta \leq \frac{\pi}{2}\sin\theta, \quad \text{for} \quad 0 < \theta < \frac{\pi}{2},$$

we have

$$|\arg(1+w)| \leq \arcsin|w| < \frac{\pi}{2}\sin(\arcsin|w|) = \frac{\pi}{2}|w|$$

since $0 < \arcsin|w| < \frac{\pi}{2}$. □

For simplicity we shall assume in the rest of this section that the two parameters $h = 1$ and $\bar{z} = 0$.

**Lemma 11.5.2** *Let*

$$R = \max\{3\sqrt{2}(2+\pi)n/4\pi, 1 + \max_{0\le i\le n} |c_i|\frac{n}{n-1}\} + \sqrt{2}$$

*and* $B(R) = \{z \in \mathbb{C} \mid |z| \le R\}$. *Then there exist no completely labelled 2-simplices outside* $B(R) \times [-1, \infty)$.

Proof: Let

$$r = \max\{3\sqrt{2}(2+\pi)n/4\pi, 1 + \max_{0\le i\le n} |c_i|\frac{n}{n-1}\}.$$

Then $R = r + \sqrt{2}$. Notice that the mesh size of any 2-simplex in $\mathcal{T}$ is less than or equal to $\sqrt{2}$. This means that if one vertex of a 2-simplex lies outside of $B(R) \times [-1, \infty)$, then all its facets must be outside of $B(r) \times [-1, \infty)$. So in order to show that there are no completely labelled simplices outside of $B(R) \times [-1, \infty)$, it is sufficient to prove that for any 1-simplex of $\mathcal{T}$ outside $B(r) \times [-1, \infty)$ with vertices $(z, k)$ and $(z', l)$ we have $|\arg(w/w')| < 2\pi/3$ where $w = f_k(z)$ and $w' = f_l(z')$.

For any $z \in \mathbb{C}$ with $|z| > r$, we rewrite $f(z)$ as

$$f(z) = z^n(1 + c_{n-1}/z + \cdots + c_0/z^n) = z^n(1 + h(z)),$$

where $h(z) = c_{n-1}/z + \cdots + c_0/z^n$. Then we have

$$|h(z')| \le |c_{n-1}|/r + \cdots + |c_0|/r^n < \max\{|c_i|\}/(r-1) \le (n-1)/n,$$

and

$$\begin{aligned} |h(z) - h(z')| &\le |c_{n-1}||1/z - 1/z'| + \cdots + |c_0||1/z^n - 1/(z')^n| \\ &\le \max\{|c_i|\}|z - z'|(r^{-2} + 2r^{-3} + \cdots + nr^{-(n+1)}) \\ &< \max\{|c_i|\}\sqrt{2}/(r-1)^2 \\ &\le \sqrt{2}(n-1)/n(r-1) \\ &\le \sqrt{2}(n-1)4\pi/[3n\sqrt{2}(2+\pi)(n-1)] \\ &= 4\pi/[3(2+\pi)n]. \end{aligned}$$

It follows that

$$\begin{aligned} \left|\frac{h(z)-h(z')}{1+h(z')}\right| &\le \frac{4\pi}{3(2+\pi)n}/(1 - \frac{n-1}{n}) \\ &= \frac{4\pi}{3(2+\pi)} \\ &< 1. \end{aligned}$$

Applying Lemma 11.5.1, we have

$$\begin{aligned} \left|\arg \frac{f(z)}{f(z')}\right| &\le \left|\arg \frac{z^n}{(z')^n}\right| + \left|\arg(1 + \frac{h(z)-h(z')}{1+h(z')})\right| \\ &\le n\left|\arg \frac{z}{z'}\right| + \frac{\pi}{2}\left|\frac{h(z)-h(z')}{1+h(z')}\right| \\ &< \frac{n\sqrt{2}}{3\sqrt{2}(2+\pi)n/4\pi} + \frac{\pi}{2}\frac{4\pi}{3(2+\pi)} \\ &= \frac{2\pi}{3}. \end{aligned}$$

We have proved the case in which $k \geq 0$ and $l \geq 0$. When $k = l = -1$, this follows from Lemma 11.4.2. □

**Lemma 11.5.3** *If $\sigma =< (z_1, k), (z_2, k), (z_3, k) >$ is a completely labelled simplex with $k \geq 0$, then some root of $f$ is closer to one vertex of the simplex than $\frac{3n\delta}{4}$, where $\delta = diam(\sigma)$.*

Proof: It is trivially true for $n = 1$. Now we discuss the case in which $deg(f) = n > 1$. As before, we can write $f(z) = (z - \xi_1)(z - \xi_2) \cdots (z - \xi_n)$. Suppose to the contrary that the conclusion is false. Then we would have

$$|z_l - \xi_i| > \frac{3n\delta}{4}$$

for all $l \in I_3$ and all $i \in I_n$. Hence for every $i \in I_n$ we have

$$|(z_2 - z_1)/(z_1 - \xi_i)| < \delta/(3n\delta/4) = \frac{4}{3n} < 1.$$

Now it follows from Lemma 11.5.1 that

$$\begin{aligned} |\arg \tfrac{f(z_2)}{f(z_1)}| &= |\arg \tfrac{(z_2-\xi_1)\cdots(z_2-\xi_n)}{(z_1-\xi_1)\cdots(z_1-\xi_n)}| \\ &\leq \textstyle\sum_{i=1}^{n} |\arg \tfrac{z_2-\xi_i}{z_1-\xi_i}| \\ &= \textstyle\sum_{i=1}^{n} \arg |(1 + \tfrac{z_2-z_1}{z_1-\xi_i})| \\ &\leq \textstyle\sum_{i=1}^{n} \tfrac{\pi}{2} |\tfrac{z_2-z_1}{z_2-\xi_i}| \\ &< n \tfrac{\pi}{2} \tfrac{4}{3n} \\ &= \tfrac{2\pi}{3}. \end{aligned}$$

Similarly, we can show that

$$\begin{aligned} |\arg \tfrac{f(z_1)}{f(z_3)}| &< \tfrac{2\pi}{3}, \\ |\arg \tfrac{f(z_2)}{f(z_3)}| &< \tfrac{2\pi}{3}. \end{aligned}$$

Notice that for $k \geq 0$, $(z_1, k)$, $(z_2, k)$ and $(z_3, k)$ are labelled by $f(z)$. It follows that $\sigma$ is not completely labelled. This contradicts the fact that $\sigma$ is a completely labelled simplex. □

For each $i \in I_n$ and each $j \in \mathbf{N}$, let $y_{i,j}$ be a vertex of $\sigma_{i,j}$ and let $y_{i,j} = (z_{i,j}, k_{i,j})$. Then the sequence $\{z_{i,j} \mid j \in \mathbf{N}\}$ has a cluster point in $B(R)$ since $B(R)$ is compact.

**Lemma 11.5.4** *For each $i \in I_n$, let $z^{*i}$ be a cluster point of the sequence $\{z_{i,j} \mid j \in \mathbf{N}\}$. Then $f(z^{*i}) = 0$, i.e., $z^{*i}$ is a root of $f(z)$.*

Proof: Let $K$ be an arbitrary positive integer. Note that the number of simplices in $B(R) \times [-1, K]$ is finite. By Theorem 11.4.5 there exists an integer $i(K)$ such that for any $j \geq i(K)$, $\sigma_{i,j}$ lies above $C_K$. So for $j \geq i(K)$, we have $k_{i,j} \geq K$. Since $z^{*i}$ is a cluster point of the sequence $\{z_{i,j} \mid j \in \mathbb{N}\}$, there is a subsequence of this sequence converging to $z^{*i}$. Without loss of generality, we may assume the sequence $\{z_{i,j} \mid j \in \mathbb{N}\}$ itself converges to $z^{*i}$. Because $f$ is continuous and $B(R)$ is compact, it is sufficient to show that for any given $\epsilon > 0$, there exists a natural number $J$ such that $j > J$ we have $|f(z_{i,j})| < \epsilon$.

Let $J = \lceil \log_2(2\sqrt{2}n^2MR^{n-1}/(\sqrt{3}\epsilon) \rceil$ where $M = \max_{0 \leq i \leq n} |c_i|$ and $R$ is the number in Lemma 11.5.2. Let $< (z_1, k), (z_2, l) >$ be a facet of a completely labelled 2-simplex above $C_J$. Then we have

$$\begin{aligned}
|f(z_1) - f(z_2)| &\leq |z_1^n - z_2^n| + |c_{n-1}||z_1^{n-1} - z_2^{n-1}| + \cdots + |c_1||z_1 - z_2| \\
&\leq \sqrt{2}2^{-J}M(nR^{n-1} + (n-1)R^{n-2} + \cdots + 1) \\
&\leq \sqrt{2}2^{-J}Mn^2R^{n-1} \\
&\leq \frac{\sqrt{2}Mn^2R^{n-1}\sqrt{3}}{2\sqrt{2}n^2MR^{n-1}}\epsilon \\
&= \frac{\sqrt{3}}{2}\epsilon.
\end{aligned}$$

Now it follows from Lemma 11.2.2 that for all $j \geq J$, we have

$$|f(z_{i,j})| \leq \epsilon.$$

□

By noticing that the number of roots of $f(z)$ is finite, one can easily prove the following lemma.

**Lemma 11.5.5** *For each $i \in I_n$, the sequence $\{z_{i,j} \mid j \in \mathbb{N}\}$ converges to a root of $f(z)$.*

Up to now we have given a constructive proof for the fundamental theorem of algebra. However, the above results do not prove that all of the roots are the limits of some sequences. In the rest of this section we are going to show that the $n$ sequences $\{\sigma_{i,j} \mid j \in \mathbb{N}\}$ for $i = 1, \cdots, n$, indeed, converge to all roots of $f(z)$. To do so, we need several auxiliary results. We first introduce a combinatorial Stokes' theorem due to Fan [1967].

Consider a house which naturally has rooms and doors. Some of the doors are *internal*, i.e., doors connecting two rooms, while the remaining doors are *external*, i.e., doors connecting a room with the outside of the house. We make the following assumptions:

(a) Each room has 0, 1 or 2 doors;

(b) For every internal door, one side is painted white and the other side black;

(c) For every external door, both sides are painted with the same color, white or black;

(d) For any room with two doors, the sides of the two doors facing the interior of the room are painted different colors, one white and one black.

A room is considered as *good*, if it has exactly one door. If the side of the unique door facing the interior of a good room is white (black), we say that this is a *good room with a white (black) door.* Under the above four assumptions, the next theorem is proved in Fan [1967].

**Theorem 11.5.6 (Fan Combinatorial Theorem)** *The number of good rooms with a white door minus the number of good rooms with a black door is equal to the number of white external doors minus the number of black external doors.*

Proof: The proof is via a *pairing process.* Let us first go through every door exactly once. To start a path, we either enter a room from the outside of the house through an external door, or go out of a good room through its unique door. Every time when we enter a room through a door and see a second door in the room, we go out of the room through the second door. This will lead us either to the outside of the house or to a new room. According to this rule, our path will be stopped either when we get to the outside of the house through an external door, or when we enter a good room. When we are stopped in either of these two situations, we start a new path. Because of Assumptions (b), (c) and (d) above, if both ends of a path are external doors, then these two external doors must be of different colors. For the same reason, if one end of a path is a white (black) external door and the other end is a good room, then this must be a good room with a white (black) door. Also, if both ends of a path are good rooms, then one of them is a good room with a white door, and the other is a good room with a black door.

After going through every door exactly once by walking on various paths, we pair off the two ends of each path. Then every white (black) external door is paired either with a black (white) external door or with a good room with a white (black) door. Also each good room with a white (black) door is paired either with a good room with a black (white) door, or with a white (black) external door. Clearly we are done. □

We now identify the rooms and the doors on $Q_m(0,h)$. An almost completely labelled 2-simplex $\sigma$ will be called a room. A completely labelled 2-simplex will be called a good room. A facet $\tau$ of a room 2-simplex will be called a door if $\tau$ is almost completely labelled. If a door of a room lying on $\partial Q_m(0,h)$ is labelled with $(1,2)$ in counterclockwise order, we paint both

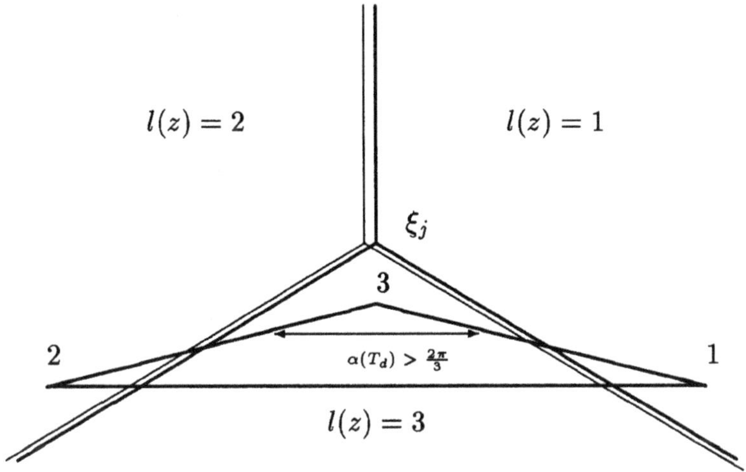


*Figure 11.13.* Approximately equal sectors.

sides of the door with white color, otherwise, we paint with black color. We paint a door of a room with white (black) color if the door is labelled with $(1, 2)$ ( $(2, 1)$) in counterclockwise order when we look the door from the interior of the room. It is clear from Lemma 11.4.1 that there are exactly $n$ white external doors and no black external doors. So there are exactly $n$ good rooms with a white door.

Now we are ready to discuss the case in which the roots of the polynomial $f(z)$ are distinct. For such a polynomial, let $P(f) = \frac{1}{2}\max_{j\neq l}|\xi_j - \xi_l|$. Then the open neighborhoods $B(z_j, \rho) = \{z \in \mathbb{C} \mid |z - \xi_j| < \rho\}$ are disjoint for $0 < \rho < P(f)$. Let $\alpha(T_d)$ denote the supremum of all angles in 2-simplices of the triangulation $T_d$.

**Lemma 11.5.7** *Let $f(z)$ be a polynomial with all roots distinct. For any $0 < \rho < P(f)$, there exists exactly one 2-simplex inside each $B(\xi_j, \rho)$ completely labelled by $f(z)$ for each triangulation $T_d$ with $d$ big enough and $\alpha(T_d) < \frac{2\pi}{3}$.*

Proof: It follows from Lemma 11.5.3 that all 2-simplices completely labelled by $f$ lie inside some $B(\xi_j, \rho)$ and arbitrarily close to $\xi_j$ for $mesh(T_d)$ small enough. Since all of the roots are simple, the mapping defined by $f$ is conformal (see Ahlfors [1979]) near $\xi_j$. So by choosing a small enough neighborhood of $\xi_j$, the regions labelled 1, 2, and 3 are approximately equal sectors rotated by a factor that takes account of $\prod_{l\neq j}(\xi_j - \xi_l)$. This is illustrated in Figure 11.13, where open boundaries are indicated by thin lines and closed boundaries by thick lines.

By an argument similar to the proof of Lemma 11.4.1, for a small enough $mesh(T_d)$, there is a circuit around the boundary of $B(\xi_j, \rho)$ with exactly one edge labelled $(1,2)$ in counterclockwise order and no edge labelled $(2,1)$ in counterclockwise order. Hence by Theorem 11.5.6, there exists one more 2-simplex labelled $(1,2,3)$ in counterclockwise order than 2-simplices labelled $(1,3,2)$ in counterclockwise order inside $B(\xi_j,\rho)$. However, Figure 11.13 shows that it is impossible to have a 2-simplex labelled $(1,3,2)$ near $\xi_j$ when $\alpha(T_d) < 2\pi/3$. □

This leads to the following lemma.

**Lemma 11.5.8** *Let $f(z)$ be a polynomial with all roots distinct. For each root $\xi_i$ of $f$, $i \in I_n$, there exists at most one sequence $\{z_{i,j} \mid j \in \mathbb{N}\}$ convergent to $\xi_i$.*

Proof: It follows from Lemma 11.5.7 that there exists a positive integer $K$ such that for all $j > K$ there is exactly one completely labelled 2-simplex inside $B(\xi_i,\rho)$ for all levels $C_j$. This means that there exists exactly one sequence $\{z_{i,j} \mid j \geq K\}$ convergent to $\xi_i$. □

**Lemma 11.5.9** *The roots of a polynomial $f$ is a continuous function of the coefficients of $f$.*

Proof: We can write $f(z) = z^n + c_{n-1}z^{n-1} + \cdots + c_1 z + c_0 = (z-\xi_1)^{i_1}\cdots(z-\xi_l)^{i_p}$. For any $l \in \{1,2,\cdots,p\}$, there exists $\epsilon > 0$ such that $\xi_l$ is the unique root of $f$ in $\{z \in \mathbb{C} \mid |z-\xi_l| \leq \epsilon\}$. By using the Argument Principle (see e.g., Ahlfors [1979, p.152]) we have

$$i_l = \frac{1}{2\pi i}\int_{|z-\xi_l|=\epsilon} \frac{f'(z)}{f(z)}dz$$

In the above equation, the left side is a positive integer and the right side is clearly a continuous function of the parameters of $f$. This means that if the parameters $c_{n-1}$, $\cdots$, $c_0$ change slightly, the value of the left side remains intact. □

**Theorem 11.5.10** *The $n$ sequences $\{z_{i,j} \mid j \in \mathbb{N}\}$ for $i = 1, \cdots, n$ converge to the $n$ roots $\xi_1, \cdots, \xi_n$ of $f(z)$.*

Proof: We have proved that the theorem is true when all roots of $f(z)$ are distinct. It remains to prove the theorem when there exist multiple roots. We will reduce the general case to the special case of polynomials with all the roots distinct.

Note that $f(z) + \epsilon$ has the derivative $f'(z)$ for all $\epsilon > 0$. Let $\omega_1, \cdots, \omega_{n-1}$ be the roots of $f'(z)$. Then $f(z)+\epsilon$ has all roots distinct if $\epsilon \neq f(\omega_j)$ for $j = 1, \cdots, n-1$. Hence, avoiding this finite set of values, we can choose $\epsilon_0 > 0$ such that $f(z) + \epsilon$ has all roots distinct for $0 < \epsilon \leq \epsilon_0$. Let the $n$

sequences described in the previous section using $f(z)+\epsilon$ for labeling be denoted by $\{z_{i,j}(\epsilon) \mid j \in \mathbb{N}\}$ where $i = 1, \cdots, n$. The index $i$ is assigned using the previous ordering of the $n$ a.c.l. 1-simplices on $\partial Q_m(0,h)$. Since all roots of $f(z)+\epsilon$ are distinct, we know that

$$\lim_{j\to\infty} z_{i,j}(\epsilon) = \xi_i(\epsilon)$$

for all $i = 1, \cdots, n$, where $\xi_1(\epsilon), \cdots, \xi_n(\epsilon)$ are the $n$ distinct roots of $f(z)+\epsilon$.

Now we discuss the relation between $\{z_{i,j}(\epsilon) \mid j \in \mathbb{N}\}$ and $\{z_{i,j} \mid j \in \mathbb{N}\}$. We will show that given any positive integer $J$, there exists $\epsilon_0 > \epsilon_1 > 0$ such that $z_{i,j}(\epsilon) = z_{i,j}$ for $1 \le j \le J$ and $0 < \epsilon \le \epsilon_1$. As shown before, the sequence $\{z_{i,j} \mid j \in \mathbb{N}\}$ is determined uniquely by the labels assigned by $f(z)$. For the finite points $z_{i,1}, \cdots, z_{i,J}$, where $i = 1, \cdots, n$, the labels induced by $f(z)$ and $f(z)+\epsilon$ will be the same for any $0 < \epsilon < \epsilon_1$, since the regions of the $w$-plance in Figure 11.1 yielding labels 1, 2, and 3 are open to the right. The above argument also implies that

$$\lim_{j\to\infty} \lim_{\epsilon\to 0+} z_{i,j}(\epsilon) = \lim_{j\to\infty} z_{i,j}.$$

By using Lemma 11.5.9 we can write

$$\xi_i = \lim_{\epsilon\to 0+} \xi_i(\epsilon)$$

for all $i \in I_n$. We have to show that

$$\xi_i = \lim_{j\to\infty} z_{i,j}$$

for all $i \in I_n$. This will be achieved if we can prove the following equations

$$\begin{aligned} \xi_i &= \lim_{\epsilon\to 0+} \xi_i(\epsilon) \\ &= \lim_{\epsilon\to 0+} \lim_{j\to\infty} z_{i,j}(\epsilon) \\ &= \lim_{j\to\infty} \lim_{\epsilon\to 0+} z_{i,j}(\epsilon) \\ &= \lim_{j\to\infty} z_{i,j}. \end{aligned}$$

It appears quite difficult to verify directly the validity of the interchange of the above double limit.

For each $i$ and $\epsilon$, $1 \le i \le n$ and $0 \le \epsilon < \epsilon_0$, let $z_{i,k}$ and $z_{i,k}(\epsilon)$ be the first $z_{i,j}$ and $z_{i,j}(\epsilon)$ that reach the level $C_k$, $k = 0, 1, \cdots$. These are subsequences that converge to the same limits. That is

$$\lim_{k\to\infty} z_{i,k} = \lim_{j\to\infty} z_{i,j},$$

$$\lim_{k\to\infty} z_{i,k}(\epsilon) = \lim_{j\to\infty} z_{i,j}(\epsilon) = \xi_i(\epsilon),$$

and

$$\begin{aligned}\xi_i &= \lim_{\epsilon\to 0^+} \xi_i(\epsilon) \\ &= \lim_{\epsilon\to 0^+} \lim_{k\to\infty} z_{i,k}(\epsilon) \\ &= \lim_{j\to\infty} \lim_{\epsilon\to 0^+} z_{i,j}(\epsilon) \\ &= \lim_{j\to\infty} z_{i,j}.\end{aligned}$$

The interchange of the above double limit is valid, since for these sequences, the convergence of $\lim_{k\to\infty} z_{i,k}(\epsilon)$ is uniform in $\epsilon$ over $0 < \epsilon < \epsilon_0$. This follows from the fact that the sequence reaches level $C_k$ in $k$ steps no matter what $\epsilon$ is. So far we have established the following:

$$\xi_i = \lim_{k\to\infty} z_{i,k} = \lim_{j\to\infty} z_{i,j}$$

for all $i \in I_n$.

Now let $\xi$ be a root of $f(z)$ with $\mu$ multiplicity. Choosing a proper $\epsilon_0$ and using the Argument Principle again, we have

$$\mu = \frac{1}{2\pi i}\int_{|z-\xi|=\epsilon_0} \frac{f'(z)}{f(z)}dz.$$

Note that $f(z)+\epsilon$ has the derivative $f'(z)$ for all $\epsilon > 0$. When we choose $\epsilon > 0$ small enough, we can have

$$\mu = \frac{1}{2\pi i}\int_{|z-\xi|=\epsilon_0} \frac{f'(z)}{f(z)+\epsilon}dz.$$

Note that all roots of $f(z)+\epsilon$ are distinct. Hence there are exactly $\mu$ sequences converging to the $\mu$ distinct roots of $f(z)+\epsilon$ in $\{z \in \mathbb{C} \mid |z-\xi| < \epsilon_0\}$. Finally, by using the fact that given any positive integer $J$, there exists $\epsilon_0 > \epsilon_1 > 0$ such that $z_{i,j}(\epsilon) = z_{i,j}$ for $1 \le j \le J$ and $0 < \epsilon \le \epsilon_1$. This shows that there exist exactly $\mu$ sequences converging to the same root $\xi$ of $f(z)$. Now the proof is completed. □

## 11.6. Complexity analysis and numerical examples

In this section we discuss the complexity of Kuhn's algorithm and give one of its applications and some numerical examples. Smale [1981] proved that for the case all $|c_i| < 1$, the cost of Newton's method of finding an approximate zero of a complex polynomial with $f(z) = z^n + c_{n-1}z^{n-1} + \cdots + c_0$ grows as $[100(n+2)]^9/\mu^7$, where $\mu$ stands for the probability allowing the conclusion to fail, $0 < \mu < 1$. The following result was given by Wang [1982]. See Kuhn, Wang and Xu [1984], Wang, Xu and Gao [1994] for related results.

**Theorem 11.6.1** *The total number of f-evaluations needed to find all roots of a complex polynomial* $f(z) = z^n + c_{n-1}z^{n-1} + \cdots + c_0$ *is at most*

$$\pi[5M^2 + 28n(1+0.75n)^2\lceil\log_2(3\sqrt{2}n/(4\epsilon))\rceil],$$

*where* $M = \sqrt{2} + \max\{3\sqrt{2}n/\pi, 1 + \frac{3}{2}\max|c_k|\}$ *and* $\epsilon > 0$ *is the accuracy demand of approximating roots.*

Proof: Using Kuhn's algorithm, we know that the computation proceeds in the half-space $\mathbb{C}\times[-1,\infty)$ triangulated. In fact, the region possible for the computation consists of a base-cylinder of radius $M$ between $C_{-1}$ and $C_0$, and $n$ cylinder-flights of steps with an infinite total height, The volume of the the base-cylinder is $\pi M^2$, and the number of tetrahedra contained in the base-cylinder is at most $5\pi M^2$. The volume of every small cylinder between $C_d$ and $C_{d+1}$ for $d \geq 0$, is $\pi(1+0.75n)^2 2^{1-2d}$, and the number of tetrahedra contained in it is at most $14\pi(1+0.75n)^2 2^{1-2d}/2^{-2d} = 28\pi(1+0.75n)^n$. Since the number of cylinder-flights of steps is $n$ and any computation over the sufficient height $\lceil\log_2(3\sqrt{2}n/(4\epsilon))\rceil$ is unnecessary, the total number of tetrahedra contained in the region possible for the computation and under the sufficient height is at most

$$\pi[5M^2 + 28n(1+0.75n)^2\lceil\log_2(3\sqrt{2}n/(4\epsilon))\rceil].$$

Notice that every $f$-evaluation is carried out by going through a tetrahedron, and every tetrahedron allows the computation to go through at most one time. The theorem then follows.

When doing so, with the fact that $|w| < 1$ implies $|\arg(1+w)| < \frac{\pi}{2}|w|$, we also improve the estimate of Kuhn's algorithm. The result is that, if $\sigma =< (z_1, k), (z_2, k), (z_3, k) >$ is a completely labelled simplex with $k \geq 0$, then some root of $f$ is closer to one vertex of the simplex than $\frac{3n\delta}{4}$, where $\delta = diam(\sigma)$, by Lemma 11.5.3. □

The above result demonstrates that Kuhn's algorithm is a good algorithm in the sense of complexity theory.

Kuhn's algorithm can be easily adjusted to compute zero points of any continuous function $g : \mathbb{C} \mapsto \mathbb{C}$. Now we consider Kuhn's algorithm starting on $Q_m(\bar{z}, h)$. The following theorems can be found in Wang [1986a-b] and proofs are left as exercises.

**Theorem 11.6.2** *Suppose that* $g : \mathbb{C} \mapsto \mathbb{C}$ *is a continuous function, and there are an integer* $n$, *a complex number* $\bar{z}$ *and a non-zero complex* $a$ *such that*

$$\lim_{|z|\to\infty} \frac{af(z)}{(z-\bar{z})^n} = 1.$$

*Then each of the* $n$ *sequences generated by Kuhn's algorithm has a subsequence which converges to a zero of* $f$.

**Theorem 11.6.3** *Let $g$, $n$, $\bar{z}$, $a$ be the same as in the above theorem, and there exists $r > 0$ such that $|z - \bar{z}| \geq r$ implies*

$$|1 - ag(z)/(z - \bar{z})^n| < \sqrt{3}/2.$$

*Then the number of zeroes of $g$ inside $G = \{z \in \mathbb{C} \mid |z - \bar{z}| \leq r\}$ is at least $n$.*

Kuhn's algorithm is programmed and implemented in Chen [1990] for complex polynomials as well as for general continuous complex functions. There are many numerical examples in this book. Here we quote several numerical results from there.

Example 1. Let $f(z) = z^{25} + 12z^6 + 245z^5 + 78z^4 + 6z^3 + 123z^2 + 245z + 1450$. The approximate roots are

$$\begin{aligned}
z_{1,9} &= 0.5887324 \mp 1.2331593i, \\
z_{2,6} &= 1.040037 \mp 0.79339091i, \\
z_{3,5} &= 1.246989 \mp 0.5447362i, \\
z_{4,24} &= 1.37473 \pm 0.1918665i, \\
z_{7,16} &= -1.291491 \mp 0.2972462i, \\
z_{8,25} &= 0.8639934 \pm 1.0091883i, \\
z_{10,23} &= 0.2313792 \pm 1.3511377i, \\
z_{11,22} &= -0.1398061 \pm 1.3367487i, \\
z_{12,20} &= 0.6740672 \pm 1.1199976i, \\
z_{13,21} &= -0.4302953 \pm 1.207803i, \\
z_{14,18} &= -0.9724528 \pm 0.9356104i, \\
z_{15,19} &= -1.192324 \pm 0.6394602i, \\
z_{17} &= -1.290797 - 2.5431315 \times 10^{-6}i.
\end{aligned}$$

Example 2. Let $f(z) = (z - 5i)^3(z + 5i)^2(z - 0.6)(z + 0.6) = z^7 - 5iz^6 + 49.64z^5 - 248.2iz^4 + 607z^3 - 3035iz^2 - 225z + 1125i$. The approximate roots are

$$\begin{aligned}
z_1 &= 3.878425 \times 10^{-4} - 5i, \\
z_2 &= 0.6 + 3.104409 \times 10^{-10}i, \\
z_3 &= 3.878425 \times 10^{-4} + 5i, \\
z_4 &= 6.208817 \times 10^{-10} + 5i, \\
z_5 &= -3.878406 \times 10^{-4} + 5i, \\
z_6 &= -0.6 - 3.104409 \times 110^{-11}i, \\
z_7 &= -3.878422 \times 110^{-4} - 5i.
\end{aligned}$$

Example 3. Let $f(z) = z^2 + z + 0.4e^{0.8z}$. In Kuhn's algorithm we set $\bar{z} = 60$ and $n = 10$. The approximate roots are

$$\begin{aligned} z_{1,2} &= -5.703023 \mp 2.643392i, \\ z_{3,8} &= -7.462748 \pm 9.480998i, \\ z_{4,10} &= -8.561687 \pm 16.89101i, \\ z_{5,9} &= -9.353496 \pm 24.49504i, \\ z_{6,7} &= -9.967876 \pm 32.18917i. \end{aligned}$$

Example 4. Let $f(z) = z - \pi + i + i\cos z$. In Kuhn's algorithm we set $\bar{z} = 7 + 4i$ and $n = 6$. The approximate roots are

$$\begin{aligned} z_1 &= 10.5217 + 2.812313i, \\ z_2 &= 3.141663 + 1.616138i, \\ z_3 &= -4.23851 + 2.812289i, \\ z_4 &= 7.589524 - 2.232563i, \\ z_5 &= 3.141493 - 2.483527 \times 10^{-9}i, \\ z_6 &= 20.27307 - 3.545788i. \end{aligned}$$

Finally, we refer to Kojima, Nishino and Arima [1979] for a related method for computing all the roots of a polynomial.

# CHAPTER 12

## Gröbner Bases for Solving Multivariate Polynomial Equations

### 12.1. Introduction

In this chapter we give an introduction to Gröbner basis theory which was developed by Buchberger [1965, 1985] and has recently received considerable attention in the field of symbolic computation and computational algebra. In algebra the famous *ideal membership problem* states that given $f$ in the set $k[x_1, \cdots, x_n]$ of all polynomials in $n$ variables with coefficients in a field $k$ and an ideal $I = < f_1, \cdots, f_m >$ of $k[x_1, \cdots, x_n]$, determine whether $f$ is in $I$ or not. Buchberger [1965] proposed a remarkable algorithm for solving this problem completely. Briefly, his algorithm can be described as follows. Given a nonzero ideal $I$, first find a standard basis $G$, called a Gröbner basis, for the ideal $I$. Then by a reduction process, Buchberger has shown that $f$ belongs to $I$ if and only if $f$ can be reduced to zero by members of $G$. The whole process can be carried out within a finite number of steps. Here we should be aware that this short description does not properly convey the true significance of Buchberger's contribution. The reader is referred to the excellent books by Cox, Little and O'Shea [1996] and Adams and Loustaunau [1994] for more details on the impact of Buchberger's work. We also refer to Wu [1984] and Chou [1984, 1988] for Wu's method on the automated proving of elementary geometry theorems based on systems of polynomial equations. Wu's method is somehow related to Buchberger's method but different. In this chapter we will particularly apply Gröbner basis theory to solve systems of multivariate polynomial equations over the complex number field $\mathbb{C}$. First we will be able to determine whether a system of polynomials has a solution or not. Second we will be able to find all solutions of the polynomial system when the system has a finite number of solutions. If the system has an infinite number of solutions, we can apply a homotopy algorithm to find as finitely many solutions as we wish to have. Third we will be able to solve a new class of complementarity problems in which the underlying functions are polynomials.

This chapter is organized as follows. In Section 2 basic algebraic concepts and the famous Hilbert basis theorem will be presented. In Section 3 we introduce a multi-variable division algorithm. In Section 4 we present the concept of a Gröbner basis and discuss its important properties. In Section 5 the fundamental algorithm of Buchberger is introduced. In Section 6 we introduce the famous Hilbert Nullstellensatz and answer the first two problems posed above. Finally in Section 7 we give an application to Gröbner bases by solving a new class of complementarity problems.

## 12.2. Basic concepts and Hilbert's basis theorem

In this chapter we study polynomials over a field. Loosely speaking, a field is a set where one can define addition, subtraction, multiplication, and division with usual properties. Most familiar fields are the real numbers $\mathbb{R}$, the rational numbers $\mathbb{Q}$, and the complex numbers $\mathbb{C}$. A standard reference on algebra is the book by van der Waerden [1970]. Now we introduce the formal definition of a field.

**Definition 12.2.1** *A group $(G, \cdot)$ consists of a nonempty set $G$ and a binary operation $\cdot$ on $G$ satisfying the following axioms:*
*(A) $a \cdot (b \cdot c) = (a \cdot b) \cdot c$ for all $a$, $b$, $c \in G$; (Associativity)*
*(B) There exists an element $e \in G$ such that $e \cdot a = a \cdot e = a$ for all $a \in G$; (Identity)*
*(C) For all $a \in G$, there exists an element $a^{-1} \in G$ such that $a \cdot a^{-1} = a^{-1} \cdot a = e$. (Inverse)*

**Definition 12.2.2** *An Abelian or commutative group $(G, \cdot)$ is a group which satisfies the axiom:*
*(D) $a \cdot b = b \cdot a$ for all $a$, $b \in G$. (Commutativity)*

**Definition 12.2.3** *A ring $(R, +, \cdot)$ consists of a nonempty set $R$ and two binary operations $+$ and $\cdot$ defined on $R$ satisfying that $(R, +)$ is an Abelian group, axioms (A) and (B) hold for $\cdot$, and the following axiom holds:*
*(E) $a \cdot (b + c) = (a \cdot b) + (a \cdot c)$, and $(a + b) \cdot c = (a \cdot c) + (b \cdot c)$ for all $a$, $b$, $c \in G$. (Distributivity)*

A ring is a *commutative ring* if axiom (D) holds for $\cdot$. As usual, for rings we denote the identity element with respect to $\cdot$ by 1, the identity element with respect to $+$ by 0, the inverse of $a$ with respect to $\cdot$ by $a^{-1}$, and the inverse of $a$ with respect to $+$ by $-a$.

**Definition 12.2.4** *A field $(F, +, \cdot)$ consists of a nonempty set $F$ and two binary operations $+$ and $\cdot$ defined on $F$ satisfying that $(F, +)$ and $(F \backslash \{0\}, \cdot)$ are Abelian groups, and axiom (E) holds.*

Note that a field is in fact a commutative ring in which every nonzero element has a multiplicative inverse. One can easily check that the integer numbers $\mathbb{Z}$ is not a field with respect to the usual operations $\cdot$ and $+$. As usual, we will simply use a single notation $F$ to denote a group $(F, \cdot)$, a ring $(F, +, \cdot)$, or a field $(F, +, \cdot)$ by omitting the binary operations. We will also denote $c \cdot d$ by $cd$ for any elements $c$ and $d$ in $F$.

From now on, let $k$ be a field with two binary operations $+$ and $\cdot$ defined on $k$. A *polynomial* $f(x_1, \cdots, x_n)$ in $n$ variables $x_1, \cdots, x_n$ with coefficients in the field $k$ is a finite sum of terms of the form $ax_1^{\alpha_1} \cdots x_n^{\alpha_n}$, where $a \in k$, and all of the exponents $\alpha_i$, $i \in I_n$ are non-negative integers. We call $x_1^{\alpha_1} \cdots x_n^{\alpha_n}$ a *monomial* (or *power product*) and $ax_1^{\alpha_1} \cdots x_n^{\alpha_n}$ a *term*. For example, $f(x_1, x_2, x_3) = 4x_1^3x_2^4 - 3x_1^4x_2^2x_3^5 + x_1x_2 + 1$ is a polynomial in three variables, whereas $f(x_1, x_2) = x_1^2x_2^3 - 4x_1x_2^{-2}$ is not a polynomial. The set of all polynomials in $x_1, \cdots, x_n$ with coefficients in $k$ is denoted by $k[x_1, \cdots, x_n]$ and the set of monomials in $n$ variables by $\Pi^n$. Sometimes we denote a monomial $x_1^{\alpha_1} \cdots x_n^{\alpha_n}$ by $X^\alpha$ where $X = (x_1, \cdots, x_n)$ and $\alpha = (\alpha_1, \cdots, \alpha_n)$. Let us now recall some notions for polynomials in one variable. For any polynomial $f(x) = a_nx^n + a_{n-1}x^{n-1} + \cdots + a_1x + a_0 \in k[x]$, with $a_0, a_1, \cdots, a_n \in k$ and $a_n \neq 0$, the *degree* of $f(x)$ is $n$, denoted by $deg(f(x)) = n$. The *leading term* of $f(x)$ is $a_nx^n$, denoted by $Lt(f(x))$, and the *leading coefficient* of $f(x)$ is $a_n$, denoted by $Lc(f(x))$. Now we introduce a basic algebraic concept of this chapter.

**Definition 12.2.5** *A subset $I$ of $k[x_1, \cdots, x_n]$ is an ideal if it satisfies:*
*(a) $0 \in I$;*
*(b) If $f, g \in I$, then $f + g \in I$;*
*(c) If $f \in I$ and $g \in k[x_1, \cdots, x_n]$, then $gf \in I$.*

Let $f_1, \cdots, f_m$ be polynomials in $k[x_1, \cdots, x_n]$. Then we call the set

$$< f_1, \cdots, f_m > = \{\sum_{i=1}^{m} g_if_i \mid g_1, \cdots, g_m \in k[x_1, \cdots, x_n]\}$$

the *ideal generated* by $f_1, \cdots, f_m$. Now we are ready to present the fundamental theorem of this section.

**Theorem 12.2.6 (Hilbert Basis Theorem)**
*(i) For any ideal $I$ of $k[x_1, \cdots, x_n]$ there exist polynomials $f_1, \cdots, f_m \in k[x_1, \cdots, x_n]$ such that $I = < f_1, \cdots, f_m >$.*
*(ii) If $I_1 \subseteq \cdots \subseteq I_m \subseteq \cdots$ is an ascending chain of ideals of $k[x_1, \cdots, x_n]$, then there exists a positive integer $L$ such that $I_L = I_{L+1} = I_{L+2} = \cdots$.*

An ideal $I$ of any ring satisfying (i) of Theorem 12.2.6 is said to be *finitely generated* or to have a *finite set of generators*. Any commutative ring satisfying (ii) is called a *Noetherian ring*. As a matter of fact, we will first prove that (i) and (ii) in Theorem 12.2.6 are equivalent.

**Lemma 12.2.7** *The two statements (i) and (ii) of Theorem 12.2.6 imply each other for any commutative ring $R$.*

Proof: We first show that (i) implies (ii). Let $I_1 \subseteq \cdots \subseteq I_m \subseteq \cdots$ be an ascending chain of ideals in $R$. Consider the set $I = \cup_{i=1}^{\infty} I_i$. It is easy to see that $I$ is still an ideal by noticing the monotonicity of the ideals $I_i$. By statement (i), we have $I =< f_1, \cdots, f_m >$ for some $f_1, \cdots, f_m \in R$. Note that $f_i \in I$ for all $i \in I_m$. It means that for each $f_i$ there exists an integer $L_i$ such that $f_i \in I_{L_i}$. Let $L$ be the largest number among all $L_i$. Then we have $f_i \in I_L$ for all $i \in I_m$ and hence $I \subseteq I_L$. Therefore, statement (ii) is implied.

Secondly we prove (ii) implies (i). Suppose to the contrary that there exists an ideal $I$ in $R$ which is not finitely generated. Let $f_1 \in I$. Then there exists an element $f_2 \notin < f_1 >$. This implies that $< f_1 >$ is a proper subset of $< f_1, f_2 >$. By continuing to proceed in this way, we can generate a strictly ascending chain of ideals in $R$ which contradicts statement (ii). Thus we proved (ii) implies (i). □

We can now prove a more general version of Hilbert's basis theorem which says that if $R$ is a Noetherian ring, so is $R[x_1, \cdots, x_n]$. Note that a field $k$ is a trivial Noetherian ring.

**Theorem 12.2.8 (Hilbert Generalized Basis Theorem)** *If $R$ is a commutative ring such that every ideal of $R$ is finitely generated, then every ideal in $R[x_1, \cdots, x_n]$ is finitely generated.*

Proof: Using induction it evidently suffices to show the theorem for the case of one variable. We have to show that any ideal $I$ in $R[x]$ has a finite set of generators. Let $I$ be any ideal in $R[x]$. For each $l \in \mathbb{Z}_+$, let $J_l$ be the set of elements $r_l$ in $R$ such that there exists an element $f_l(x)$ in $I$ having the form

$$f_l(x) = r_l x^l + r_{1-1} x^{l-1} + \cdots + r_1 x + r_0.$$

It is clear that $J_l$ is an ideal in $R$. Since $f_l(x) \in I$ implies that $x f_l(x) = r_l x^{l+1} + r_{l-1} x^l + \cdots + r_1 x^2 + r_0 x \in I$, it implies that $r_l \in J_{l+1}$ and hence $J_l \subseteq J_{l+1}$. Hence $J = \cup_{i=1}^{\infty} J_i$ is an ideal in $R$. This ideal has a finite set of generators. By Lemma 12.2.7, there exists an integer $L > 0$ such that $J_l = J_L$ for all $l \geq L$. By the assumption, each $J_l$ is finitely generated, i.e., $J_l =< r_1^l, \cdots, r_{d_l}^l >$. For each $i = 1, \cdots, L$ and each $j = 1, \cdots, d_i$, let $f_j^i(x)$ be a polynomial in $I$ with $deg(f_j^i(x)) = i$ and $Lc(f_j^i(x)) = r_j^i$. In the following, we omit $x$ in polynomials $f(x)$ and $g(x)$. Define

$$I^* =< f_1^1, \cdots, f_{d_1}^1, f_1^2, \cdots, f_{d_2}^2, \cdots, f_1^L, \cdots, f_{d_L}^L > .$$

We claim that $I = I^*$. Obviously, $I^* \subseteq I$. Let $f$ be any polynomial in $I$ with $deg(f) = l$ and $Lc(f) = r$. Clearly, $f \in I^*$ when $l = 0$ or $f = 0$. Now suppose that $l > 0$ and all polynomials in $I$ with degree at most $l - 1$ are contained by $I^*$. We need to consider the following two cases.

(i) In the case $l \leq L$, we have $r = \sum_{j=1}^{d_l} a_j r_j^l$ for some $a_j \in R$, since $r \in J_l$. Consider the polynomial $g = \sum_{j=1}^{d_l} a_j f_j^l$. It is easy to see that $deg(g) = l$ and $Lc(g) = r$. Thus $g \in I^*$. Moreover, it is also easy to check that $deg(f-g) \le l-1$ and by induction $f-g \in I^*$. Since $I^*$ is an ideal and $f = g + (f-g)$ with $f-g, g \in I^*$, we have $f \in I^*$.

(ii) In the case $l > L$, we have $J_l = J_L$, $r \in J_L$ and $r = \sum_{j=1}^{d_L} a_j r_j^L$ for some $a_j \in R$. Consider the polynomial $g = \sum_{j=1}^{d_L} a_j x^{l-N} f_j^L$. It is easy to see that $deg(g) = l$ and $Lc(g) = r$. Thus $g \in I^*$. Moreover, it is also easy to check that $deg(f-g) \le l-1$ and by induction $f-g \in I^*$. Since $I^*$ is an ideal and $f = g + (f-g)$ with $f-g, g \in I^*$, we have $f \in I^*$. □

## 12.3. Multi-variable division algorithm

In the following algorithms we often need to specify an order on the monomials. We first introduce a specific order on the set of monomials $\Pi^n$.

**Definition 12.3.1** *A linear order $<$ on $\Pi^n$ is a term order if the following conditions are satisfied:*
*(a) $1 \le X^\alpha$ for all $X^\alpha \in \Pi^n$ where $1 = x_1^0 \cdots x_n^0$;*
*(b) If $X^\alpha < X^\beta$, then $X^\alpha X^\gamma < X^\beta X^\gamma$, for all $X^\alpha$, $X^\beta$, $X^\gamma \in \Pi^n$.*

A wide variety of term orders are available. Here we discuss the following two often used term orders.

**Definition 12.3.2** *The lexicographic term order is given as follows: For any $X^\alpha$, $X^\beta \in \Pi^n$ with $\alpha = (\alpha_1, \cdots, \alpha_n)$ and $\beta = (\beta_1, \cdots, \beta_n)$, $X^\alpha < X^\beta$ if there exists an index $j \in I_n$ such that $\alpha_j < \beta_j$ and $\alpha_i = \beta_i$ for all $i \in I_{j-1}$.*

This order is shortly denoted by lex. Consider the three-variable terms with $x > y > z$. We have

$$1 < z < z^2 < \cdots < y < yz < yz^2 < \cdots < y^2 < y^2 z < \cdots < x < xz < \cdots < xy < xyz < \cdots.$$

**Definition 12.3.3** *The degree lexicographic term order is given as follows: For any $X^\alpha$, $X^\beta \in \Pi^n$ with $\alpha = (\alpha_1, \cdots, \alpha_n)$ and $\beta = (\beta_1, \cdots, \beta_n)$, $X^\alpha < X^\beta$ if $\sum_{i=1}^n \alpha_i < \sum_{i=1}^n \beta_n$, or $\sum_{i=1}^n \alpha_i = \sum_{i=1}^n \beta_n$ and there exists an index $j \in I_n$ such that $\alpha_j < \beta_j$ and $\alpha_i = \beta_i$ for all $i \in I_{j-1}$.*

This order will be denoted by deglex. In case of three variables with $x > y > z$, we have

$$1 < z < y < x < z^2 < yz < y^2 < xz < xy < x^2 < z^3 < yz^2 < y^2 z < y^3 < xz^2 < xyz < xy^2 < x^2 z < x^2 y < x^3 < \cdots.$$

From now on, when we fix a term order on $k[x_1, \cdots, x_n]$ and choose a nonzero $f \in k[x_1, \cdots, x_n]$, we will often write $f$ with $X^{\alpha_1} > X^{\alpha_2} > \cdots > X^{\alpha_m}$ as

$$f = a_1 X^{\alpha_1} + a_2 X^{\alpha_2} + \cdots + a_m X^{\alpha_m}$$

where $a_i \in k$ with $a_i \neq 0$ for all $i \in I_m$. The three parameters $a_1$, $X^{\alpha_1}$, and $a_1 X^{\alpha_1}$ are called the *leading coefficient*, *leading power product*, and *leading term* of $f$, respectively. They are denoted by $Lc(f) = a_1$, $Lp(f) = X^{\alpha_1}$ and $Lt(f) = a_1 X^{\alpha_1}$. We say that $f$ in $k[x_1, \cdots, x_n]$ is *divisible* by $g$ in $k[x_1, \cdots, x_n]$ if $f = gh$ for some $h \in k[x_1, \cdots, x_n]$.

Now we have the following lemma.

**Lemma 12.3.4** *For each nonempty subset $A$ of $\Pi^n$, there exists an element $X^\alpha \in A$ such that $X^\alpha \leq X^\beta$ for all $X^\beta \in A$. That is, every term order on $\Pi^n$ is a well-ordering.*

Proof: Suppose to the contrary that there exists a nonempty subset $B$ of $\Pi^n$ having no smallest elements. This would imply that there exist $X^{\alpha_i} \in B$, $i \in \mathbb{N}$ such that

$$X^{\alpha_1} > X^{\alpha_2} > \cdots. \tag{12.3.1}$$

This leads to a chain of ideals in $k[x_1, \cdots, x_n]$

$$< X^{\alpha_1} > \subset < X^{\alpha_1}, X^{\alpha_2} > \subset \cdots. \tag{12.3.2}$$

Notice that $< X^{\alpha_1}, \cdots, X^{\alpha_i} > \neq < X^{\alpha_1}, \cdots, X^{\alpha_{i+1}} >$ for all $i \in \mathbb{N}$. If not, we would have

$$X^{\alpha_{i+1}} = \sum_{h \in I_i} u_h X^{\alpha_h} \tag{12.3.3}$$

where $u_h \in k[x_1, \cdots, x_n]$ for all $h \in I_i$. Since we can express $u_h$ as a linear combination of power products, this means that every term in the right-hand side of equation (12.3.3) is divisible by $X^{\alpha_h}$ for some $h \in I_i$. On the other hand, $X^{\alpha_{i+1}}$ must appear as a power product of a term on the right-hand side of equation (12.3.3). This concludes that $X^{\alpha_{i+1}}$ is divisible by some $X^{\alpha_h}$ for $h \in I_i$. So there exists an $X^\beta \in \Pi^n$ such that $X^{\alpha_{i+1}} = X^{\alpha_h} X^\beta$. It follows from Definition 12.3.1 that $X^\beta \geq 1$ and $X^{\alpha_{i+1}} = X^{\alpha_h} X^\beta \geq X^{\alpha_h}$, contradicting (12.3.1). So relation (12.3.2) is true. Since this chain of ideals is strictly ascending, this contradicts Theorem 12.2.6. □

The goal of this section is to divide $f \in k[x_1, \cdots, x_n]$ by nonzero polynomials $f_1, \cdots, f_m \in k[x_1, \cdots, x_n]$ such that

$$f = u_1 f_1 + \cdots + u_m f_m + r$$

where $u_i \in k[x_1, \cdots, x_n]$ for all $i \in I_m$ and $r \in k[x_1, \cdots, x_n]$ with $Lp(r) \leq Lp(f)$. To do so, we will introduce a multi-variable division algorithm which is also one of the basic ingredients in this chapter.

**Definition 12.3.5** *Given a term order on* $k[x_1, \cdots, x_n]$, $f$, $g$, $h \in k[x_1, \cdots, x_n]$ *with* $g \neq 0$, $f$ *is reduced to* $h$ *modulo* $g$, *denoted by* $f \xmapsto{g} h$, *if there exists a nonzero term* $Y$ *in* $f$ *which is divisible by* $Lp(g)$ *and* $h = f - \frac{Y}{Lt(g)}g$.

Note that in this definition the term $Y$ in $f$ is replaced by terms strictly smaller than $Y$. For example, let $f = 3x^3y^2 + 3xy^4 + 6xy^2 + x - y^4 + 4y^2 - 5$ and $g = 7xy^3 + 4xy + 1$ be polynomials in $\mathbb{Q}[x, y]$. If the term order is lex with $x > y$, then we have $Lt(g) = 7xy^3$, $Y = 3xy^4$ and $f \xmapsto{g} h$ with $h = 3x^3y^2 + 30xy^2/7 + x - y^4 + 4y^2 - 3y/7 - 5$.

**Definition 12.3.6** *Given a term order on* $k[x_1, \cdots, x_n]$, $f$, $f_1, \cdots, f_m$, $h \in k[x_1, \cdots, x_n]$ *with* $f_i \neq 0$ *for all* $i \in I_m$, $f$ *is reduced to* $h$ *modulo* $F = \{f_1, \cdots, f_m\}$, *denoted by* $f \xmapsto{F} h$, *if there exist a set of elements* $\{i_1, \cdots, i_t\} \subseteq I_m$ *and a set of polynomials* $\{h_{i_1}, \cdots, h_{i_{t-1}}\} \subseteq k[x_1, \cdots, x_n]$ *such that*

$$f \xmapsto{f_{i_1}} h_{i_1} \xmapsto{f_{i_2}} \cdots \xmapsto{f_{i_{t-1}}} h_{i_{t-1}} \xmapsto{f_{i_t}} h.$$

For example, let $f = x^3y^2 + xy^3 - x + y^3 + y + 1$, $f_1 = x^2y + 1$ and $f_2 = xy - 1$ be polynomials in $\mathbb{Q}[x, y]$, and let $F = \{f_1, f_2\}$. Again we use lex with $x > y$. Then we have

$$f \xmapsto{f_1} xy^3 - xy - x + y^3 + y + 1 \xmapsto{f_2} xy^3 - x + y^3 + y.$$

Hence $f \xmapsto{F} xy^3 - x + y^3 + y$.

From the above definition, one can prove the following result.

**Lemma 12.3.7** *Let* $F$ *be a set of nonzero polynomials in* $k[x_1, \cdots, x_n]$. *For any* $f$, $g$, $h \in k[x_1, \cdots, x_n]$ *with* $h \neq 0$, *if* $f \xmapsto{F} g$, *then* $hf \xmapsto{F} hg$.

**Definition 12.3.8** *A polynomial* $r \in k[x_1, \cdots, x_n]$ *is irreducible modulo a set of nonzero polynomials* $F = \{f_1, \cdots, f_m\}$ *in* $k[x_1, \cdots, x_n]$ *if* $r$ *cannot be reduced modulo* $F$.

So it is equivalent to saying that $r$ is irreducible if either $r = 0$ or there exists no term in $r$ which is divisible by any of the $Lp(f_i)$'s, $i \in I_m$. If $f \xmapsto{F} r$ and $r$ is irreducible, then we call $r$ the *reminder* of $f$ with respect to $F$ and write as $f \xmapsto{F}_* r$.

Now we can describe the multi-variable division (MVD) algorithm as follows.

The MVD Algorithm

**Step (0)** Given $f$, $f_1, \cdots, f_m \in k[x_1, \cdots, x_n]$ with $f_i \neq 0$ for all $i \in I_m$. Set $u_i = 0$ for all $i \in I_m$, $r = 0$ and $h = f$.

**Step (1)** If $h = 0$, then output $u_1, \cdots, u_m$, $r$ and stop. Otherwise, set $d = 1$.

**Step (2)** If $Lp(f_d)$ divides $Lp(h)$, then set $u_d = u_d + \frac{Lt(h)}{Lt(f_d)}$, $h = h - \frac{Lt(h)}{Lt(f_d)} f_d$, and go to Step (1). Otherwise, set $d = d + 1$.

**Step (3)** If $d \geq m + 1$, go to Step (4). Otherwise, go to Step (2).

**Step (4)** Set $r = r + Lt(h)$ and $h = h - Lt(h)$. Go to Step (1).

Let us illustrate the algorithm by an example. Let $f = x^3y^2 + x^2y^4 + 3xy + 1$, $f_1 = x + y^2 - 1$ and $f_2 = y^6 + 1$ be in $\mathbb{Q}[x, y]$, and let $F = \{f_1, f_2\}$. We use lex with $x > y$. By implementing the algorithm, we finally get $u_1 = x^2y^2 + xy^2 - y^4 + y^2 + 3y$, $u_2 = 1$ and $r = -2y^4 - 3y^3 + y^2 + 3y$ such that $f = u_1f_1 + u_2f_2 + r$ with $Lp(r) < Lp(f) = x^3y^2$, $Lp(f_1)Lp(u_1) = Lp(f)$, and $Lp(f_2)Lp(u_2) < Lp(f)$.

Now we are ready to prove the following theorem.

**Theorem 12.3.9** *Let $F = \{f_1, \cdots, f_m\}$ be a set of nonzero polynomials in $k[x_1, \cdots, x_n]$. Given $f \in k[x_1, \cdots, x_n]$, within a finite number of iterations the MVD algorithm will terminate with polynomials $u_1, \cdots, u_m$, $r \in k[x_1, \cdots, x_n]$ such that*

$$f = u_1f_1 + \cdots u_mf_m + r$$

*with $r$ a reminder of $f$ with respect to $F$ and*

$$Lp(f) = \max\{Lp(u_1)Lp(f_1), \cdots, Lp(u_m)Lp(f_m), r\}.$$

Proof: We first show that the MVD algorithm will stop within a finite number of iterations. The key observation is that at each stage of the algorithm we redefine $h$, either the leading term of $h$ is subtracted off or $h$ becomes 0. That is, the algorithm generates a sequence $h_1$, $h_2$, $\cdots$ of the $h$'s, where $h_{j+1}$ is obtained from $h_j$ by subtracting off $Lt(h_j)$ and possibly some smaller terms: $h_{j+1} = h_j - (Lt(h_j)$+lower terms). This is because we compute $h_{j+1}$ from $h_j$ by subtracting off $\frac{Lt(h_j)}{Lt(f_d)} f_d = Lt(h_j)+$ lower terms (at Step (2) in case some $Lp(f_d)$ divides $Lp(h_j)$) or subtracting off $Lt(h_j)$ (at Step (4) in case no $Lp(f_d)$ divides $Lp(h_j)$). So we have that for all $j$, $Lp(h_{j+1}) < Lp(h_j)$. If the algorithm never stopped, then it would generate an infinite strictly decreasing sequence of $Lp(h_j)$'s. Since the term order is a well-ordering by Lemma 12.3.4, such sequence cannot occur. Thus after finitely many iterations the algorithm terminates with $h = 0$.

To show the second part, we have to prove that at each stage of the algorithm it holds

$$f = u_1f_1 + \cdots + u_mf_m + h + r. \tag{12.3.4}$$

It is obviously true at Step (0) since $u_i = 0$ for all $i \in I_m$, $r = 0$ and $h = f$. Now suppose that equation (12.3.4) holds at some stage of the algorithm.

If the next operation is Step (2), then some $Lp(f_d)$ divides $Lp(h)$. In this case we have

$$u_d f_d + h = (u_d + \frac{Lt(h)}{Lt(f_d)})f_d + (h - \frac{Lt(h)}{Lt(f_d)}f_d).$$

This shows that $u_d f_d + h$ remains unchanged. Since all other terms in equation (12.3.4) are not affected by this operation, equation (12.3.4) is preserved. On the other hand, if the next operation is Step (4), then we have

$$h + r = (h - Lt(h)) + (r + Lt(h)).$$

Again equation (12.3.4) still holds. Note that at the first part of this proof it is shown that $h$ becomes 0 at the last stage. Moreover, terms are added to $r$ only when they are irreducible modulo $F$. It is immediate that $r$ is a reminder of $f$ with respect to $F$ and $Lp(r) \leq Lp(f)$. On the other hand, for each $i$ we obtain $u_i$ by adding terms $\frac{Lt(h)}{Lt(f_i)}$, where $\frac{Lt(h)}{Lt(f_i)}f_i$ cancels the leading term of $h$. It is easy to see that $Lp(u_i)Lp(f_i) \leq Lp(f)$ for all $i \in I_m$. □

From the above theorem it is clear that $f - r \in I =< f_1, \cdots, f_m >$. Hence if $r = 0$, then $f \in I$. However, it may well be the case that $f$ is in $I$, but the reminder of $f$ with respect to $F$ is not zero. This problem will be resolved in the following sections.

## 12.4. Gröbner bases

In this section we introduce the most important concept of this chapter, namely, a Gröbner basis, and discuss some of its properties.

**Definition 12.4.1** *Given a term order on $\Pi^n$, a finite subset of nonzero polynomials $G = \{g_1, \cdots, g_m\}$ of an ideal $I$ is a Gröbner basis for $I$ if for every nonzero polynomial $f \in I$, $Lp(f)$ is divisible by at least one $Lp(g_i)$, $i \in I_m$.*

From the definition we could expect that an ideal $I$ of $k[x_1, \cdots, x_n]$ may have a different Gröbner basis if we use a different term order. Now we are going to introduce the following definition. For a subset $I$ in $k[x_1, \cdots, x_n]$, we denote by $Lt(I)$ the ideal generated by the leading terms of elements of $I$, that is

$$Lt(I) =< Lt(f) \mid f \in I > .$$

We also call $Lt(I)$ the *leading term ideal* of $I$. Note that in general $Lt(I) \neq I$. Now we have the following alternative characterizations of Gröbner bases.

**Lemma 12.4.2** *Given a nonzero ideal $I$ of $k[x_1, \cdots, x_n]$, a set of nonzero polynomials $G = \{g_1, \cdots, g_m\}$ in $I$ is a Gröbner basis for $I$ if and only if $f \stackrel{G}{\longmapsto}_* 0$ for every nonzero $f \in I$.*

Proof: Suppose that $G$ is a Gröbner basis for $I$. Let $f$ be an arbitrary nonzero element in $I$. Then it follows from Theorem 12.3.9 that there exists $r \in k[x_1, \cdots, x_n]$ being a reminder of $f$ with respect to $G$ such that $f \stackrel{G}{\longmapsto}_* r$. Therefore $f - r \in I$ and $r - f \in I$ as well. Clearly, if $r = 0$, we are done. Now suppose to the contrary that $r \neq 0$. Then we have $r = f + (r - f) \in I$ since $r - f$, $f \in I$. By hypothesis $r$ must be divisible by some of $Lp(g_i)$, $i \in I_m$. This contradicts the fact that $r$ is a reminder of $f$ with respect to $G$. Hence $r$ must be 0. That is $f \stackrel{G}{\longmapsto}_* 0$.

Suppose that for any $f \in I$, $f \stackrel{G}{\longmapsto}_* 0$. This means that

$$f = u_1 g_1 + \cdots + u_m g_m \tag{12.4.5}$$

where $u_i \in k[x_1, \cdots, x_n]$ for all $i \in I_m$ and

$$Lp(f) = \max\{Lp(u_1)Lp(g_1), \cdots, Lp(u_m)Lp(g_m)\}.$$

We can rewrite equation (12.4.5) as

$$f = u_1(Lt(g_1) + \text{lower terms}) + \cdots + u_m(Lt(g_m) + \text{lower terms}).$$

It is clear that $Lp(f)$ is divisible by some $Lp(g_i)$, $i \in I_m$. Hence $G$ is a Gröbner basis for $I$. □

It follows immediately from the lemma that $G = \{g_1, \cdots, g_m\}$ is a Gröbner basis for $I$ if and only if for every polynomial $f \in I$, $f = \sum_{i=1}^{m} u_i g_i$ with $Lp(f) = \max\{Lp(u_1)Lp(g_1), \cdots, Lp(u_m)Lp(f_m)\}$. Moreover, it is easy to verify that if $G = \{g_1, \cdots, g_m\}$ is a Gröbner basis for $I$, then $I =< g_1, \cdots, g_m >$.

**Lemma 12.4.3** *Given a nonzero ideal $I$ in $k[x_1, \cdots, x_n]$, a set of nonzero polynomials $G = \{g_1, \cdots, g_m\}$ in $I$ is a Gröbner basis for $I$ if and only if $Lt(I) = Lt(G)$.*

Proof: Suppose that $Lt(I) = Lt(G)$. Let $f$ be an arbitrary nonzero polynomial in $I$. Since $Lt(f)$ is in $Lt(I)$, we have

$$Lt(f) = u_1 Lt(g_1) + \cdots + u_m Lt(g_m)$$

where $u_i \in k[x_1, \cdots, x_n]$ for all $i \in I_m$. Clearly, $Lp(f)$ is divisible by some $Lp(g_i)$, $i \in I_m$. Hence $G$ is a Gröbner basis for $I$.

Suppose that $G$ is a Gröbner basis for $I$. Clearly, $Lt(G)$ is a subset of $Lt(I)$. We still have to show the reverse direction. For any $f \in I$, $Lt(f) \in Lt(I)$. Then it follows from Lemma 12.4.2 that

$$f = u_1 g_1 + \cdots + u_m g_m \qquad (12.4.6)$$

where $u_i \in k[x_1, \cdots, x_n]$ for all $i \in I_m$ and

$$Lp(f) = \max\{Lp(u_1)Lp(g_1), \cdots, Lp(u_m)Lp(g_m)\}.$$

Let $L = \{i \in I_m \mid Lp(u_i)Lp(g_i) = Lp(f)\}$. Then we have from equation (12.4.6)

$$Lt(f) = \sum_{i \in L} Lt(u_i)Lt(g_i).$$

Hence $Lt(f) \in Lt(G)$ which implies $Lt(I) \subseteq Lt(G)$. □

**Lemma 12.4.4** *Given a nonzero ideal $I$ of $k[x_1, \cdots, x_n]$, a set of nonzero polynomials $G = \{g_1, \cdots, g_m\}$ in $I$ is a Gröbner basis for $I$ if and only if for any nonzero polynomial $f \in k[x_1, \cdots, x_n]$, $f \overset{G}{\longmapsto}_* q$ and $f \overset{G}{\longmapsto}_* r$ imply $q = r$, i.e., the reminder of $f$ with respect to $G$ is unique.*

We leave the proof to the interested reader. In the following it will be shown that every ideal $I$ of $k[x_1, \cdots, x_n]$ has a Gröbner basis.

**Theorem 12.4.5** *There exists a Gröbner basis for every ideal $I$ of $k[x_1, \cdots, x_n]$.*

Proof: Let $P = \{Lt(f) \mid f \in I\}$. Consider the ideal $Lt(I)$. We first show that $f \in Lt(I)$ if and only if every term of $f$ is divisible by some element of $P$. Suppose that every term $Y$ of $f$ is divisible by some term $Z$ of $P$. Clearly $f \in Lt(I)$. Conversely, if $f \in Lt(I)$, then

$$f = u_1 Y^1 + \cdots + u_t Y^t$$

where $u_i \in k[x_1, \cdots, x_n]$ for all $i \in I_t$ and $Y^i \in P$ for all $i \in I_t$. Obviously, every term $X$ of $f$ is divisible by some term $Y$ of $P$. Now we can apply Hilbert Basis Theorem (i.e., Theorem 12.2.6) to the leading term ideal $Lt(I)$ since $Lt(I)$ is an ideal in $k[x_1, \cdots, x_n]$. Hence there exists a finite set $\{Lt(g_1), \cdots, Lt(g_m)\}$ such that $Lt(I) =< Lt(g_1), \cdots, Lt(g_m) >$ with $g_i \in I$ for all $i \in I_m$. Let $G = \{g_1, \cdots, g_m\}$. It is clear that $Lt(G) = Lt(I)$. Then it follows from Lemma 12.4.3 that $G$ is a Gröbner basis for $I$. □

## 12.5. Buchberger's algorithm

In this section we introduce the fundamental algorithm for finding a Gröbner basis for any ideal $I$ in $k[x_1, \cdots, x_n]$. Before describing the algorithm, we first introduce a basic concept which will be used in the algorithm.

**Definition 12.5.1** *The S-polynomial of the nonzero polynomials* $f$, $g \in k[x_1, \cdots, x_n]$ *is given by*

$$SP(f,g) = LCM(Lp(f), Lp(g))\left(\frac{f}{Lt(f)} - \frac{g}{Lt(g)}\right)$$

*where* $LCM(Lp(f), Lp(g))$ *is the least common multiple of* $Lp(f)$ *and* $Lp(g)$. *In other words, if* $Lp(f) = X^\alpha$ *and* $Lp(g) = X^\beta$, *then* $LCM(Lp(f), Lp(g)) = X^\gamma$ *with* $\gamma = (\gamma_1, \cdots, \gamma_n)$ *and* $\gamma_i = \max\{\alpha_i, \beta_i\}$ *for each* $i \in I_n$.

For example, let $f = 4x^3y^2 + xy^3 + x^3 + 2$ and $g = xy^3 + 3xy + 4$ be in $\mathbb{Q}[x, y]$. We use deglex with $x > y$. Then we have $Lt(f) = 4x^3y^2$, $Lt(g) = xy^3$, $LCM(Lp(f), Lp(g)) = x^3y^3$, and

$$SP(f,g) = xy^4/4 - 11x^3y/4 - 4x^2 + y/2.$$

Note that the leading term $Lt(f)$ is canceled in $SP(f,g)$. As will be seen later, the $SP(f,g)$ is in fact designed to cancel leading terms.

**Lemma 12.5.2** *Let* $f_1, \cdots, f_m \in k[x_1, \cdots, x_n]$ *be given such that* $Lp(f_i) = X^\alpha \neq 0$ *for all* $i \in I_m$. *If* $f = \sum_{i \in I_m} c_i f_i$ *with* $c_i \in k$ *for all* $i \in I_m$ *and* $Lp(f) < X^\alpha$, *then* $f$ *is a linear combination of* $SP(f_i, f_j)$, *for* $i, j \in I_m$, *with coefficients in* $k$. *Moreover,* $Lp(SP(f_i, f_j)) < X^\alpha$ *for all* $i, j \in I_m$.

Proof: Each $f_i$ can be written as $f_i = a_i X^\alpha +$lower terms, $0 \neq a_i \in k$. Since $f = \sum_{i \in I_m} c_i f_i$ and $Lp(f) < X^\alpha$, it follows that $\sum_{i=1}^m c_i a_i = 0$. Because of $LCM(f_i, f_j) = X^\alpha = Lp(f_i) = Lp(f_j)$ we have

$$SP(f_i, f_j) = \frac{f_i}{a_i} - \frac{f_j}{a_j}.$$

It is clear that $Lp(SP(f_i, f_j)) < X^\alpha$. By using $c_1a_1 + \cdots c_m a_m = 0$, we can expand $f$ as follows.

$$\begin{aligned} f &= c_1 f_1 + \cdots + c_m f_m \\ &= c_1 a_1 (\tfrac{f_1}{a_1}) + \cdots + c_m a_m (\tfrac{f_m}{a_m}) \\ &= c_1 a_1 (\tfrac{f_1}{a_1} - \tfrac{f_2}{a_2}) + (c_1 a_1 + c_2 a_2)(\tfrac{f_2}{a_2} - \tfrac{f_3}{a_3}) + \cdots \\ &\quad + (c_1 a_1 + \cdots + c_{m-1} a_{m-1})(\tfrac{f_{m-1}}{a_{m-1}} - \tfrac{f_m}{a_m}) + (c_1 a_1 + \cdots + c_m a_m)\tfrac{f_m}{a_m} \\ &= c_1 a_1 SP(f_1, f_2) + (c_1 a_1 + c_2 a_2) SP(f_2, f_3) + \cdots \\ &\quad + (c_1 a_1 + \cdots + c_{m-1} a_{m-1}) SP(f_{m-1}, f_m). \end{aligned}$$

□

Now we are ready to prove the following fundamental theorem of Buchberger [1965, 1985].

**Theorem 12.5.3 (Buchberger Theorem)** *Let $G = \{g_1, \cdots, g_m\}$ be a set of nonzero polynomials in $k[x_1, \cdots x_n]$. Then $G$ is a Gröbner basis for the ideal $I =< g_1, \cdots, g_m >$ if and only if for all $i, j \in I_m$ with $i \neq j$, $SP(g_i, g_j) \overset{G}{\longmapsto}_* 0$.*

Proof: Suppose that $G$ is a Gröbner basis for the ideal $I =< g_1, \cdots, g_m >$. By noting that $SP(g_i, g_j) \in I$, it follows from Lemma 12.4.2 that

$$SP(g_i, g_j) \overset{G}{\longmapsto}_* 0$$

for all $i \neq j$.

Suppose that $SP(g_i, g_j) \overset{G}{\longmapsto}_* 0$ for all $i \neq j$. We will use Lemma 12.4.3 to prove that $G$ is a Gröbner basis. It is clear that $Lt(G) \subseteq Lt(I)$. It remains to show that $Lt(I) \subseteq Lt(G)$. To do so, it suffices to prove that $Lt(f) \in Lt(G)$ for any nonzero polynomial $f \in I$. We can write $f$ as

$$f = u_i g_1 + \cdots + u_m g_m \tag{12.5.7}$$

where $u_i \in k[x_1, \cdots, x_n]$ for all $i \in I_m$. Let $X = \max\{Lp(u_1)Lp(g_1), \cdots, Lp(u_m)Lp(g_m)\}$. Note that there may be many ways of expressing $f$ in the form (12.5.7). Since any term order is a well-ordering, we can always choose an expression for $f$ such that $X$ is the smallest one. Clearly, if $Lp(f) = X$, then $Lt(f) \in Lt(G)$ and we are done. Now suppose to the contrary that $Lp(f) < X$. Let $L = \{i \in I_m \mid Lp(u_i)Lp(g_i) = X\}$. Then we can rewrite $f$ as

$$f = \sum_{h \in L} Lt(u_h)g_h + \sum_{h \in L}(u_h - Lt(u_h))g_h + \sum_{i \notin L} u_i g_i. \tag{12.5.8}$$

Note that $Lp(Y) < X$ for every term $Y$ appearing in the second and third sum in the above equation. By assumption the leading power product of the first sum is also strictly less than $X$.

Let $u_i = c_i X^{\alpha_i}$+lower terms. Define

$$q = \sum_{h \in L} Lt(u_h)g_h = \sum_{h \in L} c_h X^{\alpha_h} g_h.$$

Then let $f_i = X^{\alpha_i} g_i$ for all $i \in L$. Since $Lp(f_i) = X$ for all $i \in L$ and $Lp(q) < X$, it follows from Lemma 12.5.2 that

$$q = \sum_{i,j \in L} c(i,j) SP(f_i, f_j) \tag{12.5.9}$$

where $c(i,j) \in k$ for all $i, j \in L$. Let $X(i,j) = LCM(Lp(g_i), Lp(g_j))$. Because $X = LCM(f_i, f_j) = X$ for all $i, j \in L$, we have

$$\begin{aligned} SP(f_i, f_j) &= X\left(\frac{X^{\alpha_i} g_i}{Lt(X^{\alpha_i} g_i)} - \frac{X^{\alpha_j} g_j}{Lt(X^{\alpha_j} g_j)}\right) \\ &= X\left(\frac{g_i}{Lt(g_i)} - \frac{g_j}{Lt(g_j)}\right) = \frac{X}{X(i,j)} SP(g_i, g_j). \end{aligned} \tag{12.5.10}$$

It is easy to check that $\frac{X}{X(i,j)} \in \Pi^n$. By hypothesis, $SP(g_i, g_j) \xrightarrow{G}_* 0$ for all $i \neq j$, this implies that $SP(f_i, f_j) \xrightarrow{G} 0$ for all $i \neq j$, by Lemma 12.3.7. Now it follows from Theorem 12.3.9 and Lemma 12.5.2 that

$$SP(f_i, f_j) = \sum_{k \in I_m} u(i,j,k) g_k \tag{12.5.11}$$

where $u(i,j,k) \in k[x_1, \cdots, x_n]$ and

$$X > Lp(SP(f_i, f_j)) = \max_{k \in I_m} Lp(u(i,j,k)) Lp(g_k). \tag{12.5.12}$$

By substituting the expression (12.5.11) into (12.5.9), and (12.5.9) into (12.5.8), we know that $Lp(Y) < X$ for every term $Y$ appearing in the first sum of the form (12.5.8). This is a contradiction. Hence $Lp(f) = X$, $Lt(f) \in Lt(G)$ and $Lt(I) \subseteq Lt(G)$. □

This theorem gives an easy criterion for checking whether a set of nonzero polynomials is a Gröbner basis or not. We give an example. Let $f_1 = x + y^2 - 1$ and $f_2 = y^6 + 1$ be in $\mathbb{Q}[x,y]$, and let $F = \{f_1, f_2\}$. Use lex with $x > y$. Then we have that $SP(f_1, f_2) = -x + y^8 - y^6$. Moreover,

$$SP(f_1, f_2) \xrightarrow{f_1} y^8 - y^6 + y^2 - 1 \xrightarrow{f_2} -y^6 - 1 \xrightarrow{f_2} 0.$$

Hence $F$ is a Gröbner basis for $I = < f_1, f_2 >$.

Now we can formally introduce Buchberger's algorithm. Given a set of nonzero polynomials $G = \{f_1, \cdots, f_m\}$ in $k[x_1, \cdots, x_n]$, this algorithm will find a Gröbner basis for the ideal $< f_1, \cdots, f_m >$ within a finite number of steps.

Buchberger's Algorithm

**Step (0)** Let $F = \{f_1, \cdots, f_m\}$ be a set of nonzero polynomials in $k[x_1, \cdots, x_n]$. Set $G = F$ and $\mathcal{F} = \{\{f_i, f_j\} \mid f_i, f_j \in G \text{ with } f_i \neq f_j\}$.

**Step (1)** If $\mathcal{F} \neq \emptyset$, go to Step (2). Otherwise, output $G$ and stop.

**Step (2)** Choose any $\{p, q\}$ from $\mathcal{F}$. Set $\mathcal{F} = \mathcal{F} \setminus \{\{p, q\}\}$. Implement the MVD algorithm and obtain $SP(p, q) \xrightarrow{G}_* h$. If $h \neq 0$, then go to Step (3). Otherwise, go to Step (1).

**Step (3)** Set $\mathcal{F} = \mathcal{F} \cup \{\{h, f\} \mid f \in G\}$ and $G = G \cup \{h\}$. Go to Step (2).

We will demonstrate the algorithm by an example. Let $f_1 = x^2y + 1$ and $f_2 = y^2 + 1$ be in $\mathbb{Q}[x,y]$, and let $F = \{f_1, f_2\}$. We use deglex with

$x > y$. We first show that $F$ is not a Gröbner basis for $I =< f_1, f_2 >$. Take for example $f = -x^2 + y$ from $I$. Note that $f = yf_1 - x^2 f_2$. Since $Lp(f) = x^2$, $Lp(f_1) = x^2y$, and $Lp(f_2) = y^2$, it is clear that $Lp(f)$ is not divisible by $Lp(f_1)$ or $Lp(f_2)$. By implementing the above algorithm, we get $G = \{g_1, g_2, g_3\}$ with $g_1 = f_1$, $g_2 = f_2$ and $g_3 = f$ which is a Gröbner basis for $I =< f_1, f_2 >$.

Now we prove the convergence of the algorithm.

**Theorem 12.5.4** *Given a set of nonzero polynomials $F = \{f_1, \cdots, f_m\}$ in $k[x_1, \cdots, x_n]$, Buchberger's algorithm will find a Gröbner basis for the ideal $I =< f_1, \cdots, f_m >$ within a finite number of steps.*

Proof: We first prove that the algorithm will stop after a finite number of steps. Suppose to the contrary that the algorithm would never stop. Note that the sequence $G_1$, $G_2$, $\cdots$ of $G$'s generated by the algorithm is strictly ascending, since $G_i$ is obtained from $G_{i-1}$ by adding $h$ at Step (3) with $h$ being a reminder of $SP(p, q)$ of $G_{i-1}$ and $h \notin < G_{i-1} >=< g_i \mid g_i \in G_{i-1} >$. By assumption, this sequence must be infinite. That is,

$$G_1 \subset G_2 \subset \cdots \subset G_{i-1} \subset G_i \subset \cdots.$$

This means that

$$< G_1 > \subset < G_2 > \subset \cdots \subset < G_{i-1} > \subset < G_i > \subset \cdots.$$

Thus we obtain an infinite strictly ascending chain of ideals in $k[x_1, \cdots, x_n]$ which contradicts Hilbert's basis theorem. Hence the algorithm must stop with a finite set of nonzero polynomials $G = \{g_1, \cdots, g_s\}$ in $k[x_1, \cdots, x_n]$.

Now we have to show that $G$ is a subset of $I$ at each stage of the algorithm. This is true at Step (0). Moreover, $G = F$ initially. Suppose that $G \subseteq I$ at some stage of the algorithm. This implies $p$, $q$, $SP(p, q) \in I$ from Step (2). Since $h$ is a reminder of $SP(p, q)$ with respect to $G$, we have $h \in I$. Note that $G$ is enlarged at Step (3) by adding $h$. Thus, $G \cup \{h\} \subseteq I$. Because $F \subseteq G \subseteq I$, we have

$$I =< f_1, \cdots, f_m > \subseteq < g_1, \cdots, g_s > \subseteq I.$$

So $I$ is finitely generated by $G$. Since $SP(g_i, g_j) \xrightarrow{G}_* 0$ for all $g_i$, $g_j \in G$ with $g_i \neq g_j$ by the construction of the algorithm, it follows from Theorem 12.5.3 that $G$ is a Gröbner basis for the ideal $I =< f_1, \cdots, f_m >$. □

The following result can easily be shown. This theorem gives an affirmative answer to the ideal membership problem.

**Theorem 12.5.5** *Given any element $f$ and any nonzero ideal $I = < f_1, \cdots, f_m >$ of $k[x_1, \cdots, x_n]$, then $f \in I$ if and if the reminder of $f$ with respect to a Gröbner basis $G$ for $I$ is zero.*

The reader may well find that for the same ideal $I$ there might exist several Gröbner bases. In order to get a unique Gröbner basis for any ideal, we need to slightly modify the definition of Gröbner bases.

**Definition 12.5.6** *A Gröbner basis $G$ for an ideal $I$ is a minimal Gröbner basis for $I$ if for every $f \in G$, $Lc(f) = 1$ and $Lp(f)$ is not divisible by $Lp(g)$ of any $g \in G$ with $g \neq f$.*

**Definition 12.5.7** *A Gröbner basis $G$ for an ideal $I$ is a reduced Gröbner basis for $I$ if for every $g \in G$, $Lc(g) = 1$ and $g$ is a reminder of $g$ with respect to $G \setminus \{g\}$.*

This definition says that $G$ is a reduced Gröbner basis if the leading coefficient of every element $g$ of $G$ is 1 and no nonzero term of every element $g$ of $G$ is divisible by the leading product of any other element of $G$. The following theorem states that every nonzero ideal has a unique reduced Gröbner basis.

**Theorem 12.5.8** *Let $I$ be a nonzero ideal of $k[x_1, \cdots, x_n]$. Then for a given term order, $I$ has a unique reduced Gröbner basis.*

Proof: It follows from Theorem 12.4.5 that there exists a Gröbner basis for $I$, say $F = \{f_1, \cdots, f_m\}$. We will construct a reduced Gröbner basis from $F$. First, divide every $p$ in $F$ by $Lc(p)$. Hence $Lc(p) = 1$ for all $p \in F$. Second, if there exists a $p$ in $F$ of which $Lp(p)$ is divisible by $Lp(q)$ of some $p \neq q \in F$. Then delete $p$ from $F$. The newly obtained $F$ is still a Gröbner basis for $I$ from Definition 12.4.1. Repeating this we finally get a minimal Gröbner basis $F = \{f_1, \cdots, f_s\}$ for $I$. Now we proceed with the following procedure:

**Step (0)** Let $F = \{f_1, \cdots, f_s\}$. Set $G = \emptyset$ and $i = 1$.

**Step (1)** If $i = s + 1$, then output $G = \{g_1, \cdots, g_s\}$ and stop. Otherwise, set $H = G \cup \{f_t \mid t \in I_s \setminus I_i\}$, implement the MVD algorithm and get $f_i \xrightarrow{H}_* g_i$. Set $G = G \cup \{g_i\}$ and $i = i + 1$. Go back to Step (1).

We first show that $Lp(g_i) = Lp(f_i)$ and $Q_i = \{g_1, \cdots, g_i, f_{i+1}, \cdots, f_s\}$ in the procedure is a minimal Gröbner basis for $I$ for every $i \in I_s$. When $i = 1$ in the procedure, $g_1$ is a reminder of $f_1$ with respect to $\{f_2, \cdots, f_s\}$. By assumption $F$ being a minimal Gröbner basis, we know that $Lp(g_1) = Lp(f_1)$. Moreover,

$$Lt(Q_1) = < Lt(g_1), Lt(f_2), \cdots, Lt(f_s) > = Lt(F).$$

Hence $Q_1$ is a minimal Gröbner basis for $I$. The general step can be shown in the same way. Since for each $i$, $g_i$ is a reminder of $f_i$ with respect to $\{g_1, \cdots, g_{i-1}, f_{i+1}, \cdots, f_s\}$, we know that $Q_s = G = \{g_1, \cdots, g_s\}$ must be a reduced Gröbner basis for $I$ by noting that $g_j$ for $j \leq i$ is also a reminder of $g_j$ with respect to $Q_i \setminus \{g_j\}$.

Now we proceed to prove the uniqueness. Suppose that there are two different reduced Gröbner bases $F = \{f_1, \cdots, f_m\}$ and $G = \{g_1, \cdots, g_s\}$ for $I$. We may assume that $m \leq s$. We first show that $m = s$ and $Lp(f_i) = Lp(g_i)$ by renumbering elements $g$ of $G$ for all $i \in I_m$. Since $f_1 \in I$ and $G$ is a Gröbner basis for $I$, there exists some $j \in I_s$ such that $Lp(f_1)$ is divisible by $Lp(g_j)$. We may assume $j = 1$ (otherwise, we can renumber all $g \in G$). On the other hand, since $g_1 \in I$ and $F$ is a Gröbner basis for $I$, there exists $j \in I_m$ such that $Lp(g_1)$ is divisible by $Lp(f_j)$. This means that $Lp(f_1)$ is divisible by $Lp(f_j)$. Because $F$ is a reduced Gröbner basis for $I$, $j$ must be equal to 1. Hence $Lp(f_1) = Lp(g_1)$. Suppose that for all $1 < i < j < m$ we have $Lp(f_i) = Lp(g_i)$. Since $f_j \in I$ and $G$ is a Gröbner basis for $I$, there exists some $l \in I_s$ such that $Lp(f_j)$ is divisible by $Lp(g_l)$. Because $Lp(f_i) = Lp(g_i)$ for all $i < j$, and $F$ is a reduced Gröbner basis, $l$ cannot be less than $j$. We may assume $l = j$ (otherwise, we can renumber all $g \in G$). On the other hand, since $g_j \in I$ and $F$ is a Gröbner basis for $I$, there exists $l \in I_m$ such that $Lp(g_j)$ is divisible by $Lp(f_l)$. This means that $Lp(f_j)$ is divisible by $Lp(f_l)$. Because $F$ is a reduced Gröbner basis for $I$, $l$ must be equal to $j$. Hence $Lp(f_j) = Lp(g_j)$. Repeat this process until $j = m$. We have shown that $s = m$ and $Lp(f_i) = Lp(g_i)$ for all $i \in I_m$. Finally we still have to prove that $g_i = f_i$ for all $i$. Take any $i \in I_m$. If $f_i \neq g_i$, then $h = g_i - f_i \in I$ since $f_i, g_i \in I$. Then there exists some $k \in I_m$ such that $Lp(g_i - f_i)$ is divisible by $Lp(f_k)$. Note that $Lp(g_i - f_i) < Lp(f_i)$. This implies $k \neq i$. We know that some nonzero term of $g_i$ or $f_i$ is divisible by $Lp(f_k)$. But this cannot be true since both $F$ and $G$ are reduced Gröbner bases for $I$. Therefore $f_i = g_i$ for all $i$. □

Consider the previous example where $f_1 = x^2y + 1$ and $f_2 = y^2 + 1$ are in $\mathbb{Q}[x, y]$. We use deglex with $x > y$. We have shown that $G = \{g_1, g_2, g_3\}$ with $g_1 = f_1, g_2 = f_2$ and $g_3 = -x^2 + y$ is a Gröbner basis for $I =< f_1, f_2 >$. But $G$ is not a reduced Gröbner basis since $Lp(g_1) = x^2y$ is divisible by $Lp(g_3) = x^2$. However, $H = \{g_2, g_3\}$ is a reduced Gröbner basis for $I$ as one can easily see.

It is worth mentioning that there are many computer algebra systems such as Maple and Mathematica which have a program package for implementing Buchberger's algorithm or some variation of it.

## 12.6. Hilbert's zero point theorem

The goal of this section is to determine whether the system of polynomial equations

$$f_i(x_1, \cdots, x_n) = 0, \; i = 1, \cdots, m$$

has a solution in $\mathbb{C}^n$ or not, where $f_i(x_1, \cdots, x_n) \in \mathbb{C}[x_1, \cdots, x_n]$ for all $i \in I_m$. If the system has a solution, we will determine whether it has a finite number of solutions or not. When the system has a finite number of solutions, we will be able to compute all solutions of this system.

We first review several concepts. Given a field $k$ and a positive integer $n$, the $n$-dimensional *affine space* over $k$ is defined to be the set

$$k^n = \{(a_1, \cdots, a_n) \mid a_i \in k, \text{ for all } i \in I_n\}.$$

It is clear that if $k = \mathbb{R}$, then $k^n = \mathbb{R}^n$ is the $n$-dimensional Euclidean space. Given a field $k$ and a subset $S$ of $k[x_1, \cdots, x_n]$, we call the set

$$V(S) = \{(a_1, \cdots, a_n) \in k^n \mid f(a_1, \cdots, a_n) = 0, \text{ for all } f \in S\}$$

the *affine variety* defined by $S$. Let $V$ be an affine variety. We define

$$I(V) = \{f \in k[x_1, \cdots, x_n] \mid f(a_1, \cdots, a_n) = 0, \text{ for all } (a_1, \cdots, a_n) \in V\}.$$

It is easy to show that $I(V)$ is an ideal of $k[x_1, \cdots, x_n]$. Hence $I(V)$ is also called the ideal of $V$. Let $I$ be an ideal in $k[x_1, \cdots, x_n]$. We define the *radical* of $I$ to be the set

$$\sqrt{I} = \{f \mid f^m \in I \text{ for some positive integer } m\}.$$

A field $k$ is said to be *algebraically closed* if every non-constant polynomial in $k[x]$ has a zero point (or root) in $k$. In Chapter 11 we saw that every non-constant polynomial in $\mathbb{C}[x]$ has a root in $\mathbb{C}$. Thus, the complex number field $\mathbb{C}$ is algebraically closed. However, the real number field $\mathbb{R}$ is not algebraically closed. Given an algebraically closed field $k$, an ideal $I$ in $k[x_1, \cdots, x_n]$ is called a *zero-dimensional ideal* if the number of solutions in $k^n$ of the system of equations

$$f(x_1, \cdots, x_n) = 0, \text{ for all } f \in I,$$

is finite.

Now we can introduce the following famous algebraic theorem. Since the proof is very lengthy, we omit it here. The proof can be found in van der Waerden [1970, Vol.II, pp.156-157].

**Theorem 12.6.1 (Hilbert Weak Nullstellensatz)** *Let $k$ be an algebraically closed field and $I$ be an ideal in $k[x_1, \cdots, x_n]$. Then $V(I) = \emptyset$ if and only if $I = k[x_1, \cdots, x_n]$.*

Nullstellensatz means zero point theorem or solution-set theorem. By applying this theorem, we have

**Theorem 12.6.2** *Let $k$ be an algebraically closed field. Given a set of nonzero polynomials $\{f_1, \cdots, f_m\}$ in $k[x_1, \cdots, x_n]$, there is no solution to the system $f_1 = 0, \cdots, f_m = 0$ if and only if 1 is the reduced Gröbner basis for the ideal $I = < f_1, \cdots, f_m >$.*

Proof: By Theorem 12.6.1, $V(I) = \emptyset$ if and only if $I = k[x_1, \cdots, x_n]$. On the other hand, $I = k[x_1, \cdots, x_n]$ is equivalent to the statement that 1 is the reduced Gröbner basis for the ideal $I$. □

To determine whether the set $V(S)$ for a given set $S$ in $k[x_1, \cdots, x_n]$ is finite or not, we need a strong form of Hilbert Nullstellensatz. The following proof is called "Rabinowitsch trick". See van der Waerden [1970, Vol.II, p.157].

**Theorem 12.6.3 (Hilbert Strong Nullstellensatz)** *Let $k$ be an algebraically closed field. If $I$ is an ideal in $k[x_1, \cdots, x_n]$, then $I(V(I)) = \sqrt{I}$.*

Proof: We first prove that $\sqrt{I} \subseteq I(V(I))$. Consider any $f \in \sqrt{I}$. This means that $f^m \in I$ for some positive integer $m$. On the other hand $f^m$ vanishes on $V(I)$. So $f$ also vanishes on $V(I)$. Therefore $f \in I(V(I))$.

Conversely, consider any $f \in I(V(I))$. By definition, $f$ vanishes on $V(I)$. We have to show that there exists a positive integer $m$ such that $f^m \in I$. This will be done by applying Theorem 12.6.1. By Hilbert Basis Theorem, there exist a finite set of nonzero polynomials $\{f_1, \cdots, f_t\}$ such that $I = < f_1, \cdots, f_t >$. Now we have to show that there exist a positive integer $m$ and polynomials $u_1, \cdots, u_t$ in $k[x_1, \cdots, x_n]$ such that

$$f^m = u_1 f_1 + \cdots + u_t f_t.$$

Consider the ideal

$$J = < f_1, \cdots, f_t, 1 - yf > \subseteq k[x_1, \cdots, x_n, y],$$

where $f_1, \cdots, f_t$ are as given above. We will show that $V(J) = \emptyset$.

Take any $(a_1, \cdots, a_n, a_{n+1}) \in k^{n+1}$. If $f_i(a_1, \cdots, a_n) = 0$ for all $i \in I_t$ and $f(a_1, \cdots, a_n) = 0$, we have $1 - yf(a_1, \cdots, a_n) = 1 - a_{n+1} f(a_1, \cdots, a_n) = 1 \neq 0$. This means that $(a_1, \cdots, a_{n+1}) \notin V(J)$. If $f_i(a_1, \cdots, a_n) \neq 0$ for some $i \in I_t$, it is clear that $(a_1, \cdots, a_{n+1}) \notin V(J)$.

Therefore $V(J) = \emptyset$. By applying Theorem 12.6.1, we know that $J = k[x_1, \ldots, x_n, y]$. Hence $1 \in J$. Thus, there exist polynomials $v_i$ for all $i \in I_t$

and $w$ in $k[x_1, \cdots, x_n, y]$ such that

$$1 = \sum_{i=1}^{t} v_i(x_1, \cdots, x_n, y) f_i + w(x_1, \cdots, x_n, y)(1 - yf).$$

Let $y = 1/f$. Then we have

$$1 = \sum_{i=1}^{t} v_i(x_1, \cdots, x_n, 1/f) f_i.$$

By multiplying both sides with $f^m$ for a big enough positive integer $m$, we have

$$f^m = \sum_{i=1}^{t} u_i f_i$$

where $u_i \in k[x_1, \cdots, x_n]$ for all $i \in I_t$. Hence $f^m \in I$.

Since $f^m \in I$, by definition $f^m \in \sqrt{I}$. Thus, $I(V(I)) \subseteq \sqrt{I}$. □

Given $f$, $g \in k[x_1, \cdots, x_n]$ and an ideal $I$ in $k[x_1, \cdots, x_n]$, $f$ is said to be *congruent* to $g$ modulo $I$, denoted by $f \equiv g(mod\, I)$, if $f - g \in I$. Notice that the congruence relation is an equivalence relation. So for each $f$, we can define an equivalence class. This class $[f : I] = \{g \in k[x_1, \cdots, x_n] \mid f - g \in I\}$ is called a *coset* of $I$. Each element of $[f : I]$ is called a representative of $[f : I]$. Clearly, for any $g \in [f : I]$, we have $[f : I] = [g : I]$. All such cosets will be denoted by $k[x_1, \cdots, x_n]/I$. Observe that $k[x_1, \cdots, x_n]/I$ is a commutative ring and gives a partition of $k[x_1, \cdots, x_n]$. Let $I$ be an nonzero ideal in $k[x_1, \cdots, x_n]$ and let $G = \{g_1, \cdots, g_m\}$ be a Gröbner basis for $I$. For any $f \in k[x_1, \cdots, x_n]$, we denote by $N_G(f)$ the reminder of $f$ with respect to $G$. It is easy to show that $f \equiv g(mod\ I)$ if and only if $N_G(f) = N_G(g)$. Thus, $\{N_G(f) \mid f \in k[x_1, \cdots, x_n]\}$ is the set of coset representatives for $k[x_1, \cdots, x_n]/I$. Now we have the following lemma from Adams and Loustaunau [1991, p.58].

**Lemma 12.6.4** *Let $I$ be a nonzero ideal in $k[x_1, \cdots, x_n]$ and let $G = \{g_1, \cdots, g_m\}$ be a Gröbner basis for $I$. Then a basis for $k[x_1, \cdots, x_n]/I$ consists of the cosets of all the monomials in $\Pi^n$ which are not divisible by any $Lp(g_i)$ for $i \in I_m$.*

Proof: For any $f \in k[x_1, \cdots, x_n]$, we have $[f : I] = [N_G(f) : I]$. Since $N_G(f)$ is a reminder of $f$ with respect to $G$, it is by definition a linear combination of monomials in $\Pi^n$ which are not divisible by any $Lp(g_i)$ for $i \in I_m$. The cosets of such monomials are linearly independent since $N_G(f)$ is unique for any $f$ by Lemma 12.4.4. □

The dimension of $k[x_1, \cdots, x_n]/I$ is defined to be the number of basis vectors for $k[x_1, \cdots, x_n]/I$. We give an example to illustrate the lemma. Let

$f_1 = x + y^2 - 1$ and $f_2 = y^6 + 1$ be in $\mathbb{Q}[x, y]$, and let $F = \{f_1, f_2\}$. Use lex with $x > y$. As shown in the previous section, $F$ is a Gröbner basis for $I =< f_1, f_2 >$. So a basis for $\mathbb{Q}[x, y]/I$ consists of the cosets of $1, y, y^2, y^3, y^4$, and $y^5$. Obviously, $dim(\mathbb{Q}[x, y]/I) = 6$.

We are now ready to prove the following theorem which helps solve our second problem. Here the proof is adapted from Adams and Loustaunau [1994]. The interested reader is referred to van der Waerden [1970, Vol. II, chapter 16] and Buchberger [1985] for more details.

**Theorem 12.6.5** *Let $k$ be an algebraically closed field, let $I$ be a nonzero ideal in $k[x_1, \cdots, x_n]$ and let $G = \{g_1, \cdots, g_m\}$ be a reduced Gröbner basis for $I$. Then the following statements are equivalent:*
*(i) For each $i \in I_n$, there exists $j \in I_m$ such that $Lp(g_j) = x_i^d$ for some non-negative integer $d$.*
*(ii) The dimension of the set $k[x_1, \cdots, x_n]/I$ is finite.*
*(iii) The set $V(I)$ is finite.*

Proof: We first prove that (*i*) implies (*ii*). It follows from Lemma 12.6.4 that the basis of $k[x_1, \cdots, x_n]/I$ consists of the cosets of monomials in $\Pi^n$ which are not divisible by any $Lp(g_i)$ for $i \in I_m$. Because for each $i \in I_n$, there exists $j \in I_m$ such that the leading power product of $g_j$ consists only of the variable $x_i$, this implies that there are only a finite number of monomials which are not divisible by any $Lp(g_i)$ for $i \in I_m$. Hence the dimension of $k[x_1, \cdots, x_n]/I$ is finite.

Secondly we show that (*ii*) implies (*iii*). Take any $i \in I_n$. Since by hypothesis the dimension of $k[x_1, \cdots, x_n]/I$ is finite, the monomials $1$, $x_i$, $x_i^2$, $\cdots$ of $x_i$ are linearly dependent modulo $I$. Thus, there exist a positive integer $t$ and coefficients $a_j \in k$ for all $j \in I_t \cup \{0\}$ such that

$$f(x_i) = c_0 + c_1 x_i + \cdots + c_t x_i^t \in I.$$

Notice that there are only finitely many roots for $f(x_i) = 0$ in $k$. Therefore, there are only finitely many distinct values for the $i$-th coordinate of points in $V(I)$. So the set $V(I)$ is finite.

Finally, we have to show that (*iii*) implies (*i*). When $V(I) = \emptyset$, it follows from Theorem 12.6.1 that $I = k[x_1, \cdots, x_n]$ and $G = \{1\}$, so (*i*) is satisfied. When $V(I)$ is nonempty, take any $i \in I_n$. Let $a_{i,1}$, $a_{i,2}$, $\cdots$, $a_{i,l_i}$ be the distinct values of the $i$-th coordinate of the points in $V(I)$. So for each $j \in I_{l_i}$, there exists $0 \neq f_j \in k[x_i]$ such that $f_j(a_{i,j}) = 0$. Let $f = f_1 f_2 \cdots f_{l_i}$. Clearly $f \in k[x_i] \subseteq k[x_1, \cdots, x_n]$. Then $f \in I(V(I))$. By Theorem 12.6.3 there exists a positive integer $d$ such that $f^d \in I$. Therefore $Lp(f^d) = x_i^q$ for some positive integer $q$. By Definition 12.4.1, $Lp(f^d)$ must be divisible by $Lp(g)$ for some $g \in G$. Since $Lp(f^d)$ consists only of $x_i$, $Lp(g)$ must also consists only of $x_i$. Since this is true for any $i \in I_n$, we proved (*i*). □

**Theorem 12.6.6** *Let $k$ be an algebraically closed field, let $I$ be a zero-dimensional ideal in $k[x_1, \cdots, x_n]$ and let $G = \{g_1, \cdots, g_m\}$ be a reduced Gröbner basis for $I$ with lex term order by specifying $x_1 < x_2 \cdots < x_n$. Then we can renumber the elements of $G$ such that $g_1$ contains only the variable $x_1$, $g_2$ contains only the variables $x_1$, $x_2$ and $Lp(g_2)$ is the power of $x_2$, $g_3$ contains only the variables $x_1$, $x_2$, $x_3$ and $Lp(g_3)$ is the power of $x_3$, and so on until $g_n$.*

Proof: By applying Theorem 12.6.5 (i), we can renumber the elements of $G$ such that $Lp(g_i)$ consists only of $x_i$. Since we use lex term order with $x_1 < x_2 < \cdots < x_n$, this implies that only $x_1$, $x_2$, $\cdots$, $x_i$ appear in $g_i$. □

Let us now return to the problems posed at the beginning of this section. Given a set of nonzero polynomials $\{f_1, \cdots, f_m\}$ from $\mathbb{C}[x_1, \cdots, x_n]$, we want to solve the system of equations

$$f_i(x) = 0, \; x \in \mathbb{C}^n$$

for all $i \in I_m$. By using lex term order, we first obtain a reduced Gröbner basis $G = \{g_1, \cdots, g_t\}$ for the ideal $I =< f_1, \cdots, f_m >$. Then one of the next three cases must occur.

Case (1): If $G = \{1\}$, then the above problem has no solution at all.

Case (2): If we can renumber the elements of $G$ such that $g_1$ contains only the variable $x_1$, $g_2$ contains only the variables $x_1$, $x_2$ and $Lp(g_2)$ is the power of $x_2$, $g_3$ contains only the variables $x_1$, $x_2$, $x_3$ and $Lp(g_3)$ is the power of $x_3$, and so on until $g_n$. Then the above problem has a finite number of solutions. By using Kuhn's method in Chapter 11, we can find all roots of $g_1(x_1) = 0$. Then for each value $\alpha$ of $x_1$ we can find all roots of $g_2(\alpha, x_2) = 0$ by Kuhn's method. We can continue in this way until $g_n = 0$. Thus, all solutions of the above problem will be found.

Case (3): All other cases imply that the above problem has an infinite number of solutions. In this case, we can apply a homotopy algorithm to find as finitely many solutions as we wish to have. The interested reader is referred to Kojima and Mizuno [1983] for more technique details.

Finally it may be worth pointing out that it is the first time to combine Buchberger's method with Kuhn's method so that all solutions of a zero-dimensional ideal can be computed. Many authors seem to be unaware of the existence of Kuhn's method.

## 12.7. Applications: the complementarity problems

In this section we discuss an application of Gröbner bases to the complementarity problem. Recall this problem has been defined in Chapter 1.

Given a function $f : \mathbb{R}^n \mapsto \mathbb{R}^n$, the complementarity problem is to find a point $x \in \mathbb{R}^n$ such that

$$x^\top f(x) = 0, \; x \geq 0^n, \; f(x) \geq 0^n. \tag{12.7.13}$$

It is easy to see that equation (12.7.13) has a solution if and only if there exists $(x, y) \in \mathbb{R}^{2n}$ such that

$$\begin{aligned} x_1 y_1 &= 0, \\ x_2 y_2 &= 0, \\ &\vdots \\ x_n y_n &= 0, \\ f(x) - y &= 0^n, \\ x \geq 0^n, & \quad y \geq 0^n. \end{aligned} \tag{12.7.14}$$

Now we will focus on this problem by specifying that $f$ is a polynomial function. That is $f_i \in \mathbb{R}[x_1, \cdots, x_n]$ for all $i \in I_n$. A complementarity problem is said to be *zero-dimensional* if the system of polynomial equations

$$f(x) = 0^n, \; x \in \mathbb{C}^n$$

has a finite number of solutions. It is easy to see that when $f(x) = Mx + q$ with $M$ an $n \times n$ real matrix and $q \in \mathbb{R}^n$, then this linear complementarity problem is zero-dimensional under the condition that $M$ is regular or $f(x) = q$. Let $g_i = x_i y_i$ and $g_{n+i} = f_i - y_i$ for all $i \in I_n$. Then we have the following result. We leave the proof as an exercise.

**Theorem 12.7.1** *Suppose that we are given a zero-dimensional complementarity problem with $f = (f_1, \cdots, f_n)$ where $f_i \in \mathbb{R}[x_1, \cdots, x_n]$ for all $i \in I_n$. Let $H = \{h_1, \cdots, h_m\}$ be a reduced Gröbner basis for $I = < g_1, g_2, \cdots, g_{2n} >$ with lex term order by specifying $x_1 < x_2 < \cdots < x_n < y_1 < \cdots < y_n$. Then we can renumber the elements of $H$ such that $h_1$ contains only the variable $x_1$, $h_2$ contains only the variables $x_1$, $x_2$ and $Lp(h_2)$ is the power of $x_2$, $h_3$ contains only the variables $x_1$, $x_2$, $x_3$ and $Lp(h_3)$ is the power of $x_3$, and so on until $h_n$ contains only the variables $x_1$, $\cdots$, $x_n$ and $Lp(h_n)$ is the power of $x_n$; furthermore, $h_{n+1}$ contains only the variables $x_1$, $\cdots$, $x_n$, $y_1$ and $Lp(h_{n+1})$ is the power of $y_1$, and so forth until $h_{2n}$.*

Hence given any zero-dimensional polynomial complementarity problem with $f$, we can compute the affine variety $V(S) = V(\{g_1, \cdots, g_{2n}\})$ by using the above theorem and the procedure described at the end of the previous section. Next we check all elements in $V(S)$. Remember that $V(S)$ is a finite set. If there is an element $(x^*, y^*)$ in $V(S)$ such that $x^*$, $y^* \in \mathbb{R}^n_+$, then $x^*$

is a solution of the complementarity problem. We illustrate the above theorem by an example. Consider a polynomial complementarity problem with $f = (f_1, f_2)$, where $f_1 = x_2 + x_1^2 - 1$ and $f_2 = x_2 x_1^3 + 1$. Let $F = \{f_1, f_2\}$. We will show that this problem is zero-dimensional. By using term order lex with $x_1 < x_2$, we obtain a reduced Gröbner basis $G = \{g_1, g_2\}$ for $I = < f_1, f_2 >$ in $\mathbb{C}[x_1, x_2]$ with $g_1 = x_1^5 - x_1^3 - 1$ and $g_2 = x_2 + x_1^2 - 1$. Let $p_1 = y_1 x_1$, $p_2 = y_2 x_2$, $p_3 = -y_1 + x_1^5 - x_1^3 - 1$, and $p_4 = -y_2 + x_2 + x_1^2 - 1$. Using lex with $x_1 < x_2 < y_1 < y_2$, we obtain 4 elements $h_1$, $h_2$, $h_3$ and $h_4$ from the reduced Gröbner basis for $< p_1, p_2, p_3, p_4 >$ as follows:

$$\begin{aligned} h_1 &= x_1^6 - x_1^4 - x_1, \\ h_2 &= x_2^2 + x_2 x_1^2 - x_2, \\ h_3 &= y_1 - x_1^5 + x_1^3 + 1, \\ h_4 &= y_2 - x_2 - x_1^2 + 1. \end{aligned}$$

It should be noted that linear complementarity problems have been intensively studied in Cottle, Pang and Stone [1992], and Kojima, Megiddo, Noma and Yoshise [1991].

CHAPTER 13

# Intersection Theory

## 13.1. Introduction

In this chapter we deal with intersection theory. This theory studies various conditions under which the collection of sets covering a particular set has a nonempty intersection. Such existence results are called intersection theorems, which are often used to derive fixed point theorems. Those theorems are also used to prove the existence of solutions to mathematical programming, game theoretic and equilibrium problems. The most well-known intersection theorem is the lemma of Knaster, Kuratowski and Mazurkiewicz [1929] (KKM lemma) on the unit simplex $S^n$. This lemma states that $n$ closed subsets covering the $(n-1)$-dimensional unit simplex $S^n$ and satisfying some boundary condition have a nonempty intersection. A related theorem was due to Scarf [1967b, 1973]. The reformulation of this theorem was also given by Fan [1968]. Further generalizations of intersection theorems on the unit simplex can be found in Shapley [1973], Freidenfelds [1974], Gale [1984], Freund [1984b], Ichiishi [1988], and Joosten and Talman [1993]. The above theorems all build upon bounded sets. Helly [1923] proved the following different type of intersection theorem on an unbounded set. Let $C^i$, $i \in I_l$ be convex sets of $\mathbb{R}^n$ with $l \geq n+1$. If any $n+1$ of them have a nonempty intersection, then $\cap_{h=1}^{l} C^i \neq \emptyset$. Helly's theorem has been generalized in many ways. The interested reader should consult Stoer and Witzgall [1970]. In this chapter we focus on the results along the line of KKM lemma.

This chapter is organized as follows. In Section 2 we introduce an intersection theorem of van der Laan, Talman and Yang [1994]. In Sections 3 and 4 we derive the intersection theorems of KKM, Sperner, Scarf, Shapley, Ichiishi, and Freund from that theorem. In Section 5 we prove an intersection theorem with a continuum of intersection points due to Herings and Talman [1998]. In Section 6 we introduce the multi-permutation-based intersection theorem of van der Laan et al on the simplotope and apply this theorem to the exchange economy model with indivisibilities and money introduced in Chapter 2. Moreover, Gale's lemma will be derived from this theorem. Finally in Section 7 we prove Helly's theorem.

## 13.2. A general intersection theorem

Consider an arbitrary nonempty polytope $P$

$$P = \left\{x \in \mathbb{R}^n \mid a^{i\top} x \leq \alpha_i,\ i \in I \text{ and } d^{h\top} x = \delta_h,\ h \in I_{n-l}\right\},$$

where $I$ is a finite set of positive integers and $0 \leq l \leq n$. Throughout the chapter we assume that none of the constraints defining $P$ is redundant, and none of the constraints $a^{i\top} x \leq \alpha_i$, $i \in I$, is an implicit equality. For $T \subseteq I$, we define

$$F(T) = \left\{x \in P \mid a^{i\top} x = \alpha_i \text{ for } i \in T\right\},$$

with $F(\emptyset) = P$. In case $F(T)$ is nonempty, $F(T)$ is a face of $P$. When, for some $i \in I$, $T = \{i\}$, $F(T)$ is a facet of $P$ and we denote this facet by $F_i$. For $x \in bd(P)$, the set $I_x$ is defined as

$$I_x = \left\{i \in I \mid a^{i\top} x = \alpha_i\right\},$$

i.e., $I_x$ is the set of indices for which the corresponding constraint is binding at $x$. Clearly, for any $T \subseteq I_x$, $x$ belongs to $F(T)$. Furthermore, with respect to the polytope $P$, we define

$$V = \{x \in \mathbb{R}^n \mid x = \sum_{h=1}^{n-l} \nu_h d^h,\ \nu_h \in \mathbb{R} \text{ for } h \in I_{n-l}\},$$

with $V = \{\mathbf{0}\}$ for $l = n$. Now define

$$V^* = \{x \in \mathbb{R}^n \mid x^\top y = 0, \quad \text{for all} \quad y \in V\},$$

as the subspace orthogonal to $V$.

Moreover, for $T \subseteq I$, let $A(T)$ be the cone defined by

$$A(T) = V + \left\{x \in \mathbb{R}^n \mid x = \sum_{i \in T} \lambda_i a^i,\ \lambda_i \geq 0, \text{ for } i \in T\right\}$$

with $A(\emptyset) = \{\mathbf{0}\}$ when $l = n$, and let $A^*(T)$ be the *polar cone* of $A(T)$ defined by

$$A^*(T) = \left\{y \in \mathbb{R}^n \mid y^\top x \leq 0 \text{ for all } x \in A(T)\right\}.$$

By definition we have that $A(T) \cap A^*(T) = \{\mathbf{0}\}$, for all $T \subseteq I$.

Finally, for some finite nonempty set $J$, let a collection of vectors $c^j$, $j \in J$, in $\mathbb{R}^n$, be given. For a nonempty set $T \subseteq J$, we define

$$C(T) = \left\{ x \in \mathbb{R}^n \mid x = \sum_{j \in T} \mu_j c^j, \ \sum_{j \in T} \mu_j = 1 \text{ and } \mu_j \geq 0, \ j \in T \right\},$$

i.e., $C(T)$ is the convex hull of the vectors $c^j$, $j \in T$. We assume that $\mathbf{0} \in C(J)$.

**Definition 13.2.1** *For some nonempty subset $T$ of $J$, the collection of vectors $\{c^j \mid j \in T\}$ is balanced when $\mathbf{0} \in int(C(T))$, i.e., if the system of linear equations $\sum_{j \in T} \mu_j c^j = \mathbf{0}$ has a solution $\mu_j^* > 0$ for all $j \in T$.*

Without confusion, we say that the set $T$ is balanced if the corresponding collection of vectors is balanced.

Now we introduce the following lemma which can be seen as a dual version of Fan's coincidence theorem (see Fan [1972, Theorem 5, p.108]). It might be worth noting that Fan's coincidence theorem is already used in Ichiishi [1981]. The proof here is similar to the one given by Ky Fan.

**Lemma 13.2.2** *Let $D$ be a nonempty, convex and compact subset of $\mathbb{R}^n$. Let $\Psi$ and $\Phi$ be two upper semi-continuous point-to-set mappings from $D$ to $\mathbb{R}^n$ such that for any $x \in D$ both $\Psi(x)$ and $\Phi(x)$ are nonempty, convex and compact subsets of $\mathbb{R}^n$. Suppose that for each $x \in D$ and each $c \in \mathbb{R}^n$ satisfying*

$$c^\top x = \max\{c^\top y \mid y \in D\}, \tag{13.2.1}$$

*there exist $v \in \Psi(x)$ and $w \in \Phi(x)$ such that $c^\top v \geq c^\top w$. Then there is at least one point $x^* \in D$ such that $\Psi(x^*) \cap \Phi(x^*) \neq \emptyset$.*

Proof: Suppose to the contrary that for any $x \in D$, $\Psi(x) \cap \Phi(x) = \emptyset$. Then there exists a hyperplane strictly separating $\Psi(x)$ and $\Phi(x)$. So for each $x \in D$ there exist $c(x) \in \mathbb{R}^n$ and a real number $\gamma(x)$ such that $\Psi(x) \subseteq G(x) = \{y \in R^n \mid c^\top(x)y < \gamma(x)\}$ and $\Phi(x) \subseteq H(x) = \{y \in R^n \mid c^\top(x)y > \gamma(x)\}$. We shall prove that this leads to a contradiction. For each $c \in \mathbb{R}^n$, define the subset $D(c)$ of $D$ by

$$D(c) = \{x \in D \mid c^\top v < c^\top w \ \text{ for all } v \in \Psi(x) \ \text{ and all } \ w \in \Phi(x)\}.$$

The upper semi-continuity of $\Psi$ and $\Phi$ implies that there exists a neighborhood $O(x)$ of $x$ such that for all $\underline{x} \in O(x)$, $\Psi(\underline{x}) \subseteq G(x)$ and $\Phi(\underline{x}) \subseteq H(x)$. This means that $x \in int(D(c(x)))$. Furthermore, $\{int(D(c(x))) \mid x \in D\}$ is a collection of open subsets of $D$ covering $D$. Since $D$ is a compact set, there exists a finite subset of $\{int(D(c(x))) \mid x \in D\}$, say $\{int(D(c(x))) \mid x \in L\}$ covering $D$, where $L$ is a finite subset of $D$. Then by applying the unity partition theorem (see e.g., Munkres [1975]), there exist a family of continuous functions $\{\beta_x \mid x \in L\}$ being a partition of unity associated with $\{int(D(c(x))) \mid x \in L\}$. Define the continuous function $q : D \mapsto \mathbb{R}^n$ by

$$q(z) = \sum_{x \in L} \beta_x(z)c(x).$$

Then for all $z \in D$, it holds that

$$q^\top(z)v < q^\top(z)w \tag{13.2.2}$$

for all $v \in \Psi(z)$ and all $w \in \Phi(z)$. Since $q$ is a continuous function from a nonempty compact convex set $D$ to $\mathbb{R}^n$, it follows from Stationary Point Theorem I that there exists $z^* \in D$ such that

$$q^\top(z^*)z^* \geq q^\top(z^*)z,$$

for all $z \in D$. On the other hand, by the assumption there exist $v^* \in \Psi(z^*)$ and $w^* \in \Phi(z^*)$ such that $q^\top(z^*)v^* \geq q^\top(z^*)w^*$. This contradicts (13.2.2). □

Given the polytope $P$ and the collection of vectors $\{c^j \mid j \in J\}$ in $\mathbb{R}^n$ as described before, take a collection of sets $\{C^j \mid j \in J\}$ being a closed covering of the polytope $P$. For $x \in P$, we define $J_x = \{j \in J \mid x \in C^j\}$. In the next theorem we give a sufficient condition to guarantee that there is a balanced set of indices in $J$ for which the intersection of sets labelled by these indices is nonempty. The following theorem is introduced in van der Laan, Talman and Yang [1994].

**Theorem 13.2.3** *Let the collection of vectors $\{c^j \mid j \in J\}$ in $\mathbb{R}^n$ be such that $C(J) \subseteq V^*$ and let $\{C^j \mid j \in J\}$ be a collection of closed sets covering the polytope $P$ such that for every $x \in bd(P)$,*

$$C(J_x) \cap A^*(I_x) \neq \emptyset.$$

*Then there exists a balanced set $T^* \subseteq J$ for which $\cap_{j \in T^*} C^j \neq \emptyset$.*

Proof: Define two point-to-set mappings $\Psi$ and $\Phi$ from $P$ to $\mathbb{R}^n$ by

$$\Psi(x) = \{\underline{0}\}, \quad \Phi(x) = C(J_x).$$

It is clear that $\Phi$ and $\Psi$ are nonempty-valued, convex-valued, compact-valued, and upper semi-continuous. Choose any $x \in bd(P)$ and $c \in \mathbb{R}^n$ for which $c^\top x = max\{c^\top y \mid y \in P\}$. Then $c \in A(I_x)$. By the boundary condition, there exists $z \in C(J_x) \cap A^*(I_x)$. That is to say, there exists $z \in \Phi(x)$ for which

$$c^\top y = 0 \geq c^\top z$$

for all $y \in \Psi(x)$. Moreover, choose any $x \in int(P)$ and $c \in \mathbb{R}^n$ for which $c^\top x = max\{c^\top y \mid y \in P\}$. Then $c \in V$. By the assumption $C(J) \subseteq V^*$, it follows that for all $z \in \Phi(x)$,

$$c^\top y = 0 \geq c^\top z$$

for all $y \in \Psi(x)$. Thus all conditions of Lemma 13.2.2 are satisfied. So, there exists $x^* \in P$ such that $\Psi(x^*) \cap \Phi(x^*) \neq \emptyset$. That is, $\mathbf{0} \in C(J_{x^*})$. One may choose a subset $T^*$ of $J_{x^*}$ such that $\mathbf{0} \in int(C(T^*))$. So $T^*$ is balanced and $\cap_{j \in T^*} C^j \neq \emptyset$. □

The novelty of this theorem lies in the fact that the boundary condition is formulated with respect to every point on the boundary. While in all other existing intersection results the boundary condition is of the form that every face of the set under consideration is covered by the union of some specified elements of the family of sets forming the covering.

## 13.3. Intersection theorems on the unit simplex

In this section we apply Theorem 13.2.3 to prove several well-known intersection results on the unit simplex $S^n$. For $S \subseteq I_n$, let the $n$-vector $m^S$ be defined by $\sum_{i \in S} \frac{1}{|S|} e(i)$. Observe that $m^S = e(i)$ if $S = \{i\}$. Now, take $l = n - 1$, $d^1 = m^{I_n}$, $\delta_1 = 1/n$, $I = I_n$, $a^i = m^{I_n} - e(i)$ and $\alpha_i = 1/n$ for $i \in I_n$. Observe that $a^i \in V^*$ for all $i \in I_n$. Now, $S^n$ can be rewritten as

$$S^n = \left\{ x \in \mathbb{R}^n \mid a^{i\top} x \leq \alpha_i,\ i \in I_n \text{ and } d^{1\top} x = \delta_1 \right\}.$$

Notice that for $T \subset I_n$, $F(T) = S^n(T)$. We first derive the following KKM lemma.

**Theorem 13.3.1 (KKM Lemma)** *Let $\{C^j \mid j \in I_n\}$ be a collection of closed sets covering the unit simplex $S^n$ satisfying that for every $T \subset I_n$, the face $S^n(T)$ is contained in $\cup_{j \notin T} C^j$. Then $\cap_{j \in I_n} C^j \neq \emptyset$.*

Proof: Take $J = I_n$ and for $j \in I_n$, take $c^j = a^j$. Again, $C(J) \subseteq V^*$ and $\mathbf{0} \in C(J)$. From the boundary condition we have that, $J_x \cap (I_n \setminus T) \neq \emptyset$ if $x$ is a point in the boundary of $S^n$ such that $I_x = T$. Since for any pair $a^i$ and $a^j$, $i \neq j$, it holds that $a^{i\top} a^j < 0$, we have that for every $T \subseteq I_n$, $a^j \in A^*(T)$ if $j \notin T$. According to the boundary condition, $J_x$ contains at least one element $j \notin I_x$. Hence $C(J_x) \cap A^*(I_x) \neq \emptyset$ if $x \in bd(S^n)$ and the boundary condition of Theorem 13.2.3 is satisfied. So, there is a balanced set $T^* \subseteq I_n$ for which $\cap_{j \in T^*} C^j \neq \emptyset$. Again the set $I_n$ is the unique balanced set. □

The next lemma can be seen as a dual of KKM lemma and can be found in Scarf [1967b, 1973], Fan [1968] and Freidenfelds [1974]. It is said that this lemma is already established in Sperner [1928]. So we call it Sperner's intersection lemma.

**Theorem 13.3.2 (Sperner's Intersection Lemma)** *Let $\{C^j \mid j \in I_n\}$ be a collection of closed sets covering the unit simplex $S^n$ satisfying that for every $h \in I_n$ the facet $S^n_h$ is a subset of $C^h$. Then $\cap_{j\in I_n} C^j \neq \emptyset$.*

Proof: Take $J = I_n$ and for $j \in J$, take $c^j = -a^j$. Clearly, $C(J) \subseteq V^*$ and $\mathbf{0} \in C(J)$. Since the sets $C^j$, $j \in I_n$ are closed, it follows from the boundary condition that $I_x \subseteq J_x$ if $x \in bd(S^n)$. Hence, if $x \in bd(S^n)$, then $C(J_x) \cap A^*(I_x) \neq \emptyset$. For instance take $b = -\sum_{i\in T} \frac{1}{|T|} a^i$, where $T = I_x \neq I_n$. Then $b \in A^*(T) \cap C(T) \subseteq A^*(I_x) \cap C(J_x)$, since $T \subseteq J_x$. Since $S^n(T)$ is empty if $T = I_n$, the boundary conditions of Theorem 13.2.3 are satisfied and hence there is a balanced set $T^* \subseteq I_n$ for which $\cap_{j\in T^*} C^j \neq \emptyset$. However, by definition of the vectors $c^j$, $j \in I_n$, the set $I_n$ is the unique balanced set.
□

Freund [1984b, 1986] obtained the following result by relaxing the boundary condition of Sperner's intersection lemma.

**Theorem 13.3.3** *Let $\{C^j \mid j \in I_n\}$ be a collection of closed sets covering the unit simplex $S^n$. Then there exists an $x^* \in S^n$ such that $\{j \in I_n \mid x^*_j > 0\} \subseteq \{k \in I_n \mid x^* \in C^k\}$.*

Proof: Let $\bar{C}^j = C^j \cup S^n_j$ for each $j \in I_n$. Then by Sperner's intersection lemma, there is an $x^* \in \cap_{j\in I_n} \bar{C}^j$. It is clear that $x^*_j > 0$ implies $x^* \in C^j$.
□

The lemma of Sperner is a special case of a theorem of Scarf [1967b]. Also this theorem follows from Theorem 13.2.3. To state Scarf's theorem, let $B$ be an $n \times k$ matrix, $k > n$, satisfying that $b^i = e(i)$, $i \in I_n$, with $b^j$ the $j$th column of $B$, $j \in I_k$.

**Theorem 13.3.4 (Scarf's Intersection Lemma)** *Let $\{C^j \mid j \in I_k\}$ be a collection of closed sets covering the unit simplex $S^n$ satisfying that for every $h \in I_n$ the facet $S^n_h$ is a subset of $C^h$. Let $c \in \mathbb{R}^n_+ \setminus \{\mathbf{0}\}$ be given and assume that the set of solutions $\{y \in \mathbb{R}^k_+ \mid By = c\}$ is nonempty and bounded. Then there exists a set $T^* \subseteq I_k$ such that the system of equations $\sum_{j\in T^*} \mu_j b^j = c$ has a positive solution and $\cap_{j\in T^*} C^j \neq \emptyset$.*

Proof: Take $J = I_k$ and for $j \in J$, take $c^j = b^j - \gamma_j c$, where $\gamma_j = n(m^{I_n})^\top b^j$. So $\gamma_j = 1$ for $j \in I_n$. Without loss of generality we may assume that $c \in S^n$. Then, $\mathbf{0} \in C(J)$ and $(m^{I_n})^\top c^j = 0$, $j \in I_k$, and hence $C(J) \subseteq V^*$. Let $x \in bd(S^n)$ and let $T = I_x$. From the boundary condition we have that $C(T) \subseteq C(J_x)$. So the boundary condition of Theorem 13.2.3 is satisfied if $A^*(T) \cap C(T) \neq \emptyset$. First suppose that $c_h = 0$ for all $h \in T$. Then, for every $i, j \in T$, it holds that $c^{j\top} a^i = (e(j) - c)^\top (m^{I_n} - e(i)) =$

$-e^\top(j)e(i) \leq 0$, so that $c^j \in A^*(T) \cap C(T)$ for any $j \in T$. Now suppose that $c_h > 0$ for some $h \in T$. Then take $y = \sum_{j \in T} \frac{c_j}{\sum_{h \in T} c_h} c^j$. Clearly, $y \in C(T)$. Moreover, for every $i \in T$, $y^\top a^i = -y_i \leq 0$, and so $y \in A^*(T)$. Hence, the conditions of Theorem 13.2.3 are satisfied and there is a balanced set $T^* \subseteq I_k$ satisfying $\cap_{j \in T^*} C^j \neq \emptyset$. Balancedness of $T^*$ implies that there exist $\lambda_j^* > 0$ for $j \in T^*$ satisfying $\sum_{j \in T^*} \lambda_j^* = 1$ such that $\sum_{j \in T^*} \lambda_j^* c^j = \mathbf{0}$, i.e., $\sum_{j \in T^*} \lambda_j^* b^j - \alpha^* c = \mathbf{0}$ with $\alpha^* = \sum_{j \in T^*} \lambda_j^* \gamma_j$. Since $b^i = e(i)$ for $i \in I_n$, we have that $c = \sum_{i \in I_n} c_i b^i$, from which it follows that $\sum_{j \in T^*} \lambda_j^* b^j - \sum_{i \in I_n} \alpha^* c_i b^i = \mathbf{0}$. From the boundedness of the set $\{y \in \mathbb{R}_+^k \mid By = c\}$ it follows that $\mathbf{0} \notin \{By \mid y \in \mathbb{R}_+^k \setminus \{\mathbf{0}\}\}$. Hence $\alpha^* > 0$ and $\sum_{j \in T^*} \mu_j^* b^j = c$, where $\mu_j^* = \frac{\lambda_j^*}{\alpha^*} > 0$ for all $j \in T^*$. □

The next intersection theorem on the unit simplex is the intersection theorem of Shapley [1973]. To prove this theorem, we need the concept of coalitional balancedness (see Section 2.2).

**Theorem 13.3.5 (Shapley's Intersection Lemma)** *Let $\mathcal{N}$ be the collection of all nonempty subsets of $I_n$. Let $\{C^S \mid S \in \mathcal{N}\}$ be a collection of closed sets covering the unit simplex $S^n$ satisfying that for every $T \subset I_n$, it holds that $S^n(T) \subseteq \cup_{S \subseteq I_n \setminus T} C^S$. Then there is a coalitionally balanced family $\mathcal{B} = \{B_1, \ldots, B_k\}$ of elements of $\mathcal{N}$ for which $\cap_{j=1}^k C^{B_j} \neq \emptyset$.*

Proof: Take $J = \mathcal{N}$ and for any set $S \in \mathcal{N}$, take $c^S = m^{I_n} - m^S$. Then $C(J) \subseteq V^*$ and $\mathbf{0} \in C(J)$. From the boundary condition we have that for every $x \in bd(S^n)$ the set $J_x$ contains some set $S \subseteq I_n \setminus T$ if $T = I_x$. Moreover, for any $T \subset I_n$, ${c^S}^\top a^j \leq 0$ for every $j \in T$ and $S \subseteq I_n \setminus T$. Hence, for any $x$ on the boundary of $S^n$, $C(J_x) \cap A^*(I_x) \neq \emptyset$. From Theorem 13.2.3 it follows then that there exists a family $\mathcal{B} = \{B_1, \ldots, B_k\}$ satisfying that the collection of vectors $c^{B_j}$, $j \in I_k$, is balanced and $\cap_{j=1}^k C^{B_j} \neq \emptyset$. By definition of the vectors $c^S$, $S \in \mathcal{N}$, the family $\mathcal{B}$ is coalitionally balanced and hence the theorem holds. □

A typical application of Shapley's intersection lemma concerns the proof of the nonemptiness of the core of a coalitionally balanced NTU game. The following intersection theorem on the unit simplex was due to Ichiishi [1988] and can be considered as the dual of the Shapley Lemma.

**Theorem 13.3.6 (Ichiishi's Intersection Lemma)** *Let $\mathcal{N}$ be the collection of all nonempty subsets of $I_n$. Let $\{C^S \mid S \in \mathcal{N}\}$ be a collection of closed sets covering the simplex $S^n$ satisfying that for every $T \subset I_n$ it holds that $S^n(T) \subseteq \cup_{T \subseteq S} C^S$. Then there is a coalitionally balanced family $\mathcal{B} = \{B_1, \ldots, B_k\}$ of elements of $\mathcal{N}$ for which $\cap_{j=1}^k C^{B_j} \neq \emptyset$.*

Proof: For $J = \mathcal{N}$, take $c^S = m^S - m^{I_n}$ for every set $S \in \mathcal{N}$. Then again $C(J) \subseteq V^*$ and $\mathbf{0} \in C(J)$. From the boundary condition we have that $J_x$ contains some set $S$ having $T$ as a subset if $x \in S^n(T)$. Moreover, for any $T \subset I_n$, $c^{S\top} a^j \leq 0$ for every $j \in T$ and $S \supseteq T$. Hence, the boundary condition of Theorem 13.2.3 is satisfied and there exists a family $\mathcal{B} = \{B_1, \ldots, B_k\}$ satisfying that the collection of vectors $c^{B_j}$, $j \in I_k$, is balanced and $\cap_{j=1}^k C^{B_j} \neq \emptyset$. By definition of the vectors $c^S$, then also the family $\mathcal{B}$ is coalitionally balanced and hence the theorem holds. □

The next lemma is introduced in Joosten and Talman [1993] and can be proved directly by applying Theorem 13.2.3.

**Theorem 13.3.7** *Let $\{C^{ij} \mid i \in I_n,\ j \in I_n\}$ be a collection of closed sets covering the unit simplex $S^n$ satisfying that if $x \in bd(S^n)$, then $x \in C^{ij}$ for some $j \in I_n$ with $x_j > 0$ or $x \in \cup_{j \in I_n} C^{ij}$ for all $i \in I_n$ for which $x_i = 0$. Then there exists a nonempty set $I^* \subseteq I_n$ and, with $k = |I^*|$, a permutation $\pi(I^*) = (\pi(1), \ldots, \pi(k))$ of the elements of $I^*$, such that $\cap_{i \in I^*} C^{i\pi(i)} \neq \emptyset$.*

Proof: Take $J = I_n \times I_n$ and for $(i,j) \in J$, take $c^{ij} = e(i) - e(j)$. Clearly, $C(J) \subseteq V^*$ and $\mathbf{0} \in C(J)$. Without loss of generality we may assume that $C^{ii} = \emptyset$ for every $i \in I_n$. Suppose, there is an $x \in bd(S^n)$ such that $x_j = 0$ if $x \in C^{ij}$. Because of the boundary condition we then have $x \in \cup_{j \in I_n} C^{ij}$ for all $i \in I_n$ for which $x_i = 0$. As shown in Joosten and Talman [1993] then there exists a set $I^* \subseteq \{i \in I_n \mid x_i = 0\}$ and a permutation $\pi(I^*)$ such that $x \in \cap_{i \in I^*} C^{i\pi(i)}$ and hence the theorem holds. Otherwise, for every $x \in bd(S^n)$, $x \in C^{ij}$ for some pair $(i,j) \in J$ with $x_j > 0$. Then, for such a pair $(i,j)$, $(c^{ij})^\top a^h \leq 0$ for every $h \in I_x$ since $j \neq h$. Hence, $A^*(I_x) \cap C(J_x) \neq \emptyset$ if $x \in bd(S^n)$. So, the boundary condition of Theorem 13.2.3 is satisfied and there is a balanced set $T^* \subseteq J$ for which $x \in \cap_{(i,j) \in T^*} C^{ij} \neq \emptyset$. Clearly, by definition of the vectors $c^{ij}$, $(i,j) \in J$, balancedness of $T^*$ implies that there exists a set $I^* \subseteq I_n$ and a permutation $\pi(I^*)$ such that $\{(i, \pi(i)) \mid i \in I^*\} \subseteq T^*$. □

## 13.4. Intersection theorems on polytopes

In this section Theorem 13.2.3 is applied to generalize several intersection theorems on the unit simplex to an arbitrary polytope $P$. In these generalizations without loss of generality we will ignore the equality constraints on $P$. Hence $V = \{\mathbf{0}\}$, and $C(J) \cap V = \{\mathbf{0}\}$ is satisfied if $\mathbf{0} \in C(J)$. Now, the polytope under consideration equals

$$P = \left\{x \in \mathbb{R}^n \mid a^{i\top} x \leq \alpha_i,\ i \in I\right\}$$

with $I$ a set of at least $n+1$ elements. Since $P$ is bounded we have that $\mathbf{0}$ lies in the convex hull of the vectors $a^j$, $j \in I$. Now, we first generalize

Sperner's intersection lemma to the polytope. To prove this generalization, we state the following lemma, in which for $T \subseteq I$, $\tilde{A}(T) = co(\{a^j \mid j \in T\})$.

**Lemma 13.4.1** *For any $T \subseteq I$, $\tilde{A}(T) \cap -A^*(T) \neq \emptyset$.*

Proof: Suppose $\mathbf{0} \in \tilde{A}(T)$. Then the lemma holds since $\mathbf{0} \in A^*(T)$. Now, suppose that $\mathbf{0} \notin \tilde{A}(T)$. Since $\tilde{A}(T) \subseteq A(T)$ and $A(T) \cap A^*(T) = \{\mathbf{0}\}$, it follows that $\tilde{A}(T) \cap A^*(T) = \emptyset$. Hence, for every $x \in \tilde{A}(T)$, $x^\top a^i > 0$ for at least one $i \in T$. Therefore, $\alpha^* > 0$ with $\alpha^*$ the optimal value of the primal linear programming problem

$$\text{(P)} \quad \min \alpha, \text{ s.t. } x \in \tilde{A}(T) \text{ and } \alpha \geq x^\top a^i, \text{ for all } i \in T.$$

Now, let $\nu^*$ be the optimal value of the dual linear programming problem

$$\text{(D)} \quad \max \nu, \text{ s.t. } -y \in \tilde{A}(T) \text{ and } \nu \leq -y^\top a^i, \text{ for all } i \in T.$$

Then according to the primal-dual theory we have that $\nu^* = \alpha^* > 0$. Let $y^*$ be any solution to (D). Then, $-y^{*\top} a^i \geq \nu^* > 0$ for all $i \in T$ and hence $y^* \in A^*(T)$. Since $-y^* \in \tilde{A}(T)$, this proves the lemma. □

We can now prove the following generalization of Sperner's intersection lemma.

**Theorem 13.4.2** *Let $\{C^j \mid j \in I\}$ be a collection of closed sets covering the polytope $P$ satisfying that for every $h \in I$ the facet $F_h$ is a subset of $C^h$. Then there exists a set $T^* \subseteq I$ such that $\{a^j \mid j \in T^*\}$ is balanced and $\cap_{j \in T^*} C^j \neq \emptyset$.*

Proof: Take $J = I$ and $c^h = -a^h$, $h \in J$. Clearly, $\mathbf{0} \in C(J)$. For a boundary point $x$ in the interior of $F(T)$, we have that $-\tilde{A}(T) \subseteq C(J_x)$ since $x \in C^h$ for every $h \in T$. From Lemma 13.4.1 it follows that there exists a $y \in -\tilde{A}(T)$ satisfying $y \in A^*(T)$. So, $y \in C(J_x) \cap A^*(I_x)$, since $-\tilde{A}(T) \subseteq C(J_x)$ and $I_x = T$. Hence the boundary condition of Theorem 13.2.3 is satisfied and there exists a balanced set $T^* \subseteq I_m$ for which $\cap_{j \in T^*} C^j \neq \emptyset$. Clearly, if $T^*$ is balanced, then also $\{a^j \mid j \in T^*\}$ is balanced. □

For example, let the polytope be the $n$-dimensional unit cube $U^n$. We denote the $2n$ facets of $U^n$ by $F_{+i} = \{x \in U^n \mid x_i = 1\}$ and $F_{-i} = \{x \in U^n \mid x_i = 0\}$, for $i = 1, \ldots, n$. In this case $I = I_n \cup -I_n$ and the polytope is covered by $2n$ closed sets, denoted by $C^{+i}$ and $C^{-i}$ for $i = 1, \ldots, n$, such that $C^h$ contains $F_h$ for every $h \in I_n \cup -I_n$. Under this boundary condition we have the following corollary which is similar to a result stated in Freund [1986].

**Corollary 13.4.3** *Let $P$ be the $n$-dimensional unit cube $U^n$ and let $\{C^h \mid h \in I_n \cup -I_n\}$ be a collection of closed sets covering $U^n$ such that for* *every $h \in I_n \cup -I_n$ the facet $F_h$ is a subset of $C^h$. Then there is an index $i \in I_n$ such that*

$$C^{+i} \cap C^{-i} \neq \emptyset.$$

Proof: We have that $a^{+j} = e(j)$ and $a^{-j} = -e(j)$ for every $j \in I_n$. But for any $T \subseteq I_n \cup -I_n$, the system $\sum_{h \in T} \nu_h a^h = \mathbf{0}$ can only have a positive solution if for some $i \in I_n$ both $+i$ and $-i$ belong to T. □

The next result on the cube is introduced in Freund [1986].

**Theorem 13.4.4 (Freund's Intersection Lemma on Cube)** *Let $\{C^h \mid h \in I_n \cup -I_n\}$ be a collection of closed sets covering the unit cube $U^n$ such that for every $T \subseteq I_n \cup -I_n$, with $T \cap \{j, -j\} \neq \emptyset$ for each $j \in I_n$, it holds $\{x \in U^n \mid x_j = 0, j \notin T, x_j = 1, -j \notin T\} \subseteq \cup_{k \in T} C^k$. Then there is an index $i \in I_n$ such that*

$$C^{+i} \cap C^{-i} \neq \emptyset.$$

Proof: Let $a^{+j} = e(j)$ and $a^{-j} = -e(j)$ for every $j \in I_n$. Set $J = I_n \cup -I_n$ and $c^j = -a^j$ for $j \in J$. Suppose that $x$ is an arbitrary interior point of a proper face $F(T)$ of $U^n$. This implies that

$$F(T) = \{x \in U^n \mid x_j = 0, j \notin S, x_j = 1, -j \notin S\}$$

where $S = J \setminus \{-i \mid i \in T\}$. Observe that if $i \in T$, then $-i \notin T$ and that $|T| \leq n$. Notice that $I_x = T \subseteq S$ and $J_x \subseteq S$. It is easy to check that $C(S) \subseteq A^*(I_x)$. So the boundary condition of Theorem 13.2.3 is satisfied. But for any $T^* \subseteq I_n \cup -I_n$, the system $\sum_{h \in T^*} \nu_h a^h = \mathbf{0}$ can only have a positive solution if for some $i \in I_n$ both $+i$ and $-i$ belong to $T^*$. □

In Theorem 13.4.2 it is assumed that for every $h \in I$ the facet $F_h$ lies in $C^h$. The next corollary follows immediately from Theorem 13.4.2 by taking the sets covering $P$ equal to $C^j \cup F_j$, $j \in I_m$.

**Corollary 13.4.5** *Let $\{C^j \mid j \in I\}$ be an arbitrary collection of closed sets covering the polytope $P$. Then there exist a set $T^* \subseteq I$ and an $x^* \in P$, such that $\{a^j \mid j \in T^*\}$ is balanced and for every $j \in T^*$, $x^* \notin C^j$ implies $x^* \in F_j$.*

We remark that in Corollary 13.4.5 we allow that some of the sets $C^j$ are empty. In particular, suppose that $P$ is covered by just one set $C^k$ for some $k \in I_m$. Now, let $x^*$ be a solution to the linear programming problem

$$\begin{array}{ll} \min & x^\top a^k \\ \text{s.t.} & a^{i\top} x \leq b_i \text{ for all } i \in I \end{array} \tag{13.4.3}$$

and let $K$ be the set of indices given by $K = \{i \in I \mid a^{i\top} x^* = b_i\}$. Then it follows from the dual of (13.4.3) that there exist a set $J^* \subseteq K$ and positive numbers $\nu_i^*$ for $i \in J^*$ such that $\sum_{i \in J^*} \nu_i^* a^i = -a^k$ and hence with $\nu_k^* = 1$ we have that $T^* = J^* \cup \{k\}$ satisfies the conditions with $x^*$ in the intersection of the sets $C^j \cup F_j$, $j \in T^*$.

In the next theorem we generalize the KKM lemma on the unit simplex to the polytope $P$. To do so, we first observe that the face $S^n(T)$ of the unit simplex $S^n$ is just the convex hull of the vertices $e(i)$, $i \notin T$. Hence $S^n(T)$ can be written as $\Delta^{I_n \setminus T}$ with $\Delta^S$ defined as the convex hull of the vertices $e(i)$, $i \in S$. In doing so, the boundary condition of the KKM lemma can be rewritten as $\Delta^S \subseteq \cup_{j \in S} C^j$ for any $S \subseteq I_n$. This gives an important insight in the difference between Sperner intersection lemma and the KKM lemma. In Sperner's intersection lemma, each facet $S_h^n$ is covered by $C^h$ and so the number of sets $C^h$ is equal to the number of facets. In generalizing the Sperner's intersection lemma to the polytope, we indeed had that in Theorem 13.4.2 the number of sets covering the polytope equals the number of constraints. However, in the KKM lemma we have that each face, being the convex hull of a number of vertices, is covered by the union of sets corresponding to the vertices carrying the face. So, the number of sets equals the number of vertices. For $n = 2$ or $|I| = n + 1$ the number of vertices equals the number of facets. However, for $n > 2$ and $|I| > n + 1$, this is in general not the case. So, generalizing the KKM lemma to the polytope, the number of sets covering $P$ should be equal to the number of vertices. Therefore, for a given polytope $P = \{x \in \mathbb{R}^n \mid a^{i\top} x \leq \alpha_i,\ i \in I\}$, let $t$ be the number of vertices and let $\{v^j,\ j \in I_t\}$ be the set of vertices of $P$. Then we have the following generalization of the KKM lemma, in which the boundary condition says that every face is covered by the sets labelled by the vertices of that face.

**Theorem 13.4.6** *Let $\{C^j \mid j \in I_t\}$ be a collection of closed sets covering the polytope $P$ with vertices $v^j$, $j \in I_t$, satisfying that every face $F(T)$, $T \subseteq I$, is covered by $\cup\{C^h \mid v^h \in F(T)\}$. Then for any $c \in P$ there exists a set $T^* \subseteq I_t$ such that $c$ lies in the convex hull of the vectors $v^j$, $j \in T^*$, and $\cap_{j \in T^*} C^j \neq \emptyset$.*

Proof: Take $J = I_t$ and for $j \in J$, take $c^j = c - v^j$. Clearly, $\mathbf{0} \in C(J)$. Let $x$ be a boundary point of $P$ and let $v^h$ be a vertex of $F(T)$, where $T = I_x$, i.e., $x$ lies in the interior of $F(T)$. Then we have that $v^{h\top} a^i = \alpha_i$ for any $i \in T$. Hence, for every $i \in T$ it holds that $c^{h\top} a^i \leq 0$ because $c^\top a^i \leq \alpha_i$. Thus, for any vertex $v^h$ of $F(T)$ we have that $c^h \in A^*(T)$. Moreover, by the boundary condition, $x \in C^h$ for at least one set $C^h$ corresponding to a vertex $v^h$ of $F(T)$ and therefore $c^h \in C(J_x)$ for at least one $v^h$ in $F(T)$. Hence, the boundary condition of Theorem 13.2.3 is satisfied and there exists a balanced set $T^* \subseteq J$ for which $\cap_{j \in T^*} C^j \neq \emptyset$. Clearly, the balancedness of $T^*$ implies that $c$ lies in the convex hull of the vertices $v^j$, $j \in T^*$. □

Notice that Theorem 13.4.6 is a special case of Theorem 3.1 in Ichiishi and Idzik [1991]. It can be shown that under the conditions of Theorem 3.1 of Ichiishi and Idzik the conditions of Theorem 13.2.3 are satisfied and therefore that Theorem 3.1 of Ichiishi and Idzik also follows from Theorem 13.2.3. Ichiishi and Idzik also proved that their Theorem 3.1 is a special case of the theorem below, stated as Theorem 3.4 in Ichiishi and Idzik [1991]. Below we show that also this latter theorem is a special case of Theorem 13.2.3.

**Theorem 13.4.7 (Ichiishi-Idzik's Intersection Theorem)** *For $k > n$, let $B$ be an $n \times k$ matrix with columns $b^j$, $j \in I_k$, and for $t$, $n \leq t \leq k$, let $W$ be the convex hull of the vectors $b^j$, $j \in I_t$. Let $\{C^j \mid j \in I_k\}$ be a collection of closed sets covering the set $W$. For some $c \in W$, assume that the set of solutions $\{y \in \mathbb{R}^k_+ \mid By = c\}$ is bounded. Furthermore, assume that $\mathbf{0} \notin$ aff $(W)$, $b^j \in$ aff $(W \cup \{\mathbf{0}\})$ for all $j \in I_k$, and that for every proper face $F$ of $W$ it holds that $F \subseteq \cup\{C^j \mid b^j \in$ aff $(F)+$ aff $(\{\mathbf{0}, c\})$, $j \in I_k\}$. Then there exists a set $T^* \subseteq I_k$ such that the system of equations $\sum_{j\in T^*} \mu_j b^j = c$ has a positive solution and $\cap_{j\in T^*} C^j \neq \emptyset$.*

Proof: Since $\mathbf{0} \notin W$, without loss of generality we may assume that

$$\text{aff } (W) = \{x \in \mathbb{R}^n \mid \sum_{j=1}^{n} x_j = 1\}$$

and hence $l = n - 1$ and $d^1 = m^{I_n}$, so that $V = \{x \in \mathbb{R}^n \mid x = \nu m^{I_n}, \nu \in \mathbb{R}\}$. Now, take $J = I_k$, and for $j \in J$, take $c^j = -b^j + \gamma_j c$, where $\gamma_j = \sum_{i=1}^n b^j_i$. Since $c \in W$ we have that $\mathbf{0} \in C(J)$. Moreover, $m^{I_n \top} c = 1/n$ and hence for all $j \in I_k$ we have that $m^{I_n \top} c^j = 0$ and therefore $c^j \in V^*$. So $C(J) \subseteq V^*$. Now, let $x$ be a point in the interior of some face $F$ of $W$. Then, $I_x \subseteq I$ is the set of elements corresponding to its binding constraints, i.e., $F = \{x \in W \mid a^{i\top} x = \alpha_i, i \in I_x\}$, with the appropriate vectors $a^i \in \mathbb{R}^n$ and numbers $\alpha_i \in \mathbb{R}$, $i \in I_x$. According to the boundary condition there exists an element $h \in I_k$ such that both $c^h \in C(J_x)$ and $b^h = d^h + \delta_h c$, for some $d^h \in \text{aff}(F(I_x))$ and $\delta_h \in \mathbb{R}$. By substituting $b^h$ we now obtain that $c^h = -d^h + (\gamma_h - \delta_h)c$. By definition, $\sum_{j=1}^n c^h_j = 0$, whereas both $d^h$ and $c$ are in $W$. Hence, $0 = \sum_{j=1}^n c^h_j = \sum_{j=1}^n -d^h_j + (\gamma_h - \delta_h)\sum_{j=1}^n c_j = -1 + \gamma_h - \delta_h$ and so $c^h = c - d^h$. Moreover, $a^{i\top} c \leq \alpha_i$ for all $i \in I$ and $a^{i\top} d^h = \alpha_i$ for all $i \in I_x$. So, for all $i \in I_x$, $a^{i\top} c^h = a^{i\top} c - a^{i\top} d^h \leq 0$ and thus $c^h \in A^*(I_x)$. Therefore, $c^h \in A^*(I_x) \cap C(J_x)$ and the conditions of Theorem 13.2.3 are satisfied. Hence, there exists a balanced set $T^* \in I_k$ such that $\cap_{j\in T^*} C^j \neq \emptyset$. Balancedness of $T^*$ implies that there exist $\lambda^*_j > 0$ for $j \in T^*$ satisfying $\sum_{j\in T^*} \lambda^*_j c^j = \mathbf{0}$. Hence, $\sum_{j\in T^*} \lambda^*_j b^j = \alpha^* c$ with $\alpha^* = \sum_{j\in T^*} \lambda^*_j \gamma_j$. Since the set of solutions $\{y \in \mathbb{R}^k_+ \mid By = c\}$ is bounded, we must have that $\alpha^* > 0$. Hence, $\sum_{j \in T^*} \mu^*_j b^j = c$, where $\mu^*_j = \frac{\lambda^*_j}{\alpha^*} > 0$ for every $j \in T^*$. □

In case the polytope is the product space of $N$ unit simplices, i.e., the simplotope $\mathcal{S} = \Pi_{j=1}^N S^{n_j}$, intersection theorems are obtained in Freund [1986], van der Laan, Talman and Van der Heyden [1987], van der Laan and Talman [1993], and Talman [1994]. These intersection results are generalizations of the intersection theorems of Sperner, Scarf, Shapley, Freund and Ichiishi, respectively. They can be also proved by Theorem 13.2.3. We leave them to the interested reader.

## 13.5. Intersection theorems with a continuum of intersection points

In this section we introduce an intersection theorem due to Herings and Talman [1998] on the unit cube $U^n$. The proof is based on Theorem 8.4.3. In the following it will be assumed that for any $j \in I_n$, $j + 1 = 1$ when $j = n$, while $j - 1 = n$ if $j = 1$.

**Theorem 13.5.1 (Herings-Talman's Intersection Theorem)** *Let $C^1, \cdots, C^n$ be closed subsets of $U^n$ such that $U^n = \cup_{i \in I_n} C^i$. Moreover, if for $x \in U^n$, $x_j = 0$ or $x_{j+1} = 1$, then $x \in C^j$. Then there exists a connected set $S$ in $U^n$ such that $1^n, 0^n \in S$ and $S \subseteq \cap_{i \in I_n} C^i$.*

Proof: Define $J(x) = \{j \in I_n \mid x \in C^j\}$ and $p(x) = \{j \in I_n \mid x_i = 0\}$ for each $x \in U^n$. Let $d(x)$ and $s(x)$ denote the number of elements in $J(x)$ and $p(x)$, respectively. Define the point-to-set mapping $\eta : U^n \mapsto \mathbb{R}^n$ by

$$\eta(x) = co(\{e(j) - \frac{1}{n}1^n \mid j \in J(x)\}).$$

We will show that $\eta$ satisfies all the conditions (i.e., Assumption 8.2.1) of Theorem 8.4.3. It is easy to see that $\eta(x)$ is nonempty, convex and compact for every $x \in U^n$. Moreover, $\eta$ is upper-semi continuous. Thus, Assumption 8.2.1 (a) is satisfied.

If for some $j \in I_n$, $x_j = 0$ then $x \in C^j$ and so $e(j) - \frac{1}{n}1^n \in \eta(x)$. If for some $j \in I_n$, $x_j = 1$ then $x \in C^{j-1}$ and so $e(j-1) - \frac{1}{n}1^n \in \eta(x)$. The following three cases have to be considered.

Case (1): If $x = 1^n$, then consider a point $z \in \eta(x)$ with $z = \sum_{i \in I_n} \frac{1}{n}(e(j) - \frac{1}{n}1^n) = 0^n$. Hence $z_j \leq 0$ for all $j \in I_n$.

Case (2): If for all $j \in I_n$, $0 \leq x_j < 1$, then define $z = \sum_{j \in J(x)} \frac{1}{d(x)}(e(j) - \frac{1}{n}1^n)$. It is easy to see that $z \in \eta(x)$ and $x_j = 0$ implies $z_j = \frac{1}{d(x)} - \frac{1}{n} \geq 0$.

Case (3): If there exist $k \in I_n$ with $x_k = 1$ and $l \in I_n$ with $x_l < 1$, then choose some $q \in I_n$ such that $x_q = 1$ and $x_{q-1} < 1$. Let

$$\begin{aligned} z &= \sum_{j\in p(x)} \tfrac{1}{n}(e(j) - \tfrac{1}{n}1^n) + (1 - \tfrac{s(x)}{n})(e(q-1) - \tfrac{1}{n}1^n) \\ &= \sum_{j\in p(x)} \tfrac{1}{n}e(j) + (1 - \tfrac{s(x)}{n})e(q-1) - \tfrac{1}{n}1^n. \end{aligned}$$

Then it is easily verified that $z \in \eta(x)$. Furthermore, $x_j = 0$ implies $z_j \geq \frac{1}{n} - \frac{1}{n} = 0$ and $x_j = 1$ implies $z_j = -\frac{1}{n} < 0$.

So Assumption 8.2.1 (b) is satisfied.

Since we have for any $x \in U^n$, $z^\top 1^n = 0$ for every $z \in \eta(x)$, so $\eta$ satisfies Assumption 8.2.1 (c).

It is clear that if $0^n \in \eta(x^*)$ for some $x^* \in U^n$, then by the construction of $\eta$, $x^*$ must belong to $C^j$ for every $j \in I_n$. That is, $x^* \in \cap_{i\in I_n} C^i$. Now it follows from Theorem 8.3.4 that there exists a connected set $S$ of $U^n$ such that $1^n$, $0^n \in S$ and $0^n \in \eta(x^*)$ for all $x^* \in S$. This completes the proof. □

Define the set $T^n = \{x \in \mathbb{R}^n_+ \mid \sum_{i=1}^n x_i \leq 1\}$. Now we have the following intersection result.

**Theorem 13.5.2** *Let $C^1, \cdots, C^n$ be closed subsets of $T^n$ satisfying that $\cup_{i\in I_n} C^i = T^n$, and $x_i = 0$ implies $x \in C^i$ for every $x \in T^n$. Then there exists a connected subset $S$ of $T^n$ such that $0^n \in S$, $S \cap S^n \neq \emptyset$, and $S \subseteq \cap_{i\in I_n} C^i$.*

In Freidenfelds [1974] the following intersection is proved in a constructive manner.

**Theorem 13.5.3 (Freidenfelds' Intersection Theorem)** *Let $C^1, \cdots, C^n$ be closed subsets of $S^n$ such that $\cup_{i\in C^i} = S^n$ and $x_i = 0$ implies $x \in C^i$ for every $x \in S^n$. Define $A = \sum_{i\in I_n} C^i$ and $A_k = \cap_{h\in I_n\setminus\{k\}} C^h$ for every $k \in I_n$. Then for each $h \in I_n$, there exists a connected subset $S$ of $A_h$ such that $e(h) \in S$ and $S \cap A \neq \emptyset$.*

## 13.6. An m.p.b. intersection theorem and equilibria with indivisibilities

In this section we introduce a multi-permutation-based (m.p.b.) intersection theorem on the simplotope due to van der Laan, Talman and Yang [1997]. This theorem will be used to prove the existence of an equilibrium in the exchange economy model presented at Section 2.8. In the following, let $a^i$ denote the vector $\frac{1}{n}1^n - e(i)$ in $\mathbb{R}^n$ for each $i \in I_n$. Let $\mathcal{S}$ denote the simplotope $\prod_{h=1}^m S^n$. Now we present the intersection theorem of van der Laan, Talman and Yang.

**Theorem 13.6.1** *For each $i \in I_n$ and each $j \in I_m$, let the collection of closed sets $\{ C_i^{(j,1)}, \cdots, C_i^{(j,n)} \}$ be a covering of the simplotope $\mathcal{S}$ such that if $p$ lies on the boundary of $\mathcal{S}$ then for some $k \in I_n$ it holds that $p \in C_i^{(j,k)}$ and $p_{j,k} > 0$. Then there exist $\pi \in \prod_{h=1}^m \Phi$ and $p^* \in \mathcal{S}$ such that*

$$p^* \in \cap_{j=1}^m \cap_{i=1}^n C_i^{(j,\pi_j(i))}.$$

Proof: For each $i \in I_n$, we define

$$C_i^{(i_1,\cdots,i_m)} = \cap_{j=1}^m C_i^{(j,i_j)}.$$

Clearly, $C_i^{(i_1,\cdots,i_m)}$ is a closed set, and the collection of sets $\{ C_i^T \mid T \in I_n^m \}$ is a covering of $\mathcal{S}$. Furthermore, it is not difficult to show that if $p$ lies on the boundary of $\mathcal{S}$ then for some $T = (i_1, \cdots, i_m) \in I_n^m$ it holds that $p \in C_i^T$ and $p_{j,i_j} > 0$ for every $j \in I_m$. Now let $\bar{\mathcal{S}}$ denote the set $\mathcal{S} \times S^n$. For each $(i_1, \cdots, i_{m+1}) \in I_n^{m+1}$, define

$$C^{(i_1,\cdots,i_{m+1})} = C_{i_{m+1}}^{(i_1,\cdots,i_m)} \times S^n.$$

Clearly, $C^{(i_1,\cdots,i_{m+1})}$ is a closed set, and the collection of sets $\{ C^T \mid T \in I_n^{m+1} \}$ is a covering of $\bar{\mathcal{S}}$. If $p$ lies on the boundary of $\bar{\mathcal{S}}$ then for some $T = (i_1, \cdots, i_{m+1}) \in I_n^{m+1}$ it holds that $p \in C^T$ and $p_{j,i_j} > 0$ for every $j \in I_{m+1}$.

For each $(i_1, \cdots, i_{m+1}) \in I_n^{m+1}$, define a vector $c^{(i_1,\cdots,i_{m+1})} \in \prod_{h=1}^{m+1} \mathbb{R}^n$ by

$$c^{(i_1,\cdots,i_{m+1})} = (a^{i_1}, \cdots, a^{i_{m+1}}).$$

Let the set $V = \prod_{h=1}^{m+1} V^h$ be given by for $h = 1, \cdots, m+1$,

$$V^h = \{ x_h \in \mathbb{R}^n \mid \sum_{j=1}^n x_{h,j} = 1,\ x_{h,j} \geq -1/n \text{ for every } j \in I_n \}.$$

For $x \in V$ the point $p(x)$ is defined as the projection of $x$ on $\bar{\mathcal{S}}$, i.e., $p(x) = (p_1(x), \cdots, p_{m+1}(x))$ with the projection $p_h(x)$ of $x_h$ in $V^h$ on $S^n$ given by

$$p_{h,j}(x) = \begin{cases} 0 & \text{if } x_{h,j} < 0 \\ x_{h,j} / \sum_{\{ l \mid x_{h,l} \geq 0 \}} x_{h,l} & \text{if } x_{h,j} \geq 0. \end{cases}$$

Now let the point-to-set mapping $F$ from $V$ to the collection of subsets of $\prod_{h=1}^{m+1} \mathbb{R}^n$ be given by

$$F(x) = co(\{ c^{(i_1,\cdots,i_{m+1})} \mid p(x) \in C^{(i_1,\cdots,i_{m+1})}, \\ x_{j,i_j} \geq 0 \text{ for every } j \in I_{m+1} \}).$$

It is easy to see that $F$ is upper semi-continuous. Moreover, $\cup_{x\in V}F(x)$ is compact, and for each $x \in V$ the set $F(x)$ is nonempty, convex and compact. Let $Y$ be a compact, convex set in $\prod_{h=1}^{m+1}\mathbb{R}^n$ containing $\cup_{x\in V}F(x)$. Then we define the point-to-set mapping $G$ from $Y$ to the collection of subsets of $V$ by

$$G(y) = \{\, x^* \in V \mid x_h^\top y_h \leq (x_h^*)^\top y_h \text{ for all } x_h \in V^h \text{ and } h \in I_{m+1}\}.$$

Again, $G$ is upper semi-continuous. Moreover, for any $y \in Y$ the set $G(y)$ is nonempty, compact and convex and $\cup_{y\in Y}G(y)$ is bounded. For $(x, y) \in V \times Y$, let $\phi(x, y)$ be defined as

$$\phi(x, y) = G(y) \times F(x),$$

then $\phi$ is an upper semi-continuous mapping from the set $V \times Y$ into the collection of nonempty subsets of $V \times Y$ satisfying for every $(x, y) \in V \times Y$ that the set $\phi(x, y)$ is nonempty, convex and compact. According to Kakutani's fixed point theorem, there exists an $(x^*, y^*) \in V \times Y$ such that

$$x^* \in G(y^*) \text{ and } y^* \in F(x^*).$$

So it holds that

$$x_h^\top y_h^* \leq (x_h^*)^\top y_h^*$$

for every $x_h \in V^h$ and $h \in I_{m+1}$. Let $\beta_h$ be equal to $(x_h^*)^\top y_h^*$. Then by taking $x_h$ equal to $e/n$, it follows that $\beta_h \geq 0$, since $\sum_{j=1}^n y_{h,j}^* = 0$ for any $h \in I_{m+1}$. When we take $x_h$ successively equal to $e(j)$ for every $j \in I_n$, we obtain

$$y_{h,j}^* \leq \beta_h, \quad \text{for every } h \in I_{m+1} \text{ and } j \in I_n.$$

On the other hand, if for some $h \in I_{m+1}$ and $j \in I_n$ it holds that $x_{h,j}^* > -1/n$, by taking $x_h$ equal to $x_h^* + \lambda(x_h^* - e(j))$ for arbitrarily small $\lambda > 0$, we obtain that $y_{h,j}^* \geq \beta_h$. Hence $y_{h,j}^* = \beta_h \geq 0$ when $x_{h,j}^* > -1/n$.

Let the collection $\mathcal{T}^*$ of elements of $I_n^{m+1}$ be defined by

$$\mathcal{T}^* = \{T = (i_1, \cdots, i_{m+1}) \in I_n^{m+1} \mid p(x^*) \in C^T, \\ x_{j,i_j}^* \geq 0 \text{ for every } j \in I_{m+1}\}.$$

Suppose that $\mathcal{T}^* = \{T^1, \cdots, T^l\}$, where $T^k = (i_1^k, \cdots, i_{m+1}^k)$. Since $y^* \in F(x^*)$ there exist some non-negative numbers $\mu_1, \cdots, \mu_l$ with sum equal to 1 such that

$$y^* = \sum_{k=1}^{l} \mu_k c^{T^k}.$$

Suppose that $x^*_{h,j} = -1/n$ for some $h \in I_{m+1}$ and $j \in I_n$. Then it implies that $j \neq i^k_h$ for every $k = 1, \cdots, l$ and hence $y^*_{h,j} \geq 0$. Since $\sum_{j=1}^n y^*_{h,j} = 0$ for any $h \in I_{m+1}$, we have that $y^* = \mathbf{0}$. So,

$$\sum_{k=1}^{l} \mu_k c^{T^k} = \mathbf{0}. \tag{13.6.1}$$

Now for each $k \in I_m$, define

$$S_k = \{ (i^h_k, i^h_{m+1}) \mid h = 1, \cdots, l \}.$$

It is clear that $S_k$ is a subset of $I_n \times I_n$. It follows from (13.6.1) that for every $k \in I_m$

$$\sum_{(i,j) \in S_k} \mu^k_{(i,j)} (a^i, a^j) = \mathbf{0}$$

and that

$$\sum_{(i,j) \in S_k} \mu^k_{(i,j)} = 1$$

for certain $\mu^k_{(i,j)} \geq 0$ for $(i, j) \in S_k$. Moreover, it holds that for each $i \in I_n$, $\sum_j \mu^k_{(i,j)} = 1/n$ and that for each $j \in I_n$, $\sum_i \mu^k_{(i,j)} = 1/n$. From this property it follows that the $n \times n$ matrix $U(k)$ with entries $\nu^k_{(i,j)}$ defined by $\nu^k_{(i,j)} = n\mu^k_{(i,j)}$ if $(i, j) \in S_k$ and $\nu^k_{(i,j)} = 0$ if $(i, j) \notin S_k$ is a doubly stochastic matrix and therefore $U(k)$ is a convex combination of permutation matrices according to the theorem of Birkhoff and von Neumann. So, there exists a permutation $\pi_k = (\pi_k(1), \cdots, \pi_k(n))$ of $(1, \cdots, n)$ such that $\nu^k_{(\pi_k(j),j)} > 0$ and hence $(\pi_k(j), j) \in S_k$ for every $j \in I_n$. Since $p(x^*) \in \cap_{h=1}^l C^{T^h}$, it implies that

$$\cap_{k=1}^m \cap_{j=1}^n C_j^{(i^j_1(k), \cdots, i^j_{k-1}(k), \pi_k(j), i^j_{k+1}(k), \cdots, i^j_m(k))} \neq \emptyset,$$

where $(i^j_1(k), \cdots, i^j_{k-1}(k), \pi_k(j), i^j_{k+1}(k), \cdots, i^j_m(k), j) \in \mathcal{T}^*$ for every $k \in I_m$ and $j \in I_n$. For each $i \in I_n$, since

$$C_i^{(\pi_1(i), \cdots, \pi_m(i))} = \cap_{j=1}^m C_i^{(j, \pi_j(i))}$$

we obtain that

$$\cap_{j=1}^m \cap_{i=1}^n C_i^{(j, \pi_j(i))} \neq \emptyset.$$

This completes the proof. □

Now we have the following multi-permutation-based generalization of Brouwer's fixed point theorem.

**Theorem 13.6.2** *Let $f^i : \prod_{h=1}^m S^n \mapsto \prod_{h=1}^m \mathbf{R}^n$ be continuous functions for all $i \in I_n$. Then there exists a pair of vectors $(x, \pi) \in \prod_{h=1}^m S^n \times \prod_{h=1}^m \Phi$ such that*

$$f^i_{j,\pi_j(i)}(x) \leq x_{j,\pi_j(i)}$$

*for all $i \in I_n$ and all $j \in I_m$.*

Proof: For each $i \in I_n$ and each $(j,k) \in I_m \times I_n$, define

$$C_i^{(j,k)} = \{x \in \prod_{h=1}^m S^n \mid f^i_{j,k}(x) \leq x_{j,k}\}.$$

Clearly the result follows immediately from Theorem 13.6.1. □

We note that the intersection lemma of Gale [1984] on the unit simplex $S^n$ follows immediately from Theorem 13.6.1 by setting $m = 1$.

**Theorem 13.6.3 (Gale's Intersection Lemma)** *For each $i \in I_n$, let $\{C_i^{(1,j)} \mid j \in I_n\}$ be a collection of closed sets covering the unit simplex $S^n$ satisfying that for every $T \subset I_n$, it holds that $S^n(T) \subseteq \cup_{h \notin T} C_i^{(1,h)}$. Then there exists a permutation $\pi = (\pi(1), \cdots, \pi(n))$ of the elements of $I_n$ such that $\cap_{i \in I_n} C_i^{(1,\pi(i))}$.*

The next theorem can be seen as a dual of Gale's lemma.

**Theorem 13.6.4** *For each $i \in I_n$, let $\{C_i^{(1,j)} \mid j \in I_n\}$ be a collection of closed sets covering the unit simplex $S^n$ satisfying that for every $x \in S^n$, $x_h = 0$ implies $x \in C_i^{(1,h)}$. Then there exists a permutation $\pi = (\pi(1), \cdots, \pi(n))$ of the elements of $I_n$ such that $\cap_{i \in I_n} C_i^{(1,\pi(i))}$.*

Now we are ready to derive Theorem 2.8.2 by using Theorem 13.6.1.

Proof of Theorem 2.8.2: Without loss of generality, we take $M = 1$ for Assumption ($c$). Let $A^n$ be the intersection of $B^n$ and the unit $n$-cube $U^n$. We shall construct a homeomorphism $\psi$ from $\prod_{h=1}^m A^n$ into $\mathcal{S}$ such that for each $j \in I_m$ and $i \in I_n$, the collection $\{\psi(C_i^{(j,k)}) \mid k \in I_n\}$ is a covering of $\mathcal{S}$ satisfying the boundary condition of Theorem 13.6.1. Then the result immediately follows from Theorem 13.6.1. For $p = (p_1, \cdots, p_m) \in \prod_{h=1}^m A^n$ the point $\psi(p)$ is defined as $\psi(p) = (\psi_1(p_1), \cdots, \psi_m(p_m))$ where $\psi_h$ is a homeomorphism from $A^n$ into $S^n$ for each $h \in I_m$. For each $h \in I_m$, the mapping $\psi_h$ is constructed as follows. For each permutation $\rho = (i_1, i_2, \cdots, i_n)$ of $(1, \cdots, n)$ we define a subset $A^\rho$ of $A^n$ by

$$A^\rho = \{p_h \in A^n \mid p_{h,i_1} \geq p_{h,i_2} \geq \cdots \geq p_{h,i_n} = 0\}$$

and define a subset $S^\rho$ of $S^n$ by

$$S^\rho = \{\, x_h \in S^n \mid x_{h,i_1} \leq x_{h,i_2} \leq \cdots \leq x_{h,i_n} \,\}.$$

We then define $\psi_h$ from $A^\rho$ to $S^\rho$ by

$$(\psi_h(p_h))_{h,i_1} = x_{h,i_1} = \frac{1 - p_{h,i_1}}{n},$$

and

$$(\psi_h(p_h))_{h,i_k} = x_{h,i_k} = \frac{1 - p_{h,i_1}}{n} + \frac{p_{h,i_1} - p_{h,i_2}}{n-1} + \ldots + \frac{p_{h,i_{k-1}} - p_{h,i_k}}{n-k+1},$$

for $k \in I_n \backslash \{1\}$. It is easy to see that $\sum_{k=1}^{n} (\psi_h(p_h))_{h,i_k} = 1$, and $(\psi_h(p_h))_{h,i_k} \geq 0$ for any $k$. Moreover, $p_h \neq p'_h$ implies $\psi_h(p_h) \neq \psi_h(p'_h)$. Now take $x_h \in S^\rho$. We have

$$x_{h,i_1} \leq x_{h,i_2} \leq \cdots \leq x_{h,i_n}.$$

Notice that we have

$$p_{h,i_1} = 1 - n x_{h,i_1}$$

and

$$p_{h,i_k} = p_{h,i_{k-1}} - (n-k+1)(x_{h,i_k} - x_{h,i_{k-1}}),$$

for $k \in I_n \setminus \{1\}$. It follows that $p_h \in A^\rho$. Observe that $\psi_h^{-1}(e(j)) = e - e(j)$ and

$$\psi_h^{-1}(\{x_h \in S^n \mid x_{h,j} = 0\}) = \{\, p_h \in A^n \mid p_{h,j} = 1 \,\}.$$

By Assumption (c) $C_i^{(j,k)}$ does not meet $\psi^{-1}(\{x \in \mathcal{S} \mid x_{j,k} = 0\})$. This implies that $\psi(C_i^{(j,k)})$ does not meet $\{\, x \in \mathcal{S} \mid x_{j,k} = 0 \,\}$ for any $i \in I_n$, $j \in I_m$, and $k \in I_n$. Hence it follows that if $x$ is on the boundary of $\mathcal{S}$ then for some $k \in I_n$ it holds that $x \in \psi(C_i^{(j,k)})$ and $x_{j,k} > 0$. So, the boundary condition of Theorem 13.6.1 is fulfilled. We obtain the equilibrium theorem. □

In Yang [1998] the following multi-permutation-based intersection theorem on the simplotope is proved and used to show the existence of an equilibrium in an exchange economy model with bad indivisibilities.

**Theorem 13.6.5** *For each $i \in I_n$ and each $j \in I_m$, let the collection of closed sets $\{\, C_i^{(j,1)}, \cdots, C_i^{(j,n)} \,\}$ be a covering of the simplotope $\prod_{h=1}^{n} S^n$ such that if $p$ lies on the boundary of $\prod_{h=1}^{n} S^n$ with $p_{j,k} = 0$ for some $k \in I_n$, it holds that $p \in C_i^{(j,k)}$. Then there exist $\pi \in \prod_{h=1}^{m} \Phi$ and $p^* \in \prod_{h=1}^{n} S^n$ such that*

$$p^* \in \cap_{j=1}^{m} \cap_{i=1}^{n} C_i^{(j,\pi_j(i))}.$$

Moreover, we have a multi-permutation-based generalization of Brouwer's fixed point theorem.

**Theorem 13.6.6** *Let $f^i : \prod_{h=1}^m S^n \mapsto \prod_{h=1}^m \mathbf{R}^n$ be continuous functions for all $i \in I_n$. Then there exists a pair of vectors $(x, \pi) \in \prod_{h=1}^m S^n \times \prod_{h=1}^m \Phi$ such that*

$$f^i_{j,\pi_j(i)}(x) \geq x_{j,\pi_j(i)}$$

*for all $i \in I_n$ and all $j \in I_m$.*

## 13.7. Helly's intersection theorem

In this section we introduce an intersection theorem due to Helly [1923] on $\mathbf{R}^n$.

**Theorem 13.7.1 (Helly's intersection theorem)** *Let $C^1, \cdots, C^l$ be convex sets of $\mathbf{R}^n$ such that $l \geq n+1$. If any $n+1$ of them have a nonempty intersection, then*

$$\cap_{h=1}^l C^h \neq \emptyset.$$

Proof: The proof is due to Radon. We derive it by induction on $l$. For $l = n+1$, the theorem is trivially true. Suppose that the theorem holds for $l - 1 \geq n+1$ convex sets. Then by induction hypothesis there exist $l$ points

$$x^i \in \cap_{h=1,\, h\neq i}^l C^h$$

for all $i \in I_l$. Then

$$\begin{array}{rcl} \sum_{h=1}^l \beta_h x^h & = & 0^n, \\ \sum_{h=1}^l \beta_h & = & 0, \end{array} \tag{13.7.1}$$

are $n+1$ linear equations for $l > n+1$ unknown numbers $\beta_h$, $h \in I_l$. So there is a non-trivial solution

$$(\beta_1, \cdots, \beta_l) \neq (0, \cdots, 0). \tag{13.7.2}$$

Without loss of generality we may assume that

$$\beta_1 \geq 0, \cdots, \beta_t \geq 0,$$

and

$$\beta_{t+1} \leq 0, \cdots, \beta_l \leq 0.$$

Then $\sum_{i=1}^t \beta_i = -\sum_{i=t+1}^l \beta_i > 0$ because of (13.7.2) and the point

$$y = \frac{\sum_{h=1}^t \beta_h x^i}{\sum_{h=1}^t \beta_h}$$

belongs to $\cap_{h=t+1}^{l} C^h$, since it holds that $x^h \in \cap_{j=t+1}^{l} C^j$ for every $h \in I_t$. On the other hand, because of (13.7.1), we have

$$y = \frac{\sum_{h=t+1}^{l} -\beta_h x^i}{\sum_{h=t+1}^{l} -\beta_h},$$

which means that $y \in \cap_{h=1}^{t} C^h$. Consequently

$$y \in \cap_{h=1}^{l} C^i \neq \emptyset.$$

□

Helly's theorem implies the following Helly's second intersection theorem.

**Theorem 13.7.2** *Let $F = \{ C^i \mid i \in I \}$ be a family of (possibly infinitely many) convex compact sets of $\mathbb{R}^n$. If any $n+1$ members of $K$ have a nonempty intersection, then*

$$\cap_{h \in I} C^h \neq \emptyset.$$

The proof follows from an observation that if an infinite family of compact convex sets has no intersection points, then there exists a finite subfamily whose intersection is empty.

# CHAPTER 14

# Sperner Theory

## 14.1. Introduction

Sperner lemma [1928] is probably one of the most elegant and fundamental results in combinatorial topology. As we have seen, this lemma provides a very important geometric background for developing simplicial methods. Recall this lemma states that given a simplicial subdivision of the unit simplex $S^n$ and a labeling function $L$ from the set of vertices of simplices of the simplicial subdivision into the set $I_n$, there exists a completely labeled simplex, if $x_i = 0$ implies that $L(x) \neq i$ for any vertex $x \in S^n$. Scarf lemma [1967b, 1973] states a similar result if $x_i = 0$ implies that

$$L(x) = \min\{\, j \mid x_j = 0 \text{ and } x_{j+1} > 0 \,\}$$

for any vertex $x \in S^n$, where $l + 1 = 1$ if $l = n$. Cohen [1967] proved by induction a stronger version of Sperner lemma in which there exist an odd number of completely labeled simplices. Le Van [1982] gave an alternative proof of this result using topological degree theory. Shapley [1973] generalized Sperner lemma by using a set labeling rule instead of an integer labeling rule. Furthermore, existence results of completely labeled simplices have been generalized to the cube and the simplotope, while also more general labeling rules have been considered; see Tucker [1945] (in Chapter 10), Fan [1967] (in Chapter 11), Garcia [1976], van der Laan and Talman [1981b, 1982], Freund [1984b, 1986], van der Laan, Talman and Van der Heyden [1987], and Bapat [1989]. In Freund [1989], the lemmas of Sperner, Scarf, and Garcia on a full-dimensional simplex are extended to a full-dimensional polytope. In this chapter we generalize the concept of completely labeled simplices to the concept of balanced simplices. A general condition is formulated to guarantee the existence of a balanced simplex in any simplicial subdivision of an arbitrary polytope. This leads to the first theorem which implies most results mentioned above, including the lemmas of Sperner, Scarf, Shapley and Garcia, as special cases. As far as we know, no existing results can unify both Sperner and Scarf lemmas. By imposing no restrictions on the labeling rule we establish the second main theorem which unifies several main results of Freund [1989].

This chapter consists of five sections. In Section 2 we provide the basic notation. Also the concept of balancedness is introduced. In Section 3 we present and prove the main theorems. This section is based on van der

Laan, Talman and Yang [1998]. In Section 4 we show that the existing results on the unit simplex as well as their extensions on the simplotope can be derived from the first main theorem as special cases. In Section 5, it is shown that the second main theorem implies several results of Freund. Moreover, the first main theorem is used to prove some existing theorems the unit cube $U^n$.

## 14.2. Preliminaries for analysis

Let $n$ be a positive integer. For given integer $l$, $0 \le l \le n$, let $I$ be a finite set of at least $l+1$ integers. Let a polytope $P$ be given as

$$P = \{ x \in \mathbb{R}^n \mid a^{i\top} x \le \alpha_i,\ i \in I, \ \text{ and } \ d^{h\top} x = \delta_h,\ h \in I_{n-l} \}.$$

We assume throughout the chapter that $P$ is an $l$-dimensional set in $\mathbb{R}^n$ and that none of the constraints $a^{i\top} x \le \alpha_i$, $i \in I$, is an implicit equality, and no constraint is redundant.

For $T \subseteq I$, we define

$$F(T) = \{ x \in P \mid a^{i\top} x = \alpha_i \ \text{ for } \ i \in T \}$$

with $F(\emptyset) = P$. In case $F(T)$ is nonempty, $F(T)$ is clearly a face of $P$.

With respect to the polytope $P$, we define

$$V = \{ x \in \mathbb{R}^n \mid x = \sum_{h=1}^{n-l} \nu_h d^h, \nu_h \in \mathbb{R} \ \text{ for } \ h \in I_{n-l} \}.$$

with $V = \{ \mathbf{0} \}$ when $l = n$. Moreover, we define

$$V^* = \{ x \in \mathbb{R}^n \mid x^\top y = 0 \ \text{ for all } \ y \in V \}$$

as the $l$-dimensional subspace $V^*$ orthogonal to $V$.

For $T \subseteq I$, define

$$A(T) = \{ x \in \mathbb{R}^n \ \mid \ x = \textstyle\sum_{i \in T} \lambda_i a^i + \sum_{h=1}^{n-l} \nu_h d^h, \ \lambda_i \ge 0,\ i \in T \ \text{ and } \ \nu_h \in \mathbb{R},\ h \in I_{n-l} \}$$

with $A(\emptyset) = \{ \mathbf{0} \}$ when $l = n$.

Let $\mathcal{T}$ be a triangulation of $P$. The triangulation of a face $F(T)$ of $P$ induced by $\mathcal{T}$ is denoted by $\mathcal{T}(T)$, i.e.,

$$\mathcal{T}(T) = \{ \tau \subseteq F(T) \mid \tau = \sigma \cap F(T),\ \sigma \in \mathcal{T},\ dim(\tau) = dim(F(T)) \}.$$

Given a subset $B$ of $P$, define the *carrier* of $B$ as

$$Car(B) = \{ j \in I \mid a^{j\top} x = \alpha_j \ \text{ for any } \ x \in B \}.$$

For a point $v \in P$, define $Car(v) = Car(\{ v \})$. Furthermore, for some finite nonempty set $J$, let a collection of vectors $c^j$, $j \in J$, in $\mathbb{R}^n$, be given. For a nonempty set $T \subseteq J$, we define

$$C(T) = co(\{ c^j \mid j \in T \}),$$

i.e., $C(T)$ is the convex hull of the vectors $c^j$ for $j \in T$.

Finally, we assign each element of $\mathcal{T}^0$ an index from the set $J$. Let $L : \mathcal{T}^0 \mapsto J$ be such a labeling rule. For a $q$-simplex $\sigma(x^1, \cdots, x^{q+1})$ being a face of a simplex of $\mathcal{T}$, we call $L(\sigma) = \{ L(x^1), \cdots, L(x^{q+1}) \}$ the label set of $\sigma$. We are now ready to introduce the concept of balanced simplices.

**Definition 14.2.1** *A $q$-simplex $\sigma(x^1, \cdots, x^{q+1})$, being a face of a simplex of $\mathcal{T}$, is balanced if $\mathbf{0} \in C(L(\sigma))$.*

If $\mathbf{0} \in C(S)$ we also call the collection $\{ c^j \mid j \in S \}$ and sometimes $S$ itself balanced. It is clear that the set $S$ is balanced if the system of equations $\sum_{j \in S} \mu_j c^j = \mathbf{0}$ has a non-negative solution $\mu_j$, $j \in S$, with sum equal to one. In the next section we formulate a sufficient condition to guarantee the existence of a balanced simplex in $\mathcal{T}$.

## 14.3. Main integer labeling theorems

Given the polytope $P$ as defined in the previous section, a finite nonempty set $J$ and a collection of vectors $\{ c^j \mid j \in J \}$ in $\mathbb{R}^n$, let $\mathcal{T}$ be a simplicial subdivision of $P$. In the next theorem we provide a sufficient condition to guarantee that there exists at least one balanced simplex in $\mathcal{T}^+$. All results in this section are taken from van der Laan, Talman and Yang [1998].

**Theorem 14.3.1** *Let $\{ c^j \mid j \in J \}$ be a collection of vectors in $\mathbb{R}^n$ with $C(J) \cap V = \{ \mathbf{0} \}$ and let $\mathcal{T}$ be a triangulation of the polytope $P$. Let $L : \mathcal{T}^0 \mapsto J$ be a labeling rule such that for every simplex $\sigma$ of the induced triangulation $\mathcal{T}(T)$ of a proper face $F(T)$ of $P$ for some $T \subset I$, the set $A(T) \cap C(L(\sigma))$ either is empty or contains the point $\mathbf{0}$. Then there exists a balanced simplex being a face of a simplex of $\mathcal{T}$.*

Proof: Let $x$ be any point in $P$ and let $\sigma(x^1, \cdots, x^{q+1})$ be the unique simplex in $\mathcal{T}^+$ containing $x$ in its interior. Hence there exist unique positive numbers $\gamma_1, \cdots, \gamma_{q+1}$ satisfying $\sum_{i=1}^{q+1} \gamma_i = 1$ such that

$$x = \sum_{i=1}^{q+1} \gamma_i x^i.$$

Then the correspondence $\psi$ at $x$ is defined by

$$\psi(x) = co(\{ c^j \mid j = L(x^i) \text{ if } \gamma_i = \max_h \gamma_h \}).$$

Notice that $\psi(x) = C(L(\sigma))$ if $x$ is the barycenter of the simplex $\sigma$. Moreover, $\psi(x) = \{c^j\}$ where $j = L(x)$, if $x$ is a vertex of some simplex of $\mathcal{T}$. Clearly, $\psi$ is a nonempty-valued, convex-valued, and compact-valued u.s.c. correspondence from $P$ to $\mathbb{R}^n$, while $\bigcup_{x\in P}\psi(x)$ is bounded since $\psi(x) \subseteq C(J)$ for any $x \in P$. Let $\phi$ be the correspondence from a compact and convex set $Y$ containing $\bigcup_{x\in P}\psi(x)$, defined by

$$\phi(y) = \{x \in P \mid z^\top y \leq x^\top y \text{ for all } z \in P\}.$$

Clearly, $\phi$ is a nonempty-valued, convex-valued and compact-valued u.s.c. correspondence. Hence $\psi \times \phi$, defined by $(\psi \times \phi)(y, x) = \psi(x) \times \phi(y)$ for every $(y, x) \in Y \times P$, is a nonempty-valued, convex-valued and compact-valued u.s.c. correspondence from $Y \times P$ into $Y \times P$. Therefore, according to Kakutani's fixed point theorem there exists a pair of vectors $(y^*, x^*) \in Y \times P$ such that $y^* \in \psi(x^*)$ and $x^* \in \phi(y^*)$. The latter implies that

$$z^\top y^* \leq x^{*\top} y^* \text{ for all } z \in P.$$

Consequently, $x^*$ is a solution of the linear programming problem

$$\begin{array}{ll} & \max z^\top y^* \\ \text{s.t.} & a^{i\top} z \leq \alpha_i, \quad i \in I, \\ & d^{h\top} z = \delta_h, \quad h \in I_{n-l}. \end{array}$$

According to the primal-dual theory in linear programming there exist $\lambda_i^* \geq 0$ for $i \in T^*$ and $\nu_h^* \in \mathbb{R}$ for $h \in I_{n-l}$, where $T^* = \{i \in I \mid a^{i\top} x^* = \alpha_i\}$, satisfying

$$y^* = \sum_{i\in T^*} \lambda_i^* a^i + \sum_{h=1}^{n-l} \nu_h^* d^h.$$

Hence $x^* \in F(T^*)$ and $y^* \in A(T^*)$. Next, let $\sigma^*$ be any simplex of the induced triangulation $\mathcal{T}(T^*)$ of the face $F(T^*)$ of $P$ containing $x^*$. We shall show that $\sigma^*$ is balanced. Notice that since $x^* \in \sigma^*$ and $y^* \in \psi(x^*)$, we have $y^* \in C(L(\sigma^*))$. Hence $y^* \in A(T^*) \bigcap C(L(\sigma^*))$. Suppose that $T^* \neq \emptyset$. Then $\sigma^* \in \mathcal{T}(T^*)$ and therefore according to the boundary condition we have that $\mathbf{0} \in A(T^*) \bigcap C(L(\sigma^*))$. Consequently, $\sigma^*$ is balanced. Now suppose that $T^* = \emptyset$. Then $A(T^*) = V$ and therefore $y^* \in V \bigcap C(L(\sigma^*))$. However, $V \bigcap C(L(\sigma^*)) \subseteq V \bigcap C(J)$ whereas $V \bigcap C(J) = \{\mathbf{0}\}$. Hence $y^* = \mathbf{0}$ and so $\sigma^*$ is balanced. This completes the proof of the theorem. □

Although we do not require a balanced simplex to be of dimension $l$, it holds that every simplex of $\mathcal{T}$ containing a balanced simplex as a face is also balanced. Furthermore, notice that the condition $C(J) \bigcap V = \{\mathbf{0}\}$ is satisfied if $\mathbf{0} \in C(J)$ and $C(J) \subseteq V^*$. Without the boundary condition in the above theorem, we easily obtain the following result.

**Corollary 14.3.2** *Let the collection of vectors* $\{ c^j \mid j \in J \}$ *in* $\mathbb{R}^n$ *be given, let* $\mathcal{T}$ *be a triangulation of the polytope* $P$ *and let* $L : \mathcal{T}^0 \mapsto J$ *be an arbitrary labeling rule. Then there exists a simplex* $\sigma \in \mathcal{T}(T)$ *for some* $T \subseteq I$ *such that* $A(T) \cap C(L(\sigma)) \neq \emptyset$.

In order to introduce the second main theorem, we assume that with respect to the polytope $P$ the vectors $d^h$, $h \in I_{n-l}$, and $a^i$, $i \in I$, are orthogonalized and scaled such that $d^{i\top} d^j = 0$ for any $i, j \in I_{n-l}$ with $i \neq j$, $a^h \in V^*$ for every $h \in I$, and $P$ can be written as

$$P = \{ x \in \mathbb{R}^n \mid a^{i\top} x \leq 1 + a^{i\top} x^0,\ i \in I, \text{ and } d^{h\top} x = \delta_h,\ h \in I_{n-l} \}$$

for some arbitrarily chosen $x^0 \in int(P)$. Notice that the above operation is always possible. Let $X$ denote the convex hull of the vectors $a^j$, $j \in I$. Observe that if $F(T)$ is a face of $P$, then the set $co(\{a^j \mid j \in T\})$ is a face of $X$; see Grünbaum [1967, pp. 47-49]. For each $y \in X$, define

$$\begin{aligned} E(y) \;=\; & \{ (S,T) \subseteq J \times I \mid \textstyle\sum_{i \in S} \mu_i c^i + \sum_{j \in T} \nu_j a^j = y, \\ & \textstyle\sum_{i \in S} \mu_i + \sum_{j \in T} \nu_j = 1,\ \mu_i \geq 0,\ i \in S,\ \nu_j \geq 0,\ j \in T \}. \end{aligned}$$

Now we can present the second main theorem.

**Theorem 14.3.3** *Let the polytope* $P$ *be as described just above. For a finite nonempty set* $J$, *let the collection of vectors* $\{c^j \mid j \in J\}$ *in* $\mathbb{R}^n$ *be given such that* $C(J) \subseteq V^*$. *Let* $\mathcal{T}$ *be a triangulation of the* $l$*-dimensional polytope* $P$ *and let* $L : \mathcal{T}^0 \mapsto J$ *be an arbitrary labeling rule. Then for each* $y^0 \in int(X)$, *there exists a simplex* $\sigma \in \mathcal{T}^+$ *such that* $(L(\sigma), Car(\sigma)) \in E(y^0)$.

Proof: Let $x$ be any point in $P$ and let $\sigma(x^1, \cdots, x^{q+1})$ be the unique simplex in $\mathcal{T}^+$ containing $x$ in its interior. Hence there exist unique positive numbers $\gamma_1, \cdots, \gamma_{q+1}$ satisfying $\sum_{i=1}^{q+1} \gamma_i = 1$ such that

$$x = \sum_{i=1}^{q+1} \gamma_i x^i.$$

Define a correspondence $\xi : P \mapsto \mathbb{R}^n$ by

$$\xi(x) = co(\{ y^0 - c^j \mid j = L(x^i) \text{ if } \gamma_i = \max_h \gamma_h \}).$$

Consider the polytope

$$Q = \{ x \in \mathbb{R}^n \mid a^{i\top} x \leq 2 + a^{i\top} x^0,\ i \in I, \text{ and } d^{h\top} x = \delta_h,\ h \in I_{n-l} \}.$$

It is clear that $Q$ contains $P$ in its interior. Let $x$ be a point in $Q$ but not in $P$. Then let $\lambda_x \in (0,1)$ be such that $x^0 + \lambda_x (x - x^0) \in bd(P)$. Since

$x^0 \in int(P)$, such a $\lambda_x$ exists. Let $p(x) = x^0 + \lambda_x(x - x^0)$. Notice that $\{ i \in I \mid a^{i\top} x = 2 + a^{i\top} x^0 \} = Car(p(x))$ for any $x \in bd(Q)$. Now we define a correspondence $\psi : Q \mapsto \mathbb{R}^n$ by

$$\begin{aligned} \psi(x) &= \xi(x),\ x \in int(P), \\ \psi(x) &= co(\xi(x) \cup \{ y^0 - a^i \mid i \in Car(x)\}),\ x \in bd(P), \\ \psi(x) &= co(\{ y^0 - a^i \mid i \in Car(p(x)) \}),\ x \in Q \setminus P. \end{aligned}$$

Clearly, $\psi$ is a nonempty-valued, convex-valued and compact-valued u.s.c. correspondence from $Q$ to $\mathbb{R}^n$, and $\bigcup_{x \in Q} \psi(x)$ is bounded. Let $\phi$ be the correspondence from a compact, convex set $Y$ containing $\bigcup_{x \in Q} \psi(x)$, defined by

$$\phi(y) = \{ x \in Q \mid z^\top y \leq x^\top y \text{ for all } z \in Q \},\ y \in Y.$$

Clearly, $\phi$ is a nonempty-valued, convex-valued and compact-valued u.s.c. correspondence. Hence $\psi \times \phi$, defined by $(\psi \times \phi)(y, x) = \psi(x) \times \phi(y)$ for every $(y, x) \in Y \times Q$, is a nonempty-valued, convex-valued and compact-valued u.s.c. correspondence from $Y \times Q$ into $Y \times Q$. Therefore, according to Kakutani's fixed point theorem there exists a pair of vectors $(y^*, x^*) \in Y \times Q$ such that $y^* \in \psi(x^*)$ and $x^* \in \phi(y^*)$. The latter implies that

$$z^\top y^* \leq x^{*\top} y^* \text{ for all } z \in Q.$$

Consequently, $x^*$ is a solution of the linear programming problem

$$\begin{aligned} &\max\ z^\top y^* \\ \text{s.t.}\quad & a^{i\top} z \leq 2 + a^{i\top} x^0, \quad i \in I, \\ & d^{j\top} z = \delta_j, \qquad\qquad j \in I_{n-l}. \end{aligned}$$

According to the primal-dual theory in linear programming there exist $\lambda_i^* \geq 0$ for $i \in T^*$, and $\beta_j^* \in \mathbb{R}$ for $j \in I_{n-l}$, where $T^* = \{ i \in I \mid a^{i\top} x^* = 2 + a^{i\top} x^0 \}$, such that $y^* = \sum_{i \in T^*} \lambda_i^* a^i + \sum_{h \in I_{n-l}} \beta_h^* d^h$. Notice that $T^* = Car(p(x^*))$ when $T^* \neq \emptyset$.

We need to consider the following cases.

(i) In case $T^* \neq \emptyset$ and hence $x^* \in bd(Q)$, there exist non-negative numbers $\mu_j$, $j \in T^*$, with sum equal to one such that

$$\sum_{i \in T^*} \lambda_i^* a^i + \sum_{h=1}^{n-l} \beta_h^* d^h = \sum_{j \in T^*} \mu_j (y^0 - a^j).$$

Since $y^0 \in V^*$ and $a^i \in V^*$ for all $i \in I$, it implies that $\beta_h^* = 0$ for all $h \in I_{n-l}$. Hence, $\sum_{i \in T^*} \lambda_i^* + \sum_{j \in T^*} \mu_j \geq 1$ and $y^0 = \sum_{j \in T^*} (\mu_j + \lambda_j^*) a^j$. This is a contradiction with the fact that $y^0 \in int(X)$ and $F(T^*)$ is a face of $P$. Indeed, if $\sum_{i\in T^*}\lambda_i^* + \sum_{j\in T^*}\mu_j > 1$, then $y^0$ lies outside $X$, and if this sum is equal to 1, $y^0$ lies on the boundary of $X$.

(ii) In case $T^* = \emptyset$, the following subcases need to be checked. (ii-1) In case $x^* \in int(P)$, we have that $y^* \in \psi(x^*)$. Moreover, there exists a unique simplex $\sigma$ with vertices $w^1, \cdots, w^{t+1}$ containing $x^*$ in its interior. It implies that there exist non-negative numbers $\mu_j$, $j \in L(\sigma)$, with sum equal to one such that $\sum_{j\in L(\sigma)}\mu_j(y^0 - c^j) = \sum_{h=1}^{n-l}\beta_h^* d^h$. Since $y^0 \in V^*$ and $a^i \in V^*$ for all $i \in I$, it implies that $\beta_h^* = 0$ for all $h \in I_{n-l}$. So, $y^0 \in co(\{\, c^j \mid j \in L(\sigma)\})$ and therefore $(L(\sigma), Car(\sigma)) \in E(y^0)$. (ii-2) In case $x^* \in bd(P)$, we have that $y^* \in \psi(x^*)$ and that there exists a unique simplex $\sigma$ containing $x^*$ in its interior. Clearly, $Car(\sigma) = S^*$. Hence, we have

$$\sum_{i\in L(\sigma)}\mu_i(y^0 - c^i) + \sum_{j\in S^*}\nu_j(y^0 - a^j) = \sum_{h=1}^{n-l}\beta_h^* d^h$$

for some non-negative numbers $\mu_i$, $i \in L(\sigma)$, $\nu_j$, $j \in S^*$, with $\sum_{i\in L(\sigma)}\mu_i + \sum_{j\in S^*}\nu_j = 1$. Since $y^0 \in V^*$ and $a^i \in V^*$ for all $i \in I$, it implies that $\beta_h^* = 0$ for all $h \in I_{n-l}$. Hence,

$$y^0 = \sum_{i\in L(\sigma)}\mu_i c^i + \sum_{j\in Car(\sigma)}\nu_j a^j,$$

and therefore $(L(\sigma), Car(\sigma)) \in E(y^0)$. (ii-3) If $x^*$ lies in the interior of $Q$ but not in $P$, we have that $y^* \in \psi(x^*)$. It holds that $\sum_{i\in T^*}\mu_i(y^0 - a^i) = \sum_{h=1}^{n-l}\beta_h^* d^h$ for some non-negative numbers $\mu_i$ with $\sum_{i\in T^*}\mu_i = 1$, where $T^* = Car(p(x^*))$. Since $y^0 \in V^*$ and $a^i \in V^*$ for all $i \in I$, it implies that $\beta_h^* = 0$ for all $h \in I_{n-l}$. So, $y^0 = \sum_{i\in T^*}\mu_i a^i$. This contradicts the fact that $y^0 \in int(X)$ and $F(T^*)$ is a face of $P$. □

## 14.4. Applications to the unit simplex

In this section we apply Theorem 14.3.1 to obtain several well-known results on the unit simplex $S^n$. For $S \subseteq I_n$, let the $n$-vector $m^S$ be defined by $\sum_{i\in S}\frac{1}{|S|}e(i)$. Observe that $m^S = e(i)$ if $S = \{\, i \,\}$. Now, take $l = n-1$, $d^1 = m^{I_n}$, $\delta_1 = 1/n$, $I = I_n$, $a^i = m^{I_n} - e(i)$ and $\alpha_i = 1/n$ for $i \in I_n$. Observe that $a^i \in V^*$ for all $i \in I_n$. For $S \subseteq I_n$, define $A'(S) = Cone(\{\, a^i \mid i \in S\,\})$. Now, the unit simplex $S^n$ can be rewritten in the framework of this chapter as

$$S^n = \{\, x \in \mathbb{R}^n \mid a^{i\top}x \le \alpha_i, i \in I_n \text{ and } d^{1\top}x = \delta_1 \,\}.$$

We first prove Sperner lemma [1928].

**Theorem 14.4.1 (Sperner Lemma)** *Let $\mathcal{T}$ be a triangulation of $S^n$ and let $L : \mathcal{T}^0 \mapsto I_n$ be a labeling rule where $x_i = 0$ implies $L(x) \neq i$ for any vertex $x \in S^n$. Then there exists a completely labeled simplex of $\mathcal{T}$, i.e., a simplex $\sigma \in \mathcal{T}$ such that $L(\sigma) = I_n$.*

Proof: Take $J = I_n$ and for $j \in J$, set $c^j = a^j$. Clearly, $\mathbf{0} \in C(J)$ and $C(J) \subseteq V^*$. Therefore we have $C(J) \bigcap V = \{\mathbf{0}\}$. Notice that $\mathbf{0} \in C(S)$ if and only if $S = J$ and hence a balanced simplex bears all labels 1 up to $n$. To show the existence of a balanced simplex we still have to show that the boundary condition of Theorem 14.3.1 is satisfied for every simplex in a proper face $S^n(T)$ of $S^n$. So, let $\sigma \in \mathcal{T}(T)$ for some nonempty $T \subset I_n$. Then $L(\sigma) \bigcap T = \emptyset$ since for every vertex $x$ of $\sigma$ we have $x_i = 0$ and hence $L(x) \neq i$ and hence $L(x) \notin T$. Since the vectors $a^i$, $i \in S$, are linearly independent for any proper subset $S$ of $J$ we must have that $A'(L(\sigma)) \bigcap A(T) = \{\mathbf{0}\}$ and hence $C(L(\sigma)) \bigcap A(T) = \emptyset$. This completes the proof. □

The next lemma was introduced by Scarf [1967b].

**Theorem 14.4.2 (Scarf Lemma)** *Let $\mathcal{T}$ be a triangulation of $S^n$ and let $L : \mathcal{T}^0 \mapsto I_n$ be a labeling rule satisfying that $x_j = 0$ implies $L(x) = \min\{i \mid x_i = 0 \text{ and } x_{i+1} > 0\}$ for any vertex $x \in bn(S^n)$, where $l + 1 = 1$ if $l = n$. Then there exists a completely labeled simplex of $\mathcal{T}$.*

Proof: Let $J = I_n$ and $c^j = -a^j$ for all $j \in J$. Again, $C(J) \subseteq V^*$ and $\mathbf{0} \in C(S)$ if and only if $S = J$. Hence a balanced simplex must carry all labels. We still have to prove the boundary condition of Theorem 14.3.1 is fulfilled for every simplex $\sigma \in \mathcal{T}(T)$ in any proper face $S^(T)$. Suppose that $A(T) \bigcap C(L(\sigma)) \neq \emptyset$ for some nonempty subset $T$ of $J$ and some $\sigma \in \mathcal{T}(T)$. Then there exist some non-negative $\lambda_i$ for $i \in T$, a real number $\nu_1$, and non-negative $\mu_j$ for $j \in S$ where $S = L(\sigma)$ such that $\sum_{i \in T} \lambda_i a^i + \nu_1 m^{I_n} = \sum_{j \in S} \mu_j c^j$ and $\sum_{j \in S} \mu_j = 1$. Since $c^j = -a^j$ for all $j \in J$, this yields

$$\sum_{i \in T} \lambda_i a^i + \sum_{j \in S} \mu_j a^j = -\nu_1 m^{I_n}.$$

Since $m^{I_n \top} a^i = 0$ for all $i \in S \bigcup T$, it implies that $\nu_1 = 0$. It means that the vectors $a^j$, $j \in S \bigcup T$, are linearly dependent. Hence, $S \bigcup T = I_n$. Let $x^1, \cdots, x^{q+1}$ be the vertices of $\sigma$. Suppose that for some $j \in I_n$ it holds that $x^i_j > 0$ for all $i = 1, \cdots, q+1$. Then $L(x^i) \neq j$ for all $i = 1, \cdots, q+1$ and so $j \notin S$. Moreover, $j \notin T$. This contradicts the fact that $T \bigcup S = I_n$. Consequently, for every $j \in I_n$ there is at least one $i \in \{1, \cdots, q+1\}$ satisfying $x^i_j = 0$. Let $h \in I_n$ be an index such that $x^i_h > 0$ for at least one $i \in \{1, \cdots, q+1\}$ and $h + 1 \in T$, where $k + 1 = 1$ when $k = n$. Hence, $h \notin T$. Moreover, $h \notin S$ because of the fact that no vertex $x^i$ can carry label $h$ if $x^i_{h+1} = 0$. Hence, $h \notin T \bigcup S$, yielding a contradiction. Therefore, there exists a balanced simplex $\sigma$ of $\mathcal{T}$ which must then be completely labeled. □

Notice that the properness condition in Scarf lemma can be relaxed slightly. It is sufficient to require that $A(T) \bigcap C(L(\sigma)) = \emptyset$ for every $q$-simplex of $\mathcal{T}(T)$, where $q = n - 1 - |T|$. However, note that for Sperner lemma the boundary condition is equivalent to that of Theorem 14.3.1.

The third result is established in Shapley [1973]. In this theorem the vertices of a triangulation of $S^n$ are labeled with nonempty subsets of the set $I_n$. To prove Shapley lemma, we need the concept of coalitional balancedness (see Section 2.2).

**Theorem 14.4.3 (Shapley Lemma)** *Let $\mathcal{N}$ be the collection of all nonempty subsets of the set $I_n$. Let $\mathcal{T}$ be a triangulation of $S^n$ and let $L : \mathcal{T}^0 \mapsto \mathcal{N}$ be a labeling rule satisfying that $L(x) \subseteq \{i \mid x_i > 0\}$ for any vertex $x \in S^n$. Then there exists at least one face $\sigma(x^1, \cdots, x^{q+1})$ of a simplex of $\mathcal{T}$ such that the collection $\{ L(x^1), \cdots, L(x^{q+1}) \}$ is balanced.*

Proof: Let $J = \mathcal{N}$ and $c^S = m^{I_n} - m^S$ for all $S \in \mathcal{N}$. Clearly, $C(J) \subseteq V^*$ and $\mathbf{0} \in C(J)$. We next prove that the boundary conditions of Theorem 14.3.1 are fulfilled by every simplex $\sigma(x^1, \cdots, x^{q+1})$ of $\mathcal{T}(T)$ for any nonempty subset $T$ of $I_n$. Since $\sigma \in \mathcal{T}(T)$, we must have $x^i_j = 0$ for every $j \in T$, and hence according to the boundary condition $L(x^i) \bigcap T = \emptyset$ for all $i = 1, \cdots, q + 1$. Let $B_i = L(x^i)$ for $i = 1, \cdots, q + 1$ and $S = \bigcup_{i=1}^{q+1} B_i$. Then also $S \bigcap T = \emptyset$. Since the vectors $a^i$, $i \in K$, are linearly independent for each proper subset $K$ of $I_n$ we have that $A'(S) \bigcap A(T) = \{\mathbf{0}\}$. For every $i \in \{ 1, \cdots, q + 1 \}$ it holds that $L(x^i) \subseteq S$ and $c^{B_i}$ is a convex combination of the vectors $a^j$, $j \in B_i$. Hence, $C(L(\sigma)) \subseteq A'(S)$. Moreover, since for every $i \in \{ 1, \cdots, q + 1 \}$ we have $c^{B_i}_j > 0$ for any $j \notin S$, it implies that $\mathbf{0} \notin C(L(\sigma))$. Consequently, $C(L(\sigma)) \bigcap A(T) = \emptyset$ and hence the boundary condition is satisfied. This guarantees the existence of a balanced simplex according to Theorem 14.3.1. □

The next result was due to Garcia [1976].

**Theorem 14.4.4 (Garcia Lemma)** *Let $\mathcal{T}$ be a triangulation of $S^n$ and let $L : \mathcal{T}^0 \mapsto I_n$ be a labeling rule. Then there exists a simplex $\sigma \in \mathcal{T}(T)$ for some $T \subset I_n$ such that $L(\sigma) \cup T = I_n$.*

Proof: Let $J = I_n$ and let $c^j = -a^j$ for each $j \in J$. According to Corollary 14.3.2, there exists a simplex $\sigma \in \mathcal{T}(T)$ for some $T \subset I_n$ such that

$A(T) \cap C(L(\sigma)) \neq \emptyset$. Hence, the system of equations

$$\begin{gathered}\textstyle\sum_{i \in T} \lambda_i a^i + \beta m^{I_n} + \sum_{j \in L(\sigma)} \gamma_j a^j = \mathbf{0}, \\ \textstyle\sum_{j \in L(\sigma)} \gamma_j = 1, \\ \gamma_j \geq 0,\ j \in L(\sigma), \\ \beta \in \mathbb{R},\ \lambda_i \geq 0,\ i \in T,\end{gathered}$$

has a solution. Clearly, the above system has a solution only if $T \cup L(\sigma) = I_n$. It is easy to see that $T = Car(\sigma)$. Hence $Car(\sigma) \cup L(\sigma) = I_n$. □

Now we give a permutation-based generalization of Sperner lemma. A constructive proof of this result can be found in Bapat [1989].

**Lemma 14.4.5 (Bapat Lemma)** *Let $\mathcal{T}$ be a triangulation of $S^n$ and for each $i \in I_n$, let $L^i : \mathcal{T}^0 \mapsto I_n$ be a labeling rule satisfying the conditions of Sperner Lemma. Then there exist at least a simplex $\sigma \in \mathcal{T}$ with vertices $x^1, \cdots, x^n$, and a permutation $\pi = (\pi(1), \cdots, \pi(n))$ of $(1, \cdots, n)$ such that $\{L^{\pi(i)}(x^i) \mid i \in I_n\}$ is equal to $I_n$.*

Bapat lemma is closely related to Gale's intersection lemma in the previous chapter.

We remark that Sperner lemma (see van der Laan and Talman [1982]), Scarf lemma (see Freund [1986]) and Garcia Lemma (see Freund [1986] and van der Laan, Talman and Van der Heyden [1987]) have been generalized to the simplotope. These generalizations can be also derived from Theorem 14.3.1. We leave them to the interested reader.

## 14.5. Applications to polytopes

In this section we apply Theorems 14.3.1, 14.3.3 to prove existing results on polytopes. First, let us consider the $n$-dimensional cube $U^n$. The following lemmas are introduced in Freund [1984b, 1986], and van der Laan and Talman [1981b] and can be easily derived from Theorem 14.3.1 .

**Lemma 14.5.1** *Let $\mathcal{T}$ be a triangulation of the unit cube $U^n$ and let $L : \mathcal{T}^0 \mapsto I_n \bigcup -I_n$ be a labeling rule satisfying that $x_i = 1$ implies $L(x) \neq i$ and $x_i = 0$ implies $L(x) \neq -i$ for any $x \in \mathcal{T}^0$. Then there exists a complementary 1-simplex $\sigma \in \mathcal{T}^+$, i.e., a 1-simplex $\sigma$ such that $L(\sigma) = \{k, -k\}$ for some $k \in I_n$.*

**Lemma 14.5.2** *Let $\mathcal{T}$ be a triangulation of the unit cube $U^n$ and let $L : \mathcal{T}^0 \mapsto I_n \bigcup -I_n$ be a labeling rule where $L(x) = i$ implies $x_i = 1$ and $L(x) = -i$ implies $x_i = 0$ for any $x \in \mathcal{T}^0 \cap bd(U^n)$. Then there exists a complementary 1-simplex $\sigma \in \mathcal{T}^+$.*

Now we discuss three main results of Freund [1989]. Consider a full-dimensional general polytope

$$P = \{ x \in \mathbb{R}^n \mid a^{i^{\top}} x \leq 1, \, i \in I \}$$

with $|I| \geq n+1$. Since $P$ is bounded, the point $\mathbf{0}$ lies in the convex hull of the vectors $a^j$, $j \in I$. Also, $V = \{\mathbf{0}\}$. Let $X$ denote the convex hull of the vectors $a^j$, $j \in I$. Observe that the set $X$ is a full-dimensional subset of $\mathbb{R}^n$ satisfying that if $F(T)$ is a face of $P$ then $co(\{a^j \mid j \in T\})$ is a face of $X$. For $y \in X$, define

$$D(y) = \{ T \subseteq I \mid y \in co(\{a^j \mid j \in T\}) \}.$$

So, $D(y)$ is the collection of all sets $T \subseteq I$ satisfying that $y \in co(\{a^j \mid j \in T\})$. A labeling rule $L : \mathcal{T}^0 \mapsto I$ is called *dual proper* if $L(v) \in Car(v)$ for all $v \in bd(P)$ and $v \in \mathcal{T}^0$. A triangulation $\mathcal{T}$ of $P$ is *bridgeless* if for each $\sigma \in \mathcal{T}$, the intersection of all faces of $P$ that meet $\sigma$ is nonempty.

The following result is a generalization of Garcia's lemma on a full-dimensional simplex to a full-dimensional polytope, which was due to Freund [1989].

**Theorem 14.5.3 (Freund Theorem I)** *Let $\mathcal{T}$ be a triangulation of the n-dimensional polytope $P$ described above and let $L : \mathcal{T}^0 \mapsto I$ be an arbitrary labeling rule. Then for each $y \in int(X)$, there exists a simplex $\sigma \in \mathcal{T}^+$ such that $Car(\sigma) \bigcup L(\sigma) \in D(y)$.*

Proof: By taking $J = I$ and letting $c^j = a^j$ for any $j \in J$, we easily obtain the conclusion from Theorem 14.3.3. □

The next result easily follows from Freund Theorem I.

**Theorem 14.5.4 (Freund Theorem II)** *Let $\mathcal{T}$ be a bridgeless triangulation of the n-dimensional polytope $P$ just described and let $L : \mathcal{T}^0 \mapsto I$ be a dual proper labeling rule. Then for each $y \in int(X)$ there exists a simplex $\sigma \in \mathcal{T}^+$ such that $L(\sigma) \in D(y)$.*

This result extends Scarf's lemma on a full-dimensional simplex. Before introducing the next result, which is a generalization of Sperner's lemma on a full-dimensional simplex, for each $y \in int(X)$ let

$$\begin{aligned} U(y) \;=\; & \{ (S,T) \subseteq I \times I \mid \textstyle\sum_{i \in T} \mu_i a^i - \sum_{j \in S} \nu_j a^j = y, \\ & \textstyle\sum_{i \in T} \mu_i + \sum_{j \in S} \nu_j = 1, \, \mu_i \geq 0, \, i \in T, \, \nu_j \geq 0, \, j \in S \}. \end{aligned}$$

**Theorem 14.5.5 (Freund Theorem III)** *Let $\mathcal{T}$ be a triangulation of the n-dimensional polytope $P$ described above and let $L : \mathcal{T}^0 \mapsto I$ be an arbitrary labeling rule. Then for each $y \in int(X)$, there exists a simplex $\sigma \in \mathcal{T}^+$ such that $(L(\sigma), Car(\sigma)) \in U(y)$.*

Proof: By taking $J = I$ and letting $c^j = -a^j$ for any $j \in J$, we obtain the conclusion immediately from Theorem 14.3.3. □

# References


W.W. Adams and P. Loustaunau [1994], *An Introduction to Gröbner Bases*, American Mathematical Society, USA.

L.V. Ahlfors [1979], *Complex Analysis*, McGraw-Hill Book Company, New York.

E.L. Allgower and K. Georg [1980], "Simplicial and continuation methods for approximating fixed points and solutions to systems of equations", *SIAM Review* 22, pp. 28-85.

E.L. Allgower and K. Georg [1990], *Numerical Continuation Methods: An Introduction,* Springer-Verlag, Berlin.

E.L. Allgower, K. Glashoff and H.O. Peitgen [1981], *Numerical Solution of Nonlinear Equations*, eds., Lecture Notes in Mathematics 878, Springer-Verlag, Berlin.

H. Amann [1972], "On the number of solutions of nonlinear equations in ordered Banach spaces", *Journal of Function Analysis* 11, pp. 346-384.

K. Ando and S. Fujishige [1996], "On structures of bisubmodular polyhedra", *Mathematical Programming* 74, pp. 28-85.

K.J. Arrow [1963], "The role of securities in the optimal allocation of risk bearing", *Review of Economic Studies* 31, pp. 91-96.

K.J. Arrow, H.D. Block and L. Hurwicz [1959], "On the stability of the competitive equilibrium, II", *Econometrica* 27, pp. 82-109.

K.J. Arrow and G. Debreu [1954], "Existence of an equilibrium of a competitive economy", *Econometrica* 22, pp. 265-290.

K.J. Arrow and F.H. Hahn [1971], *General Competitive Analysis,* Holden-Day, San Francisco.

K.J. Arrow and L. Hurwicz [1958], "On the stability of the competitive equilibrium, I", *Econometrica* 26, pp. 522-552.

R.J. Aumann and B. Peleg [1960], "Von Neumann-Morgenstern solutions to cooperative games without side payments", *Bulletin of American Mathematical Society* 66, pp. 173-179.

R.B. Bapat [1989], "A constructive proof of a permutation-based generalization of Sperner's lemma", *Mathematical Programming* 44, pp. 113-120.

I. Bárány [1980], "Borsuk's theorem through complementary pivoting", *Mathematical Programming* 18, pp. 84-88.

I. Bárány, R. Howe and H. Scarf [1994], "The complex of maximal lattice free simplices", *Mathematical Programming* 66, pp. 272-281.

I. Bárány, H. Scarf and D. Shallcross [1998], "The topological structure of maximal lattice free convex bodies: The general case", *Mathematical Programming* 80, pp. 1-15.

P. Beato [1976], *Marginal Cost Pricing Equilibria with Increasing Returns*, Ph.D. Thesis, University of Minnesoto, Minneapolis.

J.B. Bénassy [1975], "Neo-Keynesian disequilibrium theory in a monetary economy", *Review of Economic Studies* 42, pp. 503-523.

C. Berge [1963], *Topological Spaces,* Oliver & Boyd, Edinburgh.

O. Bondareva [1962], "Theory of the core in the $n$-person game", *Vestnik Leningradskii Universitet* 13, pp. 141-142.

J.M. Bonnisseau and B. Cornet [1988], "Existence of equilibria when firms follow bounded losses pricing rules", *Journal of Mathematical Economics*, 17, pp. 119-147.

U. Borsuk [1933], "Drei Sätze über die $n$-dimensionale euklidische Sphär", *Fundamenta Mathematica* 20, pp. 177-190.

M.N. Broadie and B.C. Eaves [1987], "A variable rate refining triangulation", *Mathematical Programming* 38, pp. 161-202.

P.S. Brooks [1980], "Infinite regression in the Eaves-Saigal algorithm", *Mathematical Programming* 19, pp. 313-327.

L.E.J. Brouwer [1912], "Über Abbildung von Mannigfaltigkeiten", *Mathematische Annalen* 71, pp. 97-115.

F.E. Browder [1960], "On continuity of fixed points under deformations of continuous mappings", *Summa Brasiliensis Mathematicae* 4, pp. 183-191.

D.J. Brown, P.M. De Marzo and B.C. Eaves [1996], "Computing zeros of sections of vector bundles using homotopies and relocalization", *Mathematics of Operations Research* 21, pp. 26-43.

B. Buchberger [1965], *An Algorithm for Finding a Basis for the Residue Class Ring of a Zero-Dimensional Polynomial Ideal*, Ph.D. Thesis, University of Innsbruck, Innsbruck. (in German)

B. Buchberger [1985], "Gröbner bases: An algorithmic method in polynomial ideal theory", in: *Multidimensional Systems Theory*, ed. N.K. Bose, Reidel, Dordrecht, pp. 184-232.

J. Caristi [1976], "Fixed point theorems for mappings satisfying inwardness conditions", *Transaction of American Mathematical Society* 215, pp. 241-251.

A. Charnes, G.B. Garcia and C.E. Lemke [1977], "Constructive proofs of theorems relating to: F(x)=y, with applications", *Mathematical Programming* 12, pp. 328-343.

K.Z. Chen [1990], *Introduction to Fixed Point Theory and Methods*, Xidian University Press, Xian. (in Chinese)

S.C. Chou [1984], "Proving elementary geometry theorems using Wu's method", *Contemporary Mathematics*, 29, pp. 243-286.

S.C. Chou [1988], *Mechanical Geometry Theorem Proving*, D. Reidel Publishing Company, Dordrecht.

D.I.A. Cohen [1967], "On the Sperner lemma", *Journal of Combinatorial Theory* 2, pp.87-91.

D.I.A. Cohen [1979], "On the combinatorial antipodal-point lemmas", *Journal of Combinatorial Theory Series B* 27, pp.87-91.

O.J.C. Cornielje and G. van der Laan [1986], "The computation of quantity-constrained equilibria by virtual taxes", *Economics Letters* 22, pp. 1-6.

R.W. Cottle and G.B. Dantzig [1968], "Complementary pivot theory of mathematical programming", *Linear Algebra and Its Applications* 1, pp. 103-125.

R.W. Cottle, J.S. Pang and R.E. Stone [1992], *The Linear Complementarity Problem*, Academic Press, Boston.

D. Cox, J. Little and D. O'Shea [1996], *Ideals, Varieties, and Algorithms*, 2rd edition, Springer, New York.

I.J. Curiel and S.H. Tijs [1985], "Assignment games and permutation games", *Methods of Operations Research* 54, pp. 323-334.

Y. Dai [1991], *Path Following Algorithms for Stationary Point Problem on Polyhedra*, Ph.D. Thesis, University of Tsukuba, Tsukuba.

Y. Dai, G. van der Laan, A.J.J. Talman and Y. Yamamoto [1991], "A simplicial algorithm for the stationary point problem on an unbounded polyhedron", *SIAM Journal on Optimization* 1, pp. 151-165.

E.E.C. van Damme [1987], *Stability and Perfection of Nash Equilibria*, Springer-Verlag, Berlin.

C. Dang [1991], *The $D_1$-Triangulation in Simplicial Algorithms*, Ph.D. Thesis, Tilburg University, Tilburg.

C. Dang and H. van Maaren [1993], "A simplicial approach to integer programming", Working Paper 93-06, Department of Mathematics, Delft University of Technology, Delft.

G.B. Dantzig [1955], "Optimal solution of a dynamic Leontief model with substitution",

*Econometrica* 23, pp. 295-302.
G.B. Dantzig [1963], *Linear Programming and Extensions,* Princeton University Press, Princeton.
G. Debreu [1959], *Theory of Value,* Yale University Press, New Haven.
T.M. Doup [1988], *Simplicial Algorithms on the Simplotope,* Lecture Notes in Economics and Mathematical Systems 318, Springer-Verlag, Berlin.
T.M. Doup, G. van der Laan and A.J.J. Talman [1987], "The $(2^{n+1}-2)$-ray algorithm: a new simplicial algorithm to compute economic equilibria", *Mathematical Programming* 39, pp. 241-252.
T.M. Doup and A.J.J. Talman [1987a], "The 2-ray algorithm for solving equilibrium problems on the unit simplex", *Methods of Operations Research* 57, pp. 269-285.
T.M. Doup and A.J.J. Talman [1987b], "A new variable dimension algorithm to find equilibria on the product space of unit simplices", *Mathematical Programming* 37, pp. 319-355.
T.M. Doup and A.J.J. Talman [1987c], "A continuous deformation algorithm on the product space of unit simplices", *Mathematics of Operations Research* 12, pp. 485-521.
J.H. Drèze [1975], "Existence of an exchange equilibrium under price rigidities", *International Economic Review* 16, pp. 301-320.
D. Duffie [1987], "Stochastic equilibria with incomplete financial markets", *Journal of Economic Theory* 41, pp. 405-416.
D. Duffie [1992], *Dynamic Asset Pricing Theory,* Princeton University Press, Princeton.
D. Duffie and W. Shafer [1985], "Equilibrium in incomplete markets I: Basic model of generic existence", *Journal of Mathematical Economics* 14, pp. 285-300.
D. Duffie and W. Shafer [1986], "Equilibrium in incomplete markets II: Generic existence in stochastic economies", *Journal of Mathematical Economics* 15, pp. 199-216.
B.C. Eaves [1971a], "On the basic theory of complementarity", *Mathematical Programming* 1, . 68-75.
B.C. Eaves [1971b], "Computing Kakutani fixed points", *SIAM Journal on Applied Mathematics* 21, pp. 236-244.
B.C. Eaves [1972], "Homotopies for computation of fixed points", *Mathematical Programming* 3, pp. 1-22.
B.C. Eaves [1978], "Computing stationary points", *Mathematical Programming Study* 7, pp. 1-14.
B.C. Eaves [1984], *A Course in for Solving Equations with Deformations,* Lecture Notes in Economics and Mathematical Systems 234, Springer-Verlag, Berlin.
B.C. Eaves and R. Saigal [1972], "Homotopies for the computation of fixed points on unbounded regions", *Mathematical Programming* 3, pp. 225-237.
B.C. Eaves and H. Scarf [1976], "The solution of systems of piecewise linear equations", *Mathematics of Operations Research* 1, pp. 1-27.
B.C. Eaves and J.A. Yorke [1984], "Equivalence of surface density and average directional density", *Mathematics of Operations Research* 9, pp. 363-375.
F.Y. Edgeworth [1881], *Mathematical Psychics,* Kegan, London.
I. Ekeland [1979], "Nonconvex minimization problems", *Bulletin of American Mathematical Society* 1, pp. 443-474.
A.H. van den Elzen [1993], *Adjustment Processes for Exchange Economies and Noncooperative Games,* Lecture Notes in Economics and Mathematical Systems 402, Springer-Verlag, Berlin.
K. Fan [1967], "Simplicial maps from an $n$-pseudo-manifold into $S^m$ with the octahedral triangulation", *Journal of Combinatorial Theory* 2, pp. 588-602.
K. Fan [1968], "A covering property of simplexes equilibria", *Mathematica Scandinavica* 22, pp. 17-20.
K. Fan [1972], "A minmax inequality and applications", in: *Inequalities III,* ed., O. Oshisha, Academic Press, New York, pp. 103-113.
M.L. Fisher and F.J. Gould [1974], "A simplicial algorithm for the nonlinear complemen-

tarity problem", *Mathematical Programming* 6, pp. 281-300.
W. Forster [1980], *Numerical Solution of Highly Nonlinear Problems,* ed., North-Holland, Amsterdam.
J. Franklin [1980], *Methods of Mathematical Economics,* Springer-Verlag, New York.
J. Freidenfelds [1974], "A set intersection theorem and applications", *Mathematical Programming* 7, pp. 199-211.
H. Freudenthal [1942], "Simplizialzerlegungen von beschränkter Flachheit", *Annals of Mathematics* 43, pp. 580-582.
R.W. Freund [1984a], "Variable dimension complexes Part I: basic theory", *Mathematics of Operations Research* 9, pp. 479-497.
R.W. Freund [1984b], "Variable dimension complexes Part II: a unified approach to some combinatorial lemmas in topology", *Mathematics of Operations Research* 9, pp. 498-509.
R.W. Freund [1986], "Combinatorial theorems on the simplotope that generalize results on the simplex and cube", *Mathematics of Operations Research* 11, pp. 169-179.
R.W. Freund [1989], "Combinatorial analogs of Brouwer's fixed point theorem on a bounded polyhedron", *Journal of Combinatorial Theory* (B) 47, pp. 192-219.
R.W. Freund and M.J. Todd [1981], "A constructive proof of combinatorial lemma", *Journal of Combinatorial Theory* (A) 30, pp. 321-325.
S. Fujishige [1991], *Submodular Functions and Optimization*, North-Holland, Amsterdam.
S. Fujishige and Z. Yang [1998], "A lexicographic algebraic theorem and its applications", *Linear Algebra and Its Applications* 279, pp. 75-91.
D. Gale [1960], *The Theory of Linear Economic Models*, McGraw-Hill, New York.
D. Gale [1984], "Equilibrium in a discrete exchange economy with money", *International Journal of Game Theory* 13, pp. 61-64.
D. Gale and L.S. Shapley [1962], "College admissions and the stability of marriage", *American Mathematical Monthly* 69, pp. 9-15.
C.B. Garcia [1976], "A hybrid algorithm for the computation of fixed points", *Management Science* 22, pp. 606-613.
C.B. Garcia and F.J. Gould [1978], "A theorem on homotopy paths", *Mathematics of Operations Research* 3, pp. 282-289.
C.B. Garcia, C.E. Lemke and H.J. Lüthi [1973], "Simplicial approximation of an equilibrium point for non-cooperative $N$-person games", in: Mathematical Programming, eds., T.C. Hu and S.M. Robinson, Academic Press, New York, pp. 227-260.
C.B. Garcia and W.I. Zangwill [1979], "An approach to homotopy and degree theory", *Mathematics of Operations Research* 4, pp. 390-405.
C.B. Garcia and W.I. Zangwill [1981], *Pathways to Solutions, Fixed Points, and Equilibria,* Prentice-Hall, Englewood Cliffs.
F.J. Gould and J.W. Tolle [1974], "A unified approach to complementarity in optimization", *Discrete Mathematics* 7, pp. 225-271.
B. Grünbaum [1967], *Convex Polytopes,* John Wiley & Sons, New York.
R. Guesnerie [1975], "Pareto optimality in non-convex economies", *Econometrica* 43, pp. 1-29.
B.L. Guo and Y. Yamamoto [1996], "A note on a theorem of a continuum of zero points", Discussion Paper No. 694, Institute of Socio-Economic Planning, University of Tsukuba, Tsukuba, to appear in *Journal of the OR Society of Japan.*
T. Hansen [1968], *On the Approximation of a Competitive Equilibrium,* Ph.D. Thesis, Yale University, New Haven.
T. Hansen and H. Scarf [1969], "On the applications of a recent combinatorial algorithm", Cowles Foundation Discussion Paper No. 272, Yale University, New Haven.
P.T. Harker and J.S. Pang [1990], "Finite-dimensional variational inequality and nonlinear complementarity problems: a survey of theory, algorithms and applications", *Mathematical Programming* 48B, pp. 161-220.
O.D. Hart [1974], "On the existence of equilibrium in a securities model", *Journal of Economic Theory* 9, pp. 293-311.

P. Hartman and G. Stampacchia [1966], "On some nonlinear elliptic differential functional equations", *Acta Mathematica* 115, pp. 271-310.

E. Helly [1923], "Über Mengen konvexer Körper mit gemeinschaftlichen Punkten", *Jber. Deutsch. Math. Verein.* 32, pp.175-176.

P.J.J. Herings [1995], *Static and Dynamic Aspects of General Disequilibrium Theory,* Ph.D. Thesis, Tilburg University, Tilburg.

P.J.J. Herings [1997], "A globally and universally stable price adjustment process", *Journal of Mathematical Economics* 27, pp. 163-193.

P.J.J. Herings and A.J.J. Talman [1998], "Intersection theorems with a continuum of intersection points", *Journal of Optimization Theory and Applications* 2, pp. 311-335.

P.J.J. Herings, A.J.J. Talman and Z. Yang [1996], "The computation of a continuum of constrained equilibria", *Mathematics of Operations Research* 21, pp. 675-696.

W. Hildenbrand [1974], *Core and Equilibria of a Large Economy,* Princeton University Press, Princeton.

M.W. Hirsch [1963], "A proof of the nonretractability of a cell onto its boundary", *Proceedings of the American Mathematical Society* 14, pp. 364-365.

M.W. Hofkes [1990], "A simplicial algorithm to solve the nonlinear complementarity problem on $S^n \times \mathbb{R}^m_+$", *Journal of Optimization Theory and Applications* 67, pp. 551-565.

T.C. Hu and S.M. Robinson [1973], *Mathematical Programming,* eds., Academic Press, New York.

T. Ichiishi [1981], "On the Knaster-Kuratowski-Mazurkiewicz-Shapley theorem", *Journal of Mathematical Analysis and Applications* 81, pp. 297-299.

T. Ichiishi [1988], "Alternative version of Shapley's theorem on closed coverings of a simplex", *Proceedings of the American Mathematical Society* 104, pp. 759-763.

T. Ichiishi and A. Idzik [1991], "Closed covers of compact convex polyhedra", *International Journal of Game Theory* 20, pp. 161-169.

V.I. Istratescu [1981], *Fixed Point Theory, An Introduction,* Reidel Publishing Company, Dordrecht.

R.A.M.G. Joosten and A.J.J. Talman [1993], "A simplicial variable dimension restart algorithm to find economic equilibria on the unit simplex using $n(n+1)$ rays", Research Memorandum No. 608, FEW, Tilburg University, Tilburg.

S. Kakutani [1941], "A generalization of Brouwer's fixed point theorem", *Duke Mathematical Journal* 8, pp. 457-459.

K. Kamiya [1988], "Existence and uniqueness of equilibria with increasing returns", *Journal of Mathematical Economics* 17, pp. 149-178.

K. Kamiya [1991], "Computation of equilibria in an economy with increasing returns to scale", *Mathematical Programming* 49, pp. 253-261.

K. Kamiya and A.J.J. Talman [1991], "Variable dimension simplicial algorithm for balanced games", *Journal of the OR Society of Japan* 34, pp. 222-228.

K. Kaneko and Y. Yamamoto [1986], "The existence and computation of competitive equilibria in markets with an indivisible commodity", *Journal of Economic Theory* 38, pp. 118-136.

R. Kannan and A. Bachem [1979], "Polynomial algorithms for computing the Smith and Hermite normal forms of an integer matrix", *SIAM Journal on Computing* 8, pp. 499-507.

S. Karamardian [1977], *Fixed Points: Algorithms and Applications,* ed., Academic Press, New York.

R.B. Kellogg, T.-Y. Li and J.A. Yorke [1976], "A constructive proof of the Brouwer fixed point theorem and computational results", *SIAM Journal on Numerical Analysis* 13, pp. 473-483.

A.S. Kelso and V.P. Crawford [1982], "Job matching coalition formation and gross substitutes", *Econometrica* 50, pp. 1483-1504.

C.M. Kim [1998], "Stochastic dominance, Pareto optimality, and equilibrium asset pricing", *Review of Economic Studies* 65, pp. 341-356.

B. Knaster, C. Kuratowski and C. Mazurkiewicz [1929], "Ein Beweis des Fixpunktsatzes für n-dimensionale Simplexe", *Fundamenta Mathematicae* 14, pp. 132-137.

D.E. Knuth [1968], *The Art of Computer Programming, Vol. 1: Fundamental Algorithms*, Addison-Wesley, Reading.

M. Kojima [1978], "A modification of Todd's triangulation $J_3$", *Mathematical Programming* 15, pp. 223-227.

M. Kojima, N. Megiddo, T. Noma and A. Yoshise [1991], *A Unified Approach to Interior Point Algorithms for Linear Complementarity Problems*, Springer-Verlag, Berlin.

M. Kojima and S. Mizuno [1983], "Computation of all solutions to a system of polynomial equations", *Mathematical Programming* 25, pp. 131-157.

M. Kojima, H. Nishino and N. Arima [1979], "A PL homotopy for finding all the roots of a polynomial", *Mathematical Programming* 16, pp. 37-62.

M. Kojima and R. Saigal [1981], "On the number of solutions to a class of complementarity problems", *Mathematical Programming* 21, pp. 190-203.

M. Kojima and Y. Yamamoto [1982], "Variable dimension algorithms: basic theory, interpretation, and extensions of some existing methods", *Mathematical Programming* 24, pp. 177-215.

M. Kojima and Y. Yamamoto [1984], "A unified approach to the implementation of several restart fixed point algorithms and a new variable dimension algorithm", *Mathematical Programming* 28, pp. 288-328.

T. Koopmans and M.J. Beckman [1957], "Assignment problems and the location of economic activities", *Econometrica* 25, pp. 53-76.

M.A. Krasnosell'skii [1964], *Topological Methods in the Theory of Nonlinear Integral Equations*, Pergamon Press, Oxford.

H. Kremers and A.J.J. Talman [1994], "A new pivoting algorithm for the linear complementarity problem allowing for an arbitrary starting point", *Mathematical Programming* 63, pp. 235-252.

D.M. Kreps and R. Wilson [1982], "Sequential equilibria", *Econometrica* 50, pp. 863-894.

H.W. Kuhn [1968], "Simplicial approximation of fixed points", *Proceedings of National Academy of Science, U.S.A.* 61, pp. 1238-1242.

H.W. Kuhn [1969], "Approximate search for fixed points", *Computing Methods in Optimization Problems* 2, pp. 199-211.

H.W. Kuhn [1974], "A new proof of the fundamental theorem of algebra", *Mathematical Programming Study* 1, pp. 148-158.

H.W. Kuhn [1977], "Finding roots of polynomials by pivoting", in: *Fixed Points: Algorithms and Applications,* ed., S. Karamardian, Academic Press, New York, pp. 11-39.

H.W. Kuhn and J.G. MacKinnon [1975], "The sandwich method for finding fixed points", *Journal of Optimization Theory and Applications* 17, pp. 189-204.

H.W. Kuhn, Z. Wang and S. Xu [1984], "On the cost of computing roots of polynomials", *Mathematical Programming* 28, pp. 156-163.

M. Kurz [1982], "Unemployment equilibrium in an economy with linked prices", *Journal of Economic Theory* 26, pp. 100-123.

G. van der Laan [1980a], "Equilibrium under rigid prices with compensation for the consumers", *International Economic Review* 21, pp. 63-73.

G. van der Laan [1980b], *Simplicial Fixed Point Algorithms,* CWI, Amsterdam.

G. van der Laan [1982], "Simplicial approximation of unemployment equilibria", *Journal of Mathematical Economics* 9, pp. 83-97.

G. van der Laan [1984], "On the existence and approximation of zeroes", *Mathematical Programming* 28, pp. 1-14.

G. van der Laan [1985], "The computation of general equilibrium in economies with a block diagonal pattern", *Econometrica* 53, pp. 659-665.

G. van der Laan and L.P. Seelen [1984], "Efficiency and implementation of simplicial zero point algorithms", *Mathematical Programming* 30, pp. 196-217.

G. van der Laan and A.J.J. Talman [1979], "A restart algorithm for computing fixed points without an extra dimension", *Mathematical Programming* 17, pp. 74-84.

G. van der Laan and A.J.J. Talman [1980a], "A new subdivision for computing fixed points with a homotopy algorithm", *Mathematical Programming* 19, pp. 78-91.

G. van der Laan and A.J.J. Talman [1980b], "An improvement of fixed point algorithms by using a good triangulation", *Mathematical Programming* 18, pp. 274-285.

G. van der Laan and A.J.J. Talman [1981a], "A class of simplicial restart fixed point algorithms without an extra dimension", *Mathematical Programming* 20, pp. 33-48.

G. van der Laan and A.J.J. Talman [1981b], "Labelling rules and orientation: Sperner's lemma and Brouwer's degree", in: Numerical Solution of Nonlinear Equations, eds., E.L. Allgower, K. Glashoff and H.O. Peitgen, Lecture Notes in Mathematics 878, Springer-Verlag, Berlin, pp. 238-257.

G. van der Laan and A.J.J. Talman [1982], "On the computation of fixed points in the product space of unit simplices and an application to $N$-person games", *Mathematics of Operations Research* 7, pp. 1-13.

G. van der Laan and A.J.J. Talman [1987a], "Adjustment processes for finding economic equilibrium problems on the unit simplex", *Economics Letters* 23, pp. 119-123.

G. van der Laan and A.J.J. Talman [1987b], "Simplicial approximation of solutions to the nonlinear complementarity problem with lower and upper bounds", *Mathematical Programming* 38, pp. 1-15.

G. van der Laan and A.J.J. Talman [1993], "Intersection theorems on the simplotope", CentER Discussion Paper No. 9370, Tilburg University, Tilburg.

G. van der Laan, A.J.J. Talman and L. Van der Heyden [1987], "Simplicial variable dimension algorithms for solving the nonlinear complementarity problems on a product of unit simplices using a general labeling", *Mathematics of Operations Research* 12, pp. 377-397.

G. van der Laan, A.J.J. Talman and Z. Yang [1994], "Intersection theorems on polytopes", CentER Discussion Paper No. 9420, Tilburg University, Tilburg, to appear in *Mathematical Programming.*

G. van der Laan, A.J.J. Talman and Z. Yang [1996], "Existence and approximation of robust solutions of variational inequality problems over polytopes", Discussion Paper No. 696, Institute of Socio-Economic Planning, University of Tsukuba, Tsukuba, to appear in *SIAM Journal on Control and Optimization.*

G. van der Laan, A.J.J. Talman and Z. Yang [1997], "Existence of an equilibrium in a competitive economy of indivisibilities with money", *Journal of Mathematical Economics* 28, pp. 101-109.

G. van der Laan, A.J.J. Talman and Z. Yang [1998], "Balanced simplices on polytopes", Discussion Paper No. 131, Faculty of Business Administration, Yokohama National University, Yokohama.

C. Le Van [1982], "Topological degree and the Sperner lemma", *Journal of Optimization Theory and Applications* 37, pp. 371-377.

S. Lefschetz [1949], *Introduction to Topology,* Princeton University Press, Princeton.

C.E. Lemke [1965], "Bimatrix equilibrium points and mathematical programming", *Management Science* 11, pp. 681-689.

C.E. Lemke and J.T. Howson [1964], "Equilibrium points of bimatrix games", *SIAM Journal on Applied Mathematics* 12, pp. 413-423.

H.W. Lenstra Jr. [1983], "Integer programming with a fixed number of variables", *Mathematics of Operations Research* 8, pp. 538-548.

Q. Li, F. Janssen, Z. Yang and T. Ida [1998], "ILIN: An implementation of the integer labeling algorithm for integer programming", *IEICE Transactions on Fundamentals of Computer Sciences*, 2, pp. 304-309.

L. Lovász and H. Scarf [1992], "The generalized basis reduction algorithm", *Mathematics of Operations Research* 17, pp. 751-764.

H.J. Lüthi [1975], "A simplicial approximation of a solution for the nonlinear complementarity problem", *Mathematical Programming* 9, pp. 278-293.

R. Marshall [1920], *Principle of Economics*, Macmillan, London.
A. Mas-Colell [1974], "An equilibrium existence theorem without complete or transitive preferences", *Journal of Mathematical Economics* 1, pp. 237-246.
L. Mathiesen [1985], "Computation of economic equilibrium by a sequence of linear complementary problems", *Mathematical Programming* 23, pp. 144-162.
L. Mathiesen [1987], "An algorithm based on a sequence of linear complementary problems applies to a Walrasian equilibrium model: an example", *Mathematical Programming* 37, pp. 1-18.
O.H. Merrill [1972], *Applications and Extensions of an Algorithm that Computes Fixed Points of Certain Upper Semi-Continuous Point-to-Set Mappings*, Ph.D. Thesis, University of Michigan, Ann Arbor.
R.C. Merton [1990], *Continuous-Time Finance*, Blackwell, Malden.
M. Meyerson and A. Wright [1979], "A new and constructive proof of the Borsuk-Ulam theorem", *American Mathematical Society* 73, pp.134-136.
H. Midonick [1968], *The Treasury of Mathematics: 1,* Penguin Books, Harmondsworth.
J.R. Munkres [1975], *Topology, A First Course,* Prentice-Hall, Englewood Cliffs.
K.G. Murty [1983], *Linear Programming,* John Wiley & Sons, New York.
R.B. Myerson [1978], "Refinements of the Nash equilibrium concept", *International Journal of Game Theory* 7, pp. 73-80.
J. Nash [1950], "Equilibrium points in $N$-person games", *Proceedings of National Academy of Science U.S.A.* 36, pp. 48-49.
G.L. Nemhauser and L.A. Wolsey [1988], *Integer and Combinatorial Optimization,* John Wiley & Sons, New York.
J. von Neumann and O. Morgenstern [1944], *Theory of Games and Economic Behavior,* Princeton University Press, Princeton.
M. Newman [1972], *Integral Matrices,* Academic Press, New York.
L.T. Nielsen [1989], "Asset market equilibrium with short-selling", *Review of Economic Studies* 56, pp. 467-474.
L.T. Nielsen [1990], "Existence of equilibrium in CAPM", *Journal of Economic Theory* 52, pp. 223-231.
J.M. Ortega and W.G. Rheinbolt [1970], *Iterative Solution of Nonlinear Equations in Several Variables,* Academic Press, New York.
C.H. Papadimitriou and H. Steiglitz [1982], *Combinatorial Optimization: Algorithms and Complexity,* Prentice-Hall, Englewood Cliffs.
A. Pnueli [1968], "A method of truncated relaxation for integer programming", unpublished manuscript, IBM, Yorktown Heights.
M. Prüfer and H.W. Siegberg [1981], "Complementarity pivoting and the Hopf degree theorem", *Journal of Mathematical Analysis and Applications* 84, pp. 133-149.
M. Prüfer and H.O. Walther [1979], *Functional Differential Equations and Approximations of Fixed Points,* eds., Lecture Notes in Mathematics 730, Springer-Verlag, Berlin.
M. Quinzii [1984], "Core and competitive equilibria with indivisibilities", *International Journal of Game Theory* 13, pp. 41-60.
R. Radner [1972], "Existence of equilibrium of plans, prices and price expectations in a sequence of markets", *Econometrica* 40, pp. 289-303.
P.M. Reiser [1981], "A modified integer labeling for complementarity algorithms", *Mathematics of Operations Research* 6, pp. 129-139.
J. Renegar [1988], "Rudiments of an average case complexity theory for piecewise-linear path following algorithms", *Mathematical Programming* 40, pp. 113-163.
S. Robinson [1980], *Analysis and Computation of Fixed Points,* ed., Academic Press, New York.
J. Rosenmüller [1971], "On a generalization of the Lemke-Howson algorithm to noncooperative $N$-person games", *SIAM Journal on Applied Mathematics* 21, pp. 73-79.
K.A. Ross [1980], *Elementary Analysis: The Theory of Calculus*, Springer-Verlag, Berlin.
A.E. Roth and O. Sotomayor [1990], *Two Sided Matching,* Cambridge University Press,

Cambridge.

R. Saigal [1976], "On paths generated by fixed point algorithms", *Mathematics of Operations Research* 1, pp. 359-380.

R. Saigal [1977], "On the convergence rate of algorithms for solving equations that are based on complementarity pivoting", *Mathematics of Operations Research* 2, pp. 108-124.

R. Saigal [1979], "The fixed point approach to nonlinear programming", *Mathematical Programming Study* 10, pp. 142-157.

R. Saigal [1983], "A homotopy for solving large, sparse and structured fixed point problems", *Mathematics of Operations Research* 8, pp. 557-578.

R. Saigal and M.J. Todd [1978], "Efficient acceleration techniques for fixed point algorithms", *SIAM Journal on Numerical Analysis* 15, pp. 997-1007.

H. Scarf [1967a], "The core of an $N$ person game", *Econometrica* 37, pp. 50-69.

H. Scarf [1967b], "The approximation of fixed points of a continuous mapping", *SIAM Journal on Applied Mathematics* 15, pp. 1328-1343.

H. Scarf [1973] (with the collaboration of T. Hansen), *The Computation of Economic Equilibria*, Yale University Press, New Haven.

H. Scarf [1981a], "Production sets with indivisibilities-part I: generalities", *Econometrica* 49, pp. 1-32.

H. Scarf [1981b], "Production sets with indivisibilities-part II: the case of two activities", *Econometrica* 49, pp. 395-423.

H. Scarf [1985], "Integral polyhedra in three space", *Mathematics of Operations Research* 10, pp. 403-438.

H. Scarf [1986], "Neighborhood systems for production sets with indivisibilities", *Econometrica* 54, pp. 507-53.

H. Scarf [1991], "The origins of fixed point methods", in: *History of Mathematical Programming,* eds., J.K. Lenstra, A.H.G. Rinnooy Kan and A. Schrijver, North-Holland, Amsterdam, pp. 126-134.

H. Scarf [1994], "The allocation of resources in the presence of indivisibilities", *Journal of Economic Perspectives* 4, pp. 111-128.

H. Scarf and D. Shallcross [1993], "The Frobenius problem and maximal lattice free bodies", *Mathematics of Operations Research* 18, pp. 511-515.

A. Schrijver [1986], *Theory of Linear and Integer Programming,* John Wiley & Sons, New York.

R. Selten [1975], "Reexamination of the perfectness concept for equilibrium points in extensive games", *International Journal of Game Theory* 4, pp. 25-55.

S. Shamir [1980], "Two triangulations for homotopy fixed point algorithms with an arbitrary refinement factor", in: *Analysis and Computation of Fixed Points,* ed., S.M. Robinson, Academic Press, New York, pp. 25-56.

L.S. Shapley [1973], "On balanced games without side payments", in: Mathematical Programming, eds., T.C. Hu and S.M. Robinson, Academic Press, New York, pp. 261-290.

L.S. Shapley [1974], "A note on the Lemke-Howson algorithm", *Mathematical Programming Study* 1, pp. 175-189.

L.S. Shapley and H. Scarf [1974], "On cores and indivisibilities", *Journal of Mathematical Economics* 1, pp. 23-37.

L.S. Shapley and M. Shubik [1972], "The assignment game I: the core", *International Journal of Game Theory* 1, pp. 111-130.

J.B. Shoven and J. Whalley [1992], *Applying General Equilibrium,* Cambridge University Press, Cambridge.

S. Smale [1976], "A convergent process of price adjustment process and global Newton methods", *Journal of Mathematical Economics* 3, pp. 107-120.

S. Smale [1981], "The fundamental theorem of algebra and complexity theory", *Bulletin of American Mathematical Society* 4, pp. 1-36.

R. Smart [1974], *Fixed Point Theorems,* Cambridge University Press, Cambridge.

E. Sperner [1928], "Neuer Beweis für die Invarianz der Dimensionszahl und des Gebietes", *Abhandlungen aus dem mathematischen Seminar Universität Hamburg* 6, pp. 265-272.

J. Stoer and C. Witzgall [1978], *Convexity and Optimization in Finite Dimension,* Springer-Verlag, Berlin.

W. Takahashi [1991], "Existence theorems generalizing fixed point theorems for multi valued mappings", in: *Fixed Point Theory and Applications*, eds., M.A. Thera and J.B. Baillon, Longman Scientific & Technical, pp.397-406.

A.J.J. Talman [1980], *Variable Dimension Fixed Point Algorithms and Triangulations,* CWI, Amsterdam.

A.J.J. Talman [1994], "Intersection theorems on the unit simplex and the simplotope", in: Imperfections and Behavior in Economic Organizations, eds., R.P. Gilles and P.H.M. Ruys, Kluwer, Dordrecht.

A.J.J. Talman and G. van der Laan [1987], *The Computation and Modelling of Economic Equilibria,* eds., North-Holland, Amsterdam.

A.J.J. Talman and Y. Yamamoto [1989], "A simplicial algorithm for stationary point problems on polytopes", *Mathematics of Operations Research* 14, pp. 383-399.

A.J.J. Talman, Y. Yamamoto and Z. Yang [1995], "A new variable dimension simplicial algorithm for computing economic equilibria on $S^n \times \mathbb{R}^m_+$", *Journal of Optimization Theory and Applications* 87, pp. 679-701.

A.J.J. Talman and Z. Yang [1994], "A simplicial algorithm for computing proper Nash equilibria of finite games", CentER Discussion Paper No. 9418, Tilburg University, Tilburg.

A. Tarski [1955], "A lattice-theoretical fixed point theorem and its applications", *Pacific Journal of Mathematics* 5, pp. 285-309.

M.J. Todd [1974], "A generalized complementary pivoting algorithm", *Mathematical Programming* 6, pp. 243-263.

M.J. Todd [1976a], *The Computation of Fixed Points and Applications,* Lecture Notes in Economics and Mathematical Systems 124, Springer-Verlag, Berlin.

M.J. Todd [1976b], "On triangulations for computing fixed points", *Mathematical Programming* 10, pp. 322-346.

M.J. Todd [1976c], "Orientation in complementary pivot algorithms", *Mathematics of Operations Research* 1, pp. 54-66.

M.J. Todd [1977], "Union jack triangulations", in: *Fixed Points: Algorithms and Applications,* ed., S. Karamardian, Academic Press, New York, pp. 315-336.

M.J. Todd [1978a], "Improving the convergence of fixed point algorithms", *Mathematical Programming Study* 7, pp. 151-179.

M.J. Todd [1978b], "On the Jacobian of a function at a zero computed by a fixed point algorithm", *Mathematics of Operations Research* 3, pp. 126-132.

M.J. Todd [1980], "A quadratically-convergent fixed point algorithm for economic equilibria and linearly constrained optimization", *Mathematical Programming* 18, pp. 111-126.

M.J. Todd [1982], "On the computational complexity of piecewise linear homotopy algorithm", *Mathematical Programming* 24, pp. 216-224.

M.J. Todd [1984], "$J'$ : A new triangulation of $\mathbb{R}^n$", *SIAM Journal on Algebraic and Discrete Methods* 5, pp. 244-254.

M.J. Todd and A.H. Wright [1980], "A variable dimension simplicial algorithm for antipodal fixed point theorems", *Numerical Functional Analysis and Optimization* 2, pp. 155-186.

A.W. Tucker [1945], "Some topological properties of disk and sphere", in: *Proceedings of the first Canadian Mathematical Congress,* University of Toronto Press, Toronto, pp. 285-309.

H. Tuy [1979], "Pivotal methods for computing equilibrium points: unified approach and new restart algorithm", *Mathematical Programming* 16, pp. 210-227.

H. Uzawa [1962], "Walras' existence theorem and Brouwer's fixed point theorem", *Economic Studies Quarterly* 13, p. 1.

L. Van der Heyden [1980], "A variable dimension algorithm for the linear complementarity problem", *Mathematical Programming* 19, pp. 328-346.

L. Van der Heyden [1982], "A refinement procedure for computing fixed points", *Mathematics of Operations Research* 7, pp. 295-313.

H.R. Varian [1992], *Microeconomic Analysis,* W.W. Norton & Company, New York.

A.F. Veinott [1968], "Extreme points of Leontief substitution systems", *Linear Algebra and its Applications* 1, pp. 181-194.

R. Vohra [1988], "On the existence of equilibria in a model with increasing returns", *Journal of Mathematical Economics*, 17, pp. 179-192.

B.L. van der Waerden [1970], *Algebra*, Vols. I and II, Frederick Ungar Publishing Co., New York.

B.L. van der Waerden [1983], *Geometry and Algebra in Ancient Civilizations,* Springer-Verlag, Berlin.

L. Walras [1874], *Elements of Pure Economics,* L. Corbaz and Company, Lausanne.

Z. Wang [1982], "On the efficiency of Kuhn's root-finding algorithm", *KEXUE TONGBAO*, 27, p.1023.

Z. Wang [1986a], *The Fundamentals of Simplicial Algorithms for Fixed Points,* Zhongshan University Press, Guangzhou. (in Chinese)

Z. Wang [1986b], "A sufficient condition for the convergence of Kuhn's zero-finding algorithm", *KEXUE TONGBAO*, 31, pp. 575-576.

Z. Wang and S. Xu [1984], "Approximate zeroes and computational complexity theory", *Science Sinica (A)*, 17, pp. 566-575.

Z. Wang, S. Xu and T. Gao [1994], *Algebraic Systems of Equations and Computational Complexity Theory,* Science Press, Beijing.

J. Werner [1985], "Equilibrium in economies with incomplete financial markets", *Journal of Economic Theory* 36, pp. 110-119.

J. Werner [1987], "Arbitrage and the existence of competitive equilibrium", *Econometrica* 55, pp. 1403-1418.

P.M. White [1983], *Discrete Activity Analysis,* Ph.D. Thesis, Yale University, New Haven.

P.M. White, A.S. Caplin and L. Van der Heyden [1985], "Scarf's procedure for integer programming and a dual simplex algorithm", *Mathematics of Operations Research* 10, pp. 439-449.

R. Wilson [1971], "Computing equilibria of $N$-person games", *SIAM Journal on Applied Mathematics* 21, pp. 80-87.

A.H. Wright [1981], "The octahedral algorithm, a new simplicial fixed point algorithm", *Mathematical Programming* 21, pp. 47-69.

W.T. Wu [1984], "Basic principles of mechanical theorem proving in elementary geometries", *Journal of System Science and Mathematical Science* 4, pp. 207-235.

W.T. Wu and J.H. Jiang [1962], "Essential equilibrium points of n-person noncooperative games", *Science Sinica* 11, pp. 1307-1322.

Y. Yamamoto [1983], "A new variable dimension algorithm for the fixed point problem", *Mathematical Programming* 25, pp. 329-342.

Y. Yamamoto [1984a], "A unifying model based on retraction for fixed point algorithms", *Mathematical Programming* 28, pp. 192-197.

Y. Yamamoto [1984b], "A variable dimension fixed fixed point algorithm and the orientation of simplices", *Mathematical Programming* 30, pp. 301-312.

Y. Yamamoto [1987a], "Competitive equilibria in the market with indivisibility", in: *The Computation and Modelling of Economic Equilibria,* eds., A.J.J. Talman and G. van der Laan, North-Holland, Amsterdam, pp. 193-204.

Y. Yamamoto [1987b], "A path following algorithm for stationary point problems", *Journal of the OR Society of Japan* 30, pp. 181-198.

Y. Yamamoto [1993], "A path-following procedure to find a proper equilibrium of finite games", *International Journal of Game Theory* 22, pp. 249-259.

Y. Yamamoto and Z. Yang [1993], "The $(n+1)2^m$-ray algorithm: a new simplicial algorithm for the variational inequality problem on $\mathbb{R}^m_+ \times S^n$", *Annals of Operations*

*Research* 44, pp. 93-113.

Z. Yang [1994], "A simplicial algorithm for testing the integral property of polytopes: a revision", CentER Discussion Paper No. 9489 (November), Tilburg University, Tilburg.

Z. Yang [1995], "A constructive proof of a unimodular transformation theorem for simplices", CentER Discussion Paper No. 9503 (January), Tilburg University, Tilburg.

Z. Yang [1996], "A simplicial algorithm for computing robust stationary points of a continuous function on the unit simplex", *SIAM Journal on Control and Optimization* 34, pp.491-506.

Z. Yang [1997], "Equilibrium in an exchange economy with multiple indivisible commodities and money", Discussion Paper No.728, Institute of Policy and Planning Sciences, University of Tsukuba, Tsukuba.

Z. Yang [1998], "A multi-permutation-based intersection theorem and equilibria with bad indivisibilities", Discussion Paper No. 132, Faculty of Business Administration, Yokohama National University, Yokohama.

B. Yu and Z. Lin [1996], "Homotopy method for a class of nonconvex Brouwer fixed point problems", *Applied Mathematics and Computation* 74, pp. 65-77.

W.I. Zangwill [1977], "An eccentric barycentric fixed point algorithm", *Mathematics of Operations Research* 2, pp. 343-359.

# Index

THEORY AND DECISION LIBRARY

---

SERIES C: GAME THEORY, MATHEMATICAL PROGRAMMING AND OPERATIONS RESEARCH

*Editor:* S.H. Tijs, *University of Tilburg, The Netherlands*

---

1. B.R. Munier and M.F. Shakun (eds.): *Compromise, Negotiation and Group Decision.* 1988 ISBN 90-277-2625-6
2. R. Selten: *Models of Strategic Rationality.* 1988 ISBN 90-277-2663-9
3. T. Driessen: *Cooperative Games, Solutions and Applications.* 1988 ISBN 90-277-2729-5
4. P.P. Wakker: *Additive Representations of Preferences.* A New Foundation of Decision Analysis. 1989 ISBN 0-7923-0050-5
5. A. Rapoport: *Experimental Studies of Interactive Decisions.* 1990 ISBN 0-7923-0685-6
6. K.G. Ramamurthy: *Coherent Structures and Simple Games.* 1990 ISBN 0-7923-0869-7
7. T.E.S. Raghavan, T.S. Ferguson, T. Parthasarathy and O.J. Vrieze (eds.): *Stochastic Games and Related Topics.* In Honor of Professor L.S. Shapley. 1991 ISBN 0-7923-1016-0
8. J. Abdou and H. Keiding: *Effectivity Functions in Social Choice.* 1991 ISBN 0-7923-1147-7
9. H.J.M. Peters: *Axiomatic Bargaining Game Theory.* 1992 ISBN 0-7923-1873-0
10. D. Butnariu and E.P. Klement: *Triangular Norm-Based Measures and Games with Fuzzy Coalitions.* 1993 ISBN 0-7923-2369-6
11. R.P. Gilles and P.H.M. Ruys: *Imperfections and Behavior in Economic Organization.* 1994 ISBN 0-7923-9460-7
12. R.P. Gilles: *Economic Exchange and Social Organization.* The Edgeworthian Foundations of General Equilibrium Theory. 1996 ISBN 0-7923-4200-3
13. P.J.-J. Herings: *Static and Dynamic Aspects of General Disequilibrium Theory.* 1996 ISBN 0-7923-9813-0
14. F. van Dijk: *Social Ties and Economic Performance.* 1997 ISBN 0-7923-9836-X
15. W. Spanjers: *Hierarchically Structured Economies.* Models with Bilateral Exchange Institutions. 1997 ISBN 0-7923-4398-0
16. I. Curiel: *Cooperative Game Theory and Applications.* Cooperative Games Arising from Combinatorial Optimization Problems. 1997 ISBN 0-7923-4476-6
17. O.I. Larichev and H.M. Moshkovich: *Verbal Decision Analysis for Unstructured Problems.* 1997 ISBN 0-7923-4578-9

18. T. Parthasarathy, B. Dutta, J.A.M. Potters, T.E.S. Raghavan, D. Ray and A. Sen (eds.): *Game Theoretical Applications to Economics and Operations Research*. 1997 ISBN 0-7923-4712-9
19. A.M.A. Van Deemen: *Coalition Formation and Social Choice*. 1997 ISBN 0-7923-4750-1
20. M.O.L. Bacharach, L.-A. Gérard-Varet, P. Mongin and H.S. Shin (eds.): *Epistemic Logic and the Theory of Games and Decisions*. 1997 ISBN 0-7923-4804-4
21. Z. Yang (eds.): *Computing Equilibria and Fixed Points*. 1999 ISBN 0-7923-8395-8

KLUWER ACADEMIC PUBLISHERS – DORDRECHT / BOSTON / LONDON

www.ingramcontent.com/pod-product-compliance
Ingram Content Group UK Ltd.
Pitfield, Milton Keynes, MK11 3LW, UK
UKHW020135160926
475607UK00003B/45

* 9 7 8 0 7 9 2 3 8 3 9 5 6 *